INTERNATIONAL TRENDS IN OPTICS

INTERNATIONAL TRENDS IN OPTICS

Edited by

JOSEPH W. GOODMAN
Department of Electrical Engineering
Stanford University
Stanford, California

ACADEMIC PRESS, INC.
Harcourt Brace Jovanovich, Publishers
Boston San Diego New York
London Sydney Tokyo Toronto

This book is printed on acid-free paper. ♾

ACADEMIC PRESS, INC.
1250 Sixth Avenue, San Diego, CA 92101

United Kingdom Edition published by
ACADEMIC PRESS LIMITED
24–28 Oval Road, London NW1 7DX

Library of Congress Cataloging-in-Publication Data:

International trends in optics/edited by Joseph W. Goodman.
p. cm.
Includes bibliographical references and index.
ISBN 0-12-289690-4 (alk. paper)
1. Optics. I. Goodman, Joseph W.
QC361.I57 1991
535—dc20 90-14434
CIP

Cover design by Camille Pecoul

Printed in the United States of America

91 92 93 94 9 8 7 6 5 4 3 2 1

This book is dedicated to the proposition that international exchange of scientific and technical information contributes to mutual understanding among peoples and to world peace.

Contents

Contributors

Numbers in parentheses indicate the pages on which the authors' contributions begin.

Ivan Andonovic (77), University of Strathclyde, Electronic & Electrical Engineering Department, Royal College Building, 204 George Street, Glasgow G1 1XW, Scotland

Henri H. Arsenault (439), Centre d' Optique, Photonique et Laser, Université Laval, Quebec, QC., Canada, G1K 7P4

Alain Aspect (247), Collège de France and Laboratoire de Spectroscopie Hertzienne de l'Ecole Normale Supérieure, 24, rue Lhomond F-75231, Paris Cedex 05, France

R. H. T. Bates (423), Electrical and Electronic Engineering Department, University of Canterbury, Private Bag, Christchurch, New Zealand

Olof Bryngdahl (313), Department of Physics, University of Essen, 4300 Essen 1, Federal Republic of Germany

Katarzyna Chalasinska-Macukow (451), Warsaw University, Institute of Geophysics, 02-093 Warsaw, Poland

Pierre Chavel (499), Institut d'Optique, Centre Universitaire d'Orsay, B.P. 147, 91403 Orsay Cedex, France

Anna Consortini (267), Department of Physics, University of Florence, Via S. Marta 3, 50139 Florence, Italy

Brian Culshaw (77), University of Strathclyde, Electronic & Electrical Engineering Department, Royal College Building, 204 George Street, Glasgow G1 1XW, Scotland

J. C. Dainty (207), Blackett Laboratory, Imperial College, London SW7 2BZ, United Kingdom

R. Dändliker (57), Institute of Microtechnology, University of Neuchâtel, CH-2000 Neuchâtel, Switzerland

Yury N. Denisyuk (279), A.F. Ioffe Physical-Technical Institute, Academy of Sciences, Leningrad 194 021, USSR

J. R. Fienup (407), Optical Science Laboratory, Advanced Concepts Division, Environmental Research Institute of Michigan, P.O. Box 8618, Ann Arbor, MI 48107

Philippe Grangier (247), Institut d'Optique Théorique et Appliquée, Batiment 503, Université Paris XI, 91405 Orsay, France

Pál Greguss (195), Department of Non-ionizing Radiation, Frederic Joliot-Curie, National Research Institute for Radiobiology and Radiohygiene, Budapest 22, Pentz Karoly u. 5., Hungary

P. Hariharan (339), CSIRO Division of Applied Physics, PO Box 218, Lindfield, Australia 2070

H. P. Herzig (57), Institute of Microtechnology, University of Neuchâtel, CH-2000 Neuchâtel, Switzerland

John N. Howard (167), 7 Norman Road, Newton Highlands, MA 02161

Jean-Pierre Huignard (111), Thomson-CSF, Laboratoire Central de Recherches, Domaine de Corbeville, 91404 Orsay, France

Kenichi Iga (37), Tokyo Institute of Technology, 4259 Nagatsuta, Midoriku, Yokohama, Japan

Hong Jiang (423), Electrical and Electronic Engineering Department, University of Canterbury, Private Bag, Christchurch, New Zealand

Herwig Kogelnik (1), AT&T Bell Laboratories, Crawford Hill Laboratory, Holmdel, NJ 07733

Adolf W. Lohmann (155), Physikalisches Institut der Universität, 8520 Erlangen, Germany

Daniel Malacara (351), The Institute of Optics, University of Rochester, Rochester, NY 14627

E. W. Martin (481), High Technology Center, Boeing Aerospace & Electronics Inc., P.O. Box 3999, Seattle, WA 98124-2499

Fritz Merkle (375), European Southern Observatory, Karl-Schwarzschild-Str. 2 D-8046 Garching, Germany

D. A. B. Miller (13), AT&T Bell Laboratories, Holmdel, NJ 07733

Valentin Morozov (465), Opticomp Corporation, P.O. Box 10779, Zephyr Cove, Lake Tahoe, NV 89448

Jan Peřina (233), Joint Laboratory of Optics, Palacký University, 771 46 Olomouc, Czechoslovakia

Henri Rajbenbach (111), Thomson-CSF, Laboratoire Central de Recherches, Domaine de Corbeville, 91404 Orsay, France

Yunlong Sheng (439), Centre d'Optique, Photonique et Laser, Université Laval, Quebec, QC., Canada, G1K 7P4

Peter W. E. Smith (25), Bellcore, 331 Newman Springs Road, Red Bank, NJ 07701

S. D. Smith (481), Department of Physics, Heriot-Watt University, Riccarton, Edinburgh EH14 4AS, Scotland

Jakob J. Stamnes (141), Norwave Development, P.O. Box 144, Vinderen, 0319 Oslo 3, Norway

S. I. Stepanov (125), A.F. Ioffe Physical-Technical Institute, Leningrad, 194021, USSR

Tomasz Szoplik (451), Warsaw University, Institute of Geophysics, 02-093 Warsaw, Poland

Jumpei Tsujiuchi (297), Faculty of Engineering, Chiba University, 1-33 Yayoi-cho, Chiba 260, Japan

Yoshito Tsunoda (95), Central Research Laboratory, Hitachi Ltd., 1-280 Higashi-Koigakubo, Kokubunji, Tokyo, Japan

Christiaan H. F. Velzel (325), Philips - CFT, P.O.B. 218, 5600 MD Eindhoven, The Netherlands

Gerd Weigelt (391), Max-Planck-Institut für Radioastronomie, Auf dem Hügel 69, D-5300 Bonn 1, Federal Republic of Germany

Emil Wolf (221), Department of Physics and Astronomy, University of Rochester, Rochester, NY 14627

Zhi-Ming Zhang (185), Lab of Laser Physics & Optics, Fudan University, Shanghai 200433, China

Preface

This book has at least four separate goals. The first is to provide a broad view of work under way in the field of optics throughout the world, where breadth refers both to geography and to subject matter. While there is a strong trend toward specialization in optics, over-specialization is not healthy, and the articles in this book afford the opportunity for rather general and readable overviews of many different subjects in optics, suitable for nonspecialists.

A second goal, not unrelated to the first, is to provide a different style of technical article than is found in journals and formal reviews, one that is more informal and has some speculative aspect. Many of the authors mention unsolved problems and predict directions for the future.

A third purpose of the book derives from the rather unusual financial arrangements made with the publisher. The Editor and all authors of this book have assigned their royalties to the International Commission for Optics (ICO), which will use this modest additional income to strengthen support of its traveling lecturer program, under which developing countries can request support from the ICO to enable a visit by a distinguished lecturer in the field of optics.

A fourth purpose is to provide greater visibility for the ICO, since the organization is not as well known as it should be. With this goal in mind, a bit of background is now provided:

The ICO was formed in 1947, shortly after World War II, to promote optics on an international basis. Unlike other societies with national origins, the ICO is truly international, with a governing body (the Bureau) elected by duly appointed representatives of all member Territories once every three years. There are currently 37 member Territories from every continent except Antarctica. The ICO has no individual memberships.

The ICO has one primary barrier to maximum effectiveness: financial resources. While successful national societies in the U.S. have annual budgets of several million dollars, the ICO has an annual income of approximately $20,000, derived from fees paid by the member Territories. Clearly the program scale that the ICO is able to mount in pursuing its goal of promoting optics around the world is rather restricted.

Nonetheless, those devoting their time to the ICO believe that there is a valid role for this organization in the modern world. Usually the meetings sponsored by the ICO, and in particular its general meetings every three years, have a wider international representation than any meeting of a single national optical society. These meetings thus afford a truly unique opportunity to obtain a broad view of the state of optics around the world. The ICO has also taken a more active role in promoting topical meetings, either as a sponsor or as a co-sponsor, but seldom more than once per year. Last but not least, the ICO does serve as a co-sponsor for many other meetings in all parts of the world, in which it has less direct involvement.

In addition to its role in promoting international gatherings devoted to optics, the ICO can play other helpful roles on the optics scene in the coming decade. It can cooperate upon request, with single Territories or groups of Territories to help them organize their own societies devoted to optics, without any designs to extract financial benefits for itself. In addition, it responds to requests from developing countries to help strengthen their infrastructure in optics, through the traveling lecture program and through any other means permitted within its financial constraints.

As Editor, I wish to express my special thanks to the authors. Of the 40 invited contributors, 35 responded with manuscripts — a remarkable success rate. I have the sad duty to report that one of our authors, Richard H. T. Bates, has died since submitting his manuscript. He will be sorely missed by all his colleagues who knew him or his work.

It is my personal hope that this series can be continued in the future, perhaps every three years, which is the natural cycle of the ICO Bureau.

Joseph W. Goodman
Los Altos, California, U.S.A.

CHAPTER 1

Integrated Optics, OEICs, or PICs?

Herwig Kogelnik
AT&T Bell Laboratories
Crawford Hill Laboratory
Holmdel, New Jersey

1. Early Dreams of Integration

After early dreams of optical integration there was considerable progress in the research laboratories of the world. The absence of practical field applications, on the other hand, has been a severe test of the patience of the R&D community. Only very recently did integrated optics see its first commercial applications, but not yet at a large scale. Considerable advances in semiconductor materials processing, such as MOVPE, have led to new ideas and approaches to integration under the names of OEICs and PICs. These advances, coupled with the steady increase in the sophistication and complexity of optoelectronic systems, should make it quite safe to predict that the coming decade will see many more practical applications.

Soon after the emergence of the first practical lasers in the early 1960s, laser researchers were inspired by the highly visible success of the electronic integrated circuit and began to explore the possibility of integrating optical devices and components on a single chip [1, 2, 3, 4]. Goals and concepts crystallized and the term *integrated optics* was coined in 1969 [4]. The goals of integrated optics are modeled after those of electronic integrated circuits (ICs): They include miniaturization and compactness, improved reliability, the provision of complex circuitry at low cost, lower power consumption for active devices, and the potential for convenient mass fabrication. Early proposals envisioned that integrated optics would provide functions such as wavelength filtering, wavelength multiplexing, optical modulation, optical switching, optical processing, coupling of waveguides, heterodyne detection, nonlinear optical interactions, amplification, and many more. Research enthusiasm was high, and the first Topical Meeting on Integrated Optics was organized by the Optical Society of America in 1972. (The name of this meeting was changed to Integrated Photonics Research in 1990).

ISBN 0-12-289690-4

During the 1970s, the attention of integration-oriented research activities in lasers and optoelectronics (photonics) turned increasingly toward semiconductors. We can distinguish two focus areas in these efforts: One was to perfect InP-based materials systems for device operation in the "long-wavelength" region near 1.3 and 1.5 μm, where the loss of optical fibers is lowest. The other used the more mature GaAs-based materials technology to explore more sophisticated integrated circuitry, including junction lasers, detectors and, finally, transistors. The year 1978 saw the first demonstration of an *optoelectronic integrated circuit* (OEIC), which included both a diode laser and an electronic Gunn-diode device [5].

Many difficult technological problems had to be solved to implement and improve on those early ideas, and the research community can look with pride on many fine accomplishments. Textbooks [6, 7, 8] and review articles [9, 10] will give the reader a more detailed account of these advances.

In recent years, there has been particularly noteworthy progress in the growth and processing of semiconductor materials suitable for integration. A particularly important example is the growth of highly uniform quantum-well material for the long-wavelength region by metal-organic vapor-phase epitaxy (MOVPE). This has formed the basis for a new approach to integration which emphasizes on-chip junction lasers and on-chip optical interconnects. The term *photonic integrated circuit* (PIC) was coined for this new class of chips [11, 12].

2. Practical Applications

In spite of the remarkable R&D progress over more than two decades, the discipline of optical integration had to wait a long time for practical applications in the field. Researchers had to be patient while some humorists among them pronounced "Integrated optics is the future of optical communications—and it will remain the future of optical communications." Integrated optical devices now are being sold to the R&D community for testing and systems explorations, but a *high-volume market* for their use in commercial systems does not yet exist. However, a start was made in 1989, and there now are at least two cases where integrated optics is employed in commercial systems. The first is the distributed feedback (DFB) laser, which was conceived originally as an integrable device for integrated optics [13]. The DFB laser now is used in high-capacity links of the AT&T long-distance fiber optic network, where laser sources of high spectral purity are required. The DFB laser also is utilized increasingly in fiber-optic cable TV (CATV) links, where it exhibits superior linear amplitude modulation characteristics. The second example for the commercial application of integrated optics is the use of high-speed, guided-wave $LiNbO_3$ modulators, which are incorporated in a recent HP lightwave test set for the analysis of lightwave components and systems. The instrumentation has been demonstrated to operate at speeds up to 20 GHz [14].

All this is nice, but you will probably ask, "When will there be more commercial use, and why is it taking so long?" When we attempt to answer these questions, we should first remember that integration is envisioned to help in the handling of *complexity* in optical systems. To proceed, let us list application areas that have been proposed for integrated optics and its derivatives:

- point-to-point optical transmission
- compact disks (CDs) and optical memories
- optical sensors
- optical interconnects
- optical cross-connects
- optical processing
- photonic switching
- optical computing

Note that this listing has been sorted in order of increasing complexity of the systems. Transmission systems are relatively simple, involving just a laser, an optical fiber, and a photodetector in their simplest form. Switching and computing, on the other hand, require very complex interconnection networks and control. Consequently, integration should have greater and greater impact the further we proceed down the list. Note also that system complexity tends to increase within each of these application areas as these technologies get more mature. In transmission, for example, researchers are exploring techniques for wavelength multiplexing (i.e., sending messages over one fiber at several different wavelengths) and for coherent detection (i.e., the use of local oscillators and heterodyne receivers). Both approaches require more complexity than the early laser-fiber-detector systems.

There is a second point to be made on the listing: It also is ordered with regard to the time (or probable time) of introduction to the market. Optical transmission and compact discs already have succeeded commercially. Sensors are close to this stage. Photonic switching and optical computing still are very much a research dream at this stage.

The preceding suggests that opportunities for integration will increase as we progress in time and down the list with the commercial introduction of more complex systems. Indeed, some of the systems in the lower half of the list may have to depend on integration to become a practical reality.

A supplementary point has to do with the *single-mode* nature of integrated optics. Circuit and device properties usually are optimized if only one mode and, therefore, only one polarization are singled out in the design. This fact was an obvious obstacle to the introduction of integrated optics in the early multi-mode fiber transmission systems, which essentially were not compatible. Even the modern single-mode fiber systems are not wholly compatible because they carry two modes of different *polarization* and that polarization is not preserved along the fiber length. Integrated optics techniques exist that can control this polarization problem, but only at the cost of increased complexity.

3. The Name Game

It may not be surprising that new approaches have led to the creation of new names, as we have seen earlier, An outside observer may well be confused and ask for clear definitions of terms like *integrated optics, OEICs*, or *PICs*. The writer will not attempt to give you a clear definition. Instead, he has spoken to a considerable number of workers in the field and will try to sketch for you how the majority of "insiders" uses these and related terms. Please note that there by no means is unanimous agreement on the usage of these terms, and language usage is not expected to be rational all the time.

Most people asked were not sure whether there is a difference between *photonics* and *optoelectronics*. Some thought that photonics was broader and put more emphasis on the laser. Most consider the laser both a photonic device and an optoelectronic device. The term integrated optics now is used mostly for circuits on transparent substrates, such as glass, SiO_2, or $LiNbO_3$. Both PICs and OEICs are made on semiconductors. So far, OEICs include on-chip transistors and on-chip optoelectronic devices but no on-chip optical waveguides, while PICs emphasize the use of on-chip lasers and on-chip optical waveguides. The nature of the on-chip interconnects seems to be the main difference, with metallic connectors used for OEICs, optical guides used for PICs. In fact, one of the arguments advanced in favor of PICs is that they save on expensive off-chip laser-fiber connections.

It will be interesting to observe the name game in the future as technology advances and the PIC and OEIC approaches merge: Will a PIC become an OEIC if a single transistor is added to it? Will an OEIC become a PIC if a single optical guide is added? Or will the merger of a PIC and an OEIC produce an "EPIC?"

As a further guide to the terms integrated optics, OEIC and PIC, we will list in Table I the dominant usage for some simple circuits. More detailed illustrations for each category will be given in the following sections.

Table I

Circuit	Category
Switch array ($LiNbO_3$)	Integrated optics
Filter array ($LiNbO_3$)	Integrated optics
High-speed modulator	Integrated optics
Laser array	PIC
Laser-modulator	PIC
PIN-FET receiver	OEIC
PIN-HBT receiver	OEIC
Coherent receiver	PIC
Laser-HBT	OEIC
Laser-monitoring detector	OEIC
Tunable laser	PIC

The following illustrations are designed to provide further background on the usage of terms and to sketch recent advances in the field. For simplicity, the examples are taken from the author's home organization, but progress in integration has been worldwide with fine contributions from all continents.

4. Two Integrated Optical Circuits

Our first pair of illustrations consists of two integrated optical circuits fabricated on $LiNbO_3$ substrates, a transparent electro-optic crystal. The optical waveguide cores on these chips are made by indiffusion of titanium metal. The first example is shown in Fig. 1. It is a demultiplexing circuit

72 Gb/s 2X2 Ti:$LiNbO_3$ OTDM Switch

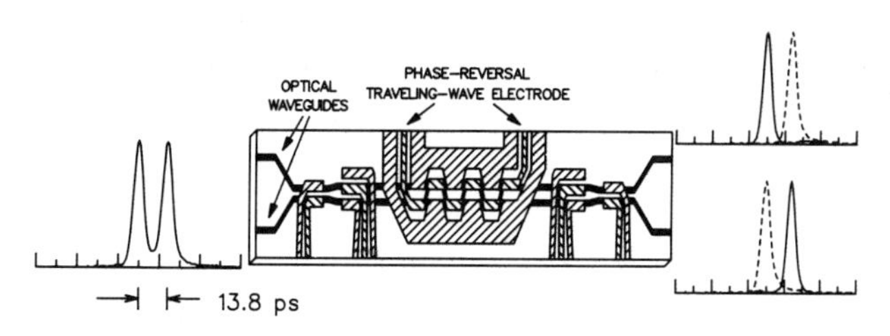

FIG. 1. Optical waveguide (black) and electrode (cross-hatched) structure of a 72 Gb/s integrated optical demultiplexing circuit made on $LiNbO_3$.

capable of operation at 72 Gb/s with a switching energy of only 5.6 pJ [15]. The figure indicates how two input pulses spaced about 14 ps apart were separated by the circuit and placed into the two output waveguides. The chip is 60 × 10 mm in size, contains two optical directional couplers, two output and two input waveguides (in black), and a special high-speed design for the control electrodes (cross-hatched). The purpose of the periodically reversed traveling-wave electrodes is to accomplish velocity matching of both the electrical and optical waves, a key requirement at these high speeds.

Figure 2 shows an optical high-speed 8 × 8 switching fabric coupled to input and output fiber arrays of eight fibers each [16]. The fabric can provide high-capacity optical interconnections from any of the input fibers to any of the output fibers. The interconnect pattern can be changed in times faster than 2.5 ns. The capacity of each interconnect has been tested to speeds up to 4 Gb/s. The switching fabric has been used as a center

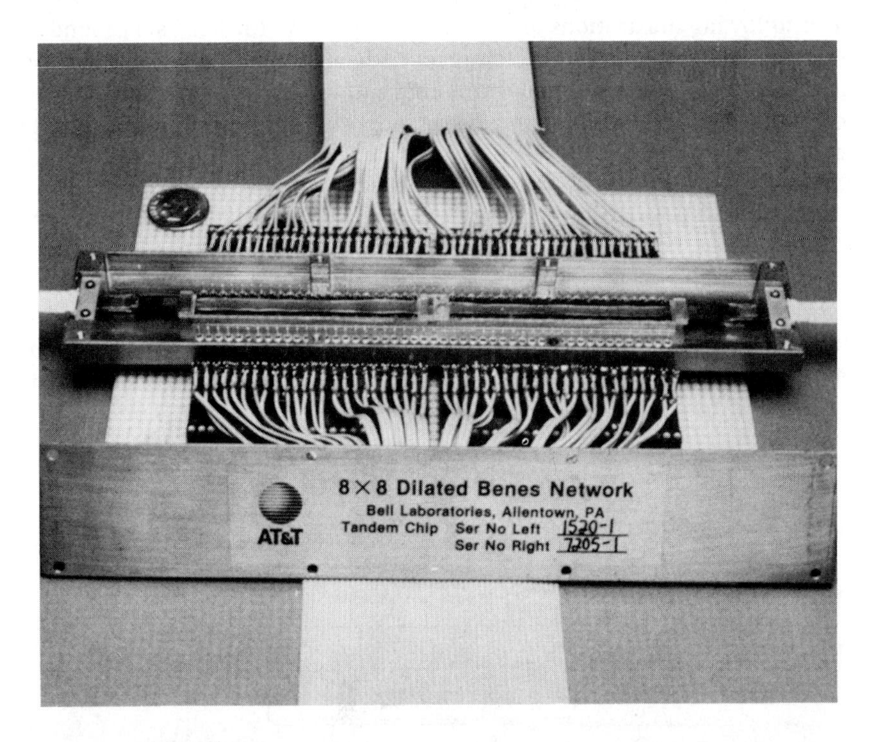

FIG. 2. Photograph of an 8 × 8 optical switching fabric implemented on two butt-coupled $LiNbO_3$ chips. The fiber arrays coupled to the input and output ports are seen on the left and right of the package.

stage in switching experiments employing the "universal time slot" concept. The fabric is designed in a "Dilated Benes" architecture consisting of 48 interconnected optical directional coupler switches. Two chips 66 × 7.6 mm in size were used; they contain 24 couplers each. The optical waveguides of the two chips are linked at the chip interface by low-loss butt coupling. The coupler switches are controlled individually by drive voltages of about 9V each. The figure shows the wiring leading to the corresponding control electrodes.

5. Two OEICs

OEICs are circuits that integrate photonic and electronic devices on a single semiconductor chip. At present, there are two dominant substrate materials, InP and GaAs. The two examples selected for our illustrations are relatively simple OEICs, containing one photodetector each and a few transistors. Both are receiver circuits made on InP for the long-wavelength region of 1.3–1.6 μm. Both combine photoreceivers with electronics and neither OEIC contains optical waveguides. Both OEICs place emphasis on reducing parasities due to high-speed metallic interconnects. The ultimate goal of this reduction of parasities is to achieve performance improvements over hybrid receiver circuits. However, today's receiver OEICs are not outperforming their hybrid counterparts yet. Difficult materials and processing challenges exist to overcome the need for compromising device optimization to allow integration.

We should also point out that OEICs of much higher complexity than those shown here already have been demonstrated. A good example is a receiver OEIC realized in the more mature GaAs IC-technology which contains four photodetectors and as many as 8,000 FETs [17].

Our first OEIC illustration is shown in Fig. 3. It is a sketch of the first long-wavelength OEIC, made in 1980 [18]. It is a p-i-n/FET receiver consisting of one photodiode and one junction FET. The figure shows how the gate electrode of the transistor (cross-hatched) is extended to form the p-i-n diode region of the photodetector. The shaded regions indicate the metallized contacts of the OEIC.

Figure 4 shows the second OEIC example, a p-i-n/HBT receiver [19]. This circuit has utilized recent advances in high-speed heterojunction bipolar transistor technologies. It is made on a semi-insulating InP substrate and contains one photodetector and three HBTs. Two transistors form a low-noise transimpedance preamplifier and the third serves as an output buffer stage. The OEIC has shown a sensitivity of -26.1 dBm at a speed of 1 Gb/s.

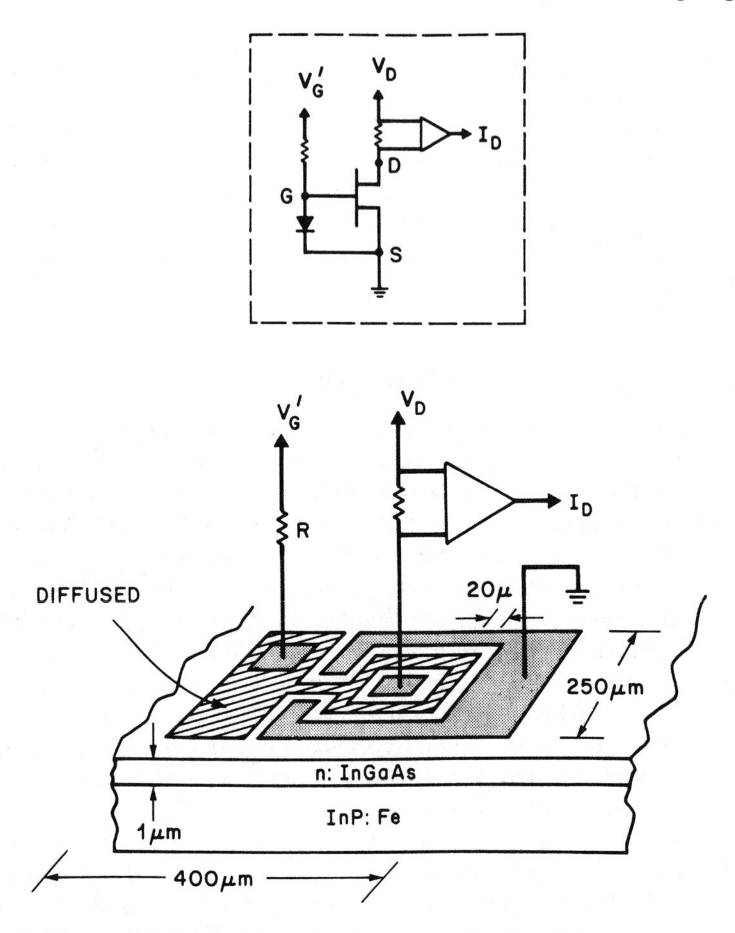

FIG. 3. Diagram of the first long-wavelength p-i-n/FET optical receiver OEIC implemented on an InP substrate.

6. Two PICs

The last pair of illustrations consists of two photonic integrated circuits, a wavelength-division multiplexing (WDM) transmitter chip and a coherent heterodyne receiver chip. Both PICs were made by metal-organic vapor-phase epitaxy on InP semiconductor substrates. Both emphasize on-chip lasers and on-chip low-loss optical guides as interconnects. The on-chip lasers are highly sophisticated devices, each of which could be regarded as a PIC all by itself. These lasers are wavelength-tunable composite junction lasers based on quantum-well structures 5–10 nm thick. The

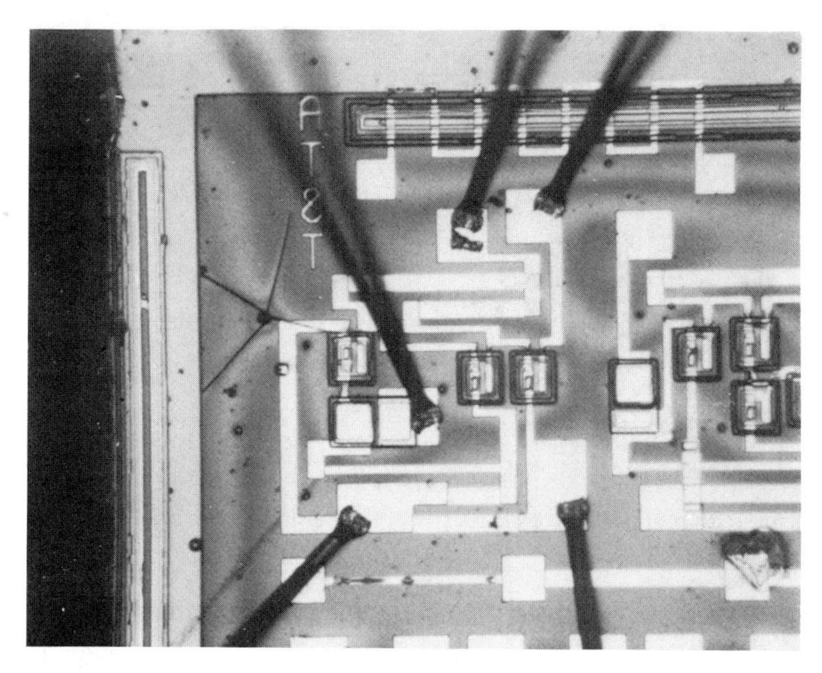

FIG. 4. Photograph of a p-i-n/HBT photoreceiver OEIC consisting of one photodetector and three HBTs. Five dark wire bonds to the circuit are seen. Parts of a neighboring test circuit also can be seen.

chips include the following integrable devices: lasers, optical amplifiers, coupler switches, filters, and detectors. The on-chip optical interconnects are lithographically self-aligned guide connections. They eliminate the need for several expensive laser chip-to-fiber connections and packages. The chips represent considerable advances in crystal growth, processing, and device design.

The *WDM transmitter PIC* is shown in Fig. 5. It was recently tested in a systems experiment transmitting information over 36 km of fiber [12]. The PIC is 3 × 4 mm in size. It contains four independently tunable lasers. For the experiment, their wavelength spacing was set at 2.5 nm. The lasers also can be modulated independently. Modulating signals at 2 Gb/s were used in the experiment for a total chip capacity of 8 Gb/s. The figure also shows the on-chip power-combining interconnect and an on-chip optical amplifier:

The *balanced heterodyne receiver PIC* is shown in Fig. 6. It consists of a tunable local laser oscillator, an adjustable directional coupler switch, and two detectors [11]. The chip size is 3 × .25 mm. This PIC replaces all of the discrete optical components required for coherent optical heterodyne

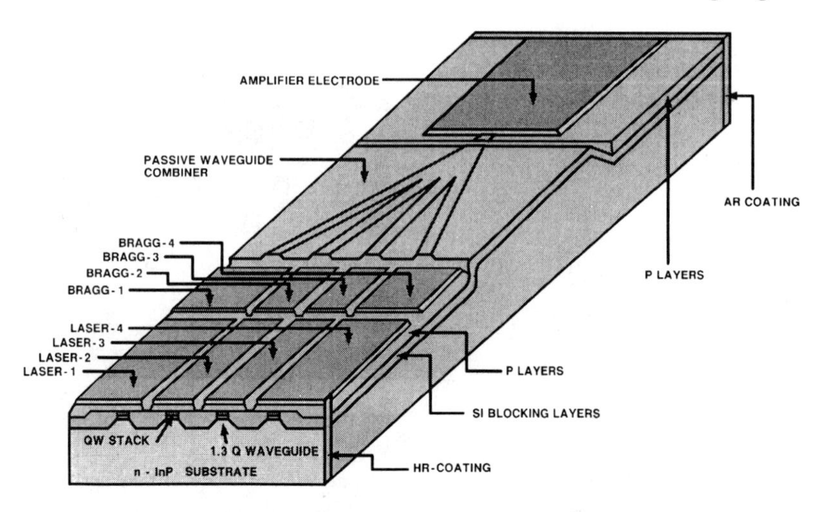

FIG. 5. Sketch of a WDM photonic integrated circuit containing four tunable lasers, a power combiner and an optical amplifier.

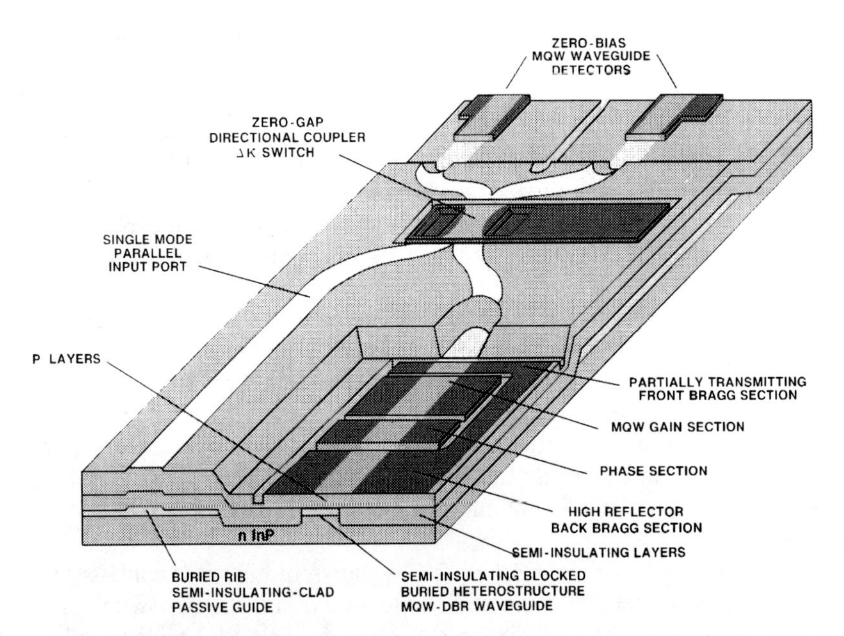

FIG. 6. Sketch of a balanced heterodyne receiver PIC containing a tunable local laser oscillator, an adjustable directional coupler and two quantum-well detectors.

reception. It was tested for frequency-shift-keying heterodyne detection and demonstrated sensitivities of -42.3 dBm and -39.7 dBm at data rates of 108 Mb/s and 200 Mb/s, respectively.

7. Outlook

After more than two decades of exploration, the field of integrated optics finally has seen the commercial application of its products, albeit not yet on a large scale. Considerable advances in materials and processing technology as well as device design, coupled with the increasing complexity of advanced photonic systems, have enhanced the expectation that further commercial applications will follow in the not-too-distant future.

References

1. D. B. Anderson, "Application of Semiconductor Technology to Coherent Optical Transducers and Spatial Filters," in "Optical and Electropt. Information Processing," (J. T. Tippet, editor), p. 221, MIT Press, Cambridge, MA, 1965.
2. R. Shubert and J. H. Harris, "Optical Surface Waves on Thin Films and Their Application to Integrated Data Processors," *IEEE Trans. MTT* **16**, 1048–1054 (Dec., 1968).
3. P. K. Tien, R. Ulrich, and R. J. Martin, *Appl. Phys. Lett.* **14**, 291–294, (May, 1969).
4. S. E. Miller, "Integrated Optics: An Introduction," *Bell Syst. Techn. J.* **48**, 2059–2069 (Sept., 1969).
5. C. P. Lee, S. Margalit, I. Ury, and A. Yariv, "Integration of an Injection Laser with a Gunn Oscillator on a Semi-Insulating GaAs Substrate," *App. Phys. Lett.* **32**, 806–807 (June, 1978).
6. R. G. Hunsperger, "Integrated Optics: Theory and Technology," Springer, Berlin, 1984.
7. L. D. Hutcheson (ed.), "Integrated Optical Circuits and Components: Design and Applications," Marcel Dekker, New York, 1987.
8. T. Tamir (ed.), "Guided-Wave Optoelectronics," Springer, Berlin, 1988.
9. M. Hirano, "OEIC and the Research Activity of the Optoelectronics Joint Research Laboratory," *Optoelectronics* **2**, 137–175 (Dec., 1987).
10. R. C. Alferness, "Waveguide Electro-optic Switch Arrays," *IEEE J. Selected Areas in Communications* **6**, 117–1130 (August, 1988).
11. T. L. Koch, F. S. Choa, U. Koren, et al., "Balanced Operation of an InGaAs/InGaAsP Multiple-Quantum-Well Integrated Heterodyne Receiver," *Digest Topical, Meetg. on Integrated Photonics Research*, Optical Soc. of Am., PD-5, 1990.
12. A. H. Gnauck, U. Koren, T. L. Koch, et al., "Four-Channel WDM Transmission Experiment Using a Photonic-Integrated-Circuit Transmitter," *Digest Conf. Opt. Fiber Communications*, Optical Soc. of Am., PD-26 1990.
13. H. Kogelnik and C. V. Shank, "Stimulated Emission in a Periodic Structure," *Appl. Phys. Lett.* **18**, 152–154 (February, 1971).
14. R. L. Jungerman, C. A. Johnsen, et al., "High Speed Optical Modulator for Application in Instrumentation," *Digest Conf. Optical Fiber Comm.*, Optical Soc. of Am., FB2 1990.
15. S. K. Korotky and J. J. Veselka, "Efficient Switching in a 72 Gb/s Ti:$LiNbO_3$ Binary Muliplexer/Demultiplexer," *Digest Conf. Optical Fiber Comm.*, Optical Soc. of Am., TUH2 1990.

16. J. E. Watson, et al., "A Low-Voltage 8 × 8 Ti:LiNbO$_3$ Switch with a Dilated-Benes Architecture," *J. Lightw. Techn.*, **8**, 794–801 (May, 1990).
17. J. D. Crow, "Optoelectronic Integrated Circuits for High-Speed Computer Networks," *Digest Conf. Optical Fiber Comm.*, Optical Soc. of Am., WJ3 1989.
18. R. F. Leheny, R. E. Nahory, M. A. Pollack, et al., "Integrated InGaAs p-i-n FET Photoreceiver," *Electronics Lett.* **16**, 353–355 (May, 1980).
19. S. Chandrasekhar, et al., "An InP/InGaAs p-i-n/HBT Monolithic Transimpedance Photoreceiver," *Photonics Technology Lett.*, **2**, (July, 1990).

CHAPTER 2

Quantum Optoelectronics for Optical Processing

D. A. B. Miller
AT&T Laboratories
Holmdel, New Jersey

Almost since the invention of the laser, we have been tantalized by the possibility of the use of optics for processing information. Linear optics, we know, offers so many surprising and useful capabilities. It can communicate and manipulate vast amounts of information with simple lenses and mirrors. It can perform sophisticated operations, such as Fourier transforms and correlations. Yet, given our current preoccupation with processing information, its actual use here has been disappointingly small. The reason for this is simple. The majority of tasks we have to perform on information (and, arguably, the most useful ones to us) involve the interaction between one piece of information and another. Fundamentally, this interaction is a nonlinear process.

For example, we may be trying to search a database for some information. We compare our starting knowledge, such as an author name, with the records in the database. This comparison is a nonlinear, logical operation, although at first sight, a fixed comparison can appear to be a linear operation. (We could do such a search by comparing with a fixed mask, using linear optics.) Looking at the records we have retrieved, we have to modify the comparison to be made, adding another key term, and so on; this involves making a new mask, an operation that, essentially, is nonlinear. Importantly, therefore, we generally do not know in advance what the comparison is to be. A simple fixed comparison is not enough. We need

ISBN 0-12-289690-4

the ability to use the output of the present operation as the input to the next. We may fool ourselves into thinking that we can do such operations only with linear optics, but we can never get to this next stage without nonlinearity. It is, arguably, the absence of adequate nonlinear physical systems—not only with optical inputs, but also with optical outputs—that has prevented us from using optics in many important information-handling tasks.

Primarily, the problem is a physical one, not one of engineering. The phenomena of nonlinear optics have been a disappointment as far as useful devices for information processing are concerned. The effects of classical nonlinear optics, such as those that give wave-mixing phenomena, largely are too weak, and most of those that are not too weak are too slow to be of serious interest. Furthermore, there are other physical issues that we have ignored too often in trying to make useful optical "gates." Issues such as good input/output isolation and avoiding any critical biasing of the devices, both attributes of transistor logic circuits, are extremely important in allowing us to make large systems [1–3].

Hence, we need a different approach. We need both strong physical effects and device configurations that operate in qualitatively different ways from those of classical nonlinear optics of bulk crystals. The approach here is to try to use a new opportunity in materials technology and device physics: namely, the increasingly sophisticated technology of layered semiconductor growth. This technology lets us make structures such as *quantum wells*, which consist of very thin layers of different semiconductors [4]. Techniques such as molecular beam epitaxy (MBE) and metal-organic chemical vapor deposition (MOCVD, also known as organometallic vapor phase epitaxy, or OMVPE can controllably grow structures with multiple layers of different semiconductor materials. This technology offers two benefits: one in new or improved physical effects through the quantum mechanics of thin layers, the second in integration of optical and electronic device structures. The combination of these benefits allows us to coin a new term to describe this growing field, hence the *quantum optoelectronics* in the title of this chapter.

The first benefit of layered semiconductor growth comes because it can grow semiconductor layers in a controlled manner with thicknesses down almost to single atomic layers (e.g., 2–3 Å). Hence, we can design and engineer using quantum mechanics. This new design freedom lets us make new kinds of structures, fundamentally different from the bulk materials we had available before. Some of these structures show physical effects that are qualitatively different. One example is the quantum-confined Stark effect, a large electroabsorptive mechanism that allows us to make new kinds of optical modulators and switches with low operating energy

[5–7]; another example is resonant tunneling [8], an electronic transport effect that can give negative resistance for electrical oscillators and switches. In other structures, we also can use the new freedom to improve existing effects quantitatively, as in the quantum-well laser diode [9]; here, the quantum mechanics improves the form of the density of states, giving us more laser gain for a given carrier density, and the small gain volumes possible with a thin-layer-gain medium mean that we can make lasers with lower thresholds.

The second benefit, integration, is profoundly important for the use of optics in processing information. This growth technology allows us to make optical structures, such as mirrors, waveguides, and Fabry-Perot resonators; electronic devices, such as transistors of various kinds; and optoelectronic devices, such as laser diodes and quantum-well modulators and switches. The flexibility of the technology, therefore, let us combine these various devices in integrated structures. One physical benefit at the microscopic level is the potential to build physical processes and devices that neither are simply optical nor electronic, but are fundamentally optoelectronic. This is an exciting area for future basic research. This integration also buys us a benefit that goes beyond physics into the realm of systems. That benefit is flexibility of functionality—in other words, we can choose what the device or system does for us. At the device level, it will allow us to make our devices operate in an acceptable way [1, 2]. For example, we will be able to make devices that isolate the input from the output. This isolation is important for designing large systems, because it decouples the system from all that follow it. Most nonlinear optical processes can run equally well backwards or forwards; hence, they have poor input/output isolation. Another example is that we will be able to avoid devices that individually require critical setting of some parameter to make them work (critical biasing); such devices are not acceptable if we are to make large systems. At the macroscopic level, the flexibility of functionality enables us to make logically complex systems; an extreme example would be an optically interconnected integrated circuit, which uses electronic logic to perform some complex function and uses optics to interconnect the information to other circuits. We should not underestimate the importance of complex functionality; it is, arguably, the ease with which electronic systems can perform many different complex functions that makes them so pervasive.

The quantum mechanics that we use in most of these structures is relatively simple in practice, being based on the "particle-in-a-box" problem familiar from elementary quantum mechanics. When we work with normal semiconductors, we treat the electrons in the material essentially as classical particles. Current flows because the classical charged masses, the electrons, move when we apply a field. In very thin layers, however, we cannot neglect the fact that electrons also are waves. The waves reflect back and forth between the "walls" of the layers. They resonate in much the same way as optical waves in a resonant cavity. Therefore, when the layers are significantly thinner than the coherence length of the waves, these electron wave interference effects become strong. Since coherence lengths of electron waves may be ~1000 Å at room temperature, this kind of interference effect can be strong under practical conditions with this growth technology. We see the consequences of this interference clearly in both electrical and optical effects in these materials.

The classic structure that shows these wave effects is a *quantum well.* A typical structure would consist of a gallium arsenide (GaAs) layer, perhaps ~100 Å thick, between aluminium gallium arsenide (AlGaAs) layers. In practice, we might make many such layers to give a multiple quantum-well structure of alternating GaAs and AlGaAs layers, each ~100 Å thick. (There also are many other semiconductors we can use to make such structures.) Both the electrons in the conduction band and their equivalent in the valence band, the positively charged holes, see the GaAs layer as a potential well. On solving Schroedinger's wave equation for this simple system, we find the "standing wave" eigenfunctions as shown in Fig. 1 for the first two levels, $n = 1$ and $n = 2$. The solutions of this wave equation are the only allowed motions of the electron (or the hole) in the direction perpendicular to the layers. One immediate consequence of these quantized levels is that the optical absorption spectrum breaks up into a series of steps, spaced essentially by the separation of these levels, as shown in

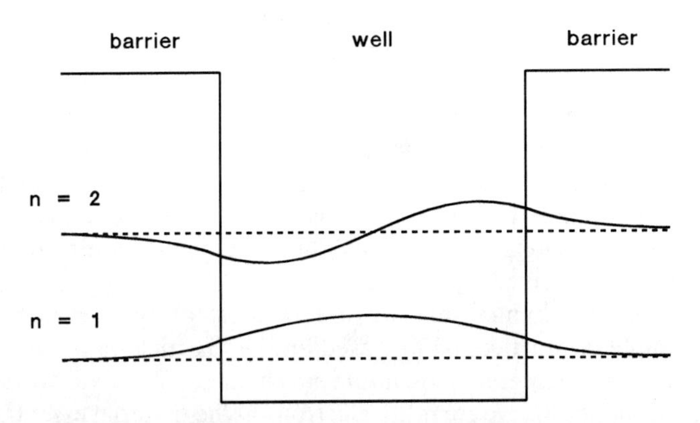

FIG. 1. Wavefunctions of the first two confined levels ($n = 1$ and $n = 2$) of an electron in a quantum well. These actually are calculated wave functions and the potential for an electron in a 95 Å GAs quantum well surrounded by $Al_{0.3}$ $Ga_{0.7}$ As barrier layers.

Fig. 2. Associated with each of these steps is a sharp peak, called the *exciton peak*. Absorption at the exciton peak creates an electron and a hole orbiting around one another like a large hydrogen atom. The quantum well squeezes the exciton, making it smaller and more stable, and allows these peaks to be seen clearly, even at room temperature. By contrast, in normal semiconductors, such peaks usually are clear only at low temperature. Thus, even in a simple absorption spectrum at room temperature, we see large changes resulting directly from the quantum mechanics of the layered structure.

The quantum-confined Stark effect [5–7] is a good example of an effect in such structures that is qualitatively different from those seen in usual materials. By applying an electric field perpendicular to the layers, we find that we can shift the exciton peaks to lower energies. This shift occurs because we are polarizing the exciton. The difference, compared to the effect seen in bulk materials, is that the walls of the quantum well hold the exciton together even with large fields, allowing large polarizations (and shifts) without destroying the particle and its associated absorption peak.

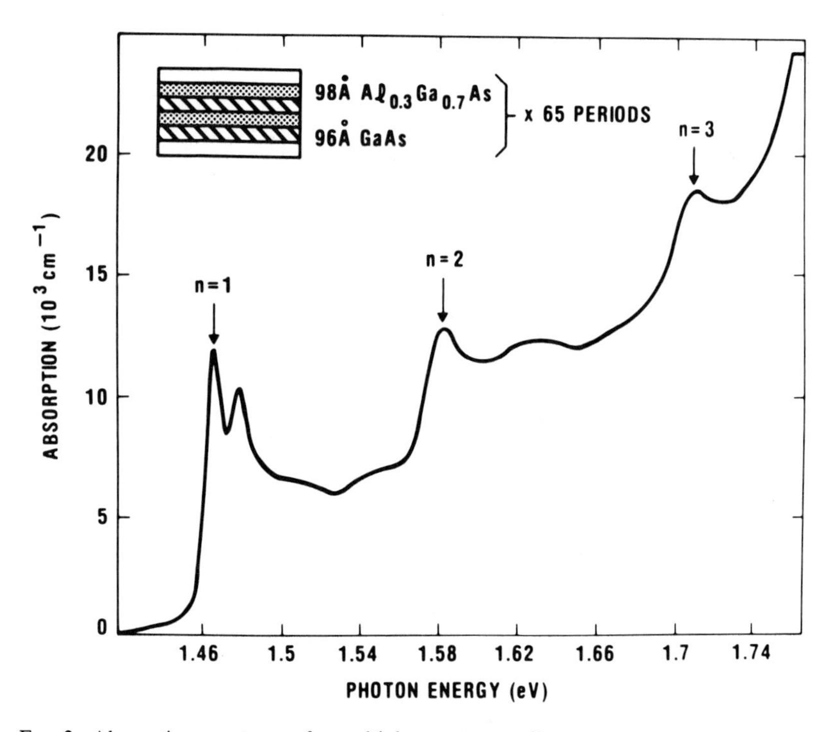

FIG. 2. Absorption spectrum of a multiple quantum-well structure at room temperature.

The necessary fields can be applied by including quantum wells in the intrinsic (i) region of a p-i-n diode and reverse-biasing the diode. This gives both a low energy absorption modulation mechanism and a simple structure capable of modulating light propagating perpendicular to the surface. This latter fact is very important because it allows us to make two-dimensional arrays of modulators and switches.

An example of the kinds of novel devices that can be made with these effects is the *self-electro-optic effect device* (SEED) [6, 7, 10, 11]. In the SEED, the quantum-well modulator is combined with photodetection. As we shine light on the photodetector, the resulting photocurrent through some circuit causes the voltage to change across the modulator, modulating an output beam. Hence, we have a simple device with both an optical input and an optical output. If integrated, such devices can be very efficient, despite the fact that they convert light to electricity and back to light. They also can be fast; a switching time of 33 ps has been reported [12], and this does not appear to represent any fundamental limit. The conversion from optics to electronics and back does not, therefore, result in a slow device. Input optical switching energies are ~1 pJ in $10 \times 10\,\mu m$ devices in the devices demonstrated at this time [13].

The simplest forms of SEED are bistable devices, without any transistors, and can use the quantum-well diode simultaneously as both detector and modulator. A simple bistable device usually is critically biased, however: It must be set near a switching point so that a small additional beam can trip it over; otherwise, we have no signal gain. A solution is the so-called *symmetric SEED* (S-SEED) [13, 14]. This device is more complex, comprising two quantum-well diodes in series with a power supply. This complexity gives us the improved device functionality that we need. The S-SEED is bistable in the ratio of the two light beams that shine, one on each diode. Hence, we can switch it at low total power and read it out with high power to get a "time-sequential" gain. At no time is the device ever biased close to its switching point, so there is no critical biasing. It also has reasonable input/output isolation. Therefore, this device is good enough to be used in optical logic systems experiments [15]. Partly because it is a relatively simple device, it can be made in large arrays, such as the 64×32 array shown in Fig. 3 [13]. Each of the S-SEEDs can be used as a two-input NOR gate. The array, therefore, corresponds to a chip with 6,144 logical connections, two inputs and one output per gate, all of which fit within the center of a typed letter "o." All these connections are externally accessible in parallel with two-dimensional arrays of light beams. To operate the entire array at once requires a total of 16,384 light beams, a formidable challenge for optics. Such a device is experimental, but working demonstration systems have been made. It can be taken as indicative of the kinds

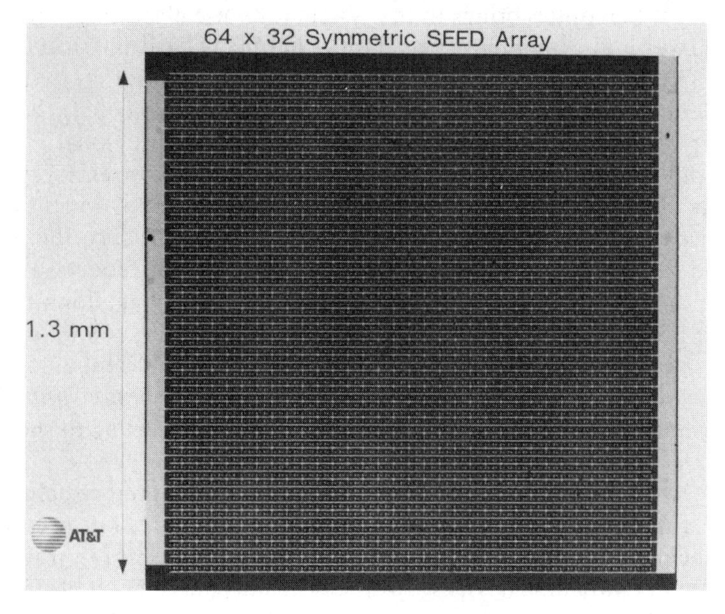

FIG. 3. 64 × 32 array of symmetric self-electro-optic effect devices for optical logic or memory.

of novel devices made possible by such technology. There are, for example, many different configurations of SEEDs (without transistors) [10, 11], including oscillators, spatial light modulators, complex logic devices [16], multi-stable systems, and optical level shifters.

There is no fundamental reason not to include more electronic components in devices such as SEEDs, of course. The more components that can be included, the more sophisticated the functionality of the systems can become. The price is that inclusion of more and different components makes integration harder. Integrating different devices usually entials compromise in each device. Fortunately, these quantum-well modulators are relatively easy to make; hence, they have good prospects for integration. For example, they already have been integrated with GaAs field effect transistors in simple demonstrations [17]. The potential exists for transistor circuits with small optical input and output pads distributed at will over the entire chip. There also are similar possibilities for GaAs/AlGaAs modulators grown on Si. Demonstrations have shown working modulators with good lifetimes after being grown on Si substrates [18]. Ultimately, it may be possible to grow on Si integrated circuits, thereby combining the complexity of silicon electronics with the connection abilities of optics.

What is it that optics brings to the system if we use electronic logic? One clear answer is in communications. Optics has many basic physical advantages [1–3]. It has as large a bandwidth as we can use, with no frequency-dependent loss or cross talk. It avoids problems of keeping constant ground potentials throughout the system. It is very good at global interconnect topologies, with many crossing "wires." It can connect very large numbers of devices. One other communication advantage of optics that has been discussed recently is that, fundamentally, it can reduce the energy required for communication of logic level signals. This reduction occurs because optical devices can be quantum devices, a fact that allows them to perform a simple effective impedance transformation [19]. As we are able to make well-integrated, small optical devices that are efficient at low total powers, we will be able to take advantage of this *quantum impedance transformation*. Devices such as S-SEEDs already are starting to show this benefit.

There are many other devices possible with such layered structures, of course; a full catalog would be too long to include here, but it certainly would include many other modulators, detectors [20], lasers [9], transistors [8], and other switching devices, both optical and electronic. It also is quite likely that the most useful devices are yet to be invented. We may see for example, devices in which the processes are not distinguishably optical or electronic even at the most basic level, perhaps involving novel, ultrafast optoelectronic processes.

Therefore, this combination of optics, electronics, and quantum physics looks to be very powerful. The new quantum-mechanically-engineered mechanisms are so strong that they make a qualitative difference to optoelectronics; we now can have both optical inputs and optical outputs that operate at the same energy densities as efficient electronic devices. Furthermore, it appears that the prospects are good for an optoelectronic integration technology with very large numbers of optical input and output devices.

What can we speculate for the future? At the very least, with the existing layered semiconductor technology, we can expect continued evolution of better quantum-mechanically-engineered structures for particular applications. This engineering is used already in the design of better modulator and switch structures, and it will remain a fertile field for invention for some time. Certainly, we can hope for more and more integration of optics and electronics. This now is largely a matter of technology, although it remains a major challenge. The need for integration is clear from a systems perspective. The abilities of optics and electronics largely are complementary. Optics excels in communication. Electronics is the only way we have of making complexity of function. We cannot neglect the importance of complexity. In fact, we also could argue that the historic inability to make optical systems that can perform complex tasks has prevented their wider use. On the other side, if we could relieve electronic systems of the communications problems that currently limit them, it is difficult for us to imagine the improvements in systems that might result. All our notions of processor architecture might change. It seems unlikely that the optimum processor architecture just happens to be the one that exactly matches the abilities and shortcomings of electronics.

Can we hope for new classes of quantum devices for optics? Perhaps we can. It is fairly clear from a fundamental point of view that, in principle, we could make better optical nonlinearities and electro-optical effects if we had the ability to make arbitrary structures that were quantum-confined in all three directions [21]. There is no clear technology to do this at the present time, however. We arguably have possibilities in semiconductors through quantum wires and quantum dots, but we know of no way of making any large number of them with the very high uniformity that we would need for a simple extension of current quantum-well ideas. One tantalizing aspect is that one interesting size of structure lies exactly between two technologies. With semiconductor quantum dots, crystallites with about 1,000 atoms would have all the advantages of quantum-confined semiconductors without the disadvantages of some smaller molecular structures. (In molecules, the optical absorption strength often is spread over many vibrational sidebands, reducing the concentration of oscillator strength that is one of the main aims of quantum confinement.) However, 1,000 atoms is too small to make reproducibly by simple lithography, but too large to make reproducibly by simple chemistry. Several novel techniques are being tried [22]. Perhaps the answer lies in organic materials, some of which are natural quantum wires, for example. Here, too, the materials issues are not clear, however. Can we make the exact material we want? Will it be stable and easy to grow? Could we use biological processes to grow very uniform mesoscopic systems?

There are no clear answers currently to these questions of future physics and technologies. It is clear, however, that even with the technology we have now, we are able to make devices that simply were not possible before. These devices may be the key to the marriage of optics and electronics, since they offer the ability to have efficient optical inputs and outputs in conjunction with electronics. The real benefit of such a union is to better system performance overall, using optics and electronics each where it is best. If this is achieved, the dogmatic arguments of optics against electronics will become thankfully irrelevant.

References

1. D. A. B. Miller, "Optical Switching Devices: Some Basic Concepts," in "Optical Computing," (B. S. Wherrett and F. A. P. Tooley, eds.) pp. 55–70, Adam Hilger, Bristol, 1989.
2. D. A. B. Miller, "Device Requirements for Digital Optical Processing," *SPIE Critical Reviews* (to be published).
3. N. Streibl, K.-H. Brenner, A. Huang, J. Jahns, J. Jewell, A. W. Lohmann, D. A. B. Miller, M. Murdocca, M. E. Prise, and T. Sizer, "Digital Optics," *Proc. IEEE* **77**, 1954–1969 (1989).
4. For an introductory discussion of quantum wells and their optical applications see D. A. B. Miller, "Optoelectronic Applications of Quantum Wells," *Optics and Photonics News* **1** (2), 7–15 (Feb., 1990).
5. For extended discussion of quantum well optical properties and applications, see D. A. B. Miller, "Quantum-Well Optoelectronic Switching Devices," *Int. J. High Speed Electron.* **1**, 19–46 (1990).
6. D. A. B. Miller, D. S. Chemla, T. C. Damen, A. C. Gossard, W. Wiegmann, T. H. Wood, and C. A. Burrus, "Electric Field Dependence of Optical Absorption near the Band Gap of Quantum-Well Structures," *Phys. Rev. B* **32**, 1043–1060 (1985).

7a. For a longer discussion of quantum well electrooptical properties and applications, see D. A. B. Miller, D. S. Chemla, and S. Schmitt-Rink, "Electric Field Dependence of Optical Properties of Semiconductor Quantum Wells," in "Optical Nonlinearities and Instabilities in Semiconductors," (H. Haug, ed.), pp. 325–359, Academic Press, San Diego, 1988.

7b. S. Schmitt-Rink, D. S. Chemla, and D. A. B. Miller, "Linear and Nonlinear Optical Properties of Semiconductor Quantum Wells," *Adv. Phys.* **38**, 89–188 (1989).

8. S. Sen, F. Capasso, and F. Beltram, "Resonant Tunneling Diodes and Transistors: Physics and Circuit Applications," in "Introduction to Semiconductor Technology: GaAs and Related Compounds," (C. T. Wang, ed.,), pp. 231–301, Wiley, New York, 1990.
9. N. K. Dutta, "Physics of Quantum Well Lasers," in "Heterojunction Band Discontinuities: Physics and Device Applications," (F. Capasso and G. Margaritondo, eds.,), pp. 565–593, North-Holland, Amsterdam, 1987.
10. D. A. B. Miller, D. S. Chemla, T. C. Damen, T. H. Wood, C. A. Burrus, A. C. Gossard and W. Wiegmann, "The Quantum-Well Self-Electro-Optic Effect Device: Optoelectronic Bistability and Oscillation, and Self-Linearized Modulation," *IEEE J. Quantum Electron.* **QE-21**, 1462–1476 (1985).
11. D. A. B. Miller, "Quantum-Well Self-Electro-Optic-Effect Devices," *Opt. Quantum Electron.* **22**, S61–S98 (1990).
12. G. D. Boyd, A. M. Fox, D. A. B. Miller, L. M. F. Chirovsky, L. A. D'Asaro, J. M. Kuo, R. F. Kopf, and A. L. Lentine, "33 ps Optical Switching of Symmetric Self Electro-Optic Effect Devices," *Appl. Phys. Lett.* **57**, 1843–1845 (1990).
13. A. L. Lentine, F. B. McCormick, R. A. Novotny, L. M. F. Chirovsky, L. A. D'Asaro, R. F. Kopf, J. M. Kuo, and G. D. Boyd, "A 2-kbit Array of Symmetric Self-Electro-Optic Effect Devices," *IEEE Photonics Tech. Lett.* **2**, 51–53 (1990).
14. A. L. Lentine, H. S. Hinton, D. A. B. Miller, J. E. Henry, J. E. Cunningham, and L. M. F. Chirovsky, "Symmetric Self-Electro-Optic Effect Device: Optical Set-Reset Latch, Differential Logic Gate, and Differential Modulator/Detector," *IEEE J. Quantum Electron.* **25**, 1928–1936 (1989).

15. M. E. Prise, R. E. LaMarche, N. C. Craft, M. M. Downs, S. J. Walker, L. A. D'Asaro, and L. M. F. Chirovsky, "A Module for Optical Logic Circuits Using Symmetric Self Electro-Optic Effect Devices," *Appl. Optics* **29**, 2164–2174 (1989).
16. A. L. Lentine, D. A. B. Miller, J. E. Henry, J. E. Cunningham, L. M. F. Chirovsky, and L. A. D'Asaro, "Optical Logic Using Electrically Connected Quantum-Well PIN Diode Modulators and Detectors," *Appl. Optics* **29**, 2153–2163 (1990).
17. D. A. B. Miller, M. D. Feuer, T. Y. Chang, S. C. Shunk, J. E. Henry, D. J. Burrows, and D. S. Chemla, "Field-Effect Transistor Self-electro-Optic Effect Device: Integrated Photodiode, Quantum-Well Modulator and Transistor," *IEEE Photonics Technol. Lett.* **1**, 62–64 (1989).
18. K. W. Goossen, G. D. Boyd, J. E. Cunningham, W. Y. Jan, D. A. B. Miller, D. S. Chemla, and R. M. Lum, "GaAs-AlGaAs Multiquantum-Well Reflection Modulators Grown on GaAs and Silicon Substrates," *IEEE Photonics Technol. Lett.* **1**, 304–306 (1989).
19. D. A. B. Miller, "Optics for Low-Energy Communication inside Digital Processors: Quantum Detectors, Sources, and Modulators as Efficient Impedance Converters," *Optics Lett.* **14**, 146–148 (1989).
20. B. F. Levine, G. Hasnain, C. G. Bethea, and N. Chand, "Broadband 8-12 μm High-Sensitivity GaAs Quantum-Well Infrared Photodetector," *Appl. Phys. Lett.* **54**, 2704–2706 (1989).
21. D. A. B. Miller, D. S. Chemla, and S. Schmitt-Rink, "Electroabsorption of Highly Confined Systems: Theory of the Quantum-Confined Franz-Keldysh Effect in Semiconductor Quantum Wires and Dots," *Appl. Phys. Lett.* **52**, 2154–2156 (1988).
22. M. L. Steigerwald and L. E. Brus, "Synthesis, Stabilization, and Electronic Structure of Quantum Semiconductor Nanoclusters," in "Annual Review of Material Science," (R. A. Huggins, J. A. Giordmaine, and J. B. Wachtman, eds.), pp. 471–95, Annual Reviews, Palo Alto, CA, 1989.

CHAPTER 3

Optics in Telecommunications: Beyond Transmission

Peter W. E. Smith
Bellcore
Red Bank, New Jersey

1. Introduction

There is at present a growing trend towards the installation of an optical fiber telecommunications network all—or most—of the way to the home [1]. One reason is simply an economic one: as shown in Fig. 1, the cost of installing a fiber telephone line is projected to drop below that of a copper line early in this decade. Another reason is that the existence of a

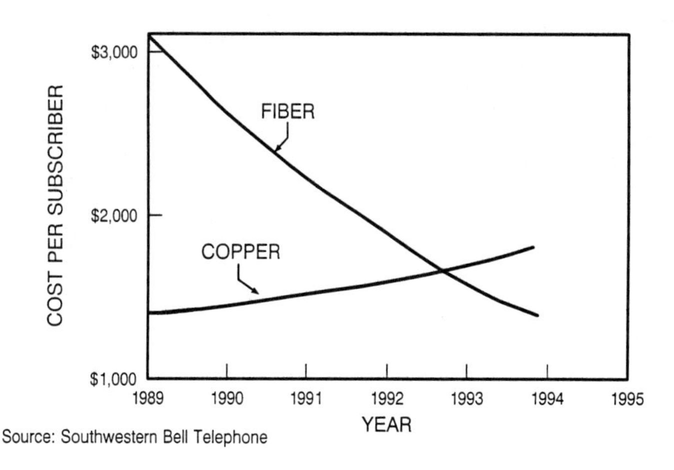

FIG. 1. Estimated cost of installing a telephone line.

ISBN 0-12-289690-4

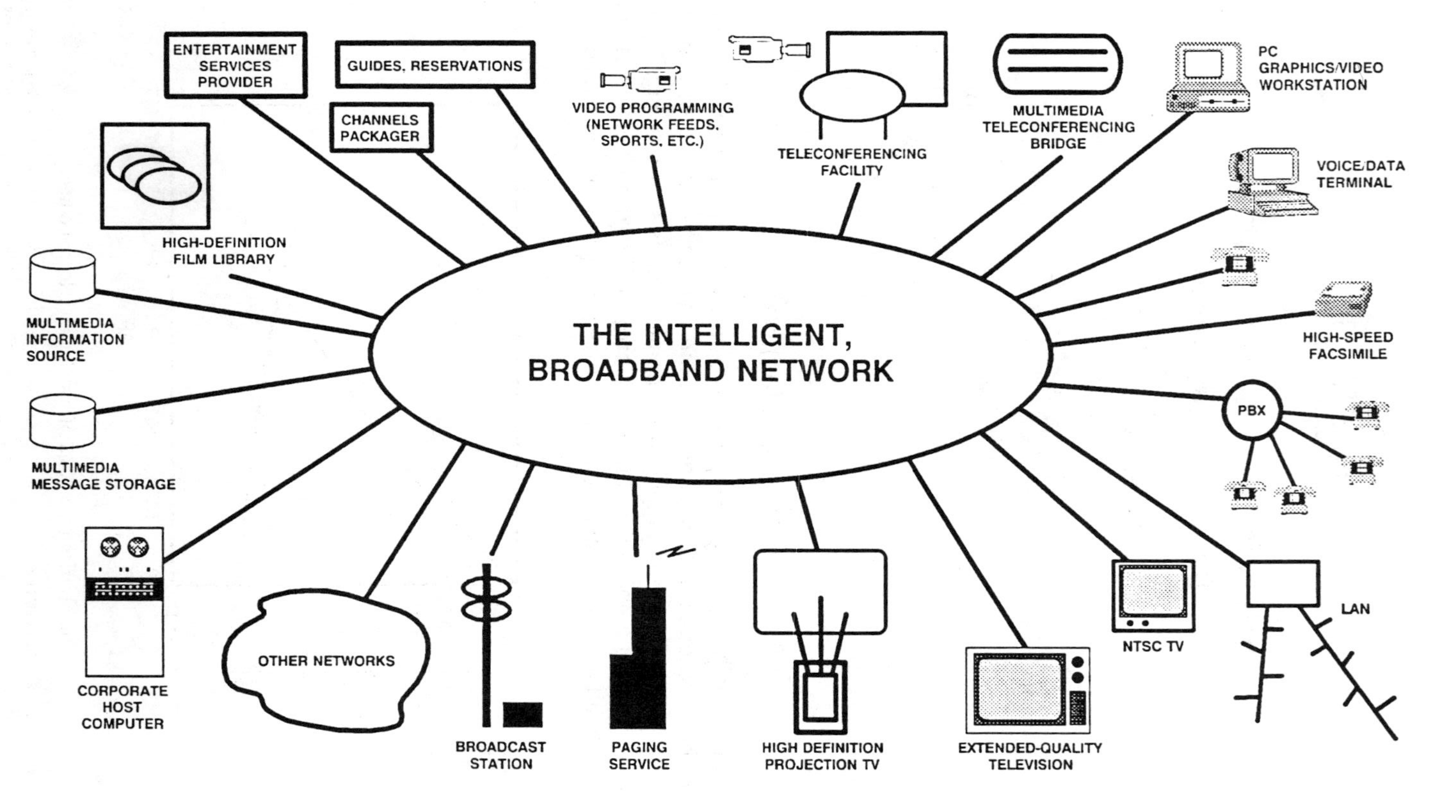

FIG. 2. Potential services in the future intelligent broadband network.

very wide bandwidth transmission channel makes it possible to envision a future high-bandwidth, intelligent telecommunication network capable of providing a wide range of "information age" services (Fig. 2). To make such a dream a reality, it will be necessary to develop switching, processing, and control technology that will be capable of handling a wide variety of broadband optical signals of different formats in a flexible and economical way. What role can we expect optics to play in providing the switching, control and "network intelligence?"

The often-quoted strengths of optical devices are high bandwidth, transparency, parallelism, and speed. The weaknesses are large size, high power requirements, and small nonlinearity (difficulty of sensing an optical signal) [2]. To compete with existing technology, optical devices and circuits must exhibit at least some of the following advantages: better performance; higher functionality; lower costs/power dissipation; ease of fabrication/packaging, etc.; new service capabilities; and higher levels of integration.

Although for most optical devices, these advantages remain to be proven it is clear that the bit-rate capacity of present-day lightwave communications systems is limited by the electronic parts of the system. As the trend towards still higher capacity systems continues, it will become more and more necessary to perform some of the signal-processing operations with photonic components to eliminate the electronic bottleneck (Fig. 3). There

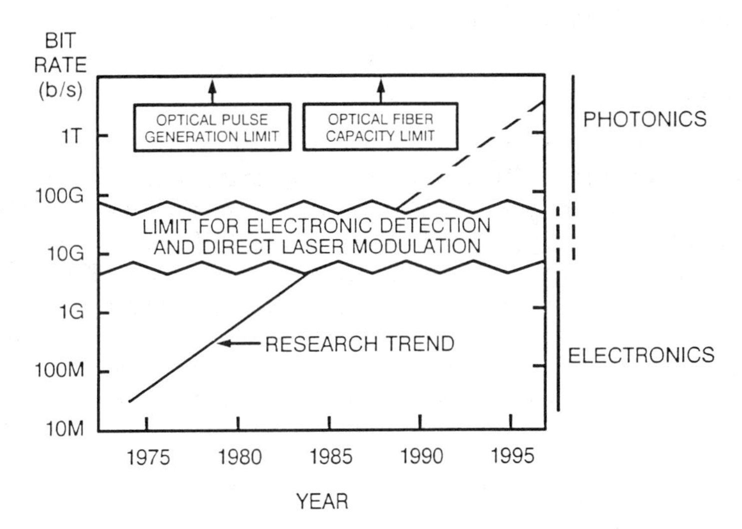

FIG. 3. Trends in lightwave communications systems. Present systems are limited not by the ability to generate short optical pulses, nor by the capacity of the optical fiber. They are limited by the electronics used for laser modulation and detection.

have been tremendous strides in recent years in the development of a photonics technology for the generation, control, and detection of optical signals. There also has been remarkable progress in the integration of photonic and electronic circuits. In this chapter, we will review these technology trends briefly and explore their implications for future telecommunications networks.

2. Technology Trends

2.1. Devices

Semiconductor diode lasers are evolving rapidly in many ways [3]. Low-cost, reliable, and long-life lasers have been demonstrated. Power levels and efficiencies are such that many applications previously requiring other types of lasers are now being considered. Tunable, narrow-linewidth lasers are making possible new applications for coherent and wavelength division systems. The recent development of low-threshold (<1 mA) lasers and arrays of surface-emitting lasers has opened up new opportunities for optical interconnects and optoelectronic integration.

Similar progress has been made in other photonic device areas. Broad-band, high-gain amplifiers have been made and tested in a variety of systems experiments. Tunable filters have made possible dense wavelength division multiplexing networks. Detectors have been developed that can be integrated efficiently with optical waveguides.

2.2. Integration

Aided by advances in materials and processing technology, optoelectronic integrated circuits (OEICs) have advanced to the point where the performance of OEICs is approaching the performance of hybrid circuits [4]. This is illustrated in Fig. 4, which shows that in 1985, simple OEIC receiver has sensitivities at least 25 dB poorer than the best hybrid receivers. By the end of 1989, however, OEIC technology had improved to the point where OEIC receiver performance was within 4 or 5 dB of the best hybrids. High-speed integrated optical transmitters and receivers have been demonstrated. Figure 5 shows an OEIC transmitter that has recently been demonstrated to operate at 10 Gbit rates. The trend to more complex circuits incorporating large numbers of individual elements (lasers, transistors, etc.) is very evident. These higher levels of integration offer major advantages in many projected applications.

One technique for optoelectronic integration is a lift-off technique recently developed at Bellcore [5]. This technique allows an optical element (such as a detector or laser) to be removed from the substrate on which it

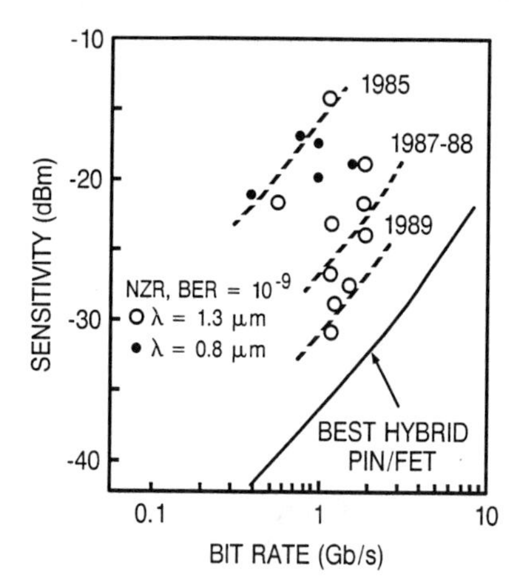

FIG. 4. Trends in OEIC receiver performance. The receivers consist of a PIN photodetector and a field effect transistor (FET).

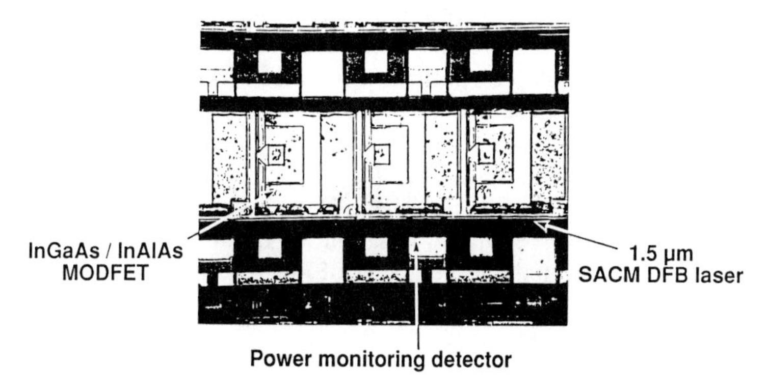

FIG. 5. 10 Gbit transmitter OEIC consisting of a MODFET transistor and a long wavelength distributed feedback laser.

was grown and processed, and attached to another substrate. In this way, lasers have been mounted on silicon and glass substrates, and detectors have been integrated with glass and $LiNbO_3$ optical waveguides. With this lift-off technique, hybrid integrated optical circuits can be made using components that are fabricated using their optimum material technology.

Self electro-optic devices (SEEDs) are optical elements whose transmission can be controlled by optical or electrical signals. Recently, large arrays of these elements have been fabricated, and have been used in a demonstration of optical computation [6]. Arrays of elements which combine optical light emitters with electronic control—the so-called VSTEP devices—also have been shown to perform optical switching and signal-processing functions [7]. The ability to fabricate large arrays of optoelectronic elements opens up the possibility of information-processing systems that exploit the parallelism of optics to make large numbers of simultaneous high-bandwidth interconnects.

2.3. Towards Applications

Active programs are under way in many institutions to study the ways in which optical interconnects can be used to overcome problems such as pin-out limitations and clock skew in conventional integrated electronic circuits. J. D. Crow at IBM has recently described an optoelectronic circuit designed for optical interconnection of multiprocessor networks [8]. It contains four lasers, four photodiodes, and 8,000 FETs; it operates at a 1 Gb/s rate.

The dense interconnection requirements of neural networks make optics a natural choice for the interconnection of the neuron elements. Optoelectronic implementations of neural networks have been demonstrated to have unique capabilities. Neural networks using holographic elements for optical interconnections have recently been used to demonstrate learning behavior for complex pattern recognition functions.

Large arrays of surface-emitting lasers have recently been developed at a number of research laboratories. At Bellcore, we have fabricated a 1,000-element array of surface-emitting lasers that can be individually electrically addressed [9] (Fig. 6). These laser arrays have been used to make a rapidly addressable, high-throughput holographic memory readout (Fig. 7). The ability to make large arrays of low-threshold, independently addressable light sources opens up exciting possibilities for many information processing applications.

A number of optical networks have been proposed and demonstrated that take advantage of the large bandwidth of optical fiber transmission by using time- and/or wavelength-division multiplexing to increase the throughput of the network. These networks offer new possibilities for optical interconnections and the switching of optical signals [10].

In the last few years, there have been many proposals published for photonic switching systems, ranging for simple arrays of electrically controlled optical routing elements to complex telecommunications packet switching networks with logic and memory circuits [11]. Figure 8 shows a representative example of the wide variety of switching systems that have recently been proposed and demonstrated. It is clear that for such systems to become practical, the technology for the optical components and optoelectronic interfaces must be economical, compact, and reliable.

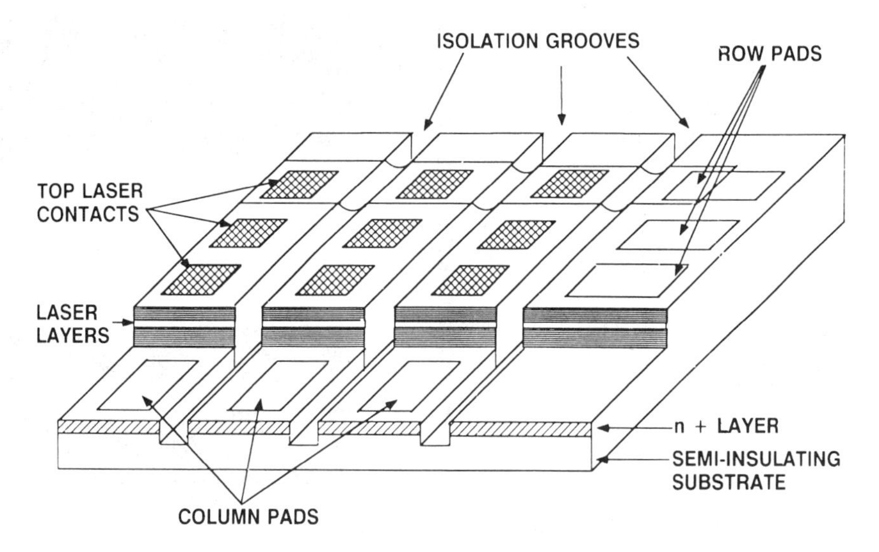

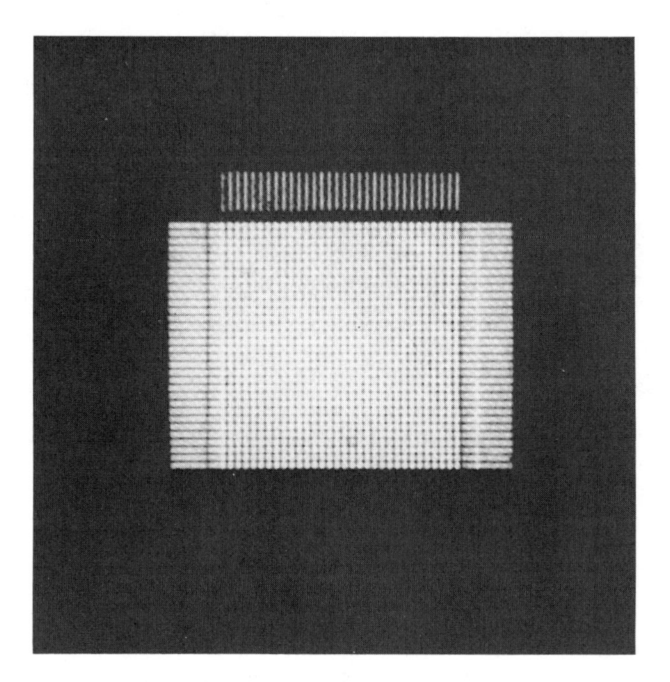

FIG. 6. Integrated matrix-addressed surface-emitting laser array.

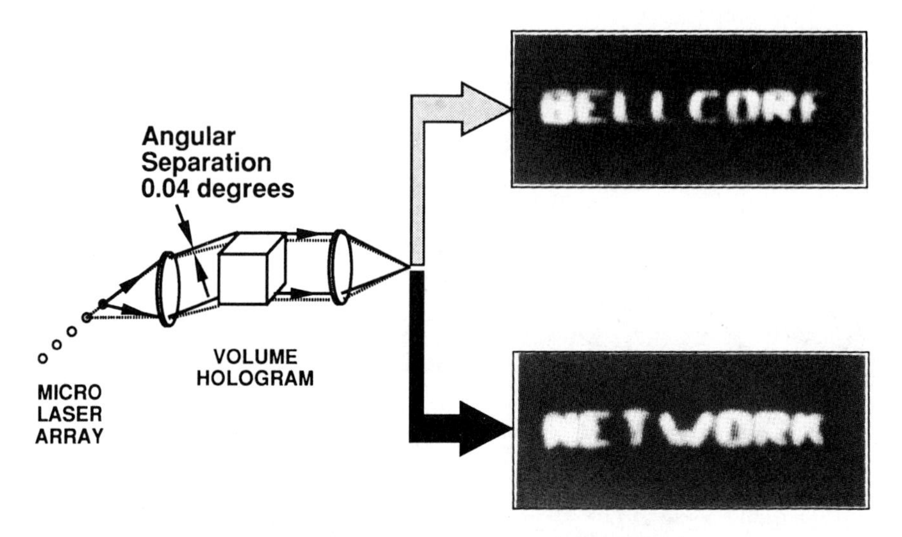

FIG. 7. Holographic memory readout using surface-emitting laser array.

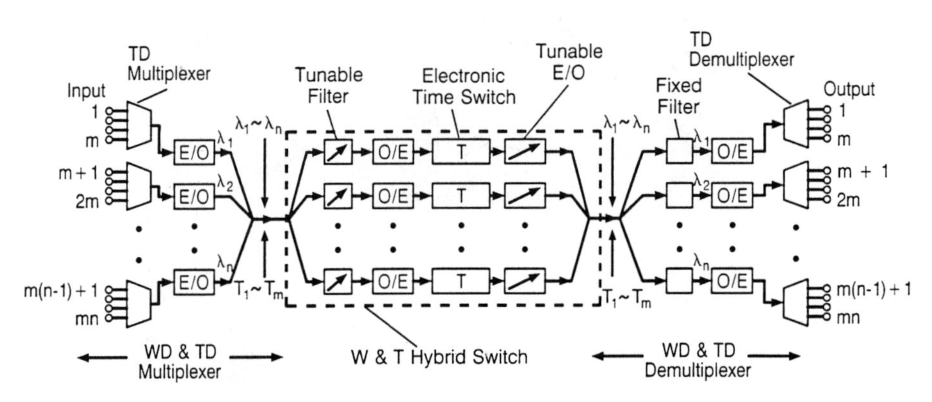

FIG. 8. Photonic wavelength-division and time-division hybrid switch. O/E and E/O refer to optics-to-electronics and electronics-to-optics conversions, respectively.

3. The Evolving Role of Optics

If we look at the evolution of ideas for the switching and control of telecommunications networks, we see that the ideas and demonstrations have evolved over the last three or four years, from simple switching elements and arrays to far more complex circuits and systems involving sophisticated signal-processing operations. The trend is clearly towards more control, more functionality, and more complex processing. This is, in part, because of continuing progress in electronic technology, and, in part, because of the increasing complexity of requirements for "information age" networks. It appears likely that the high data rates involved in many projected optical transmission applications will make conventional electronic switching systems inadequate; new switching technologies must be developed [12].

The development of small, low-power, and efficient surface-emitting semiconductor lasers, coupled with increasingly sophisticated techniques for achieving higher levels of integration of electronic and optical components, will permit future systems to be designed with a intimate mix of optical and electronic elements. These future systems will be capable of performance far superior to present-day systems because the incorporation of optical components will eliminate the electronic bottlenecks. The most attractive prospect for lightwave networks would appear to be implementing switching and control functions in a hybrid optical-electronic technology. In this way, each technology can be used for what it does most efficiently: electronics for "logic" and optics for "communications." As the capabilities for integration grow, we will see ever more complex operations performed on a single chip, with electronics providing the logic and control, and optics being used not only for transmission, but also for high-bandwidth interconnections, multiplexing, and, perhaps also, high-speed, high-density memory.

Figure 9 shows some predictions for future applications of optics. Optical fiber transmission will move closer to the customer. Optical switching systems will see wider use as the need for high-bandwidth switched services increases. Optical interconnects will become the natural pathways for information in complex hybrid optoelectronic systems. High-speed optical

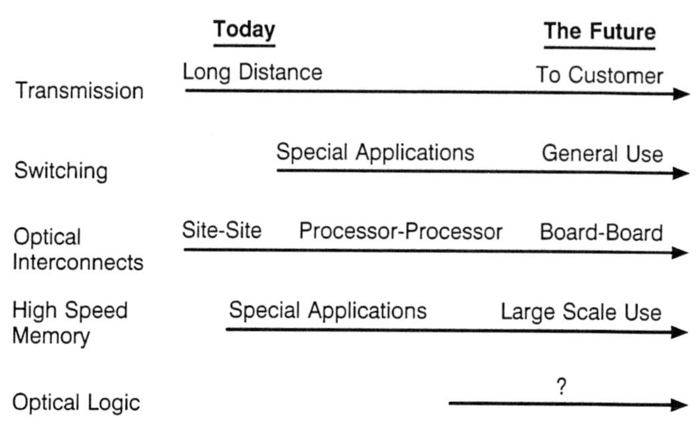

FIG. 9. Applications of optics in communications and computing.

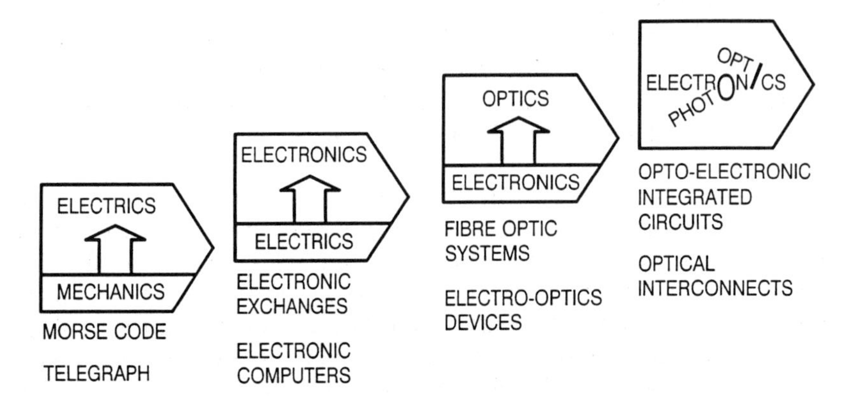

FIG. 10. The evolution of communications and computing.

memories will find progressively more use because of their high throughput capabilities. The use of all-optical elements for digital logic has a more uncertain future. Large-scale applications may not appear until the need for sub-picosecond response times is great enough to overcome the limitations of large size and high power requirements.

The key to the practical implementation of many of the current ideas for photonic technology will be the development of cheap, reliable, and simple optics-to-electronics and electronics-to-optics interfaces—to the point where optical interconnects become "easier" to implement than wires. At this point, optoelectronic circuits will afford greatly increased functionality with the potential advantages (low cost, reliability, etc.) of today's electronic chips. The evolution of the optical fiber network toward the customer (fiber to the home) brings the possibility of high volume requirements for OEIC chips. This "market pull" coupled with the "technology push" should aid in the development of a low-cost OEIC technology.

The future intelligent telecommunications network will require new technologies for the operation, control, and switching of data. The powerful data-handling capabilities made possible with new generations of optoelectronic circuits incorporating an intimate mix of optical and electronic components will provide the necessary performance to turn today's dreams into tomorrow's reality (Fig. 10).

References

1. P. W. Shumate, Jr. "Optical Fibers Reach into Homes," *IEEE Spectrum,* **43** (Feb 1989).
2. See, for example, P. W. Smith, "Photonic Switching: Present Research and Future Prospects," in "Optical Computing and Nonlinear Materials," *SPIE* **881**, 30 (1988).

3. See, for example, T. P. Lee, "Recent Advances in Long Wavelength Semiconductor Lasers for Optical Fiber Communications," *IEEE Proceedings* (1991), to be published.
4. See, for example, R. F. Leheny, "Optoelectronic Integration: a Technology for Future Telecommunication Systems," *IEEE Circuits and Devices Magazine*, 38 (May, 1989).
5. E. Yablonovich, T. Gmitter, J. P. Harbison, and R. Bhat, "Extreme Selectivity in the Lift-Off of Epitaxial GaAs Films," *Appl Phys Lett* **51**, 2222 (1987).
6. See, for example, P. Batacan, "Can Physics Make Optics Compute?" *Computers in Physics*, 9 (March/April, 1988).
7. Y. Tashiro, N. Hamao, M. Sugimoto, N. Takado, S. Asada, K. Kasahara, and T. Yanase, "Vertical to Surface Transmission Electrophotonic Device with Selectable Output Light Channels," *Appl Phys Lett* **54**, 329 (1989).
8. J. D. Crow, C. J. Anderson, S. Bermon, A. Callegari, J. F. Ewen, J. D. Feder, J. H. Grenier, E. P. Harris, P. D. Hoh, J. H. Hovel, J. H. Magerlein, T. E. McKoy, A. T. F. Pomerene, D. L. Rogers, G. J. Scott, M. Thomas, G. W. Mulvey, B. K. Ko, T. Ohashi, M. Scontras, and D. Widiger, "A GaAs MESFET IC for Optical Multiprocessor Networks," *IEEE Trans Elect Devices* **36**, 263 (1989).
9. H. J. Yoo, A. Scherer, J. P. Harbison, L. T. Florez, E. G. Paek, B. P. Van der Gaag, J. R. Hayes, A. VonLehmen, E. Kapon, Y. S. Kwon, "Fabrication of Two-Dimensional Phased Array of Vertical Cavity Surface-Emitting Lasers," *Appl Phys Lett* **56**, 1189 (1990).
10. See, for example, the special issue of *IEEE Communications Magazine* on "Lightwave Systems and Components" **27** (Oct., 1989).
11. See, for example, J. E. Midwinter and H. S. Hinton, eds., "Photonic Switching," Optical Society of America, Washington, 1989.
12. E. Nussbaum, "Communication Network Needs and Technologies—a Place for Photonic Switching?" *IEEE J Selected Areas in Comm* **6**, 1036 (1988).

CHAPTER 4

Microoptics

Kenichi Iga
Tokyo Institute of Technology
Yokohama, Japan

1. Introduction

Microoptics is a technology where very small optical devices (measured in μm to mm) are skillfully integrated. In optoelectronics, tiny optical devices support various functions, such as focusing, branching, coupling, multiplexing, demultiplexing, switching, and so on. It may be said that the optics related to mircoelectronics technology for optoelectronics systems is microoptics. Microoptics contributes towards the active operation of various optical parts of optoelectronics, including optical fiber communication systems and optical disk systems.

This chapter will review recent progress in microoptics that is applicable to various electro-optics fields. First, some basic microoptic elements will be summarized. Next, applied microoptic systems for lightwave communications, laser memory disk systems, and electro-optic equipment will be introduced. Two-dimensionally arrayed integrated optics (so-called *stacked planar optics*) using planar microlenses will be included.

2. Classifications of Microoptics

As for the optical components used in these applied optical systems, four types have been considered:

(a) Discrete microoptics, which consists of microlenses.

ISBN 0-12-289690-4

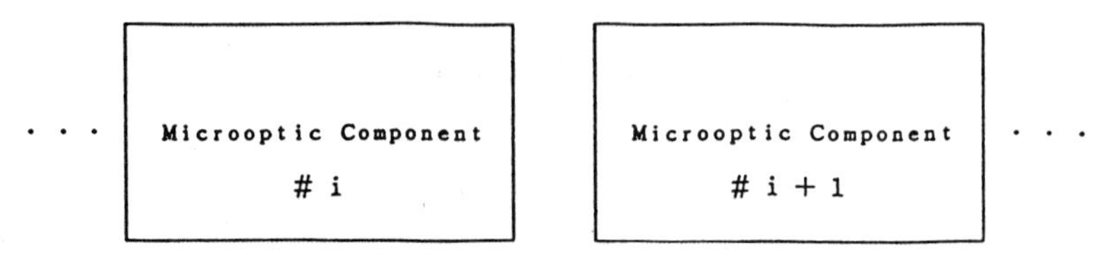

FIG. 1. A Model of a microoptic component.

(b) Optical fiber circuits made from manufactured fibers.
(c) Guided-wave optics.
(d) Stacked planar optics.

Optical alignment and complexity of the fabrication process have been problems with the first two schemes. Guided-wave integrated optics devices are limited to special devices.

The role of microoptics then is believed not to be replacing other components, such as guided-wave optic components or fiber-optic circuits, but fully utilizing these optical systems more effectively by complementing them. Hopefully, some much more modern concepts to integrate microoptic devices will be developed in this area. One such idea may be stacked planar optics, which will be detailed in the last section of this chapter.

The design of necessary optical systems as shown in Fig. 1 should be performed, [1, 2] but that will not be done here. We only will define three important parameters to characterize the microoptic systems that follow.

NA: Numerical aperture of the element defined by

$$NA = n_0 \sin \theta_{max}. \tag{1}$$

$\phi(r)$: Eigenmode, or allowable mode of the element.
n: Refractive index of the medium or mode index of the element.

Here, θ_{max} is an acceptance angle of the element and n_0 is the refractive index of the outside medium.

3. Basic Microoptics Elements

Since there are various kinds of microoptics components, their formation also is being developed as a total technology.

3.1. Microlens

In microoptics, several types of microlenses have been developed as listed in Table I. A spherical microlens is used mostly to gather light from

Table I

Microlenses	Shape	Applications	Diameter (mm)	f (mm)	NA
• Sphere Microlens		LD→Fiber	0.5–1	0.15	0.8
• Distributed-index Rod Microlens		Conjugate image Fiber components	0.5–2	1–2.8	0.35–0.46
• Triplet microlens		Disc objective		4.6	0.47
• Asperic Microlens		Disc objective	6.4	4.5	0.45
• Distributed-index +Spherical surface		Disc objective	4	3.55	0.45
• Fresnel Microlens		Disc collimator	1	2	0.25
• Planar Microlens		Multi-imager Fiber components	0.1–2	0.2–3	0.3–0.5

a laser diode by taking advantage of the high numerical aperture and the small Fresnel reflection of light from its spherical surface. Moreover, the precise fabrication of the sphere is very easy, but its mount to a definite position must be considered more carefully.

If we cut a rod having a parabolic refractive index distribution into the length of $L_p/4$ (where L_p is the *pitch*), the lens acts as a single piece of positive (focusing) lens, a so-called *distributed-index* (DI) or *gradient index* (GRIN) lens. On the other hand, if the length is $\frac{3}{4} L_p$, an erect image can be formed by a single piece of lens. In addition, it has the advantage of a flat surface, so that other optical elements—such as dielectric multilayer mirrors, gratings, optical fibers, and so on—can be directly cemented without any space in between. Another feature is its ability to constitute a conjugate image device; i.e., real images with magnification of unity can be formed by a single piece of rod microlens.

A planar microlens as shown in Fig. 2 was invented for the purpose of constructing two-dimensional lightwave components [3]. A large number of microlenses with diameters of $0.1 \sim 2$ mm can be arranged two-dimensionally on one substrate and their position can be determined by the precision of photomasks. These microlenses were intended for application to lightwave components and multi-image forming devices.

The importance of microlenses for focusing the light from a diode laser onto a laser disc could increase. Various types of microlenses are being developed. One is a molded lens with aspheric surfaces to reduce aberrations and allow diffraction-limited focusing. A number of fabrication methods for microlenses have been developed, such as the ion staffing method, the photolysis method, holography, the etching method for grating lenses, and so on.

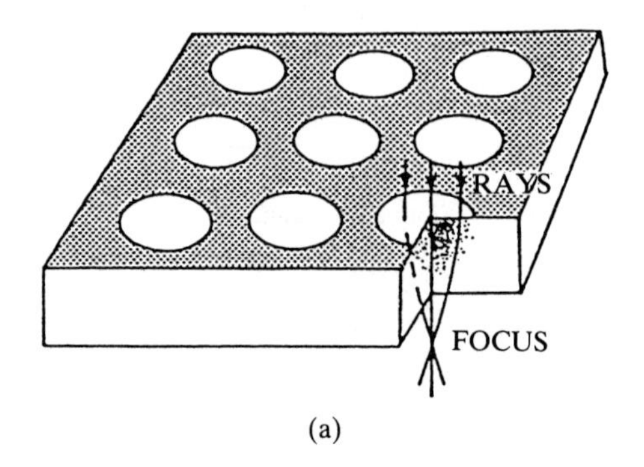

(a)

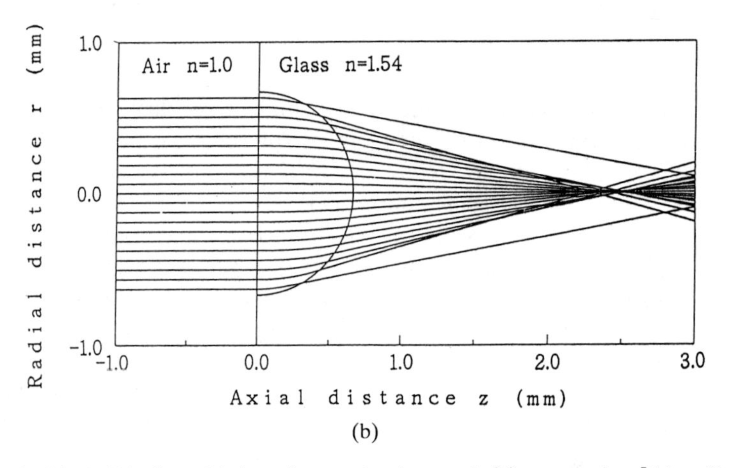

(b)

FIG. 2. (a) A Distributed-index planar microlens and (b) ray tracing [After X. Zhu, A. Akiba, and K. Iga].

3.2. *Micro-grating*

A grating normally used in monochromators is utilized to multiplex or demultiplex different wavelengths. The band elimination property is sharp, and reliable components can be fabricated. One important problem is that the angle of the exit ray is changed if the wavelength varies.

3.3. *Multilayer Mirror and Filter*

The dielectric-coated multilayer mirror or filter has none of the problems associated with grating filters. Its reflectivity and transmission can be controlled. The aging of optical properties especially must be examined, such as regarding resistance to moisture and temperature change.

3.4. Aperture Stop and Spatial Frequency Filter

An aperture stop is used to eliminate light from unwanted directions. In special cases, some types of spatial filters are inserted to cut out unwanted modes.

3.5. Optical Waveguide Devices

Devices based on optical waveguides are classified by materials, modes, etc. Here, we classify them as follows:

(*a*) *Monolithic Integrated Optical Waveguide Devices.* In 1969, the concept of integrated optics using dielectric optical waveguides was introduced [5], and an attempt began to fabricate optical circuits in a monolithic way. The basic composition was such that cores with a high refractive index were sandwiched between claddings with a low refractive index. A novel single-mode waveguide, an ARROW-type optical device using a single-mode waveguide with a thin buffer layer and thick active layer, has been developed [6]. It can be fabricated on a Si substrate with a rather thin buffer layer.

Multimode optical waveguide devices seem to have come into use quickly, but have the inconvenience of mode dependence. A branching method by means of a mode scrambler has been studied [7]. The phase space diagram was employed to characterize the device. Here is another method where light is focused and transformed through a waveguide-type lens horizontally, using a planar optical waveguide.

(*b*) *Hybrid Integrated Optical Waveguide Devices.* A new method has been proposed, where a silica glass-based optical waveguide, fabricated on a Si substrate, or a glass diffusion waveguide is used as an optical transmission circuit, and a tiny filter or reflecting mirror is inserted into a slit to make up an optical component, as shown in Fig. 3 [8]. Hybrid integrated optical waveguide devices have begun to attract practical interest.

(*c*) *Fiber-Based Optical Circuit Devices.* A fiber component consists of manufactured optical fibers. A branch is realized by polishing fibers to expose the core region and putting them close together. A directional coupler, star coupler, polarizer, Faraday rotator and other components can be constructed with optical fibers only. One merit of a fiber component is that of the mode of the component is identical to that of the fibers employed.

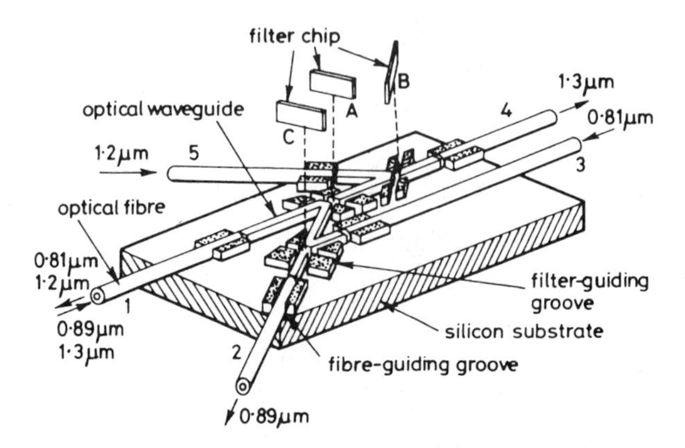

FIG. 3. Hybrid-integration of optical components [After M. Kawachi, *et al*].

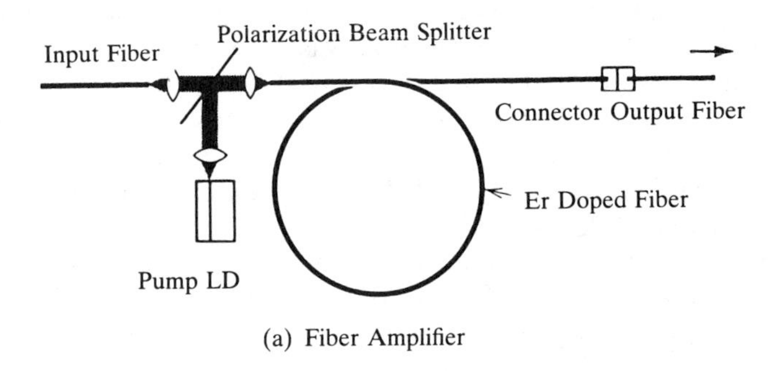

(a) Fiber Amplifier

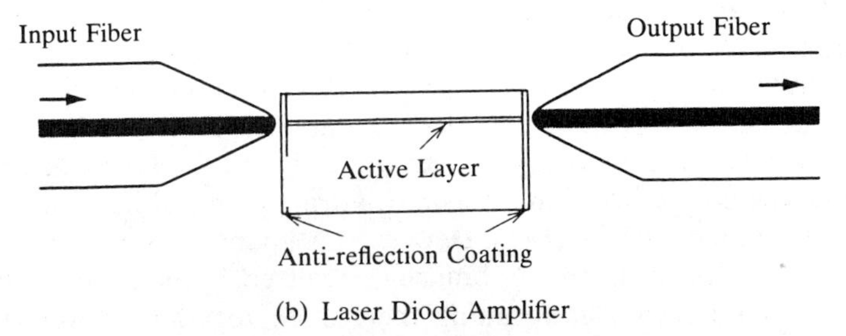

(b) Laser Diode Amplifier

FIG. 4. Optical amplifiers: (a) fiber amplifier (b) laser diode amplifier.

These devices are used to realize desired circuits using optical fibers as optical waveguides through optical fiber processing, such as elongation, grinding, adhesion, deposition, etching etc. [9].

A component using a polarization-maintaining fiber is being developed especially for optical sensing. Easy mode-matching and polarization independence are the advantages of fiber-based components. Recently, optical amplifiers using a piece of Er-doped optical fiber [10] or a semiconductor laser [11] have gained interest for regenerating repeaters, as shown in Fig. 4.

4. Applied Microoptics Systems

4.1. Lightwave Systems

4.1.1. Microoptics for Lightwave Systems

(*a*) *Focusing.* For the purpose of focusing light from a laser into an optical fiber, a microlens is employed. A *spherical lens* is the simplest one with a large enough NA to gather light from a semiconductor laser emitting light with 0.5–0.6 of NA. In the single-mode fiber application, the combination of a spherical and DI lenses or specially designed aspheric lens is considered.

(*b*) *Branching.* High-performance optical communication systems and electro-optic systems need a kind of component that divides light from one port to two or more ports. A simple *branch* consisting of a lens and prisms has been demonstrated. A microoptic branch is used inherently both for single-mode and multimode fibers, and no polarization preference exists. Another possibility is to utilize a planar waveguide, where the design must be done separately for single-mode or multimode waveguides.

(*c*) *Coupling.* A *power combiner* or *coupler* is a component that combines light from many ports into one port. In general, we can use a branching device as a combiner if it is illuminated from the rear side. A *directional coupler* is a component consisting of a branch and coupler [12]. A compact component consisting of a DI lens and a half mirror has been demonstrated.

A *star coupler* or *optical mixer* is a device for branching m ports into n ports, which sends light to many customers, as in a data highway or a local area network (LAN). A mixer made of fabricated fibers and utilizing a planar waveguide configuration has been developed.

(*d*) *Wavelength MX/DMX.* A *wavelength multiplexer/demultiplexer* (MX/DMX) is a device that combines/separates light of different wavelengths at the transmitter and is essential to communication systems using many wavelengths at the same time. Shown in Fig. 5 are (a) a device consisting of a multilayer filter and DI lenses that is good for several wavelengths and (b) one with a grating and lenses [13]. The grating DMX [14] can be used for many wavelengths, but a problem is that the direction of the beam changes when the wavelength varies, due to reasons such as a change of source temperature.

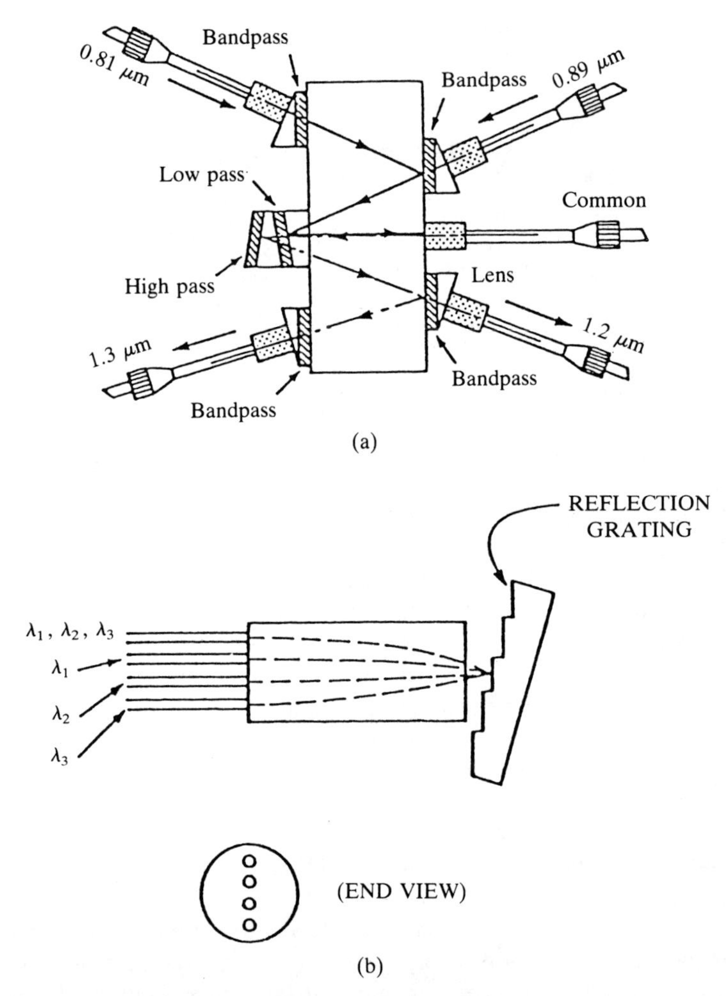

FIG. 5. (a) Wavelength demultiplexer consisting of a microlens and optical filters [After Watanabe, R., Sano, K. and Minowa, J. IOOC' 83, 30C1-2, (1983)]. (b) Wavelength demultiplexer consisting of a grating [After W. J. Tomlinson].

(*e*) *Optical Isolator.* An *optical isolator* is used in sophisticated lightwave systems where one can not allow any reflection of light that might disturb the laser oscillator. The principle of the isolator is understood as follows: The Faraday rotator rotates the polarization of incident light by 45 degrees. Then the reflected light can be cut off by the polarizer at the input end while the transmitted light passes through the analyzer at the exit end. Pb glass is used for short wavelengths and YIG($Y_2Fe_5O_{12}$) for the long-wavelength region. A device with 1 dB of insertion loss and 30 dB of isolation has been developed for wavelength multiplexing communication.

(*f*) *Functional Components.* Functional components such as switches and light modulators are important for electro-optic systems. Several types of optical switches have been considered: mechanical switch, electro-optic switch, and magnet-optic switch. A switch using the same idea as the isolator shown in Fig. 6 has been introduced to switch laser sources in the undersea cable (Trans Pacific Cable-3) for the purpose of maintaining the system when one of the laser devices fails [15].

Beam deflectors are becoming important in the field of laser printers. A rotating mirror is used in the present system, but some electro-optically controlled devices are required to simplify the system.

4.1.2. Lightwave Communication Systems

There is a wide variety of microoptic components used in lightwave communication systems. The simplest system consists of a light source such as a laser diode, an optical fiber, and a detector at the receiving end. The employed component focuses the laser light into the fiber. Needless to say, many long-haul transmission systems have been or will be installed in many

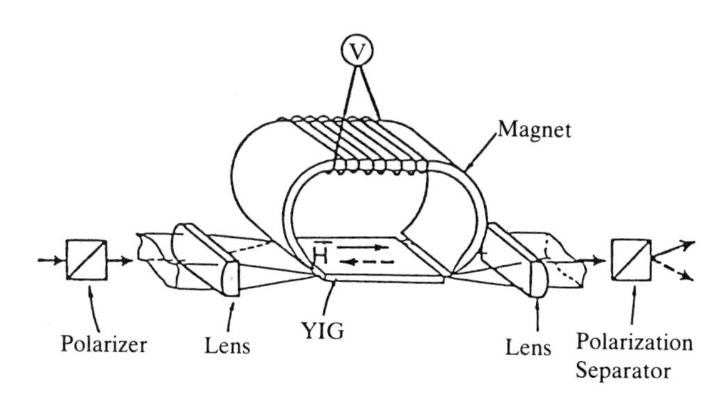

FIG. 6. Optical switch [After M. Shirasaki].

countries, and transatlantic and transpacific undersea cables are being prepared for international communication with very wide bandwidths of several thousand voice channels.

A *wavelength multiplexing* (WDM) system needs a more sophisticated combination of devices. At the sending end, a multiplexer is used to combine multiple wavelengths into a single piece of fiber; at the receiving end, a demultiplexer is utilized for the purposes of separating different signals on different wavelengths.

The local area network will be a most popular lightwave communication system. A video transmission system is an attractive medium for education and commercials. At the Tokyo Institute of Technology, TV classrooms connected two campuses 27 km apart by using eight single-mode fibers; 400 Mbits/s signals are transmitted back and forth.

4.2. Optical Disks

A light-focusing component (*focuser*) is the simplest—yet an important—component. We first consider a simple focuser composed of a single lens. The diffraction limit of the focused spot D_s is given by the following equation:

$$D_s = 1.22(\lambda/NA) \simeq 1.22 f\lambda/a, \tag{2}$$

where λ is wavelength, $2a$ the aperture stop diameter, f the focal length of the lens employed, and NA the numerical aperture of the lens.

The focuser for a laser disc system must have the smallest aberration, since the lens is used in the diffraction limit. Another important consideration is the working distance of the lens. In the laser disc application, some space must be considered between the objective lens and the focal point because there must be a clearance of about 2 mm; there also is the thickness of the disc to consider. Therefore, we need a lens with a large diameter (~5 mm) as well as a large NA (~0.5). Molded plastic aspheric lenses are used mostly in CD systems.

More than one million pieces of plastic aspheric lenses are produced per month. CD device production reached 20 million in 1988. Along with increased CD production, the monthly production of microlenses and semiconductor lasers in some companies is increasing to more than one million units.

Higher mechanical speed and more simplified design can be achieved by using a grating microlens or holographic optical element (HOE), as shown in Fig. 7 [16]. A new floating system, which does not use any focusing lenses, has been proposed [17].

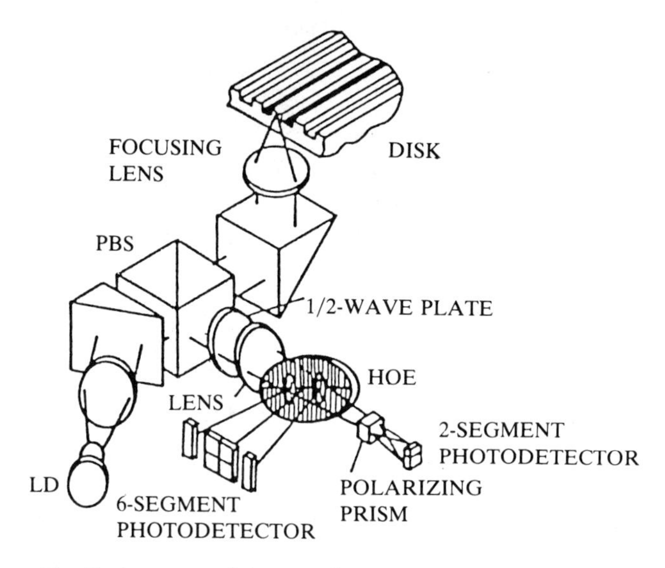

FIG. 7. A compact disk system [After Y. Kimura and Y. Ono].

4.3. Imaging Systems Image Transmission by a Distributed-Index Fiber

In this subsection, we deal with some imaging components (*imagers*) consisting of distributed index lenses, including a *simple imager*, *conjugate imager*, and *multiple imager*.

4.3.1. Image Fiber Bundle

A new type of fiber image bundle has been developed, using silica fibers instead of composite glass fibers, by taking advantage of their high temperature capability and low loss over a wide range of the spectrum. The fiber diameter is about a few microns, providing better resolution.

4.3.2. Copier

A lens array consisting of distributed index lenses has been introduced in photocopiers where *conjugate images* (erect images with unit magnification) of a document can be transferred to the image plane as illustrated in Fig. 8 [18]. This configuration is effective for eliminating a long focal length lens and the total volume can be drastically reduced. The number of lenses produced is expected to exceed 100 million.

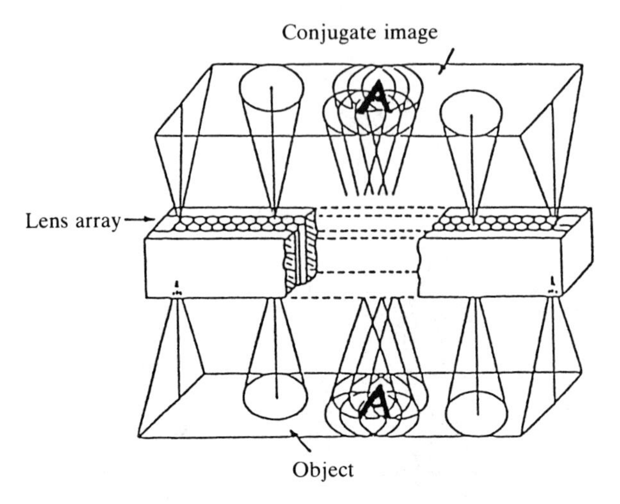

FIG. 8. 1:1 imaging system and application to a copy machine [After I. Kitano].

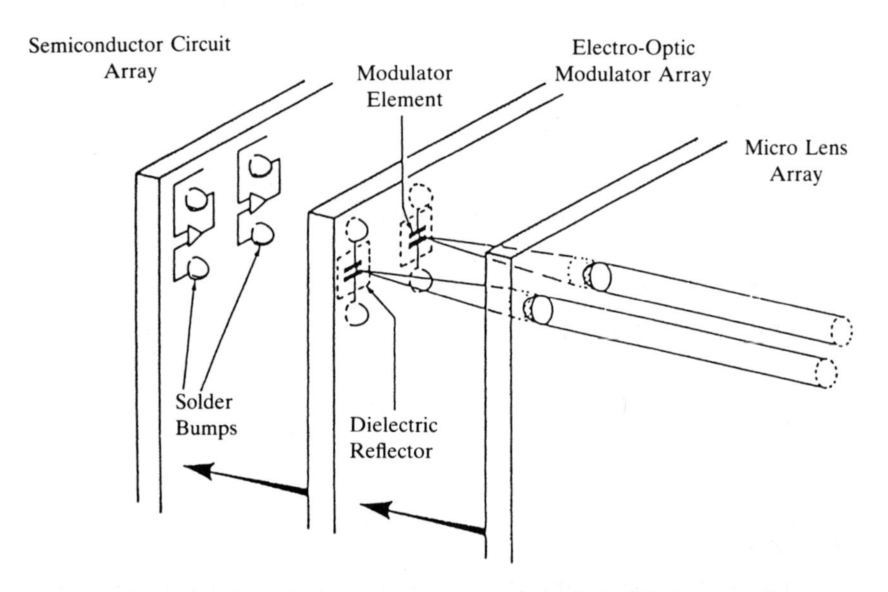

FIG. 9. Two-dimensional array of optical switches [After W. J. Stewart].

4.3.3. Auto Focuser

An auto-focusing element to provide easy focusing for a still camera is becoming popular now. Multiple images are formed by a lens array, and a CCD detector generates an error defocus signal until the main lens is automatically moved to the right position. Various types of microlens arrays are being developed. A planar microlens array can be used for this application.

4.4. Other Microoptics Systems

4.4.1. Fiber Sensor

A fiber gyro and other lightwave sensing systems are being considered for various measurement demands. Microoptic elements being employed include some types of interferometers consisting of a beam splitter and half mirrors. Single-mode components that match the mode of the single-mode and polarization-maintaining fiber employed are necessary. Making an optical circuit from manufactured fiber is an interesting method.

4.4.2. Optical Computer

A future technique may be using an optical parallel processor like a TSE computer and an optical neural computer [19], for which architectures have been presented. Some functional devices such as optical AND and OR elements based on semiconductor materials must be developed to realize this sophisticated system. A two-dimensional configuration made of planar microlenses and surface-emitting laser arrays will be very helpful.

A novel optical triode device using a microoptic concept has been proposed and fabricated [20]. Also, a 2-D arrayed microlens system has been introduced into a component for multi-matched filtering [21].

4.4.3. Optical Interconnection

In computers, there are chip-to-chip and circuit-board-to-circuit-board communications in starlike networks. If optics can be used there, such communication may be called *micro-haul communication*. The system may be constructed using a lot of tiny transmitters, waveguides or transmitting media, and receivers. Microoptics will play an important role in this system. Parallel arrayed optics, described in the next section, will be suitable for this application.

A Si/PLZT optical modulator has been demonstrated (as shown in Fig. 9), aiming at interconnection or optical parallel processing [22]. Though its modulation speed is slow, a practical parallel device has been developed.

5. Stacked Planar Optics

5.1. Concept of Stacked Planar Optics

Stacked planar optics consists of planar optical components arranged in a stack, as shown in Fig. 10 [4]. All components must have the same two-dimensional spatial relationship, which can be achieved from planar technology, with the help of photolithographic fabrication as used in electronics. Once we align the optical axis and adhere all of the stacked components, two-dimensionally arrayed components are realized; mass production of axially aligned discrete components also is possible if we

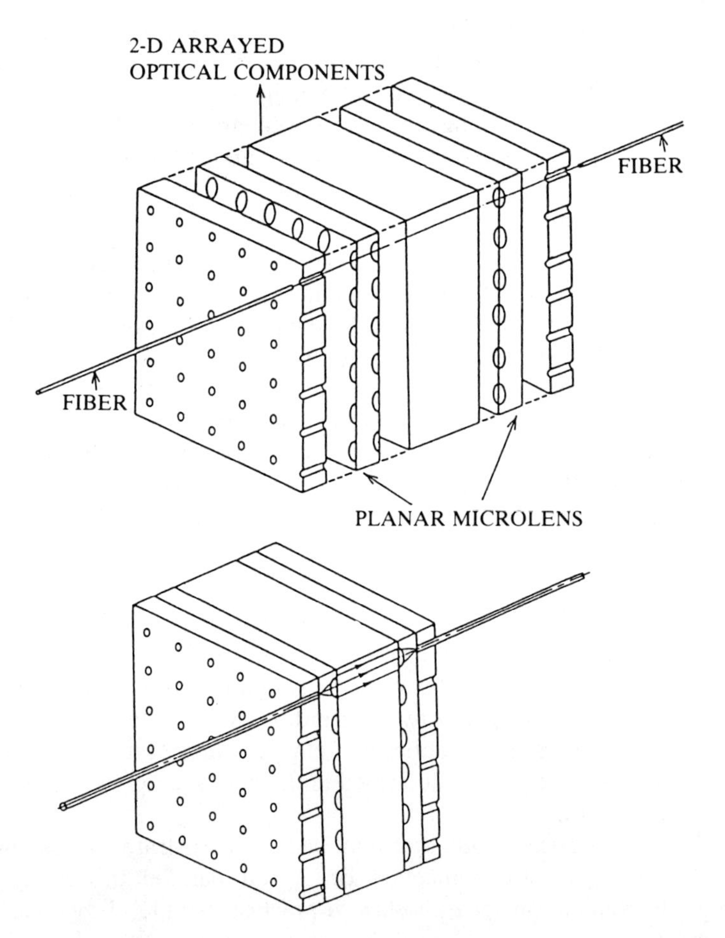

FIG. 10. Stacked planar optics [After K. Iga, Y. Kokubun, and M. Oikawa].

separate individual components. This is the fundamental concept of stacked planar optics, which may be a new type of integrated optics.

5.2. Planar Microlens Array

To realize stacked planar optics, all optical devices must have planar structure. An array of microlenses on a planar substrate is required to focus and collimate the light in optical circuits. A planar microlens as shown in Fig. 2 has been developed [3]. Recent data is mentioned in [23]; 10×10 cm square array samples now are available. The data for available planar microlenses is shown in Table II.

5.3. Surface-Emitting Lasers and Surface-Operating Devices

In the author's laboratory in 1977, we started the study of surface-emitting lasers [24], and in 1988, achieved room temperature CW operation in the GaAlAs/GaAs system [25]. Also, we fabricated a two-dimensional array of surface-emitting lasers, as shown in Fig. 11. Furthermore, research on surface-emitting laser-type optical devices,

Table II

Lens Diam 2a (μm)	10–1000
N.A. (max)	0.25
Back Focal Fb (μm)	20–2000
Maximum Index Difference Δn	0.17
Spot Diam ($\lambda = 633$ nm)	4 μm

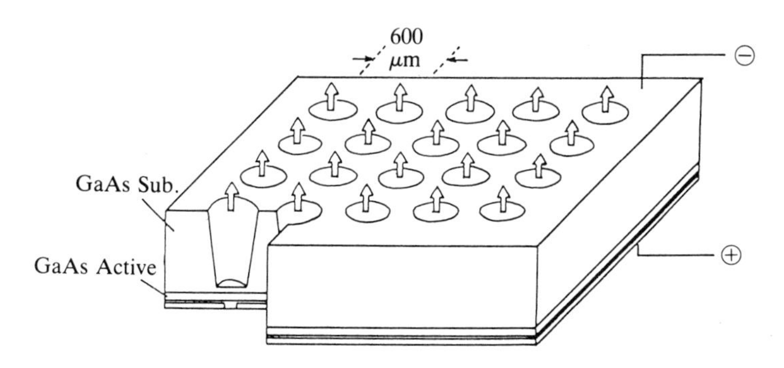

FIG. 11. Surface-emitting laser array [After F. Koyama and K. Iga].

such as optical switches introducing an ultra-miniature structure called a quantum well, has become very active.

5.4. Design Rule of Stacked Planar Optics

A possible fabrication process proposed for stacked planar optics is given in [4], and will not be detailed here. Features of stacked planar optics include:

(a) Mass production of standardized optical components of circuits, since planar devices are fabricated by planar technology.
(b) Optical alignments.
(c) Connection in tandem of optical components of different materials, such as glass, semiconductors, electro-optical crystals, etc.

Table III. Basic Optical Components for Stacked Planar Optics and Optical Circuits[a]

Basic components	Application
Coaxial imaging components	Coupler[b]
Noncoaxial imaging components (transmission-type)	Branching circuit[b] Directional coupler[b] Star coupler[c] Wavelength demultiplexer[b]
Noncoaxial imaging components (reflection-type)	Wavelength demultiplexor[b] Optical tap[b]
Collimating components	Branching insertion circuit[b] Optical switch[c] Directional coupler[b] Attenuater[b]

[a] From Iga, K., Kokubun, Y., and Oikawa, M. (1984). "Fundamentals of Microoptics." Academic Press, Orlando.

[b] Circuit integrated in a two-dimensional array.

[c] Circuit integrated in a one-dimensional array.

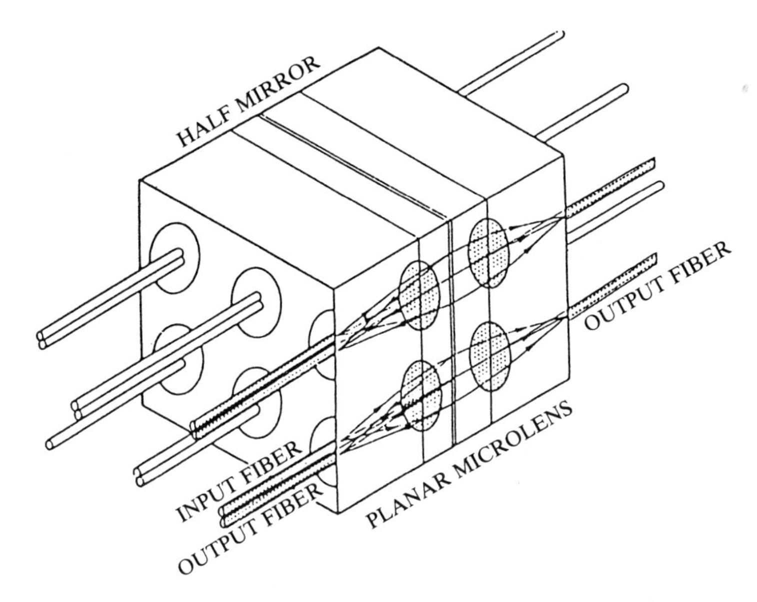

FIG. 12. Directional coupler using a stacked planar optics configuration [After K. Iga and M. Oikawa].

5.5. Applications of Stacked Planar Optics

Many kinds of optical circuits can be integrated in the form of stacked planar optics, as summarized in Table III. We have listed some components as examples of stacked planar optics. The *optical tap* is a component for monitoring part of the light being transmitted through an optical fiber. A 2×3 branching component has been fabricated with two pieces of stacked planar microlenses and a half mirror, as shown in Fig. 12. We also can construct three-dimensional optical circuits by allowing coupling among individual components in the array with a suitable design.

6. Summary

Microoptics is a concept which provides very practical schemes for a wide variety of optoelectronic systems. When one wants to design a system, one must consider the use of microoptics. Microoptics can be described as a technology for which various demands for devices in optoelectronics can be realized in practice.

A problem is finding methods for satisfying alignment requirements, and one solution may be to employ a two-dimensional microoptic component, a stacked planar optics component that consists of planar microlenses. In the future, the realization of three-dimensional large-scale integrated microoptics is desirable [26].

References

1. K. Iga, Y. Kokubun, and M. Oikawa, "Fundamentals of Micro-optics," Academic Press, New York, 1984.
2. K. Iga, "Progress of Microoptics," *Proceeding of 14th Congress of the International Commission for Optics (ICO '86)*, paper C5.4, 125–128 (August, 1986).
3. M. Oikawa, K. Iga, and S. Sanada, "Distributed-Index Planar Microlens Array Prepared from Deep Electromigration," *Electron. Lett.* **17**, no. 13, 452–454 (June, 1981).
4. K. Iga, M. Oikawa, S. Misawa, J. Banno, and Y. Kokubun, "Stacked Planar optics: an application of the Planar Microlens," *Appl. Opt.* **21**, 3456–3460 (1982).
5. S. E. Miller, "Integrated Optics—Introduction," *Bell Syst. Tech. J.* **48**, no. 7, 2059–2069 (September, 1967).
6. T. Baba, Y. Kokubun, T. Sakaki, and K. Iga, "Loss Reduction of ARROW Waveguide in Shorter Wavelength and Its Stack Configuration," *J. Lightwave Tech.* **6**, no. 9, 1440–1445 (September, 1988).
7. Y. Kokubun, S. Suzuki, and K. Iga, "Phase Space Evaluation of Distributed-Index Branching Waveguide," *J. Lightwave Tech.* **LT–4**, no. 10 1534–1541 (October, 1986).
8. M. Kawachi, Y. Yamada, M. Yasu, and M. Kobayashi, "Guided-Wave Optical Wavelength-Division Multi/Demultiplexer Using High-Silica Channel Waveguides," *Electron. Lett.* **21**, no. 8, 314–315 (April, 1985).
9. S. Ohshima, T. Ito, K. Donuma, and Y. Fujii, "Small Loss-Deviation Tapered Fibre Star Coupler with Mixer Rod," *Electron. Lett.* **20**, no. 23, 976–978 (November, 1984).
10. D. Payne and L. Reekie, "Rare-Earth Fibre Lasers and Amplifiers," (Invited), *Technical Digest of 13th European Conf. Opt. Comm. (ECOC '87),* **3**, 89–94 (1987).
11. T. Saitoh and T. Mukai, *IEEE J. Quant. Electron.* **QE–23**, no. 6, 1010 (June, 1987).
12. T. Uchida and K. Kobayashi, "Micro-optic Circuitry for Fiber Communication," in "Optical Devices and Fibers," pp. 172–189, Ohm-sha-North Holland, 1982.
13. R. Watanabe, K. Sano, and J. Minowa, "Design and Performance of Multi/Demultiplexer for Subscriber Loop System," *Technical Digest, International Conference on Intergrated Optics and Optical Fiber Communication (IOOC '83)*, paper 30C1-2, 368–369 (July, 1983).
14. W. J. Tomlinson, "Aberrations of GRIN-Rod Lenses in Multimode Optical Fiber Devices," *Appl. Opt.* **19**, no. 7, 1117–1126 (April, 1980).
15. M. Shirasaki, "Optical Switch Using Thin Plate Waveguide Polarization Rotator," in "Optical Devices and Fibers," JARECT, Vol. 11, pp. 152–166, Ohm-sha-North Holland, 1984.
16. Y. Kimura and Y. Ono, "Polarizing Holographic Optical Element for Magneto-Optical Disk Heads," *Technical Digest, 1st Microoptics Conference (MOC '87)* (Tokyo, Japan), paper F8, 162–165 (October, 1987).
17. S. Sawada, Y. Uenichi, Y. Isomura, and H. Ukita, "Laser Diode Acting as Micro-optical Head," *Microoptics News* **6**, no. 3, 200–203 (September, 1989).
18. I. Kitano, "Current Status of Gradient-Index Rod Lens," in "Optical Devices and Fibers," JARECT, Vol. 5, p. 151–166, Ohm-sha-North Holland, 1983 [also from the catalogue of SELFOC lens array by Nippon Sheet Glass Co. Ltd.].
19. K. Kyuma, J. Ohta, K. Kojima, and T. Nakayama, "Optical Neural Networks: System and Device Technologies," *Optical Computing, SPIE,* **963**, 475–484 (1988).
20. H. Tsuda and T. Kurokawa, "Optical Triode Switch Module with a Nonlinear Etalon," *2nd Microoptics Conference/8th Topical Meeting Gradient-Index Optical Imaging System (MOC/GRIN '89)* (Tokyo, Japan), paper K2, 264–267 (June, 1989).
21. M. Agu, A. Akiba, and S. Kamemaru, "Multimatched Filtering System as a Model of Biological Visual System," *SPIE* **1014**, The International Congress on Optical Science and Engineering (September, 1988).

22. W. J. Stewart, "A μ-Optic Si/PLZT Modulator Array and an Orthoconjugate Reflector," *1st Microoptics Conference (MOC '87)* (Tokyo, Japan), paper D1, 66–69 (October, 1987).
23. K. Iga and S. Misawa, "Distributed-Index Planar Microlens and Stacked Planar Optics: a Review of Progress," *Appl. Opt.* **25**, no. 19, 3388–3396 (October, 1986).
24. K. Iga, F. Koyama, and S. Kinoshita, "Surface-Emitting Lasers", *IEEE J. Quant. Electron.* **QE–24**, no. 9, 1845–1855 (September, 1988).
25. F. Koyama, S. Kinoshita, and K. Iga, "Room-Temperature Continuous Wave-Lasing Characteristics of a GaAs Vertical Cavity Surface-Emitting Laser," *Appl. Phys. Lett.* **55**, no. 3, 221–222 (July, 1989).
26. K. Iga, "Two-dimensional Microoptics," *CLEO '89,* **TUB2** (April, 1989).

CHAPTER 5

Holographic Optical Elements for Use with Semiconductor Lasers

H. P. Herzig and R. Dändliker

Institute of Microtechnology, University of Neuchâtel, Neuchâtel, Switzerland

1. Introduction

Holographic optical elements (HOE) are diffractive structures, which are fabricated by recording the interference pattern of an object beam and a reference beam. Compared with conventional refractive and reflective optics, they are thinner and lighter; they also can perform multiple functions simultaneously.

The immense progress in developing semiconductor lasers as cheap and compact monochromatic light sources has been accompanied by a parallel upswing in HOE technology. A wide range of applications has been proposed, such as compact optical disk heads, laser scanners, laser-beam shaping for space communications, fan-out elements for parallel optical processing, and fiber-optic communications.

Unfortunately, semiconductor lasers emit light at wavelengths in the near IR, typically at $\lambda = 780$ nm (GaAlAs) and $\lambda = 1{,}300$ nm (InGaAsP), whereas highly efficient holographic recording materials are only sensitive below 580 nm. Recording HOE in the visible range and reconstructing them in the IR requires careful control of all wavelength-dependent parameters, such as Bragg angle, focal length, and aberrations. There are different methods to solve this problem. For small focal lengths and numerical apertures (NA), an optimized recording geometry is sufficient to achieve a diffraction limited spot size. In other cases, aspheric waves are

ISBN 0-12-289690-4

necessary. These aspheric waves can be generated by using several recording and readout steps, or by the help of computer-generated holograms (CGH). Analytical equations, ray-tracing, and diffraction calculations are used to analyze the geometrical aberrations.

In this chapter, we shall give a short overview of the design of HOE, an analysis of the aberrations, some recent and future applications, and a comparison with diffractive optical elements (DOE). Note that we limit our discussion to phase holograms. Amplitude holograms absorb light and, therefore, are of less interest.

2. Semiconductor Laser Sources and Holographic Materials

Semiconductor lasers are compact monochromatic light sources, which are well-suited for use with HOE. Commercially available single-mode semiconductor lasers have been developed for telecommunications, emitting light typically at $\lambda = 1300$ nm (InGaAsP lasers at 1,170–1,570 nm), and for compact disk players, typically at $\lambda = 780$ nm (GaAlAs lasers at 750–850 nm). Efforts to fabricate semiconductor lasers for the visible range already have led to lasers on the market at $\lambda = 670$ nm. The output beam of single-mode semiconductor lasers is astigmatic and elliptic, but it can be collimated properly by conventional optics.

Unfortunately, there is no holographic material available that is sufficiently sensitive in the IR to record high-quality HOE. Candidates for highly efficient holographic materials are bleached photographic emulsions with reasonable sensitivity at 400–700 nm, dichromated gelatin at 350–580 nm, photopolymers at UV–650 nm and photoresists at UV–500 nm [1].

Silver halide emulsions are widely used because of their high sensitivity, and because they are commercially available. On the other hand, the efficiency is limited by scattering and absorption in the material.

Dichromated gelatine (DCG) is an ideal recording material for volume HOE with the capability of large refractive index modulation, high resolution, low absorption, and scattering. Close to 100 % efficiency is possible. On the other hand, the wet process must be controlled carefully to get reproducible results.

Promising results for photopolymers show that they might become competitive with DCG [2], although the refractive index modulation is smaller than for DCG. A great advantage of photopolymers is dry processing.

High efficiencies (> 85 %) also can be achieved with surface relief gratings in photoresist. Relief structures are advantageous for mass production. They can be replicated by embossing. Interferometrically recorded photoresist patterns also can be used as masks, which then are transformed into other materials by etching techniques (Section 6).

3. HOE Design

The HOE design includes calculation of the HOE phase structure and an analysis of its optical properties at the operating wavelength.

3.1. HOE Phase Function

Efficiency is an important factor for HOE. Therefore, we limit our considerations to phase-only elements, which can be described by an amplitude transmittance $t(x, y) = \exp[i\Phi_H(x, y)]$. For the recording, the phase function $\Phi_H(x, y)$ is defined by the wave fronts that are used to construct the HOE:

$$\Phi_H(x, y) = \Phi_O(x, y) - \Phi_R(x, y), \tag{1}$$

where $\Phi_O(x, y)$ is the phase distribution of the object wave and $\Phi_R(x, y)$ the phase distribution of the recording reference wave in the hologram plane (x, y). For the readout with a wave Φ_r, the phase distribution Φ_P of the reconstructed wave is governed by the phase-matching condition

$$\Phi_P(x, y) = \Phi_r(x, y) + m\Phi_H(x, y) \tag{2}$$

in the hologram plane, where m is the diffraction order. In general, only the first order is of interest, so that $m = 1$.

In the case of a focussing HOE in the IR (e.g., at $\lambda_r = 780$ nm) for readout with a spherical wave Φ_r, the required HOE phase function is determined by Eq. 2, where both phase functions (Φ_P and Φ_r) are of the form

$$\Phi_i = \pm\frac{2\pi}{\lambda}\sqrt{(x - x_0)^2 + (y - y_0)^2 + (z_0)^2}). \tag{3}$$

The point (x_0, y_0, z_0) is the source or focus of the spherical wave.

If only spherical waves are available for recording at the wavelength λ_R different from λ_r, Eq. 1 cannot be fulfilled exactly. Thus, aberrations occur. A perfect phase function Φ_H can be realized by using aspheric waves (Subsection 4.2).

As a guideline for HOE design, it is useful to know the basic aberrations for holographic grating lenses and the conditions for high-diffraction efficiency. In the following subsection, we shall discussed second-order aberrations, astigmatism for point sources, and the Bragg condition.

3.2. Second-Order Aberrations for Point Sources and Bragg Condition

With plane and spherical waves, only simple focussing HOE can be recorded. Nevertheless, these elements have great potential as holographic lenses. Furthermore, these considerations will help us to understand the basic problems of HOE.

The principal parameters (recording and readout angles, astigmatic focal lengths) for a HOE can be calculated analytically by using second-order approximation [3]. In the case of recording and reconstruction of point sources in a common plane of incidence (Fig. 1), one gets for the angles the grating equation

$$\sin\theta_P = \sin\theta_r + m\mu\,(\sin\theta_O - \sin\theta_R), \tag{4}$$

and for the radii of curvature ρ_i,

$$\frac{\cos^2\theta_P}{\rho_P^{\parallel}} = \frac{\cos^2\theta_r}{\rho_r} + m\mu\left(\frac{\cos^2\theta_O}{\rho_O} - \frac{\cos^2\theta_R}{\rho_R}\right), \tag{5}$$

$$\frac{1}{\rho_P^{\perp}} = \frac{1}{\rho_r} + m\mu\left(\frac{1}{\rho_O} - \frac{1}{\rho_R}\right). \tag{6}$$

θ_i are the angles of incidence in the air (refractive index $n = 1$) and the wavelength ratio is given by $\mu = \lambda_r/\lambda_R$. For further considerations, we choose the diffraction order $m = +1$. The indices refer to the waves involved, namely R to the recording reference, O to the object wave, r to the reconstructing reference, and P to the reconstructed astigmatic ray pencil, with the principle radii of curvature $\rho_P^{\parallel}$ parallel and $\rho_P^{\perp}$ perpendicular to the plane of incidence. Wavefronts without astigmatism require that $\rho_P^{\parallel} = \rho_P^{\perp}$.

To get high diffraction efficiency from thick volume holograms, the Bragg condition has to be satisfied. In the holographic emulsion with a

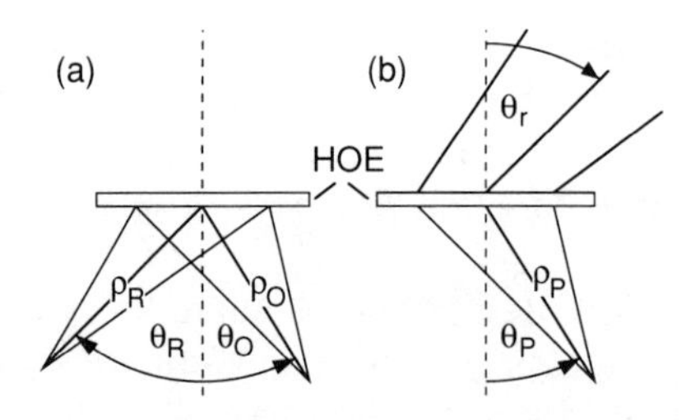

FIG. 1. (a) HOE recording geometry. (b) HOE readout geometry.

refractive index n', the wavevectors are related by

$$\mathbf{k}_P - \mathbf{k}_r = \mathbf{k}_O - \mathbf{k}_R, \tag{7}$$

with $|\mathbf{k}_i| = 2\pi n'/\lambda_i$.

Note that for the components in the hologram plane, Eqs. 4 and 7 are identical. For the components normal to the hologram, it also follows that

$$\cos\theta'_P = \cos\theta'_r + \mu(\cos\theta'_O - \cos\theta'_R), \tag{8}$$

where the relation between the angles in the medium n' and in the air $n = 1$ are determined by the law of refraction,

$$n \sin\, \theta_i = n' \sin\, \theta'_i. \tag{9}$$

If a readout geometry for a focussing HOE is determined now by the angles θ_i and radii of curvature ρ_i, we can find from Eqs. 4 to 9 the recording parameters for a readout without astigmatism ($\rho_P^{\|} = \rho_P^{\perp}$) and fulfilling the Bragg condition simultaneously.

For increasing aperture and focal length, second-order aberrations do not describe the HOE sufficiently. Higher-order aberrations become dominant. They can be analyzed by ray tracing or by using scalar diffraction theory (Subsection 3.4) to calculate light propagation.

3.3. Ray Tracing

The aberration properties of HOE can be analyzed by geometrical optics, using *ray tracing* [4], which is very common in conventional lens design. In ray tracing, the path of light through lens systems is determined with the help of elementary geometry by successive application of the law of refraction (or reflection). In holography, the law of refraction has to be replaced by the law of grating diffraction.

The holographic recording and reconstruction process is governed essentially by the condition of phase matching in the hologram plane (x, y), which is given by Eq. 2. The phase-matching condition in the hologram plane yields relations for the normal projections $\mathbf{k}_{iH}$ of the wavevectors $\mathbf{k}_i$ $(i = P, r, O, R)$ onto that plane, namely

$$\mathbf{k}_{PH} = \mathbf{k}_{rH} + m\mathbf{K}_H = \mathbf{k}_{rH} + m[\mathbf{k}_{OH} - \mathbf{k}_{RH}], \tag{10}$$

where the phase functions Φ_i and the vectors $\mathbf{k}_{\mathrm{iH}}$ are related by $\mathbf{k}_{iH}(x, y) = \mathrm{grad}[\Phi_i(x, y)]$.

The length of the wavevectors at the reconstructing wavelength is given by $|\mathbf{k}| = |\mathbf{k}_r| = 2\pi/\lambda_r$. For the component k_{Pz} of the outgoing wave normal to the hologram plane it follows that for a transmission HOE,

$$k_{Pz} = \mathrm{sign}\,(k_{rz})[|\mathbf{k}_r|^2 - |\mathbf{k}_{PH}|^2]^{1/2}, \tag{11}$$

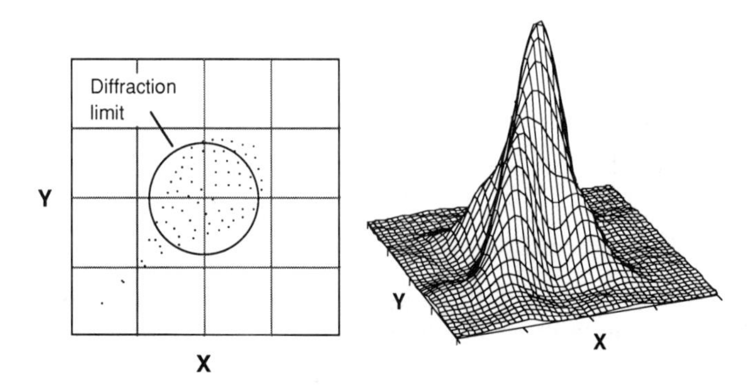

FIG. 2. Comparison of ray tracing and diffraction analysis.

where sign (k_{rz}) denotes the sign of k_{rz}. For reflection holograms, sign (k_{rz}) in Eq. 11 has to be replaced by $-$sign (k_{rz}).

Equations 10 and 11 describe the grating diffraction, which allows the tracing of finite rays through the holographic component.

The shape and the size of the image spot can be determined by tracing a set of rays corresponding to the illumination of the HOE by a wavefront of finite size. The results are usually presented as spot diagrams (Fig. 2), which are the points of intersection of the calculated rays with the image plane.

The propagation of the phase function also could be calculated with the help of geometrical ray tracing, taking into account the optical path length. If the phase Φ_A is known at a point A, the phase Φ_B at the point B on the ray is given by

$$\Phi_B = \Phi_A + snk, \tag{12}$$

where n is the refractive index and s the distance between the two points A and B on the ray.

3.4. Wave Propagation Calculated by Scalar Diffraction Theory

According to the Rayleigh-Sommerfeld theory, the complex amplitude $U(x', y', z_O)$ in the *image* plane is given by

$$U(x', y', z_O) = \frac{1}{i\lambda} \int_S U(x, y) \frac{e^{ikr}}{r} \cos(\mathbf{n}, \mathbf{r})\, dx\, dy, \tag{13}$$

where $U(x, y)$ is the complex amplitude in the hologram plane, $\mathbf{r}$ is the vector between the points (x, y) and $(x', y', z_O), r = |\mathbf{r}|$ and $(\mathbf{n}, \mathbf{r})$ is the angle between the normal $\mathbf{n}$ of the HOE plane and the vector $\mathbf{r}$.

The complex amplitude $U(x,y)$ in the HOE plane is given by

$$U(x,y) = A(x,y)\exp[i\Phi_P(x,y)], \tag{14}$$

where $A(x,y)$ is the amplitude of the illumination within the pupil and $\Phi_P(x,y)$ is obtained through Eq. 2 from the HOE phase function Φ_H and the readout wave Φ_r. In the case of homogeneous illumination, $A(x,y)$ is equal to 1 inside the pupil and 0 outside. $A(x,y)$ also can describe inhomogeneous illumination, such as readout with a gaussian beam.

A comparison of ray tracing and numerical results obtained from the diffraction integral in Eq. 13 is shown in Fig. 2.

3.5. Optimum Design

For many applications—e.g. imaging, Fourier lenses, and laser scanners—a set of continuous readout wavefronts $\Phi_r(x,y,t)$ has to be converted into another continuous set of output wavefronts $\Phi_P(x,y,t)$ [5, 6]. The parameter t may describe, for example, the direction of the wave or the position of the focus. From Eq. 2, we get the desired hologram phase function Φ_d as

$$\Phi_d(x,y,t) = \Phi_P(x,y,t) - \Phi_r(x,y,t). \tag{15}$$

The phase $\Phi_d(x,y,t)$ varies with the parameter t. We now are looking for a continuous hologram phase function $\Phi_H(x,y)$, which is as close as possible to $\Phi_d(x,y,t)$ for all t. The performance of the optical element is considered to be optimum when the value of the mean-squared difference between the desired set of holograms $\Phi_d(x,y,t)$ and the real hologram $\Phi_H(x,y)$ is minimum, i.e.

$$\int W(t)P(x,y,t)\,[\Phi_d(x,y,t) - \Phi_H(x,y) + \phi(t)]^2 dt\,dx\,dy \rightarrow \min. \tag{16}$$

The absolute phase of the desired function $\Phi_d(x,y,t)$ is not significant; therefore, an arbitrary additional phase $\phi(t)$ can be added. $W(t)$ is a weighting function and $P(x, y, t)$ is the pupil function. The t-dependence of the pupil function indicates that the readout wave may illuminate different parts of the HOE for different values of the parameters t.

Another approach is based on analytic ray tracing and relies on propagation vectors and grating vectors rather than on phase functions [7]. This means that the phases Φ_i in Eqs. 15 and 16 are replaced by the corresponding wavevectors $\mathbf{k}_{iH}(x,y) = \text{grad}[\Phi_i(x,y)]$. The function $\phi(t)$ disappears.

For either approach, variational methods are applied to find the optimum phase function $\Phi_H(x,y)$ for the given functions $W(t), P(x,y,t)$ and $\Phi_d(x,y,t)$. In many cases, the desired phase function [$\Phi_d(x,y,t)$ in Eq. 15] is not precisely known, for example, if distortions or field curvature can be tolerated. This has to be taken into account.

For holographic disk scanners, as another example, a rotating disk deflects the readout beam and the pupil function $P(x, y, t)$ is not known initially. It has to be determined through the optimization process. To solve this problem, an alternative method has been introduced [8]. The method compares local derivatives of the desired hologram phase function $\Phi_d(x, y, t)$ at the position of the pupil $P(x, y, t)$, given by the parameter t, and the real hologram phase $\Phi_H(x, y)$. This yields good solutions and also reveals basic design restrictions.

Numerical methods are used widely to calculate the optimum function. Starting with a phase function described by N-free parameters, a merit function has to be minimized [9]. Different numerical methods—such as damped least-squares, downhill simplex, simulated annealing, and others—have been reported in the literature. Their success depends, in many cases, on the choice of merit function and the parameter representation of the investigated functions Φ_H. For that purpose, an analytical approach may be helpful to analyze the basic problems, determine free parameters, and suggest the most suitable initial phase function.

4. HOE Fabrication

The HOE phase function $\Phi_H(x, y)$ is fabricated by recording the interference pattern of an object and a reference wave (Eq. 1) in a holographic emulsion. There are limitless possibilities to record the same phase function. Practically, however, the choice is restricted to easily available waves, mainly spherical waves. Another restriction is the requirement to fulfill the Bragg condition. In many cases, spherical waves may be sufficient to approximate the desired function. In other cases, aspheric waves are necessary. These aspheric waves can be generated by using several recording and readout steps, or by the help of computer-generated holograms (CGH). Less frequently, classical optical elements are used to generate aspheric waves.

4.1. *Multiple-Step Methods*

In this subsection, we assume that only spherical and plane waves are available for HOE recording. A one-step recording under optimum conditions is preferred if the aberrations can be reduced sufficiently, because every step is an additional source of noise and errors. However, the optical performance of the desired element may require more complex wavefronts, which can be generated by two or more steps [3]. Special recursive design techniques have been developed for this purpose [10].

4.2. Computer-Generated Holograms

The generation of aspheric waves by *computer-generated holograms* (CGH) is widely used [11]. CGH are binary structures, which allow the creation of optical waves from numerical data by using well-known encoding techniques (Lohmann, Lee, Burch, Arnold, etc.) Today, a CGH is usually fabricated by laser-beam writing or by *e*-beam lithography, permitting a large space band-width product [12].

A typical recording setup is shown in Fig. 3. The laser beam is split into a plane wave branch for the reference and a spherical wave branch for the object. The CGH is inserted into the plane wave branch of the recording setup. A telescopic lens system creates a 1:1 image of the CGH at the HOE. The image plane of the CGH and the HOE plane, in general, are inclined with respect to each other. The phase function Φ_{CGH} necessary to generate Φ_R in the hologram plane can be calculated, e.g. with the aid of the ray-tracing method, taking into account the optical path length between the CGH image plane and the hologram plane. The carrier frequency of the CGH, which is a binary hologram, separates the higher-order diffracted waves from the zero order. The desired aspherical wave is obtained by inserting a spatial filter for the first order in the Fourier plane.

4.3. Diffraction Efficiency

The diffraction efficiency of HOE depends, besides material properties, on the recording and the readout geometry. However, whereas aberrations and image geometry depend on the two-dimensional grating structure $\Phi_H(x, y)$ in the plane of the HOE, the diffraction efficiency is determined by the three-dimensional volume (or surface-relief) structure of the grating.

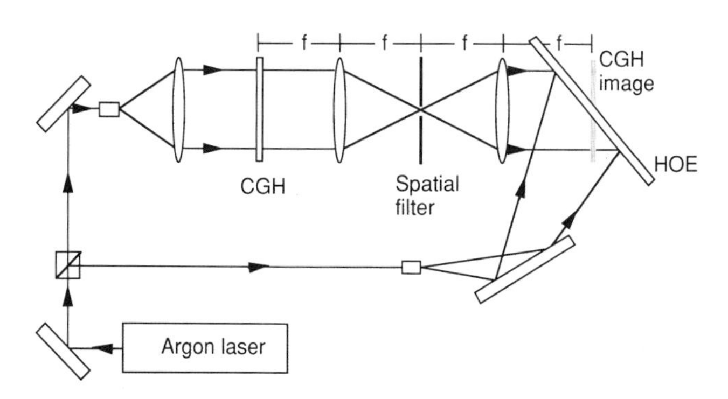

FIG. 3. Recording setup to realize a HOE with the aid of a computer-generated hologram.

Efficient HOE, fabricated by interferometric recording, are volume gratings in most cases (except photoresist HOE). Their diffraction efficiency is well described by Kogelnik's coupled wave theory [13]. This theory assumes that only the zero and the first diffraction order are significant. As an example, the diffraction efficiency η of a transmission HOE is given by

$$\eta = \frac{\sin^2(\mathrm{v}d\sqrt{1 + (\Delta \mathrm{k}_z/2\mathrm{v})^2})}{1 + (\Delta \mathrm{k}_z/2\mathrm{v})^2}, \tag{17}$$

with

$$\nu_s = \frac{\pi \Delta n}{\lambda\sqrt{\cos\theta_p \cos\theta_r}}, \qquad \nu_p = \nu_s \cos(\theta_P - \theta_r), \tag{18}$$

where ν_s and ν_p hold for the s- and p-polarization, respectively. The angles with respect to the surface normal are θ_r for the zero and θ_P for the first diffraction order. Δk_z is the mismatch of the Bragg condition in the z-direction. If the Bragg condition (Eq. 7) is fulfilled, i.e., $\Delta k_z = 0$, the diffraction efficiency is highest. The maximum, ideally $\eta = 1$, is achieved for $\nu d = \pi/2$. For $\Delta k_z \neq 0$, the diffraction efficiency is reduced. For thinner emulsions, larger Δk_z can be accepted, but this requires larger Δn for the same maximum efficiency.

4.4. Copying

Reasons for copying HOE may be mass production or changing the three-dimensional structure to get higher diffraction efficiency (Bragg, blazing). Surface relief HOE can be copied by embossing or casting. Volume phase holograms also can be copied. One possibility is to replay the HOE and to add a new reference wave. This technique has the advantage of great flexibility. Contact copying is another possibility. Here, the transmitted light and the diffracted light of the master hologram are the reference and object beam for the copy. This method is insensitive to vibrations or to variations of the illuminating beam, as long as the master and the copy do not move relative to each other.

5. Applications

5.1. Collimators for Semiconductor Lasers

The output beam of semiconductor lasers is astigmatic and elliptic. Thus, it would be possible to use HOE to collimate the light. On the other hand, inexpensive collimators with good optical output quality, which use conventional optics, already are on the market. This may be the reason why there is little motivation for this HOE application. There would be more motivation if the optics has to include several tasks, as in optical disk heads (Subsection 5.2), or if the laser beam quality is really poor, as in the case of laser diode arrays (Subsection 5.7).

5.2. Optical Disk Head

Read-only, write-once, and erasable storage systems based on optical disks have been developed successfully in recent years. Compact disk players stand today in nearly everyone's house. The optical head is a complex subsystem, which has to fulfill different tasks: focussing the beam from the diode laser for reading and writing, detecting tracking and focus errors, and detecting the reflected signal. A conventional optical head consists of different kinds of bulk optical components, such as lenses, beamsplitters, and prisms. The application of HOE for optical heads permits simplification of the system, reduction of the number of elements, and improvement of the drive performance by reducing the overall size and mass of the head.

Recent research projects have investigated HOE for use in optical disks with great success [14]. They report a reduced number of elements and better stability under adverse environmental conditions.

5.3. Spectrometer

An off-axis holographic lens (HOE) has a large chromatic dispersion. This restricts the use of a single HOE to monochromatic applications. On the other hand, it allows the construction of simple *spectrometers*, which have high spectral resolution in a limited range of wavelengths [3]. This makes them well suited to measure and monitor the longitudinal mode spectrum of semiconductor lasers.

When illuminated with a collimated readout beam from a semiconductor laser, the HOE deflects and focuses the light onto a linear CCD array. As the deflection angle depends on the wavelength (Eq. 4), the spectrum of the laser can be detected by the CCD array, as shown in Fig. 4. The spectral resolution in that case is 0.02 nm over a range of a few nanometers.

5.4. Holographic Laser Scanner

Holographic optical elements can serve as the deflecting as well as the focussing element in laser scanners. They have been incorporated into supermarket point-of-sale systems [15] and laser-beam printers [16].

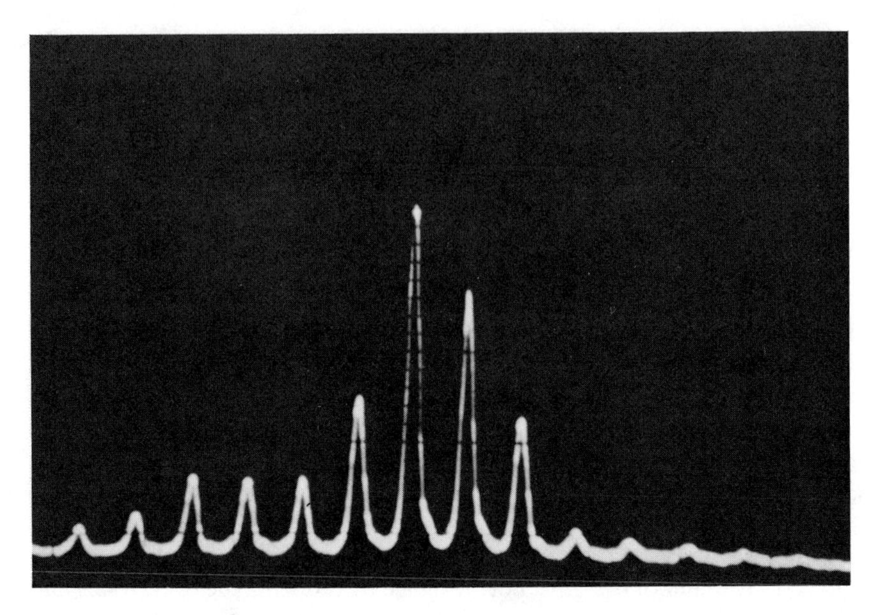

FIG. 4. Spectrum of a multimode semiconductor laser.

Rotational motion usually creates circular scan lines. For some applications, such as point-of-sale systems, curved scan lines can be accepted. However, for other applications, such as laser printers, straightness of the scan line and spot quality are far more critical. Disk scanners are relatively easy to manufacture; therefore, substantial work has been done to develop focussing disk scanners that produce straight-line images [17]. Figure 5 shows the basic arrangement for a holographic disk scanner. The disk contains a number of HOE. Each HOE deflects and focuses the light simultaneously. When the disk rotates, each HOE generates a scan line. The scanning image point permits reading or writing information. Depending on the required scan quality (spot diameter, scan line deviation), additional elements may be necessary.

The optimization of disk scanners in order to find the required phase function for straight line scanning is a complex problem. Analytical as well as numerical optimization techniques have been applied successfully (Subsection 3.5). The performances achieved are equivalent to those of the actual laser printers on the market, which deflect the laser beam with the help of a polygon mirror and use a special lens for focussing. Advantages of holographic scanners are that they can include focussing power in the hologram and they are less sensitive to mechanical wobble of the rotation axis.

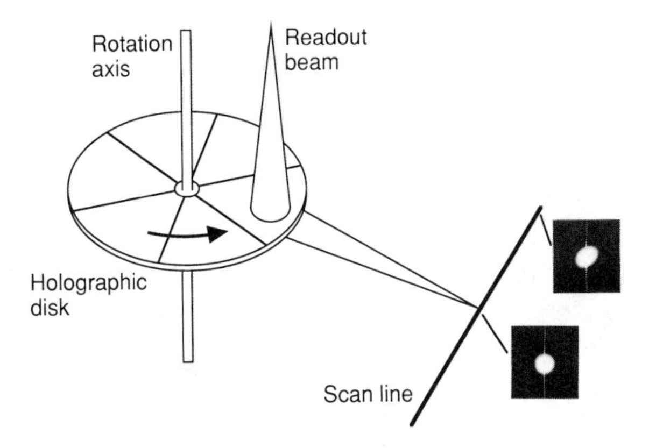

FIG. 5. Holographic scanner. Experimental results for a scan length of ±105 mm: spot diameters (at half intensity) <60 μm for a scan-line deviation of ±30 μm and <85μm for a scan-line deviation of ±8 μm.

5.5. Lenslet Arrays

Holographic lenslet arrays are proposed for free-space optical interconnection. They allow parallel communication, eliminate mechanical point-to-point contacts, and have the potential to address any location on an integrated circuit or in a logic plane.

Lenslet arrays can be fabricated by sequentially recording single holographic lenses. Arrays of spherical waves also can be generated simultaneously by illuminating a pinhole array on a mask with a laser beam. Such a mask can be fabricated by electron beam lithography. Problems occur due to undesirable intermodulations between the individual beams. Better results are achieved if the pinhole array is illuminated with a random phase distribution. This can be done with the help of a scattering plate.

Transmission holograms are recorded by interfering an object and a reference wave, where both sources have to be on the same side of the hologram. For recording holographic lenses with short focal lengths, a problem occurs because of geometry. This is overcome by using total internal reflection (TIR) near-field holography [18]. A pinhole array mask is recorded holographically by passing a collimated beam through the mask, which is placed in close proximity (typically 50–500 μm) to the photosensitive layer, as shown in Fig. 6(a). The transmitted wave interferes with the reference wave, which is fed through a prism and totally reflected at the film-air interface. For reconstruction without a prism, the readout beam can be coupled into the hologram substrate by another

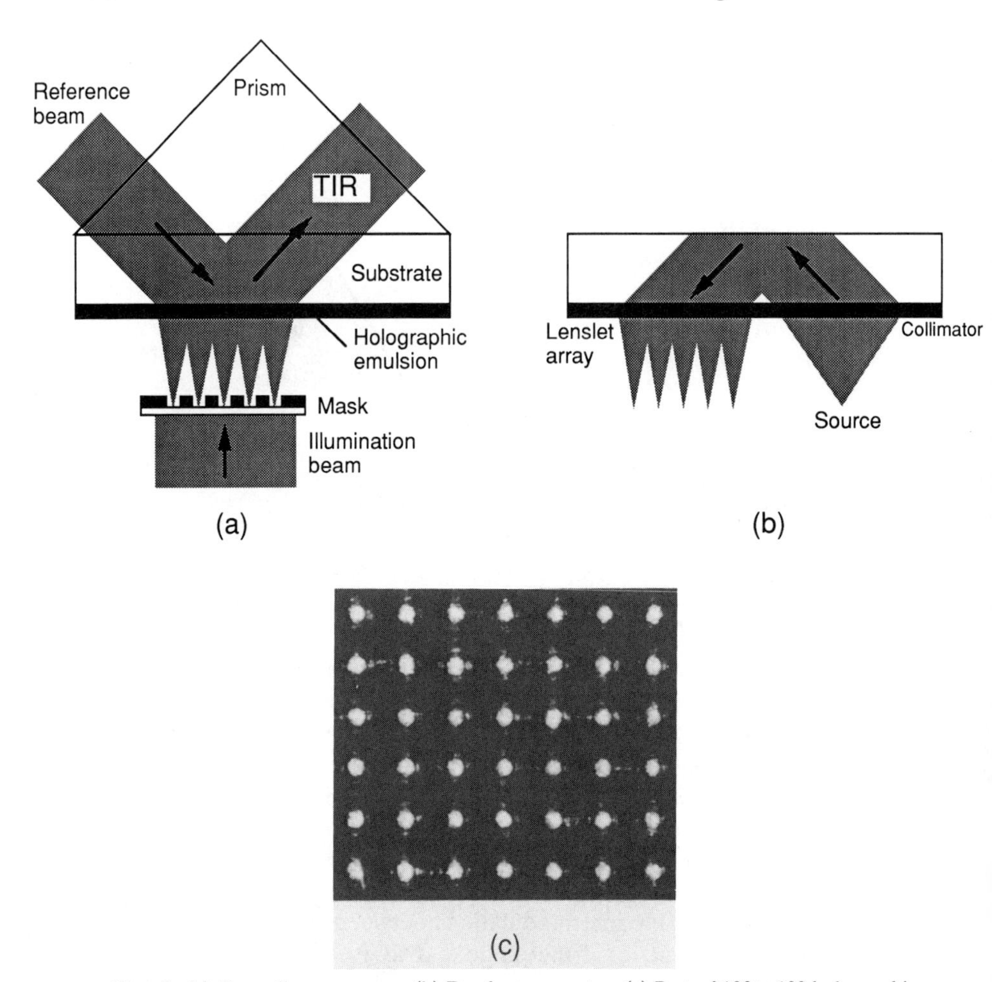

FIG. 6. (a) Recording geometry. (b) Readout geometry. (c) Part of 100×100 holographic lenslet array, with focal length $f = 400$ μm, spacing between the spots $a = 100$ μm and $\varnothing = 10$ μm.

holographic grating, which acts at the same time as a collimator, as shown in Fig. 6(b). This coupling element can be fabricated in the same photosensitive layer as the lenslet array. The collimator and lenslet array together form a compact system. Because of its symmetry, this system is quite insensitive to wavelength changes. The fabrication of a 100×100 lenslet array with a focal length of 400 μm, a diffraction-limited spot size of 10 μm, and a spacing between the spots of 100 μm, as shown in Fig. 6(c), has been demonstrated recently [19].

5.6. Fan-Out Elements

In contrast to lenslet arrays, which are multifaceted elements, *fan-out elements* are single-facet elements. Fan-out elements split a single laser beam into a one or two-dimensional array of beams. They are used in many fields, such as parallel optical processing and fiber optic communication. Binary phase gratings, also called Dammann gratings [20], represent a successful technique to fabricate fan-out elements with good uniformity of the generated array of beams, but with moderate efficiency (60–70%). More recently, efforts are concentrated on kinoforms to increase the diffraction efficiency. Today, such gratings are fabricated synthetically by using microfabrication techniques (Section 6). They can also be made holographically.

A fan-out element can be considered as the far-field hologram of a one- or two-dimensional array of coherent light sources, recorded with a plane or a spherical reference wave. The light sources of the array are characterized by their amplitudes and phases, A_i and ϕ_i, respectively. The phase grating corresponding to this hologram becomes efficient only if the intermodulation terms between the coherent sources are minimized; otherwise, part of the input energy is diffracted into undesired beams. This can be achieved by appropriate choice of the phases ϕ_i, using numerical optimization [21]. Close to 100% efficiency can be achieved for nearly any fan-out with more than six beams.

Ultimately, for this type of fan-out elements, the fabrication technology will become the limiting factor for efficiency and uniformity. Fan-out elements can be fabricated holographically by different methods. The first uses a computer-generated hologram to produce an array of coherent sources with optimized phases for minimum intermodulation (Fig. 7), which can be used to record efficient holographic optical elements as

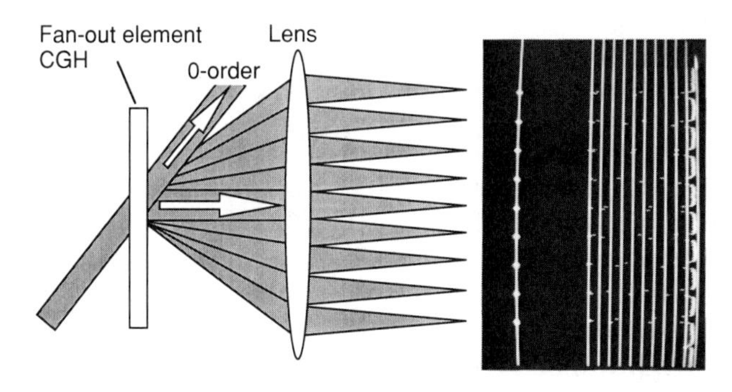

FIG. 7. Readout of the CGH fan-out element at $\lambda = 488$ nm. Spacing between the generated array of sources is $s = 500\ \mu$m and the focal length of the lens is $f = 100$ mm.

volume phase gratings in dichromated gelatine or photopolymer. A similar array of sources can be generated by an array of pinholes (electron beam mask) combined with a phase plate to produce optimized amplitude and phase distribution. Both holographic recording techniques allow the inclusion of focussing power in the fan-out element.

5.7. Beam Shaping

Beam shaping is the most general task a HOE can accomplish. Starting with a complex amplitude distribution $U_1(x, y, z = 0)$ in an object plane, one would like to generate another distribution $U_2(x, y, z = d)$ in an image plane without loss of optical power. In general, two HOE are necessary to perform this task [22]. Applications include the transformation of a gaussian beam into a flat-top profile and the far-field shaping of laser diode arrays (LDA).

Commercially available single-mode lasers emit light powers up to 50 mW. Actually, we can observe a very strong competition to fabricate high-power semiconductor lasers up to several watts. Typical applications are laser pumping and space communications. Of great interest are single-mode lasers which can be collimated. Promising results to generate coherent light beams of high power have been achieved with laser diode arrays. Unfortunately, the output beam is of poor quality, and cannot be collimated properly by conventional diode laser optics.

It has been shown that stable supermodes emitted by LDA can be shaped and collimated with very small inherent losses using two HOE [22]. In the case of a LDA with nine phase-locked stripes emitting equal intensities, 99.3% of the output power can be converted, in principle, into a single-lobed beam with a shape corresponding to the diffraction-limited emission of a single stripe.

6. Interferometric HOE Recording versus Synthetic Fabrication

We limited the scope of this chapter to HOE, fabricated by interference of a reference and an object wave. Nevertheless, perfect diffractive structures also can be generated synthetically. However, the laws of diffraction remain the same. The difference between the elements is the fabrication technology. For the future, we expect that the term *diffractive optical elements* (DOE) will be adopted to cover all these elements.

Figure 8 shows the different fabrication methods for DOE. Fabrication methods for synthetic DOE use techniques which have become standard in

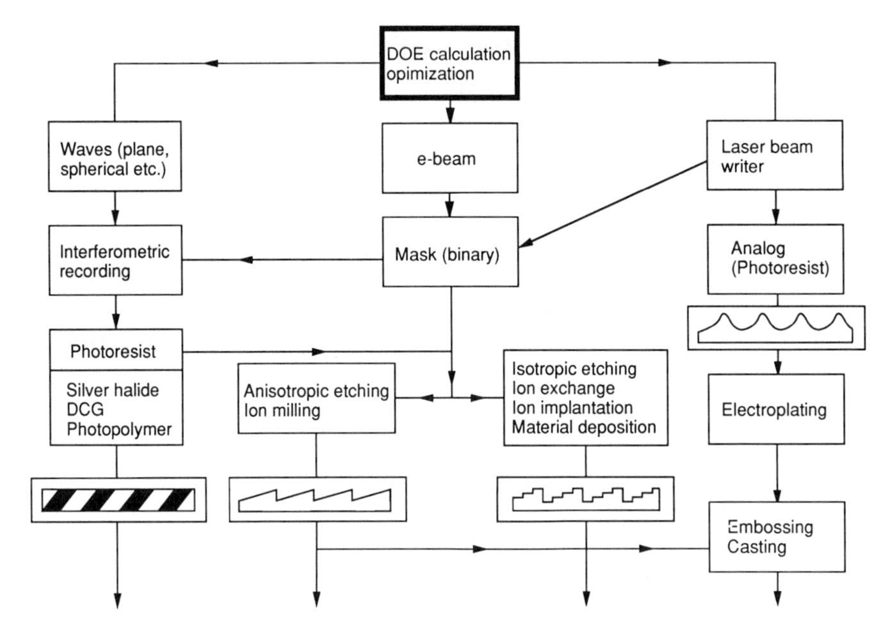

FIG. 8. Fabrication of diffractive optical elements (DOE).

VLSI technology. Masks are generated by *e*-beam lithography or by laser beam writing lithography. Then, to get high efficiency, they are transformed into surface-relief structures by etching in materials like silicon nitrite, quartz, silicon, or by thin film deposition, e.g., SiO. Typical examples are Dammann gratings. Using several masks, multilevel functions can be generated to improve the efficiency (blazing) [23]. Anisotropic etching is another method to introduce blaze effects.

Another interesting technique is the direct fabrication of the phase structure by laser beam writing lithography in photoresist [24]. The developed photoresist relief can be converted into a metalized master relief by electroplating to emboss or cast low-cost replicas. This method allows smooth relief structures, such as lenslet arrays or highly efficient fan-out elements [21].

The microfabrication techniques used for synthetic (computer-generated) elements are well established, and their feasibility has been proven under industrial conditions. However, it still is a problem to make properly blazed structures for high spatial frequencies ($>1{,}000$ lines/mm).

An advantage of interferometric recording is the high resolution offered by holographic materials ($>3{,}000$ lines/mm). On the other hand, it is difficult to control the recording process and the materials to get reproducible results.

7. Future Perspectives

Diffractive optical elements (DOE) are components that make full use of the wave nature of light. By relying on diffraction and interference rather than on reflection and refraction, unique and novel properties can be realized, complementing and exceeding the possibilities of the traditional lenses, prisms, and mirrors. The flexibility of DOE, and their potential for compact multitask elements, make them attractive for many future applications.

Today's HOE market essentially is limited to applications for laser scanning and to different kinds of displays (e.g., head-up displays). Only a few of them use semiconductor lasers as the light source. On the other hand, many applications are planned in the growing field of modern optics, including an optical disk head, fiber optic communication, space communication, and parallel optical processing. The potential of these elements is nearly unlimited. Meanwhile, lens designers, too, have discovered the appealing properties of diffractive elements when combining them with classical optics. Thus, it is difficult to imagine what still could retard the introduction of these elements into the market. The key question is the fabrication technology necessary to produce HOE and DOE in large quantities and with reliable accuracy. Actual trends in manufacturing show promising results, so that a brilliant future for diffractive optical elements can be expected.

References

1. P. Hariharan, "Optical Holography," in "Cambridge Studies in Modern Optics 2," Cambridge University Press, Cambridge, 1986.
2. W. K. Smothers, T. J. Trout, A. M. Weber, and D. J. Mickish, "Hologram Recording in Dupont's New Photopolymer Materials," *Holographic Systems, Components and Applications*, IEE, Conference Publication No. 311, Bath, UK, 184–189 (1989).
3. H. P. Herzig, "Holographic Optical Elements (HOE) for Semiconductor Lasers," *Opt. Commun.* **58**, 144–148 (1986).
4. W. T. Welford, "Aberrations of Optical Systems," Adam Hilger Ltd, Bristol, 1986.
5. J. N. Cederquist and J. R. Fienup, "Analytic Design of Optimum Holographic Optical Elements," *J. Opt. Soc. Am. A* **4**, 699–705 (1987).
6. J. Kedmi and A. A. Friesem, "Optimized Holographic Optical Elements," *J. Opt. Soc. Am. A* **3**, 2011–2018 (1986).
7. E. Hasman and A. A. Friesem, "Analytic Optimization for Holographic Optical Elements," *J. Opt. Soc. Am. A* **6**, 62–72 (1989).
8. H. P. Herzig and R. Dändliker, "Holographic Optical Scanning Elements: Analytical Method for Determining the Phase Function," *J. Opt. Soc. Am. A* **4**, 1063–1070 (1987).
9. Y. Ono and N. Nishida, "Holographic Optical Elements with Optimized Phase-transfer Functions," *J. Opt. Soc. Am. A* **3**, 139–142 (1986).
10. Y. Amitai and A. A. Friesem, "Design of Holographic Optical Elements by Using Recursive Techniques," *J. Opt. Soc. Am. A* **5**, 702–711 (1988).

11. W. H. Lee, "Computer-Generated Holograms: Techniques and Applications," in "Progress in Optics," (E. Wolf, ed.), Vol. 16, pp. 119–232, North-Holland, 1978.
12. H. Buczek and J. M. Teijido, "Application of Electron-Beam Lithography at CSEM for Fabricating Computer-Generated Holograms," *Computer-Generated Holography II, Proc. SPIE* **884**, 46–51 (1988).
13. H. Kogelnik, "Coupled Wave Theory for Thick Hologram Gratings," *Bell. Syst. Tech. J.* **48**, 2909–2947 (1969).
14. Y. Kurata, H. Yamaoka, T. Ishikawa, P. Coops, A. Duijvestijn, and P. de Zoeten, "CD Optical Pickup Using a Computer-Generated Holographic Optical element," *Optical Storage and Scanning Technology, Proc. SPIE* **1139**, 161–168 (1989).
15. G. T. Sincerbox, "Holographic Scanners: Applications, Performance, and design," in "Optical Engineering," Vol. 8: "Laser Beam Scanning," (G. F. Marshall, ed.), pp. 1–62, Marcel Dekker, New York, 1985.
16. L. Beiser, "Holographic Scanning," John Wiley & Sons, New York, 1988.
17. H. P. Herzig and R. Dändliker, "Holographic Optical Scanning Elements with Minimum Aberrations," *Appl. Opt.* **27**, 4739–4746 (1988).
18. K. Stetson, "Holography with Total Internally Reflected Light," *Appl. Phys. Lett.* **11**, 225–226 (1967).
19. D. Prongué and H. P. Herzig, "HOE for Clock Distribution in Integrated Circuits: Experimental Results," *Optical Interconnections and Network, Proc. SPIE* **1281**, 113–122 (1990).
20. H. Dammann and K. Görtler, "High-Efficiency In-Line Multiple Imaging by Means of Multiple Phase Holograms," *Opt. Commun.* **3**, 312–315 (1971).
21. H. P. Herzig and D. Prongué, "Design and Fabrication of Highly Efficient Fan-Out Elements," *Jpn. J. Appl. Phys.* **29**, L1307–L1309 (1990).
22. H. P. Herzig, R. Dändliker, and J. M. Teijido, "Beam Shaping for High Power Laser Diode Array by Holographic Optical Elements," *Holographic Systems, Components and Applications*, IEE, Conference Publication No. 311, Bath, UK, 133–137 (1989).
23. J. Jahns and S. J. Walker, "Two-Dimensional Array of Diffractive Microlenses Fabricated by Thin Film Deposition," *Appl. Opt.* **29**, 931–936 (1990).
24. M. T. Gale and K. Knop, "The Fabrication of Fine Lens Arrays by Laser Beam Writing," *Industrial Applications of Laser Technology, Proc. SPIE* **398,** 347–353 (1983).

CHAPTER 6

Fibre-Optic Signal Processing

Brian Culshaw and Ivan Andonovic
University of Strathclyde
Electronic & Electrical Engineering Department
Glasgow, Scotland

1. Introduction

There is no doubt that fibre optics has made a substantial contribution in the communications network. Signal processing systems are very similar—arguably identical—in topology. The question then arises as to whether the unique features of this transmission medium could be useful in this different context.

The term *signal processing* applies to a wide range of data manipulation and modification procedures. In the context of this chapter, it may be restricted to means whereby electronic input information, modulated on to an optical carrier, is modified whilst in the optical domain and then transmitted for further processing in an electronic form. The data is modified by transmitting it through an optical fibre network, which—by using a combination of signal division, delays, and signal recombination—will alter the characteristics of the signal modulated on to the original optical carrier.

Clearly, the data originates and terminates electronically, so for the technique to be viable, there must be specific advantages that apply when the processing is implemented within an optical transmission medium. Such advantages may include the following:

- The manipulation of data that is already optical—for example, in a local area optical fibre communications network (LAN)—may be simplified considerably if pre-processing can be effected in the optical domain.

ISBN 0-12-289690-4

- The extremely broad dispersion-free bandwidth of the optical fibre, as perceived at the modulation frequency, may be used to good effect in delay-based processors.

Processors involving delay-and-recombine elements are used widely and include the very simple delay lines used in European colour television sets and, at a more specialised level, pulse compressors and surface wave filters. A figure of merit for a delay-processing medium is the time-bandwidth (TB) product. Single-mode optical fibres—thanks primarily to the extremely high bandwidths available—offer a TB product that exceeds that of any other delay medium by up to two orders of magnitude. In effect, the product may be viewed as the number of data points that may be stored upon a transmission line and, therefore, may be available for processing at any given time.

In practice, even though the TB product is so high, it is appropriate to observe that the bandwidth is far more useful than the delay and that it is probable that fibre optic processors predominately will find applications in high bandwidth processes. They enable signals in the optical domain to be split, recombined, and modified with frequency-independent delays, without the requirement for multiple dispersion-matched, very high frequency signal channels and, more importantly, without the need for analogue-to-digital converters operating in the 100Gbits-per-second region.

The following sections contain a summary of the basic principles of fibre-optic signal processors, a brief outline of some applications, and a discussion of their future potential. The principal benefits appear to lie in the manipulation of data from multiple sources on a single fibre link in a fibre-optic LAN and in the handling of very high frequency analogue signals. The future potential will be enhanced greatly by developments occurring in optical communications. Of these, the most important probably are the distributed optical amplifier and, related to that, the concepts of non-demolition detection and similar nonlinear processes, together with the ever increasing availability of optical data manipulation components originating from the desire to realise all-optical telephone switching networks.

2. Principal Features of Fibre-Optic Signal Processing Networks

The manner in which a fibre-optic signal processing system will operate is determined by the fundamental characteristics of the transmission medium and the manner in which the medium may be interconnected.

2.1. General Observations

An optical fibre transmission path, interfaced with a fibre- or integrated-optic component or component set, may modify the properties of light propagating within it in a variety of ways:

- — by providing a delay.
- — by splitting the input optical power into two, generally unequal portions.
- — by providing an electronically or optically controllable attenuation or phase change.
- — by providing a polarization- or wavelength-selective transmission path.

In a general network (Fig. 1), the processing element may be reconfigured either optically or electronically, and may have functions within it that depend upon the levels of the optical signals transmitting along the interconnecting paths. This network is nothing more than a guided-wave processor and, as such, performs in a similar fashion to—though significantly quicker than—a wire interconnect computer system. Its characteristics are totally different from those of a spatial processor, whose properties are covered elsewhere in this volume.

An important and often neglected characteristic of optical radiation is that only its intensity can be detected electronically. Furthermore, due to the phase instability of relative optical paths in a guided-wave network,

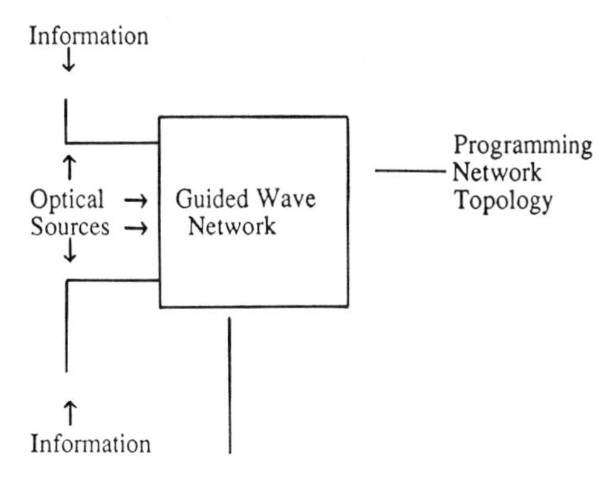

Fig. 1. Global representation of a fibre-optic signal processing network. The principal features are that the inputs and outputs are electronic, and that the signal manipulation takes place on the modulated signal in the optical domain.

phase-modulated (that is, *interferometric*) fibre-optic signal processors likely will be difficult to implement apart from when utilising very short delays and, therefore, by implication, extremely high modulation frequencies.

An immediate corollary of this is that since intensity modulation must be used to perform the processor functions, any possibility of interferometric effects must be eliminated. The coherence length of the optical source, therefore, must be extremely small when compared to the minimum modulation wavelength.

2.2. Processor Architectures

Linear processes, as seen from the preceeding discussion, are the simplest to implement. This leads to the perhaps perverse implication that analogue data is easier to handle than digital data. The constraints imposed by the availability of only positive weighting functions limit the range of operations that are available. However, by using special coding approaches, both digital and bipolar operations may be realised.

2.2.1. Programmabiliity

The parameter determining the functional role of most fibre-optical processor networks is the coupling ratio of the signal dividing and recombining elements. In integrated optics, this ratio readily is controlled electronically, but, at least for systems with large delays, the interface losses between the fibre and the integrated optic chips can compromise the network performance. All fibre components now are highly developed and used both in telecommunications and sensing networks, but apart from mechanically and thermally controlled couplers, electrically variable splitters currently are unavailable.

2.2.2. Functional Capabilities

The basic delay-split-recombine format is naturally compatible with Finite Impulse Response (FIR) and Infinite Impulse Response (IIR) filtering systems. Additionally, matrix-matrix analogue multiplication may be realised, so that a wide range of appropriate analogue processing functions is implemented readily. Additionally—given the availability of nonlinear elements—functions such as digital code correlation and related spread spectrum operations also may be implemented.

2.2.3. Functional Cascadability

A key feature of any processing network is that the output for one stage should be compatible with the input to the next and, ideally, this input should be common to all processors. The degree to which this condition may be met in practice varies with the processing configuration, but with suitable coding of the input data, cascadable networks may be implemented.

2.3. Components for Linear Processors

Virtually all key components used in fibre-optic signal processors are obtainable directly from the communications area and include:

— an optical source or sources that may be modulated at gigahertz rates and introduce substantial power into (preferably) single-mode fibres, and are incoherent over the minimum differential delay times.
— fibre-optic (or, for high frequency systems, integrated optic) delays.
— splitters and recombiners (i.e. directional couplers).

All of these are currently available, so that simple networks are a "here and now" technology [1–4]. Nonlinear operations are discussed later.

3. Some Practical Examples

3.1. Analogue Operations

3.1.1 Transversal and Recursive Filters

The basic principles of these devices are shown in Figs. 2 and 3. In both cases, the input signal is presented as analogue modulation on to the light source launched into a signal-mode optical fibre. State-of-the-art direct-drive laser modulators are capable of modulating over bandwidths of several GHz with linearities of better than 1%. The finite impulse response filter shown in Fig. 2 consists of a series of directional couplers with coupling coefficients k_i with time delays τ_i between the input to the fibre and the ith coupler. If the modulation on the input is $f(t)$, then the output

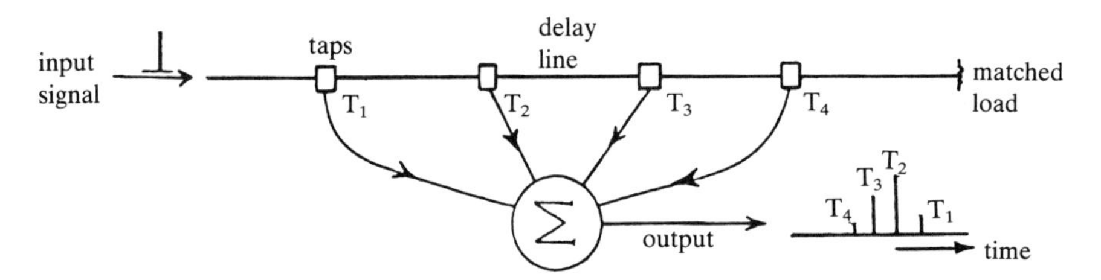

FIG. 2. Delay-line representation of a transversal filter indicating the derivation of the impulse response.

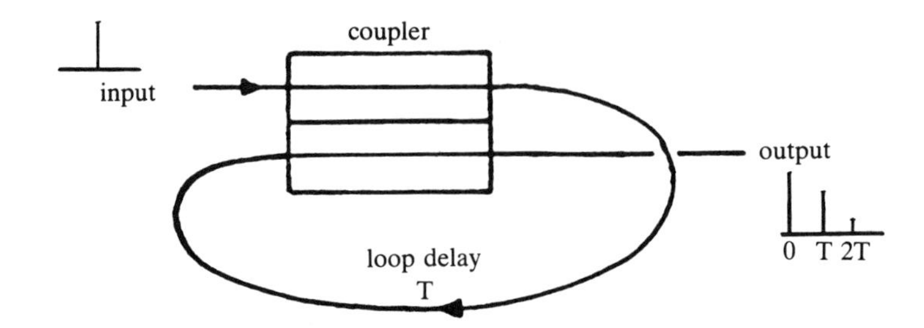

FIG. 3. Recirculating delay-line optical-fibre filter with an impulse response set for the coupling ratio equal to 0.34. Changing the coupling ratio alters the impulse response in the limit to that of a comb filter.

$f_{\text{out}}(t)$ is given by:

$$f_{\text{out}}(\tau) = \sum_i f_{\text{in}}(t - \tau_i) k_i . \Pi_j (1 - k_j) \quad (1)$$

Normally, the coupling coefficients k_i are assumed to be much less than 1, so that the continued product term may be neglected to a first approximation. The coupling coefficients k_i and the corresponding delays τ_i then become directly equivalent to the impulse response of the filter.

The exceptional feature of filters constructed using this principle is that the delays τ_i are independent of modulation frequency on the input laser over any currently usable bandwidth. Consequently, the filter response may be defined precisely.

The IIR configuration shown in Fig. 3 consists of a single adjustable directional coupler and a recirculating loop, possibly incorporating electronic or optical gain to compensate for the loss at the coupler. Depending on the coupling coefficient of the coupler, the response of this configuration can be varied from a simple (almost) dual impulse with a cosine squared characteristic, where the precisely defined time delays produce an extremely deep notch filter, to a temporal comb response producing very narrow pass-band filters.

These filters were studied primarily in the early 1980s [5]; a commercial prototype of the recirulating filter has been built and evaluated.

These are the simplest of fibre-optic signal processors and a rapid study of their characteristics soon reveals some of the major constraints. The architecture is inherently analogue, so that the accuracy of the filtering process is limited, and all the taps in the system are positive and real. The range of impulse responses available, therefore, is somewhat limited. These filters also demonstrate the advantage of the very broad band non-dispersive delay medium, as well as the inherent difficulty in realising the immense TB product mentioned in the introduction due to the necessary losses at the coupling points. However, strategically placed electronic or optical gain may be adequate to overcome this particular problem. Optical gain is preferable, since its phase and gain characteristics at the modulation frequency will be equally non-dispersive.

3.1.2. Matrix-Matrix Multiplication

The matrix multiplication process involving two $N \times N$ matrices AB may be expressed as:

$$[C] = [A][B], \tag{2}$$

where the elements of matrix $[C]$ are given by

$$c_{ij} = \sum_{k=1}^{N} a_{ik} b_{kj} \tag{3}$$

The matrix-matrix multiplication process, therefore, may be expressed in terms of the inner product between the ith row vector of A and the jth column vector of B. Therefore, an inner product processor coupled in a systolic array could form the basis of analogue matrix-matrix multiplication. The basic inner product processor is shown schematically in Fig. 4, whilst Fig. 5 shows a fibre-optic implementation of the matrix multiplication process, where the matrix B is programmed as the coupling ratios of an array of directional couplers (again assumed to be much less than 1, so that the *through* signal may be assumed to be unaltered, though the modifications may be incorporated easily to allow for the coupler attenuation). Each coupler in this configuration performs equivalently to an inner product processor and the delay times T in the diagram are the clock period of the system. This can be made to be extremely small, of course, by using short lengths of fibre or even integrated optic waveguides. The basic configuration in Fig. 5 may be extended to a general matrix [6].

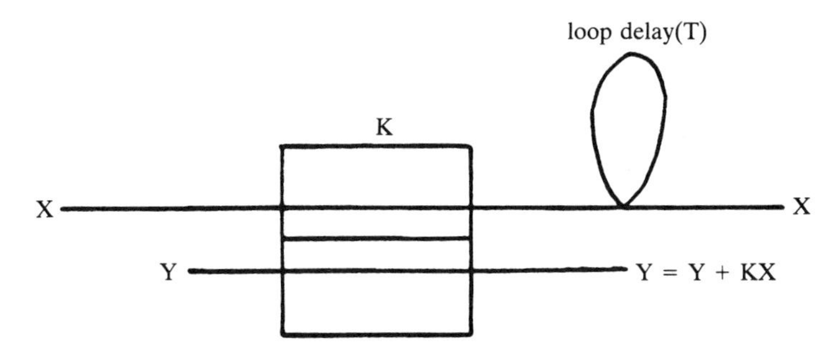

FIG. 4. The basic inner product operation.

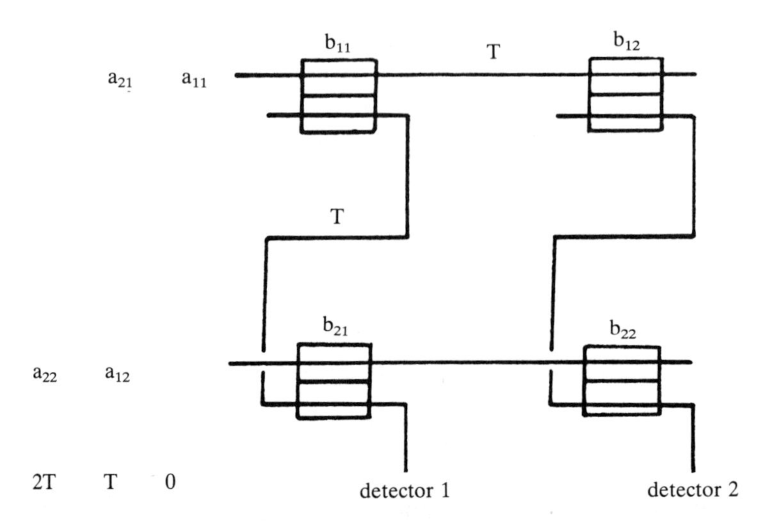

FIG. 5. Implementation of 2 × 2 fibre-optic matrix multiplier.

This class of matrix-matrix multipliers is limited by the constraint that the coupling ratio b_{kj} usually is fixed, implying that the system cannot be programmed. Of course, mechanically, thermally, or electro-optically programmable couplers may be realised using polished block, Mach Zehnder interferometric, or integrated-optic technologies, respectively. These add inevitably to the cost and complexity of the system and, at present, it seems most likely that these concepts will find a niche in applications that require relatively little programming. Such applications could be found, for example, in certain types of images processing functions, including contrast enhancement and edge extraction, where the processing can be defined by fixed constants and time delays.

3.2. Digital Processing

3.2.1. Basic Digital Operations

The preceding examples of fibre-optic signal processors have been analogue in nature. Analogue optical systems are inherently difficult to realise with any realistic accuracy, since factors such as source output powers, detector sensitivities, and transmission levels through fibre-optic links all are subject to variations with time, temperature, and the physical environment. This limitation also is well-known in the Fourier plane processing community.

Digital optics may be able to realise some, if not all, of the accuracy offered by standard digital electronic computational processors. In the context of fibre-optics systems, the term *digital optics* means the use of binary or pseudo-binary (i.e., multi level) coding as a means of representing information in a fibre-optic processor.

Possibly the most important issue in exploiting any computational technique is optimising the information coding method that is exploited. For electronic processors, binary coding usually is accepted as being the most appropriate. In the past, there have been some exploratory evaluations of the possibilities for multilevel coding in the context of optical processors. Especially for those in which the availability of nonlinear elements is limited, it is likely that multilevel coding will be the most appropriate. For all fibre processors, attention thus far has focussed on the representation of numbers in twos-complement arithmetic, coupled with the use of algorithms such as DMAC (digital multiplication by analogue convolution). This approach enables the realisation of bipolar binary operations in a format that may be cascaded from one stage to another of a hypothetical optical computer.

In two-complement arithmetic, the most significant bit is the sign bit, 0 for positive and 1 for negative. Positive numbers are represented in the normal way, with the 0 preceding the conventional number; negative numbers begin with a 1 and, thereafter, all the 1's and 0's are inverted and, finally, a 1 is added. The usefulness of this scheme may be demonstrated by illustrating a fibre-optic implementation of a binary multiplier capable of multiplying by plus or minus 1, depending upon the setting of directional couplers. Such a scheme is shown in Fig 6. Here, the centre coupler represents the value 1 in all cases. The polished coupler (*PC*1) is

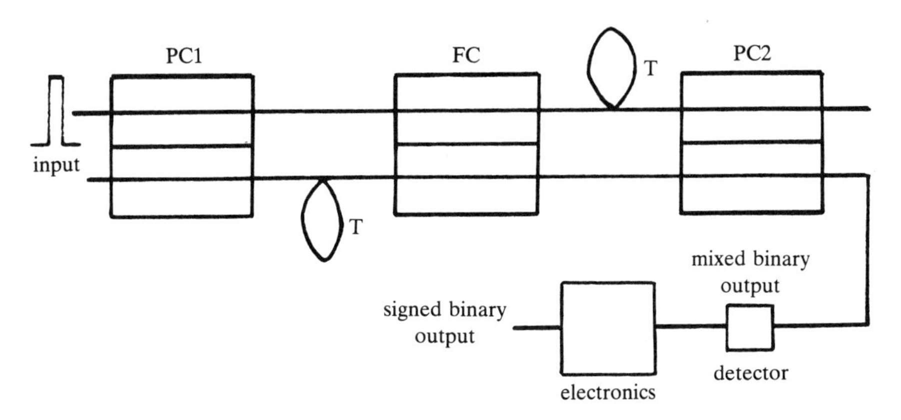

FIG. 6. The realisation of a fibre-optic bipolar tap.

	0	0	0	0	0	0	0	1	1	0	1		+13
	1	1	1	1	1	1	1	1	1	0	1	x	−3
	0	0	0	0	0	0	0	1	1	0	1		
	0	0	0	0	0	0	0	0	0	0			
	0	0	0	0	0	1	1	0	1				
	0	0	0	0	1	1	0	1					
	0	0	0	1	1	0	1						
	0	0	1	1	0	1							
	0	1	1	0	1								
	1	1	0	1									
	1	0	1										
	0	1											
	1												
Mixed binary output	3	3	3	3	3	3	2	2	2	0	1		
Signed binary output	1	1	1	1	1	0	1	1	0	0	1		

FIG. 7. Illustrating the mode of operation of the DMAC algorithm.

set to 50% or 0% when the multiplicand is -1 (11 in signed binary) or $+1$ (01 in signed binary), respectively, and the value of $PC2$ depends upon the sign bit of the multiplier. Inputs represented as on/off pulses for 1 or 0, respectively, may be coded to represent $+/-1$ and may be multiplied by $+/-1$, using the system shown in Fig. 6. This technique is described in detail in [7] and can be expanded to realise the basic binary functions of multiplication, addition, and subtraction.

The multiplication function is obtained using the DMAC algorithm. An example of the operation of this algorithm is shown in Fig. 7, which also demonstrates the use of signed to represent the numbers. One important feature of the DMAC approach is that each number must be represented by a word whose length is equal to the longest word that will be appearing in the system. Therefore, cascading (see Fig. 8) must be implemented with care to ensure that sufficient bits are used throughout the system. However, a full and self-consistent technique for multiplication may be realised based upon this algorithm, and the entire function can be implemented in guided-wave optical form, with the exception of the final multi-level-to-binary conversion step.

3.4. Code Recognition and Correlation

A particular interesting form of fibre-optical signal processor is shown in Fig. 9. This essentially is a combination of the bipolar tap outlined earlier with a carefully chosen code system and non-linear element designed to perform a code recognition function. The function of the non-linear coupler is outlined in Fig. 10. The system recognises the presence of a

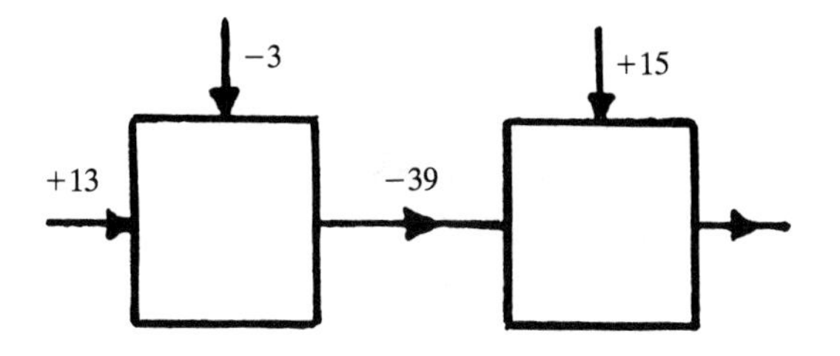

MIXED BINARY OUTPUTS

0	0	0	0	0	0	0	1	1	1	1		+15
3	3	3	3	3	3	2	2	2	0	1	x	−39
0	0	0	0	0	0	0	1	1	1	1		
0	0	0	0	0	0	0	0	0	0			
0	0	0	0	0	2	2	2	2				
0	0	0	0	2	2	2	2					
0	0	0	2	2	2	2						
0	0	3	3	3	3							
0	3	3	3	3								
3	3	3	3									
3	3	3										
3	3											
3												
12	12	12	11	10	9	6	5	3	1	1		
1	0	1	1	0	1	1	0	1	1	1		

FIG. 8. Cascading DMAC.

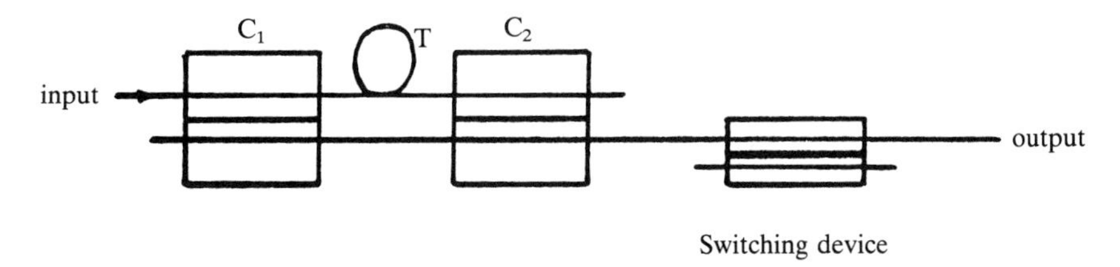

FIG. 9. A digital correlator architecture. Here, the loops represent delays of one bit interval and the second row of couplers are nonlinear couplers used as thresholding elements.

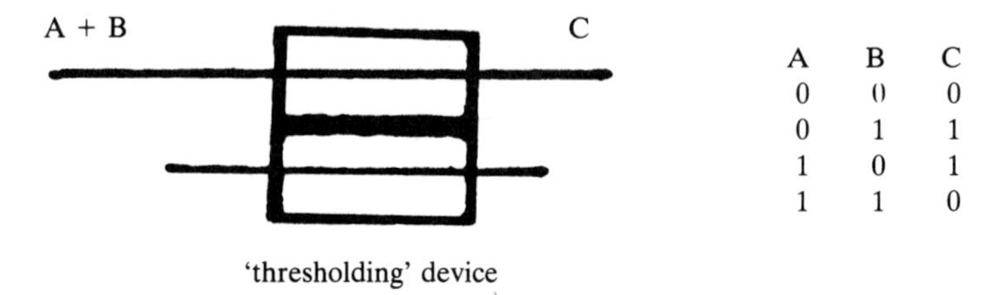

A	B	C
0	0	0
0	1	1
1	0	1
1	1	0

FIG. 10. The truth table for the nonlinear coupler shown in Fig. 9.

particular bit sequence transmitted along the optical fibre. In this system, each data bit is coded into one period of a specific code sequence, e.g., Gold sequences [8] with a 1 represented as 01 and a 0 as 11. The following table shows how the network produces an equivalent to a 1 (i.e., the sequence 01) if the tap is matched to the input code and a sequence 11 if the tap is mismatched.

Table I

Input	Bipolar Tap Weight	Output(C)
01	01	01
01	11	11
11	01	11
11	11	01

Some proposals already have been described [9] for realising the nonlinear element, and some initial experimentation currently is under way. The technique shows enormous promise for use in local area networks and high-capacity, multi-user optical transmission systems. Demonstrations of the thresholding properties of a semiconductor laser amplifier recently have been completed [10].

4. Applications Potential

The principal technical features that may make optical-fibre signal processing networks attractive are:

- the obvious compatibility with data that already is encoded optically and is transmitted along the fibre.

• the ability to cope with extremely high frequency signals in the optical domain and thereby eliminate the need for the high-frequency electronic components within the processor.

The applications then are relatively well-defined and will include such areas as:

— the distribution of triggering signals in phased-array radars.
— receiver processing for large phased arrays.
— high-precision filtering.
— code recognition/ generation in communication systems.

These applications all have the need to take one or more high-speed single-channel data inputs and combine/compare these inputs with others of very similar nature to realise the required processing function. This currently appears to be the focus of any applications analysis of fibre-optic signal processing.

5. Fibre-Optic Signal Processing—Future Prospects

There are numerous obvious difficulties with the processors described earlier, and all of these stem from the restricted range of optical properties currently available in single-mode optical fibres.

Perhaps the most important problem is the influence of loss on the maximum number of taps which may be incorporated within a system (see Fig. 11). The other constraint lies in the restricted range of non-linear elements that may be used to implement operations such as thresholding. A related difficulty stems from the very variable optical level used to represent, say, a "1." These variations, which may extend to several tens of decibels, imply that operation involving threshold detection are extremely difficult to define.

This last feature, which may be rephrased as the need to define the *data unit* within the fibre-optic network, probably is the most fundamental parameter that determines the network design. The only *natural* optical power unit in fibres is the soliton, and there are strong arguments for adopting the soliton as the unit of currency within both signal processors and communication networks. The principal advantages gained thereby include relatively stable and predictable peak power levels within the soliton itself, high information capacity, and, perhaps, a bipolar potential offered by the use of dark solitons.

Clearly, the intensity dependence of soliton propagation characteristics implies that any losses must be made up by using the appropriate amounts of gain. One approach may be to exploit non-demolition detection (see

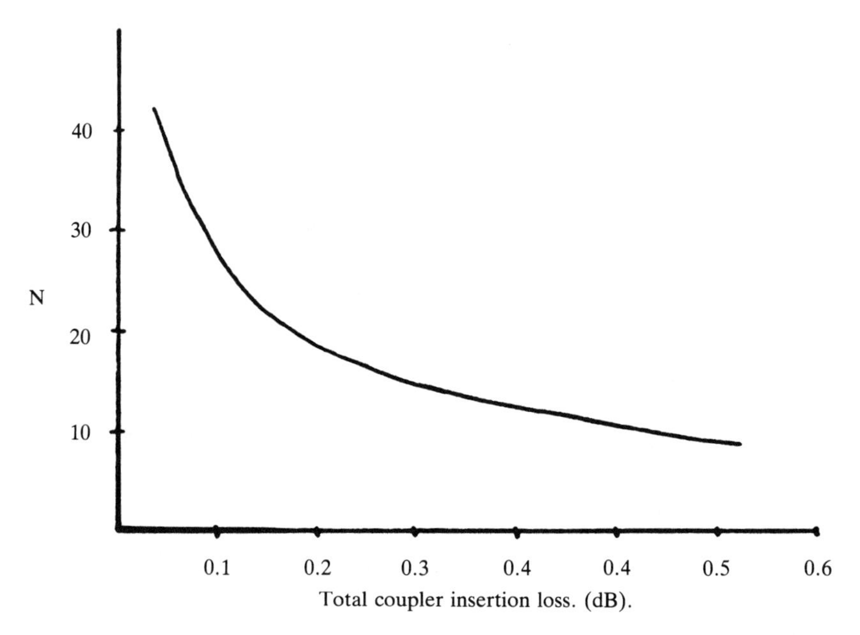

FIG. 11. The maximum number of taps (N) expressed as a function of the coupler loss in dB.

later in this section). Whilst this may be attractive, it does imply that alternate stages of the processing network operate at different wavelengths. A single-wavelength scheme offers advantages, including simple cascadability and more straightforward technology, and will require the use of optical amplifiers within the network. Both distributed (all fibre) and point (semiconductor) amplifiers have been developed with the communications industry. The all fibre-device is much simpler optically and has been shown to be capable of high gains of the order of 40dB and low noise figures [11]. The situation in signal processing is different in that the gain required is probably quite small—in the range of a few decibels—and spontaneous emission noise may detract significantly from the amplifier performance. A semiconductor amplifier will provide localized gain. These devices are operated best at around 10dB gain. Consequently, there is a trade-off, as yet not understood, between the system need for relatively constant signal levels and the inherent characteristics of readily available optical amplifiers. Fig. 12 illustrates the basic features of all-fibre and discrete amplifier networks. The former may exploit Raman or Brillioun gain mechanisms all well as rare earth doping.

An alternative gain mechanism would be to use parametric amplifiers [12]. These do not exhibit spontaneous emission noise. Parametric amplifiers have other interesting features, in that they may generate non-

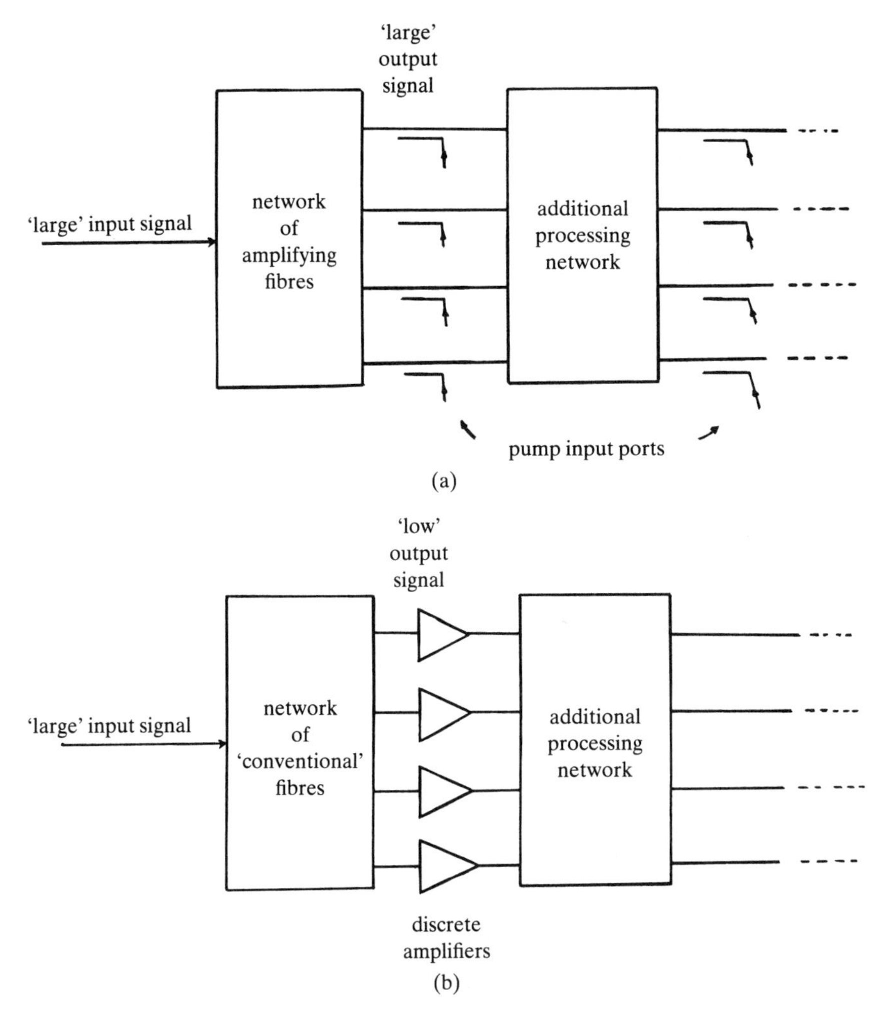

FIG. 12. The available network topologies for (a) distributed-fibre amplifiers and (b) discrete semiconductor amplifiers in a signal processing system. The discrete approach simplifies the network components, but implies that any thresholding must be implemented after amplification. The distributed network is complicated by the need for pump-port inputs, but offers stability of the optical level throughout.

classical states. A recent analysis of the possibilities offered by such states in communications has indicated that some potential benefits may accrue especially for coherent systems. The same probably is true for signal processing, since close analogies can be drawn between the two system requirements. However, even the most optimistic speculation only could see a role for squeezed light long after the soliton has become established.

Nonlinear elements obviously are important. Conventionally, optical physicists turn to crystalline materials to experimental demonstrations of non-linear phenomena. However, an alternative look at optical fibres soon demonstrates that they may be viewed as the most non-linear component available [13]. The key for any nonlinear interaction is the product of the second- or third-order nonlinear susceptibility with an interaction length. The attenuation in optical fibres at the 1.3- and 1.55-micron wavelengths is extremely low, so that interaction lengths of kilometres imply that Kerr-induced nonlinearities are observed readily in fibres. This is well-known in the context of fibre-optic gyroscopes, where nonlinearities detectable due to power levels of small fractions of a microwatt are known to cause bias offsets [14]. The fibre gyroscope—though rechristened the nonlinear loop mirror—is emerging as an important new component for communication and signal processing systems. The concept is that the Sagnac interferometer, when not subjected to rotation, reflects all the input light back towards the source. In the presence of an asymmetry, some light emerges from the other unconnected port of the input coupler. Such an asymmetry may be induced by inertial rotation, by the Faraday effect, or by induced nonlinearities. The last of these can form the elements of a non-demolition detection scheme, indicated in Fig. 13, where the presence of a signal is transferred through the nonlinear loop mirror on to the wavelength, which is used to excite the Sagnac interferometer. The signal power levels required to do this obviously depend upon the interaction length, but typi-

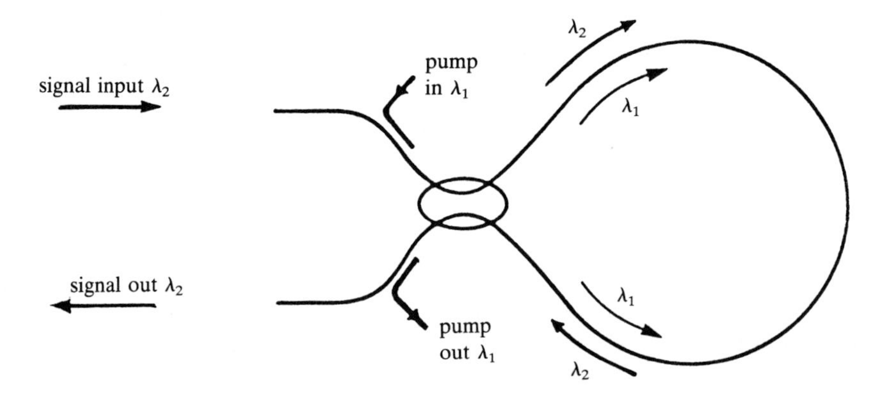

FIG. 13. The Sagnac interferometer used as a nonlinear loop mirror. The interferometer coupler is 50:50 at the pump wavelength λ_1 and is connected in the through state at the signal wavelength λ_2. The asymmetry introduced by the signal wavelength offsets the two arms of the interferometer, thereby producing an output at the wavelength λ_1. The signal wavelength λ_2 may continue throughout the rest of the system in principle without any further attenuation.

cally are in the microwatt region. The obvious disadvantages of this technique is the delay between the input signal and the arrival of its replica on the pump wavelength. This accumulates at five microseconds per kilometre and a 10-kilometre mirror may not be unreasonable. For a single-channel system, this will cause no difficulties and the nonlinear loop mirror undoubtedly will find a place in communications. However, in signal processing, there must be synchronism between a number of disparate inputs. Maintaining this to a clock (that is, one soliton) period over a length of several kilometres may prove to be impracticable.

There undoubtedly will be other important inputs to the evolution of fibre-optical signal processors. The availability of the optical head producing a soliton word from an analogue input will facilitate greatly the input/output interfaces. Optically programmable components also will play their role in system programming.

The next generation of guided-wave processors will exploit the concepts implicit in the loss-free tap and will incorporate programmable components either in all-fibre or a similar compatible technology. The applications will be in specific niches along the lines indicated in Section 4 and will involve the manipulation of very high speed, essentially serial, all-optical data inputs. The new soliton-based processing architectures have yet to be realised, but will make a significant contribution to the manipulation of optical data.

Acknowledgments

The authors would like to thank Mohammed Shabeer for his contribution to this work, and Ian Garrett, Ian Marshall, Tony Kinghorn, and George Georgioux for their various styles of stimulating input. The work has been supported, in part, by the SERC, Pilkington, and BT.

References

1. M. Shabeer, I. Andonovic, and B. Culshaw, "Fibre-Optic Systolic Array Architectures," *Proc. ICOC.* **963**, SPIE, Toulon, France (August, 1988).
2. B. Culshaw, "Optical Fibre Sensing and Signal Processing," pp. 170–187, Peter Peregrinus, London, 1984.
3. B. Moslehi, J. W. Goodman, M. Tur, and H. J. Shaw, "Fiber-Optic Lattice Signal Processing," *Proc. IEEE* **72**, no. 7, 942–953 (1984).
4. C. M. Verber, "Integrated-Optical Approaches to Numerical Optical Processing," *Proc. IEEE* **72**, no. 7, 942–953 (1984).
5. K. P. Jackson, J. E. Bowers, S. A. Newton, and C. C. Cutler, "Microbend Optical Fibre Tapped Delay Line for Gigahertz Signal Processing," *Appl. Phys. Lett.* **41** no. 2, 139–142 (1982).
6. M. Shabeer, I. Andonovic, and B. Culshaw, "Fibre-Optic Matrix Multiplier Using a Two-Dimensional Systolic-Array Architecture," *Optics Letter* **12**, 959–961 (1987).

7. M. Shabeer, I. Andonovic, and B. Culshaw, "Fibre-Optic Bipolar Tap Implementation Using an Incoherent Optical Source," *Opt. Lett* **12**, 726 (1987).
8. S. Tamur, S. Nakano, and K. Okazaki, "Optical Code-Multiplex Transmission by Gold Sequences," *Lightwave Tech* **LT-3,** 121–127 (1985).
9. D. F. Clark, I. Andonovic, and B. Culshaw, "Perturbation Analysis for the Design of an Optically Controlled Fibre-Optic Directional Coupler," *Opt. Lett.* **11** no. 6, 540–542 (1986).
10. D. Riddell, "All-optical Code Recognition," MSc Thesis, University of Strathclyde, 1989.
11. M. J. F. Digonnet (ed.), "Fibre Laser Sources and Amplifiers," *Proc SPIE* **1171**, Boston (1989).
12. R. E. Slusher and B. Yurke, "Squeezed Light for Coherent Communications," *IEE JLT* **8**, no. 3, pp. 466–477 (March, 1990)
13. I. Garrett, BTRL, first indicated to the authors this interesting feature of silica fibres.
14. R. B. Smith, "The Fibre Optic Gyroscope," *SPIE Collected Reprints Series* (1990).

CHAPTER 7

Optical Memories

Yoshito Tsunoda
Central Research Laboratory, Hitachi Ltd.,
Tokyo, Japan

1. Introduction

The present concept of optical disk memory first was proposed in 1966 [1]. Since then, considerable progress has been made in this area, together with the creation of completely new markets for optical disk memories in consumer and computer application areas. Recent rapid increases of document, image, and computer data has accelerated the development of optical disk memory products.

The first generation of optical memory was developed mainly with holographic recording technology in the late 1960s and early 1970s. A considerable number of developments have been made in both analog and digital memory applications [2]. Unfortunately, these technologies never resulted in a commercial product. The practical development of optical memories in products started at the beginning of the 1970s with bit-by-bit recording technology, as presently used in optical disk memories. Read-only-type optical disk memories, such as video disks and compact audio disks, have been investigated extensively [3]. Since laser diodes first were applied to optical video disk readout in 1976[4], there has been extensive development of laser diode pickups for optical disk memories. On the other hand, the development of user-writable optical disk memories using laser diodes became active in the late 1970s [5, 6]. Development of recording materials, including both write-once and rewritable, has been pursued

ISBN 0-12-289690-4

actively at several research institutes. Write-once optical disk products for document, image, and computer data storage were introduced to the market in the early 1980s. At present, these write-once products are used widely as external storage devices for large-scale computers, work stations, and personal computers. Rewritable optical disk products, such as magneto-optic storage devices, were introduced to the market in the late 1980s. Moreover, advanced technologies for future generations of optical disk memories have been investigated extensively. Data transfer rates and access times competitive with magnetic hard disks will be possible in the near future.

In this chapter, the present developmental status and future potential of optical disk memories for computer applications are described.

2. Read/Write Principles and Features

Read/write principles and features of several types of optical disk memories are shown in Table 1. Read-only optical disks store the data in the form of prerecorded phase pits. Data readout can be performed by the diffraction of the reading laser spot. Mass production of the data can be achieved easily using an injection molding replication process. A write-once optical disk stores the data in the form of deformations of the recording materials. A typical example is burning holes on ablative thin film materials with focused laser beams. Data can be read out by the change

Table I

Type		Principle: Write	Principle: Read	Feature
Read-Only		Phase Pit	Diffraction	Mass Productive
Write-Once		Burned Hole	Reflectivity Change	Long Archival Life
Rewritable	Magneto-Optic	Magnetization Direction	Polarization Rotation	Erasable
Rewritable	Phase-Changing	Crystal/Amorphous	Reflectivity Change	Erasable

of reflectivities of burned holes. Long archival storage life can be expected for the recorded data. There are two major approaches for rewritable optical disks. These are *magneto-optic recording* and *phase-changing recording*. In magneto-optic recording, data can be stored in the form of magnetization direction on vertical magnetic recording materials. Data readout can be performed by Kerr rotation of the reading laser beam. In phase-changing recording, data can be stored in the form of a reversible phase change between crystallized and amorphous states. Data can be read out by the change of reflectivities of the recorded pits. Erasing and rewriting the data are quite flexible for both approaches.

The common features of optical disk memories are as follows:

(1) *High-storage density.* More than 4×10^8 bits of data can be stored per square inch area. A total capacity of 10^{10} bits can be stored on a 30cm-diameter disk. This storage density is more than five times higher than that of magnetic disks.

(2) *Long archival life.* The recorded data on the read-only, write-once, and rewritable disks can be stored for more than several decades. This long archival life is one of the most important advantages of optical disk memories for large-capacity data storage.

(3) *Disk removability.* The recording surface of the disk is protected from dust and scratches by a thick transparent substrate. The laser beam is focused on the recording surface through this substrate. Consequently, the disk can be handled and removed easily without any special care.

3. Fundamental Techonologies

There are several fundamental technologies that are indispensable for the development of optical disk memories. The following are descriptions of the main technologies used.

(1) Laser diode. The laser diode is a vital device in optical disk memories. Required specifications for laser diodes include stable pulse operation with more than 20~30 mW peak power, stable far field pattern with single transversal mode, low noise, low astigmatism, and long lifetime. The most difficult problem encountered when introducing laser diodes in optical disk memories is the laser output fluctuation caused by a temperature shift or optical feedback, as shown in Fig. 1 [7]. Laser output fluctuation is related closely to the change in longitudinal mode properties. To solve this problem, a noise reduction technique called the *high-frequency current superposition method* usually has been used. An index-guided single-mode laser is driven by a high-frequency current, with the frequency more than 600 MHz. The laser diode oscillates in multiple

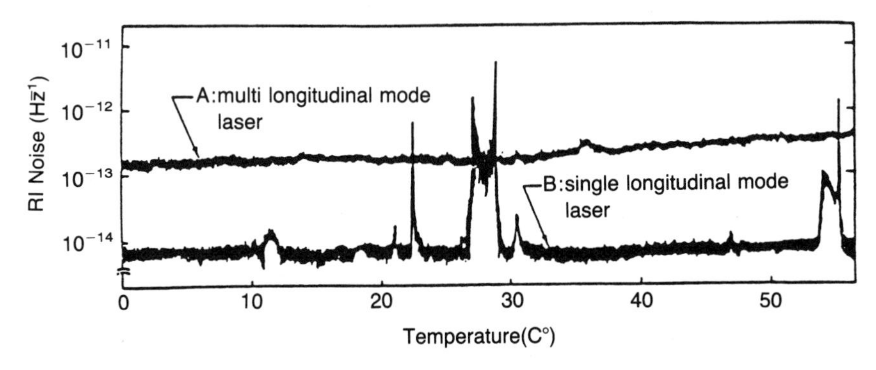

FIG. 1. Laser output fluctuation caused by a temperature shift.

modes under this condition, and the output fluctuation is suppressed drastically.

(2) *Optical Pickup.* Read-only and write-once/rewritable types of pickups appear similar in their basic optical configurations, as shown in Fig. 2. However, there are several definite differences between them. One difference is the laser power output from the pickup. With read-only pickups, about 1~2 mW of output power is required. Consequently, a relatively low power laser (~5 mW) and low efficiency optics (10~20%) are used. Write-once/rewritable pick-ups, on the other hand, require a rather high output power (5~15 mW) for writing erasing the data. Consequently, a relatively high output power laser (20~30 mW) and high efficiency optics (40~50%) are used. Another difference is the signal detection scheme for spot positioning. Focusing and tracking accuracy requirements for writing data are much more severe than those for reading data. The required accuracies for focusing are within $\pm 1.5\ \mu m$ for reading and $\pm 1\ \mu m$ for writing. Focusing error signal detection optics for write-once/rewritable pickups are much more restricted than those for read-only pickups. This restriction results from the diffraction effect caused by pre-grooves and heading sections commonly used in the write-once/rewritable disks.

(3) *Disk Substrates and Recording Materials.* Both read-only and write-once/rewritable optical disks are fabricated through the process shown in Fig. 3. First of all, prerecorded data, such as read-only data, pre-grooves, and heading-section data, are recorded on a photoresist coated master disk with an Ar or He-Cd laser. Secondly, a Ni stamper is produced through an electro-chemical plating process. Then, a large number of replicated disks are manufactured through injection molding or a photopolymer curing process. On top of the replicated disk surface, aluminum or a reflective recording layer is evaporated for read-only and

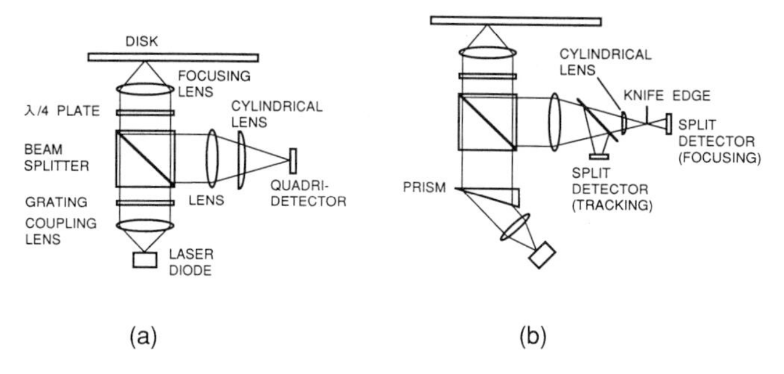

FIG. 2. Optical configurations of pickups: (a) read-only type. (b) write-once/rewritable type.

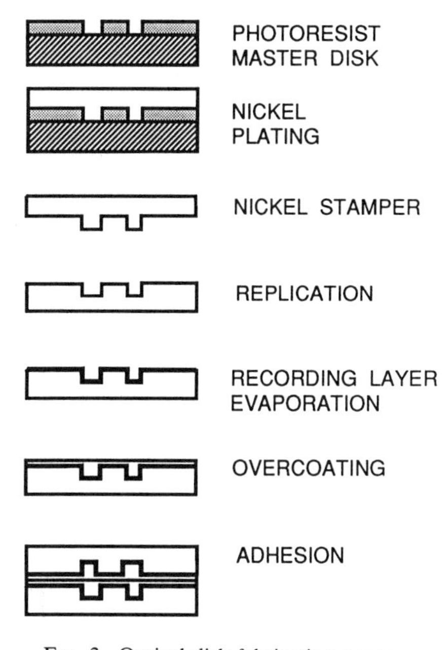

FIG. 3. Optical disk fabrication process.

write-once/rewritable optical disks, respectively. Ablative materials, phase changing materials, and organic dyes have been developed so far as write-once recording materials. Among several characteristics of the recording layer, the signal-to-noise ratio (S/N) of the layer is the most crucial. As rewritable recording materials, TbFeCo alloy commonly is used for magneto-optic recording and Te alloy for phase-changing recording.

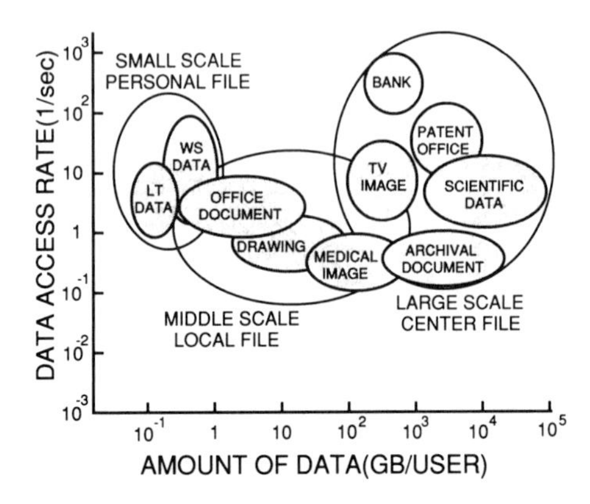

FIG. 4. Application areas of optical disk memories for computer use.

4. Application Areas

The main application areas of optical disk memories for computer use are shown in Fig. 4. Applications are categorized in three major areas on the capacity-throughput map. A large-scale central file requires the largest capacity and wide range of throughput. On-line transaction data of banks, broadcasting images, patent files, scientific data, and archival documents are typical examples of the application. Large-capacity write-once-type optical disks with 300 mm diameter and a library system are used in this application. The middle-scale local file requires intermediate capacity and rather slow throughput. Office documents, technical drawings, and medical images are typical examples of this application. Large-capacity write-once-type optical disks and middle-capacity write-once/rewritable-type disks with 130 mm diameter and their library systems are used in this application. The small-scale personal file requires rather small capacity and rather high throughput. Multimedia data for work stations and laptop computer data are typical examples of the application. Small-capacity rewritable-type optical disks with less than 130 mm diameter and small capacity read-only-type optical disks are used in this application.

5. Developmental Status of Product

5.1. Read-Only-Type Optical Disk

A typical example of the read-only-type optical disk product is the CD-ROM. More than 500 megabytes of data are prerecorded on the mass-produced 120 mm-diameter small disk, which has the same mechan-

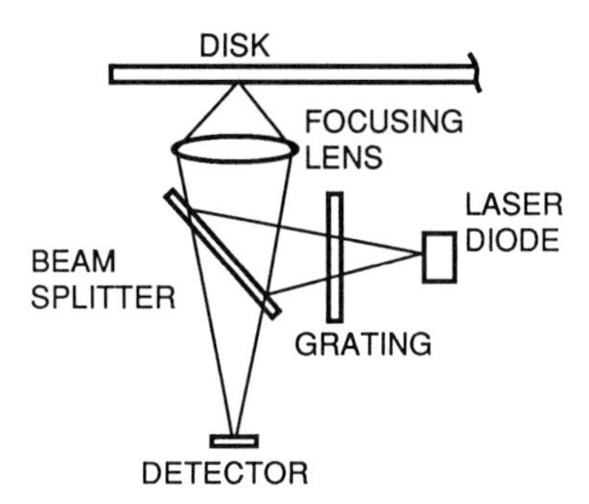

FIG. 5. Configuration of the optical pickup for CD-ROM.

ical and optical characteristics as the conventional compact audio disk. A wide variety of data—such as educational programs, manuals, dictionaries, patent files, timetables, scientific data, telephone books, and application programs—are prerecorded in electrical data form and supplied to many types of users. A great expansion of applications is expected by establishing logical format standardization. A typical configuration of the optical pickup is shown in Fig. 5. The number of optical components is decreased drastically compared with the other types of optical pickups used in write-once and rewritable optical disk memories. Tracking-error signal detection using the three-spot method and focusing-error signal detection using the astigmatic method are used. To achieve simpler pickups, the aspheric focusing lens and the holographic optical components for tracking and focusing signal detection are being used.

5.2. Write-Once-Type Optical Disk

For the write-once-type optical disk, two types of products have been developed: large-capacity optical disks with a 300 mm diameter and middle-capacity optical disks with a 130 mm diameter. The basic configurations of a large-capacity optical disk system and the optical pickup used are shown in Fig. 6 [8]. In Fig. 6(a), an optical pickup for writing and reading the data is driven by a linear actuator and can be accessed randomly to the target track. Air-sandwiched pre-grooved disks with 1.6 μm track pitch are used. A recording layer consisting of Te-Se-Pb alloy is evaporated on a pre-grooved glass substrate 300 mm in diameter. Each track is divided into 64 sectors beginning with a heading section in which track address, sector address, and other information are prerecorded in the form of phase pits. The number of tracks is about 40,000. The optical pickup consists of the

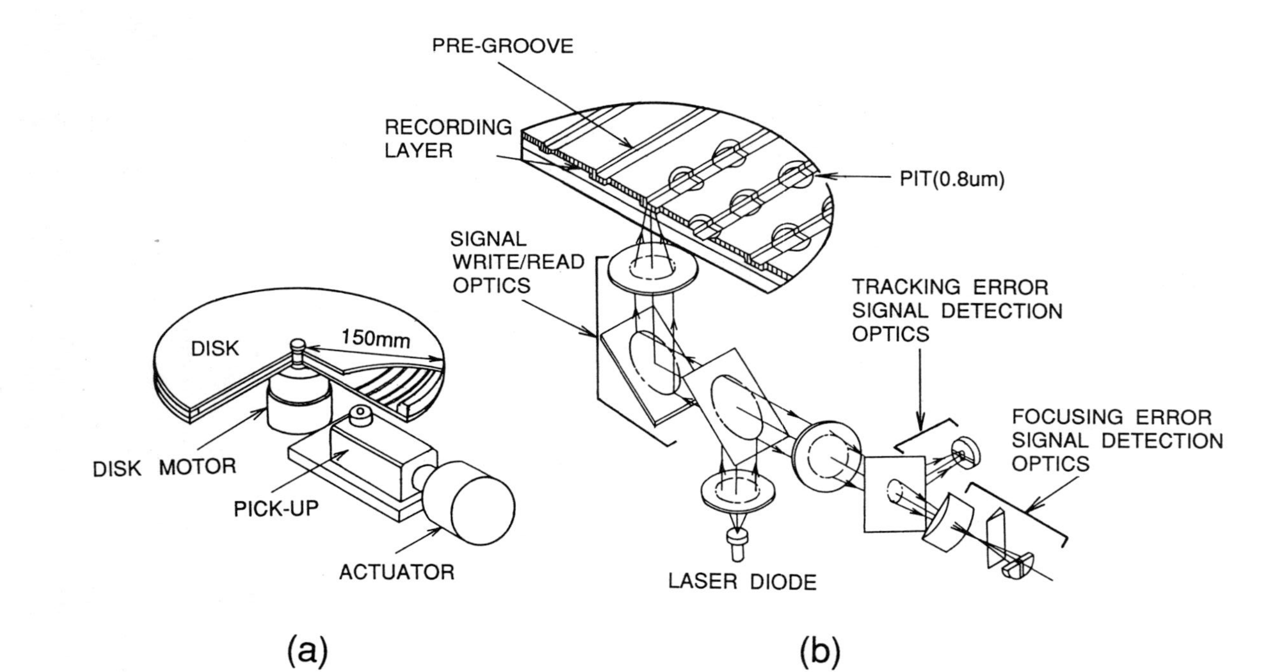

FIG. 6. Configuration of the large capacity write-once-type optical disk memory: (a) whole system. (b) optical pickup and disk.

Table II

Item	Spec
Capacity (GB/Side)	1.3
Linear Density (BPI)	19500
Track Density (TPI)	16000
Data Rate (KB/s)	440
Disk Rotation (rpm)	600
Average Access (ms)	250
Bit Error Rate (ECC)	$\leqq 10^{-12}$

data write/read optics, the focusing error signal detection optics, and the tracking error signal detection optics. Data write/read can be achieved with one diode laser, which oscillates at 20 mW for writing and 4 mW for reading. The focusing-error signal detection optics and the tracking-error signal detection optics are used to control the position of the focused laser spot precisely on the target track on the recording layer. Typical specifications of this system are shown in Table 2. The storage capacity is 1.3 gigabytes/side, the data transfer rate is 440 kirobytes/sec, and the average access time is 250 ms. Bit error rates of less than 10^{-12} have been achieved using error correction codes (ECC), so that computer data can be stored with high reliability. Moreover, due to the adoption of the glass substrate (which is mechanically and optically stable) and the long archival-life recording layer, the recorded data can be stored more than 50 years with high reliability. A middle capacity write-once optical disk product also has been developed. This product has 300 megabytes of storage capacity, 690 kilobytes/sec data transfer rate, and 100 ms average access time. The size of the disk is 130 mm in diameter. A plastic substrate and the same recording material as the large-capacity optical disk are used. Bit error rates of less than 10^{-12} and a long archival life of more than 30 years have been achieved. The recording format of the data was standardized internationally in 1987 by the International Organization for Standardization (ISO). The basic layout of one of the standardized track formats, called Composite Continuous Servo (CCS), is shown in Fig. 7 [9]. In this format, servo-tracking can be performed using pre-grooves with one byte-length mirror (non-pre-groove) area. The laser spot is focused on the land area between pre-grooves and the laser beam diffracted by the pre-grooves is detected by split detectors. Precise laser spot positioning can be done by balancing the two output signals from the split detectors. The mirror area can be used for suppressing the low-frequency tracking offset. Each track is divided into 17 sectors. Each sector has a 52-bytes-length prerecorded heading section and can store 1,024 bytes of user data.

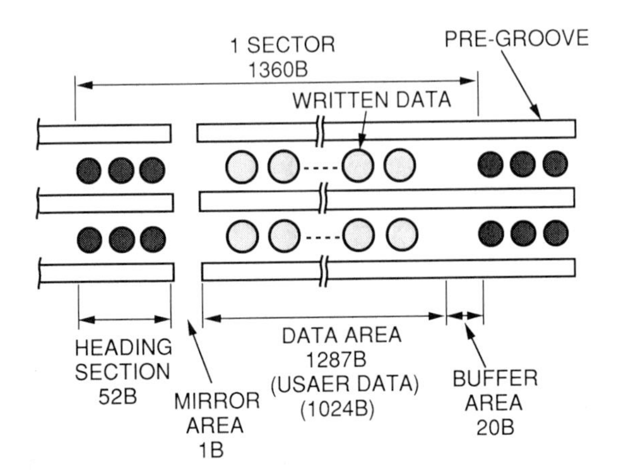

FIG. 7. Basic layout of the standardized Composite Continuous Servo (CCS) track format.

5.3. *Rewritable-Type Optical Disk*

As a rewritable-type optical disk product, middle-capacity magneto-optic disks with 130 mm diameters were introduced to the market in the late 1980s [10]. The basic configuration of the magneto-optic disk is shown in Fig. 8(a). An optical pickup consists of a laser diode, a focusing lens, optical components, and photodetectors in the same manner as the write-once pickup. A vertically magnetized thin film recording layer consisting of Tb-Fe-Co alloy is evaporated on a pre-grooved substrate 130 mm in diameter. Magnetic domains are recorded along the pre-grooves with 1.6 μm track pitch. An electromagnet is placed on the disk to supply magnetic field at the recording point. Figure 8(b) shows the write/read mechanism. The recording layer is magnetized uniformly in the same direction before recording the data. The electromagnet supplies the magnetic field in the direction opposite to the recording layer magnetization direction. When the focused laser spot irradiates and heats the recording layer up to the Curie temperature, the initial magnetization of the recording layer disappears and the opposite direction magnetization appears at the recorded domain. Erasure of the recorded domain can be done by changing the direction of the magnetic field of the electromagnet. To read out the recorded data, the Kerr effect or the Faraday effect is utilized. When a linearly polarized laser beam irradiates the recording layer, the direction of polarization can be rotated by the magnetization. This rotation can be converted to a beam intensity change through the polarization device and detected by the photodetector.

There have been several technical subjects to be solved for practical application of the magneto-optic disk. One of the most serious problems

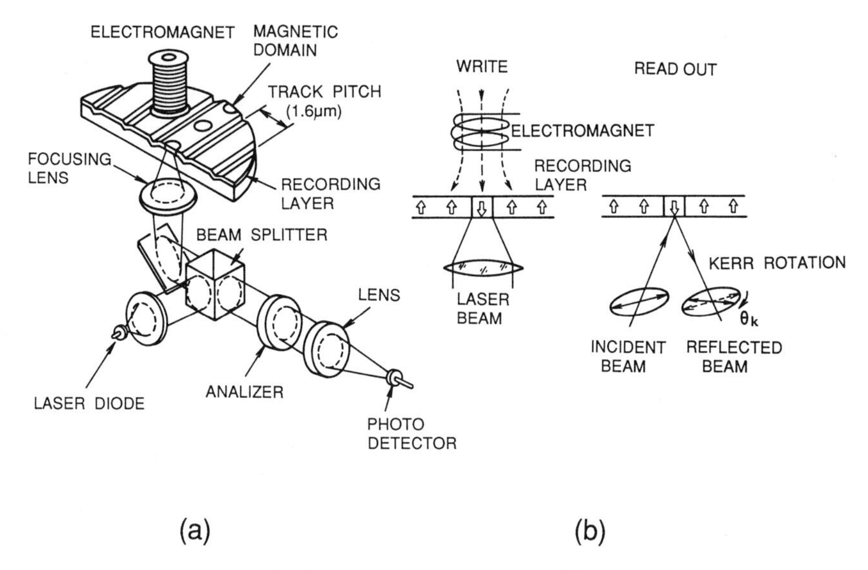

FIG. 8. Configuration of the magneto-optic disk: (a) optical pickup and disk. (b) write/read mechanism.

was improvement of the signal-to-noise ratio. The rotation angle of the polarized laser beam by the Kerr effect is less than one degree, and the amount of the detected signal is too small for reliable data readout. To solve this problem, extensive investigations have been pursued on optical pick-ups and recording materials. Consequently, a satisfactory signal-to-noise ratio for highly reliable computer data storage has been obtained. Another problem was improvement of the archival storage life of the recording materials. Improvement of the overcoating layer and the recording layer itself has resulted in an archival storage life of more than 20 years.

6. Future Technologies

Optical disk memories have made great progress in the past 20 years as a new type of computer memory. However, expanding their application areas in the future will require new technologies to improve the present storage density, data transfer rate, and access time. Typical activities for future technology development are described in this section.

6.1. High Speed and High-Density Recording

To increase the storage density and data transfer rate of optical disk memories, an increase of the linear density is the most efficient approach. Smaller spot-size recording using short wavelength laser diodes is one of the typical approaches for higher linear-density recording. GaAlAs laser diodes with 780 nm ~ 830 nm wavelength commonly are used for present optical disk products. InGaAlP laser diodes with 620 nm ~ 670 nm wavelengths are to be used for the next generation optical disk memories [11]. For optical disk memories in the distant future, second-harmonic generation (SHG) of the laser diode is promising. On the other hand, improvement of the recording scheme also is significantly effective for increasing the linear density. The basic principle of one typical high linear-density recording scheme, *pit-edge recording*, is shown in Fig. 9 [12]. In a conventional pit-position recording scheme, the codeword data is recorded at the center position of the pit, so that interference between two pits limits the storage density. In the case of pit-edge recording, the codeword data is recorded at the front and back edges of the pit. This results in almost double the linear density. Using all these technologies, it is expected that more than 10 gigabytes of data can be stored on a surface of a 300 mm-diameter optical disk and 10 megabytes/sec data transfer rate can be achieved in the future.

6.2. High Speed Overwrite

Conventional magneto-optic recording usually requires additional rotation for erasing the data before writing. To solve this problem and achieve direct overwriting, several approaches have been investigated. One of the most promising methods is an optically assisted magnetic field modulation

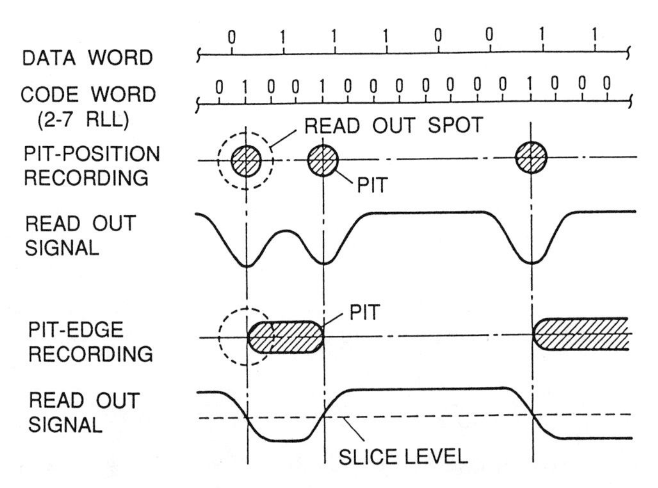

FIG. 9. Principle of the high-density pit-edge recording.

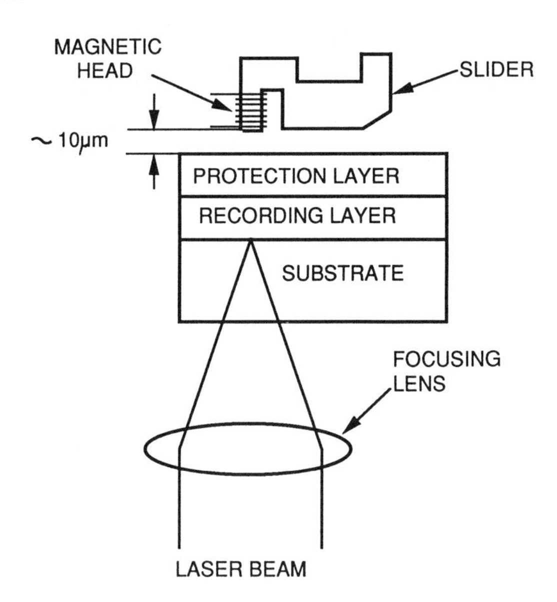

FIG. 10. Configuration of the optically assisted magnetic field modulation method.

method using an air-floating magnetic head [13]. The basic configuration of this method is shown in Fig. 10. In this method, a focused laser beam continuously irradiates the information track, and the magnetic field is modulated according to the data to be recorded. The distance between the magnetic head and the optical disk is more than 10 μm, which is 50 ~ 100 times larger than that of the magnetic disks, so the removability of the optical disks can be sustained with high reliability. A prototype model of the second-generation magneto-optic disk has been developed using pit-edge recording and the direct overwriting scheme. Typical specifications of the developed model are shown in Table 3. Storage capacity of 600

Table III

Item	Spec
Capacity (GB/Side)	0.6
Linear Density (BPI)	42300
Track Density (TPI)	18000
Data Rate (KB/s)	2200
Disk Rotation (rpm)	3000
Average Access (ms)	60
Bit Error Rate (ECC)	$\leqq 10^{-12}$

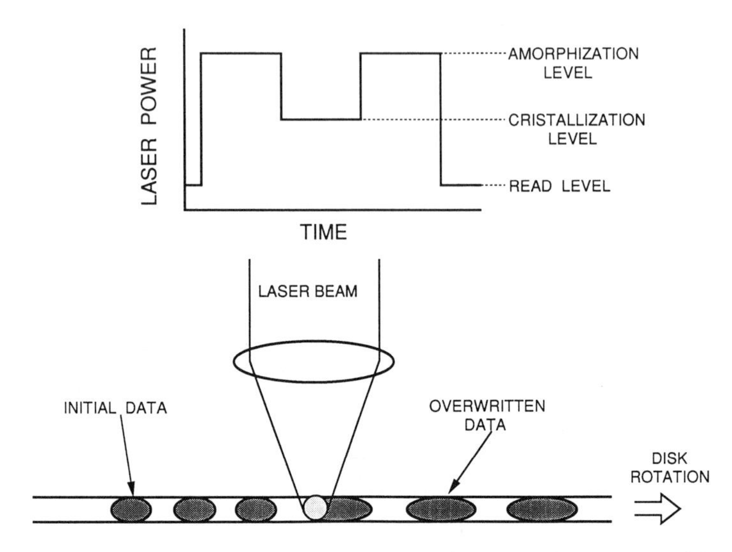

FIG. 11. Principle of high speed overwriting using phase-changing materials.

megabytes/side and a data transfer rate of 2.2 megabytes/sec have been achieved with 130 mm disk using Tb-Fe-Co alloy as the recording layer.

Another promising approach to high-speed overwriting is the use of phase-changing materials with high response for erasing and writing. As shown in Fig. 11, the output power of the laser diode is modulated with two levels according to the signals to be recorded [14]. Te alloy is one of the promising candidates of the recording materials.

6.3. *Compact Optical Pickup*

The present access times of optical disks are $3 \sim 5$ times slower than those of magnetic disks. This is mainly because of the slow access speed due to heavy and bulky optical pickup. There are two approaches to compact optical pickup. One approach is decreasing the number and size of the optical components used in the pickup; another is utilizing integrated optics, such as waveguide optics. The former approach is simple but laborious. Plastic aspheric lenses, compact actuators, and simple signal-detection optics already have been investigated. Using all these optical components, a compact optical pickup with 10 mm thickness has been developed [15]. Although the latter approach still is under basic research, a number of challenging trials have been made. Thin film optical pickup will be feasible in the future.

7. Toward the 21st Century

The present status and the future of optical disk memories have been described. As a conclusion to this chapter, the evolution of optical disk memories toward the 21st century will be discussed.

Figure 12 is the hierarchical structure of computer memories. The computer memories consist of *internal memories*, including cache memory and main memory, and *external memories*, including direct access memory and mass file memory. In the present large-scale computer, the capacities of cache memory, main memory, direct access memory, and mass file memory are one megabyte, one gigabyte, 500 gigabytes, and five terabytes, respectively. Semiconductor memories are used for cache memory and main memory. Magnetic disk memory commonly is used for direct access memory. Optical disk memory and magnetic tape memory commonly are used for mass file memory.

These capacities of computer memories drastically will increase toward the 21st century. The capacities of cache memory, main memory, direct access memory, and mass file memory at the beginning of the 21st century are expected to be 100 megabytes, 100 gigabytes, 10 terabytes, and 100 terabytes, respectively. Optical disk memories are considered to be the most likely candidate to achieve such a huge capacity for data storage. Achieving a mass file memory with 100 terabytes capacity is an especially challenging goal for optical disk memories. To achieve such a huge storage capacity, the advanced concept of a super high-density optical disk memory has to be developed. A wide variety of investigations on materials, devices, and systems are now proceeding toward this goal. Great process for optical disk memories is expected as the 21st century approaches.

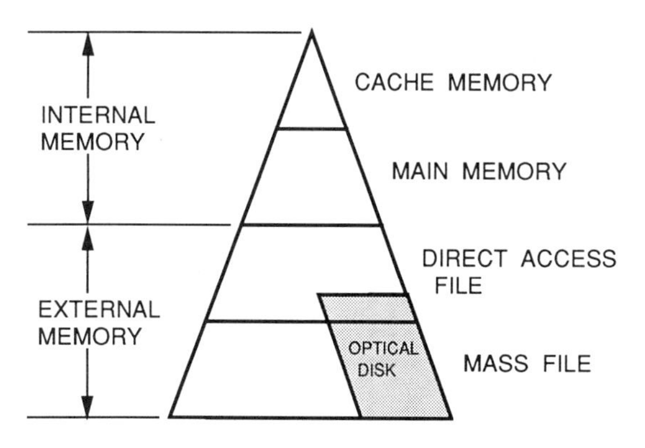

FIG. 12. Hierarchical structure of computer memories.

References

1. P. Rice, "An Experimental Television Recording and Playback System Using Photographic Disc," *J. SMPTE* **79**, 997 (1970).
2. Y. Tsunoda, K. Tatsuno, K. Kataoka, and Y. Takeda, "Holographic Video Disk— An Alternative Approach to Optical Video Disks," *Appl. Opt.* **19**, no. 6, 944 (1980).
3. K. Compaan and P. Kramer, "The Philips VLP System," *Philips Tech. Rev.* **33**, 178 (1973).
4. Y. Takeda, and Y. Tsunoda, "Use of Heterostructure Diode Lasers in Video Disk Systems," *Appl. Opt.* **17**, no. 6, 863 (1976).
5. J. P. J. Heemskerk, "Optical Recording with Diode Lasers," *1979 CLEA Digest of Technical Papers*, 17.4 (1979).
6. Y. Tsunoda and T. Kaku, "Large Capacity Optical Disk Memory with Diode Lasers," *Technical Digest on 26th Joint Conference of Japanese Appl. Phys. Soc.*, 103 (1979).
7. M. Ojima, A. Arimoto, N. Chinone, T. Gotoh, and K. Aiki "Diode Laser Noise at Video Frequencies in Optical Video Disc Players," *Appl. Opt.* **25**, no. 9, 1404 (1986).
8. Y. Tsunoda, S. Horigome, and Z. Tsutsumi, "Optical Digital Data Storage Technologies with Semiconductor Laser Head," *Digest of Technical Papers of Topical Meeting on Optical Data Storage*, MB1, (1983).
9. Y. Tsunoda, T. Kaku, S. Nakamura, and T. Maeda, "On-land Composite Pre-groove Method for High Track Density Recording," *Proc. SPIE* **695**, 224 (1986).
10. M. Ojima, A. Saito, T. Kaku, M. Ito, Y. Tsunoda, S. Takayama, and Y. Sugita, "Compact Magneto-Optical Disk for Coded Data Storage," *Appl. Opt.* **25**, 483 (1986).
11. H. Fujii, K. Kobayashi, S. Kawata, A. Gomyo, I. Hino, H. Hotta, and T. Suzuki, "High Power Operation of a Transverse Mode Stabilized AlGaInP Visible Light Semiconductor Laser," *Elect Lett.* **23**, no. 18, 938 (1987).
12. H. Sukeda, M. Ojima, M. Takahashi, and T. Maeda, "High-Density Magneto-Optic Disk Using Highly Controlled Pit-Edge Recording," *Jap. J. Appl. Phys.* **26**, Suppl. 4, 243 (1987).
13. T. Nakao, M. Ojima, Y. Miyamuta, S. Okamine, H. Sukeda, N. Ohta, and Y. Takeuchi, "High Speed Overwritable Magneto-Optic Recording," *Jap. J. Appl. Phys.* **26**, Suppl. 4, 149 (1987).
14. M. Terao, T. Nishida, Y. Miyauchi, S. Horigome, T. Kaku, and N. Ohta, "In-Se Based Phase Change Reversible Optical Recording Film," *Proc. SPIE* **695**, 105 (1986).
15. S. Nakamura, M. Ojima, T. Nakao, T. Kato, and K. Mizuishi, "Compact Two-Beam Head with a Hybrid Two-Wavelength Laser Array for Magneto-Optic Recording," *Jap. J. Appl. Phys.* **26**, Suppl. 4, 117 (1987).

CHAPTER 8

How Can Photorefractives Be Used?

Henri Rajbenbach and Jean-Pierre Huignard
Thomson-CSF
Laboratoire Central de Recherches
Orsay, France

1. Introduction

The last few years have witnessed an increasing interest in nonlinear optical interactions to process laser beams and manipulate complex time-varying coherent wavefronts. A highly nonlinear response, even to low-level illuminations, converts photorefractive materials into attractive candidates for these applications. The main purpose of this chapter is to highlight some device potentials and the chances of practical applications of photorefractives for laser beam control and image processing. The following section briefly reviews the physics of the photorefractive effect with emphasis on relevant characteristics of the most exciting materials. The coupling of optical waves then is analyzed to introduce some prototype laboratory applications, including dynamic interconnects, phase locking of lasers, homodyne detection, phase conjugation, and associative memories. The crucial challenges that remain are discussed briefly in the last section.

2. The Photorefractive Effect

The *photorefractive effect* [1] is a phenomenon in which the material local index of refraction is altered by a nonuniform illumination. Electrons or holes are photoexcited from impurities, vacancies, or defects present in the material; upon migration by thermal diffusion or by drift in an electric

ISBN 0-12-289690-4

field, they are retrapped at other locations, leaving behind a space-charge field that modulates the refractive index via the electro-optic effect. If a photorefractive crystal is illuminated with the interference pattern of two coherent light beams, a dynamic refractive index grating will result. The amplitude $\Delta n(x)$ of this change in the index of refraction is

$$\Delta n(x) = 1/2 n^3 r E(x), \quad (1)$$

where r is the effective Pockels coefficient of the crystal, n is the background refractive index, and $E(x)$ is the periodic photoinduced space-charge field. This space-charge field, which reaches values of several $\mathrm{kV \cdot cm^{-1}}$, can become phase shifted spatially with respect to the illumination pattern (Fig. 1), and the photoinduced change Δn can be as high as $\sim 10^{-3}$. In addition, numerous media display photorefractive properties at uv, visible, or infrared wavelengths. Current photorefractive materials consist of electro-optic oxides—such as $LiNbO_3$, $BaTiO_3$, $KNbO_3$, $Bi_{12}SiO_{20}$ (or BSO), $Bi_{12}\ GeO_{20}$ (or BGO), and $Ba_{1-x}\ Sr_x\ Nb_2O_6$ (or SBN)—and compound semiconductors, such as GaAs, InP, and CdTe. High values of the diffraction efficiency of the photoinduced index gratings are available in materials with high electro-optic coefficients, such as $BaTiO_3$, SBN ($r \sim 10^3$ pm/V) or $KNbO_3$ ($r \sim 60$ pm/V). In other materials having low electro-optic coefficients—such as BSO, BGO, GaAs and InP($r \sim 1$–3 pm/V)—the diffraction efficiency can be increased with an externally applied electric field until saturation occurs when all the photoexcited charges are separated by one fringe period.

The photorefractive effect essentially is a response to optical energy. The refractive index changes are due to electro-optic effects driven by space-charge fields, and the time required to record a grating depends on the efficiency of the charge generation and transport processes. The inertia in the nonlinear response of photorefractive media constitutes an important difference from other nonlinear media, where the refractive index change is of electronic origin and thus occurs instantaneously. A simple expression for the time dependence of the space-charge field during grating recording is the following [2]:

$$\Delta E_{sc} = m E_{sc}[1 - e^{-t/\tau}], \quad (2)$$

where m is the incident fringe modulation and $\tau = 10^{-3} - 1$ sec are typical recording times for elementary gratings with a CW incident intensity of $10 - 100\ \mathrm{mW \cdot cm^{-2}}$ at the blue or green line of the Argon laser. BSO and GaAs are used as fast and sensitive materials, while crystals such as $BaTiO_3$ have a large electro-optic coefficient but respond rather slowly, i.e., a response time of a few seconds.

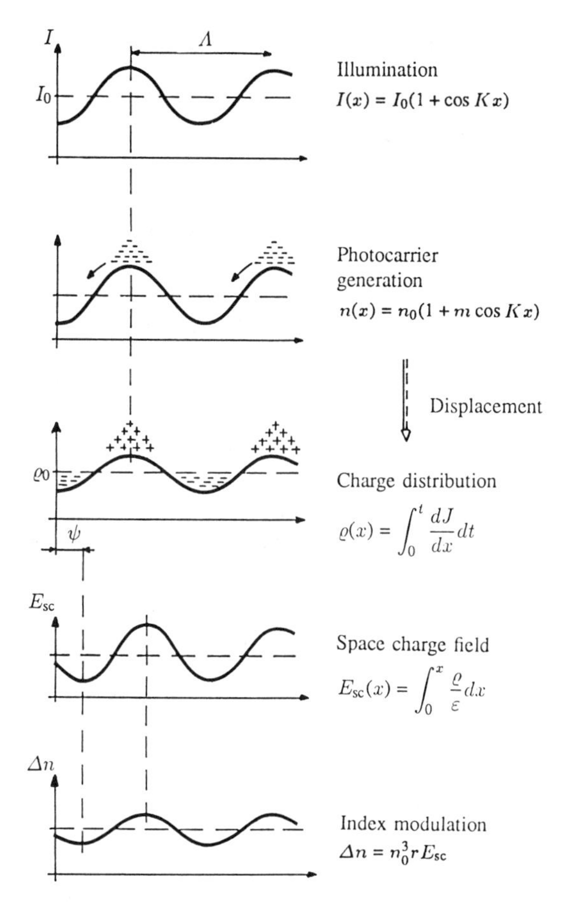

FIG. 1. Mechanism of photorefractive grating recording in electro-optic crystals: Light with a spatially periodic intensity (I) resulting from the interference of two beams in the material rearranges the charge density (ρ), which causes a periodic field (E_{sc}) according to Poisson's equation. This electric field then causes a change in the refractive index of the crystal (Δn) by the linear electro-optic effect.

The photorefractive effect can be regarded from two different and complementary perspectives, real-time holography [1] and a nonlinear optical process [3]. Real-time holography in photorefractives applies to optical storage, image correlation for pattern recognition, and real time interferometry. This last application is now close to practical development for industrial non-destructive testing. In the following section, we will focus on the nonlinear optical processing point of view and show some applications of nonlinear wave mixing in these materials.

3. Two-Wave Mixing in Photorefractives

The recording of phase-volume gratings in photorefractive media leads to a stationary energy exchange between the two interfering beams [4]. The resulting energy redistribution that has been observed in many electro-optic crystals ($LiNbO_3$, $KNbO_3$, $BaTiO_3$, BSO, and GaAs) is due to a self-diffraction process of the reference pump beam by the dynamic phase grating photoinduced in the crystal. More specifically, the self-interference of the incident beam with the diffracted beam creates a new holographic grating that can add to (or subtract from) the initial one. Since the diffracted wave is phase-delayed by $\pi/2$ with respect to the reading beam, the maximum energy transfer is obtained when the incident fringe pattern and the photoinduced index modulation are shifted by $\psi = \pi/2$. In photorefractive crystals, such a $\pi/2$ phase shift exists when the recording is by *diffusion* of photocarriers (no external applied electric field), as shown in Fig. 2. Permanent and efficient amplification of a low-intensity signal beam has been observed in crystals like $LiNbO_3$, $BaTiO_3$, and $KNbO_3$. If we now apply the coupled wave equations to the $\pi/2$ phase-shifted component of the photoinduced index modulation, the coherent interaction between the two waves of respective amplitudes R and S is described by the following coupled wave equations:

$$\frac{dS}{dz} = \frac{1}{2}\Gamma \frac{R^2 S}{R^2 + S^2} - \frac{1}{2}\alpha R; \frac{dR}{dz} = \frac{1}{2}\Gamma \frac{R S^2}{R^2 + S^2} - \frac{1}{2}\alpha S. \tag{3}$$

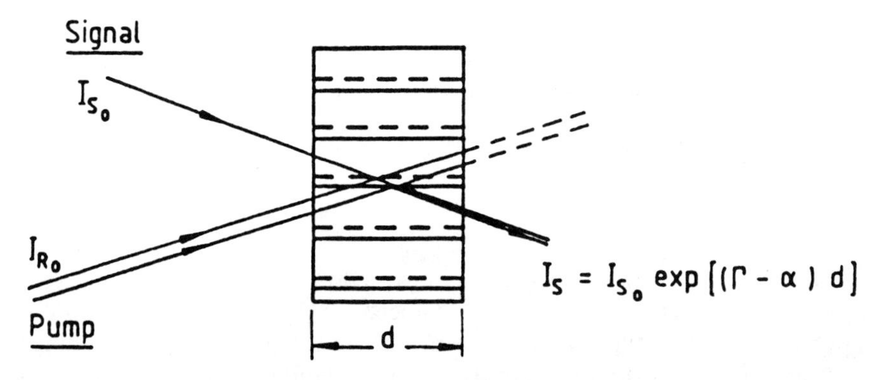

FIG. 2. Degenerate two-wave mixing (2WM) in photorefractive crystals. The continuous and dashed lines represent the maxima of the illumination and index pattern, respectively. In the diffusion mode (no applied field), the phase shift ψ is $\pi/2$, and large gains are observed in crystals such as $BaTiO_3$ and SBN.

In the case of a negligible pump depletion, the transmitted signal beam intensity resulting from the dynamic two-beam coupling takes the form

$$I_s = I_{so} \exp[(\Gamma - \alpha)d], \tag{4}$$

where Γ is the exponential gain coefficient of the interaction and d is the interaction length. Γ is related to the maximum amplitude of the photo-induced index modulation Δn_s through the relation

$$\Gamma = \frac{4\pi \Delta n_s}{\lambda \cos \theta} \sin \psi, \tag{5}$$

where ψ is the spatial phase shift of the grating and, in agreement with the previous arguments, Γ is maximum when $\psi = \pi/2$. Large values of Γ ($\Gamma \approx 20\ \text{cm}^{-1}$) may be obtained in materials having large Δn_s when recording by diffusion ($\psi = \pi/2$), and this is the case for $BaTiO_3$, SBN, $LiNbO_3$, and $KNbO_3$. However, the same 2WM experiment performed with highly photoconductive BSO or GaAs crystals leads to very low beam coupling ($\Gamma \approx 1\ \text{cm}^{-1}$) for the following reasons: (*a*) For diffusion, the required phase shift $\psi = \pi/2$ is established, but the steady-state index modulation is low, and (b) for drift, with an electric field applied to the crystal, the index modulation is much higher, but the corresponding phase shift ψ is negligible. However, efficient beam coupling can be obtained if the fringe pattern (or the crystal) is moved at a constant velocity. The speed is adjusted so that the index modulation is recorded at all times, but with a spatial phase shift with respect to the interference fringes. This interaction is called *nearly degenerate two-wave mixing* and provides efficient energy transfer in crystals like BSO and GaAs. More recently, a dramatic improvement of the two-wave mixing gain was observed in semiconductors by combining the large electrorefractive nonlinearities near the band edge of semiconductors with the sensitivity and speed of photorefractive buildup.

4. Four-Wave Mixing and Phase Conjugation

Optical phase conjugation with photorefractive crystals utilizes a four-wave mixing (4WM) interaction to reverse both the direction of propagation and the phase of an arbitrary input wavefront [5]. Phase conjugate mirrors have many applications in problems associated with passing through distorting media: The phase distortion can be removed by allowing the wavefront to travel back through these same media (Fig. 3).

A second property of phase-conjugate mirrors is their ability to generate a conjugate signal with an amplified intensity. Amplified phase conjugation

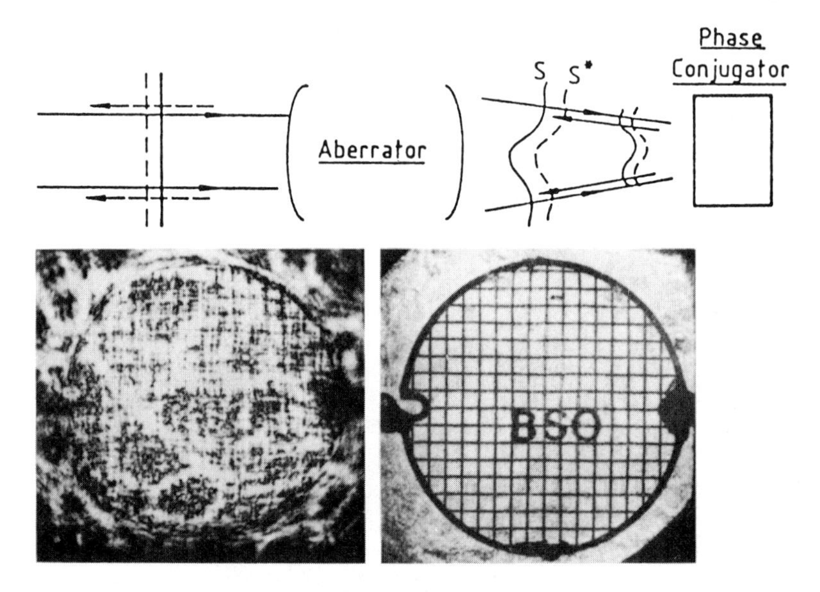

FIG. 3. Phase-distortion compensation by wavefront reflection at a phase-conjugate mirror. The distorted image (bottom left) is restored after travelling back through the same medium.

has been observed in photorefractive crystals such as $LiNbO_3$, $LiTaO_3$, $KNbO_3$, $BaTiO_3$, and SBN with typical time responses of a few seconds, and in BSO and GaAs (10–100 ms) when recording with a moving grating. The optical configuration used for phase conjugation by nearly-degenerate 4WM is presented in Fig. 4. As shown in this figure, the moving grating in the crystal is obtained by a frequency detuning of the reference beam by a small amount $\delta\omega$ with the use of a piezomirror driven by a saw-tooth voltage.

Theory shows that to reconstruct a high-quality phase-conjugate replica of an incident image using four-wave mixing, the pump beams themselves must be phase conjugates of each other. In practice, this is not so easy to do. The problem is solved in self-pumped phase conjugation, in which the pump beams themselves are self-generated from the incident image via amplified scattering and interface reflections [6].

Double-phase conjugation [7] is an even more impressive effect: Two independent inputs to opposite sides of a photorefractive $BaTiO_3$, carrying different spatial images, pump the same four-wave mixing process, resulting in the phase-conjugate reproduction of the two images simultaneously. Furthermore, the two inputs need not originate from the same laser. In fact, they even can be of different colors. In Fig. 5, a $BaTiO_3$ crystal is

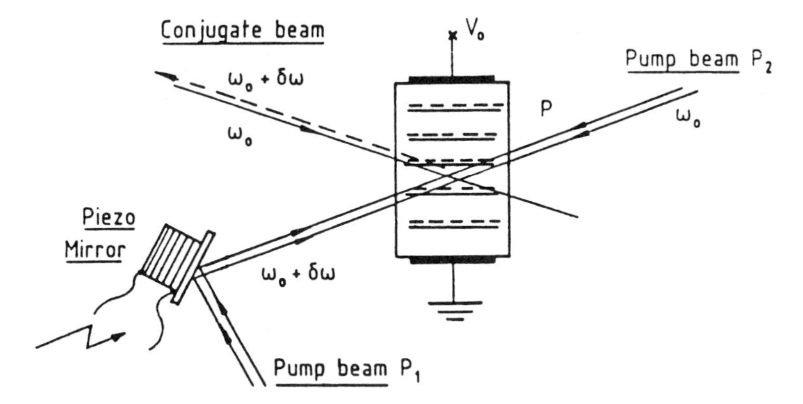

FIG. 4. Nearly degenerate four-wave mixing configuration. The pump beam is frequency-shifted to obtain amplified phase conjugation.

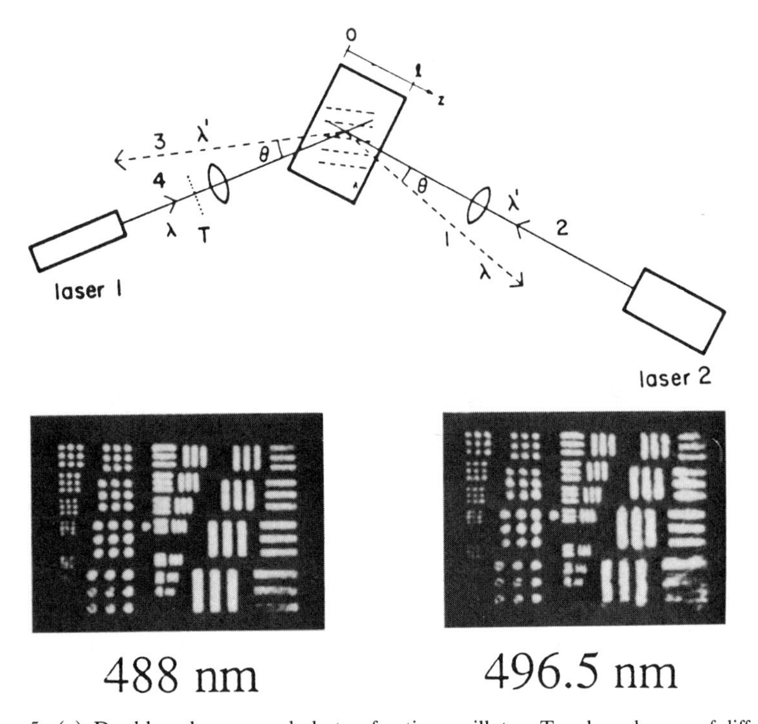

FIG. 5. (a) Double-color-pumped photorefractive oscillator. Two laser beams of different color interact to generate a common grating and two oscillations of different colors. (b) Color conversion from $\lambda = 488\,\text{nm}$ (laser 1) to $\lambda' = 496.5\,\text{nm}$ (laser 2). The spatial modulation T (resolution chart) carried by beam 4 emerges on the oscillation beam 3.

pumped by two lasers of different colors. Despite their difference in wavelength, these beams interact with each other to cause the self-generation of a common grating and two oscillation beams of different colors. This process is self-produced and permits a pictorial input (on beam 4 in Fig. 5a) to be transferred to beam 3 (at a different color). Figure 5b shows the output image 3 ($\lambda' = 496.5$ nm) when the corresponding resolution chart ($\lambda = 488$ nm) is inserted on beam 4.

5. Applications of Wave-Mixing to Optical Signal Processing

5.1. Beam Steering and Optical Interconnects

Free-space optical networks are important for highly parallel architectures and optical communications [8]. Two- and four-wave mixing can be exploited for beam steering and optical interconnects with reconfigurable capabilities. In two-wave mixing, the basic principle is shown in Fig. 6a: The pump beam interferes in the crystal with the probe beam, whose direction is selected by the spatial light modulator (an array of electro-optic shutters, for example), and after two-beam coupling, a complete energy exchange from the pump to the selected probe-beam direction can be obtained by using photorefractive crystals with large gain coefficients [9]. Therefore, we can say that the pump beam has been deflected in the direction of the probe. If another direction is selected for the probe beam, the previous grating is erased, and rewriting a new one deflects the pump beam in another direction. By means of this principle, a new type of random-access digital laser-beam deflector with large scan angles is realized.

Double phase-conjugate mirrors (DPCM) can be used to implement bidirectional transmitter-receivers, as shown in Fig. 6b [10]. In the off position, the photorefractive crystal is pumped by one (or no) light beam (A_4 in the figure). This beam is scattered in a wide angle (the fanning effect). The link is on when the two beams, A_2 and A_4, are directed simultaneously to the opposite sides of the crystal. This results in the collapse of the fanned light into two oscillating beams, A_1 and A_3, in anti-parallel directions (phase conjugation) with respect to the two input beams. While the two input beams exchange their spatial structures, the photons and their temporal (intensity and average phase) information are transmitted through the DPCM; A_1 and A_3 carry the temporal information of A_4 and A_2, respectively.

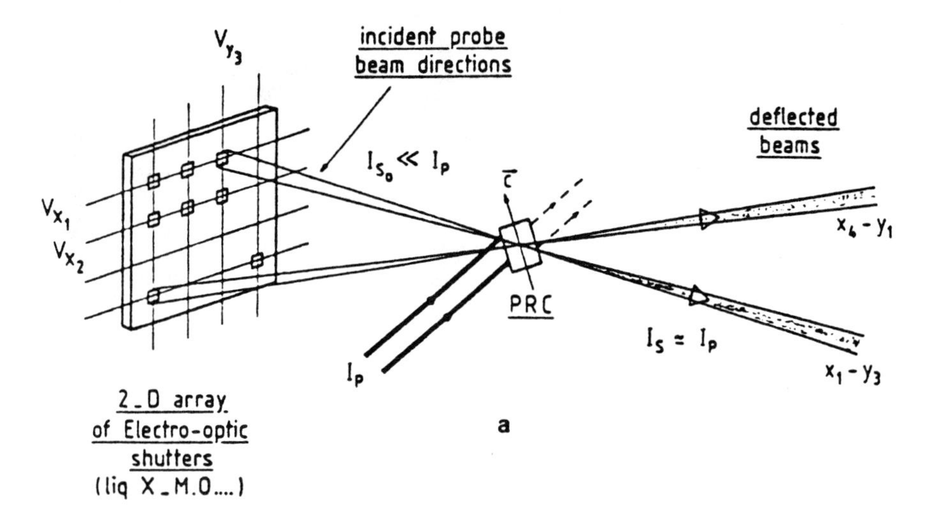

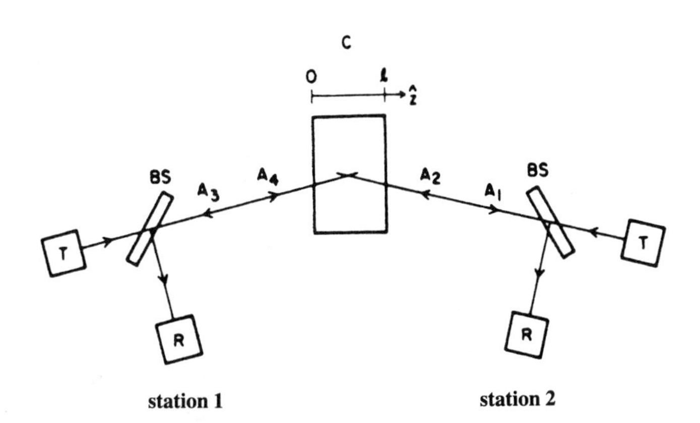

FIG. 6. Optical interconnects with photorefractives. (a) Laser beam deflection with two wave mixing. (b) Optical switch with double phase-conjugate mirror.

5.2. Phase Locking of Lasers

Two-wave mixing techniques in photorefractives offer the possibility of phase-locking lasers to form a single-mode output beam arising from the coupling of several coherent sources [11]. A schematic diagram for the coupling of two semiconductor laser arrays is shown in Fig. 7. The two

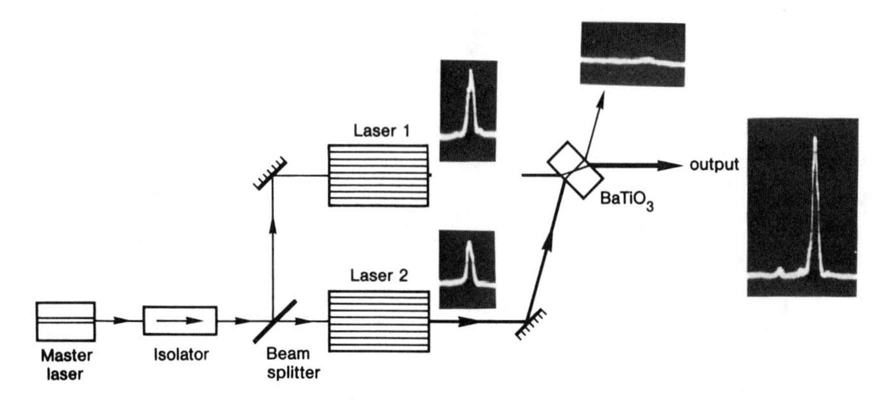

FIG. 7. Phase locking of two semiconductor laser arrays with two-wave mixing in $BaTiO_3$. Oscilloscope traces represent the far-field distribution.

arrays are injected with a single-mode low-power master laser. Their outputs become coherent and intersect in a photorefractive $BaTiO_3$ crystal. Energy, therefore, is transferred from one beam to the other, as shown by the far-field beam profiles. Moreover, this technique can be extended to combine coherently more than two lasers. In that case, the master laser injects the N slaves, the output of which stimulate a two-wave mixing process that transfers energy (but not phase) from $N - 1$ beams into the direction of the last one.

5.3. *Optical Novelty Filters*

Photorefractive crystals can be employed to construct dynamic filters that detect variations in intensity (or phase) of a two-dimensional image or data plane [12]. Self-pumped phase conjugation and two-beam coupling techniques were used to implement these *novelty filters*. In the experiment, two arms of an interferometer share a common phase-conjugate mirror that replaces the classical mirrors in a Michelson interferometer. A transmitting spatial light modulator imposes a phase image onto the optical beam in one of the interferometer arms. The phase-conjugate mirror guarantees that, in steady state, the output is zero. However, when the phase information in one arm changes suddenly, the recombined fields at the output of the interferometer no longer interfere destructively, thus yielding non-zero intensity at the output. The output returns to its original state after a delay governed by the response time of the phase-conjugate mirror.

More recently, novelty filters using signal depletion due to noise amplification (fan-out) in a crystal of $BaTiO_3$ were demonstrated. Depletion on an input signal is achieved by amplified scattering from crystal defects (Fig. 8a). As the input signal varies, the fanning hologram written in the crystal adjusts to the change according to its photorefractive response time, and the changing image components become brighter as deamplification is disturbed. Figure 8b shows edge detection of a model car obtained by moving the object (the input signal) in front of the crystal; only the edges are changed in intensity, resulting in a transient edge-enhanced output image.

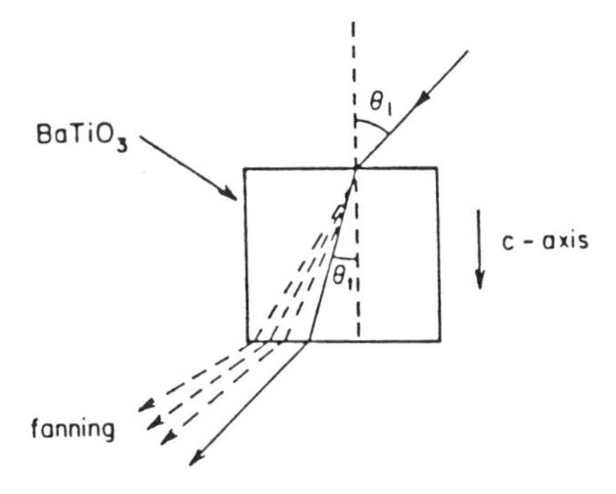

FIG. 8. (a) Optical novelty filter using amplifield scattering in $BaTiO_3$. A time-varying input signal results in a transient deamplified output. (b) Output signal showing edge enhancement of a model car. The car was moved forward slightly after the signal was depleted completely so that only the edges were changed in intensity.

5.4. Coherent Homodyne Detection of Complex Wavefronts

Consider a complex wavefront that carries temporal and spatial information (i.e., a wavefront diffracted by a target object, such as a spatial light modulator). In two-wave mixing, this wave interferes with the pump beam to create a complex dynamic hologram in the volume of the photorefractive crystal. The self-diffraction of the pump beam generates a wave perfectly in phase at any point with the incident signal wavefront (Fig. 9). Therefore, the self-diffracted wave can be considered as a local oscillator wave in a homodyne detection scheme involving an arbitrary signal wavefront [13]. This local oscillator wave has the required properties for this purpose: (a) Its phase is perfectly matched to the incident wavefront, (b) its amplitude can be much larger than that of the incident signal wave, provided that a large gain coefficient is achieved in the interaction, and (c) its optical frequency is equal to that of the incident signal wave. When this prevails, the minimum detectable signal corresponds to the quantum limit.

5.5. Associative Memories

In an *associative memory*, a stored piece of information can be retrieved completely when the input is only partially complete. The thresholding and gain properties of phase-conjugate mirrors with photorefractives provide regenerative optical feedback that enable the unlocking of database memories in which multiple objects are stored holographically by angular coding of the reference beam directions [14]. The principle of a photorefractive associative memory is shown in Fig. 10. The illumination of the hologram by part of the object generates a diffracted beam propagating in

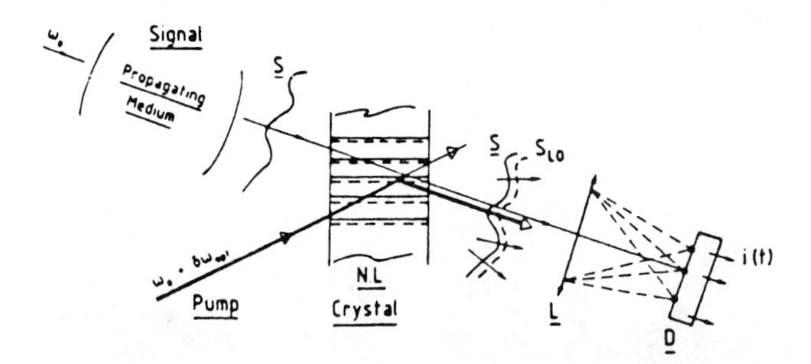

FIG. 9. Two-wave mixing in a nonlinear photorefractive crystal applied to coherent homodyne detection involving a complex wavefront. The incident arbitrary wavefront carrying spatial and temporal information (S) is perfectly phase-matched with the local oscillator wave S_{LO} that originates from the pump diffracted at the dynamic grating.

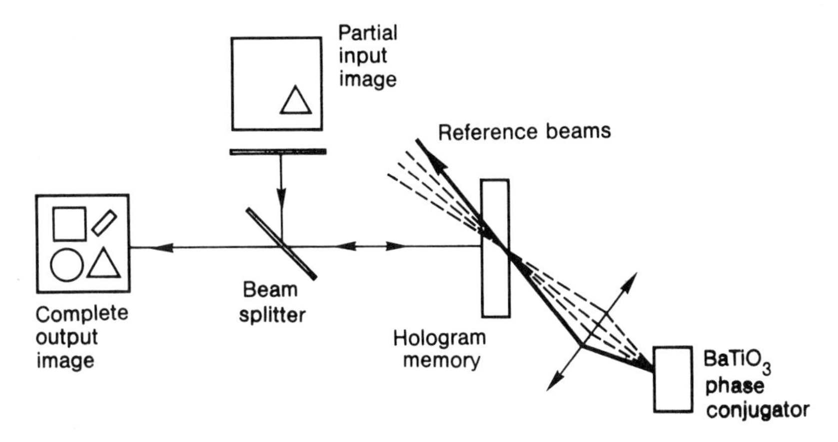

FIG. 10. Optical associative memory using a phase conjugate resonator and information storage in a volume holographic memory.

the original direction of the reference beam. This beam then is phase conjugated and amplified by four-wave mixing in a photorefractive $BaTiO_3$ crystal. Thresholding provides the necessary nonlinearity emphasizing only the strongly correlated signal. When this readout beam impinges on the hologram, it is diffracted and recreates the initial object beam. This recreated object beam contains all the information recorded originally in the hologram memory.

6. Prospects for Photorefractives

It is too early to assess fully the future of photorefractives for optical processing applications. As yet, no single material has emerged as superior to all the others, and new photorefractive materials continue to be identified. Furthermore, among all the impressive prototype demonstrations reported to date, none has made its way to the marketplace. A number of factors, however, appear to motivate continuing collaboration between researchers in crystal growth and characterization, in optical processing architectures and algorithms, and in the development of exploratory and real systems. First, the nonlinearities are so large that at modest cost, with sturdy and easy-to-use materials, many parallel processing architectures can be tested with low-power cw lasers. Second, photorefractives allow the study of complex stimulated nonlinear processes on convenient time scales. This has led to the discovery of many qualitatively new or unexpected processes, such as self-pumped phase conjugation and the coherent coupling of independent lasers. Third, photorefractive behavior can be exploited to reach important parameters of electronic and optoelectronic materials. For example, two-beam coupling behaviors provide a ready access to the mobilities, lifetimes, and impurity concentrations of semiconductors such as GaAs and InP. Being optical, this type of analysis can be nondestructive and locally selective. Last but not least, it is fun to play with photorefractive materials. Impressive and often beautiful effects occur in these materials that often behave in unexpected (if not mysterious) fashion. Photorefractive materials occupy a unique position in nonlinear optics research. They certainly have most of the characteristics needed to demonstrate exploratory devices and to stimulate further studies on materials and architectures for the use of optics in all-optical and hybrid systems. Whether or not they will find practical applications remains an open question.

References

1. See, for example, P. Günter and J. P. Huignard (eds.), "Photorefractive Materials and Their Applications I and II," Vol. 61 and 62, Springer Verlag, Berlin, 1989; A. M. Glass, *Opt. Eng.* **17**, 470 (1978).
2. G. C. Valley and M. B. Klein, *Opt Eng.* **22**, 704 (1983).
3. P. Yeh, *IEEE Jour. Quant. Elec.* **25**, 484 (1989); D. H. Auston, *Opt. Eng.* **26**, 211 (1987); J. Feinberg, *Physics Today*, **46**, Oct., 1988.
4. N. V. Kukhtarev, V. M. Markov, S. G. Odulov, M. S. Soskin, and V. L. Vinetskii, *Ferroelectrics* **22**, 949 (1979).
5. See, for example, R. A. Fisher (ed.), "Optical Phase Conjugation," Academic Press, Orlando, 1983; M. Cronin-Golomb, B. Fisher, J. O. White and A. Yariv, *IEEE Jour. Quant. Elec.*, **QE 20**, 12 (1984).
6. J. Feinberg, *Opt. Lett.* **7**, 486 (1982).
7. S. Weiss, S. Sternklar, and B. Fisher, *Opt. Lett.* **12**, 498 (1987); A. M. C. Smout and R. W. Eason, *Opt. Lett.* **12**, 498 (1987).
8. J. W. Goodman, F. J. Leonberger, S. Y. Kung, and R. A. Athale, *Proceedings of the IEEE*, **72**, 850 (1984).
9. D. Rak, I. Ledoux, and J. P. Huignard, *Opt. Commun.* **49**, 302 (1984).
10. S. Weiss, M. Segev, S. Sternklar and B. Fisher, *Appl. Opt.* **27**, 3422 (1988).
11. W. Christian, P. Beckwith, and I. McMichael, *Opt. Lett* **14**, 81 (1989); J. M. Verdiell, H. Rajbenbach, and J. P. Huignard, *Phot. Tech. Lett* (August, 1990).
12. D. Z. Anderson and J. Feinberg, *IEEE J. Quant. Electron.* **25**, 635 (1989); M. Cronin-Golomb, A. M. Biernachi, C. Lin, H. Kong, *Opt. Lett.* **12**, 1029 (1987); J. Ford, Y. Fainman, S. H. Lee, *Opt. Lett.* **13**, 856 (1988).
13. G. Hamel de Monchenaux and J. P. Huignard, *J. Appl. Phys.* **63**, 624 (1988).
14. B. H. Soffer, G. J. Dunning, Y. Owechko and E. Marom, *Opt. Lett.* **11**, 118 (1986).

CHAPTER 9

Adaptive Interferometry: A New Area of Applications of Photorefractive Crystals

S. I. Stepanov
A. F. Ioffe Physical-Technical Institute Leningrad, USSR

1. Introduction

Adaptive interferometry is a new term that appeared recently in the field of research and applications of photorefractive crystals (PRCs). What does this area have in common with such well-established fields as dynamic holography and adaptive optics? And how does it differ from conventional holographic interferometry, where adaptivity also can be traced—specifically, to the complicated shape of the object being tested?

First of all, this new field deals with the metrological aspects of interferometry, i.e., with measurements of mechanical vibrations, interferometric fiber-optic sensors, optical memory, and communication systems. The basic scheme of these devices is presented in Fig. 1a. It includes an interferometer with two arms: the signal arm, where the phase modulation to be measured is performed, and the reference arm. A conventional beamsplitter usually is used to produce interference between the signal and reference waves at the photodetector (photodiode, photomultiplier, photoresistor, etc.). The measured phase shift is converted here into amplitude modulation of the output light beams, and then to an electric signal.

Along with an extremely high sensitivity that is limited by the shot noise of the laser source [1], these optical measuring systems have serious

ISBN 0-12-289690-4

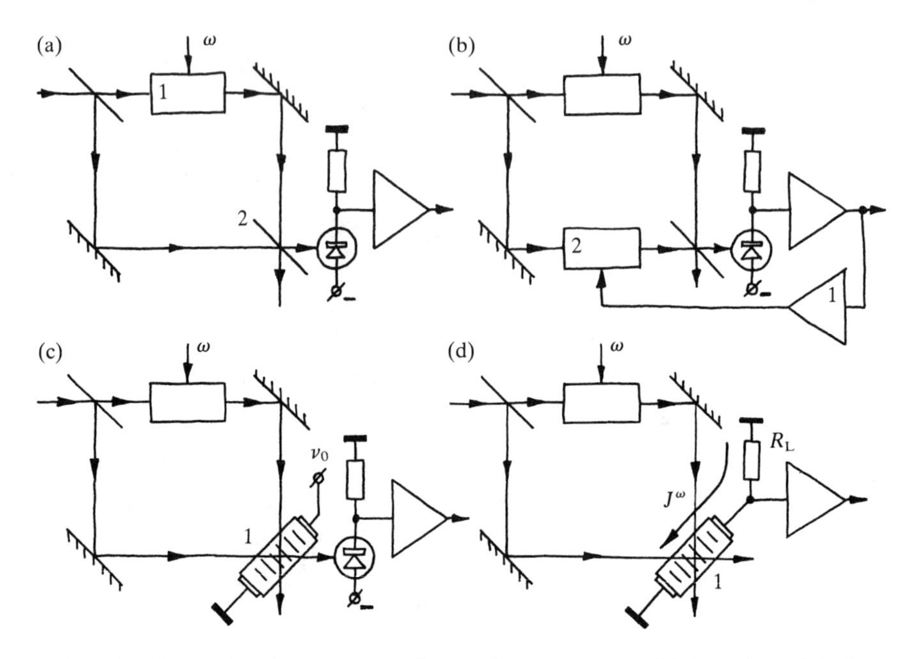

FIG. 1.a. Conventional two-arm interferometric arrangement with homodyne detection (1—part of the scheme where the signal beam is phase-modulated, 2—beamsplitter). b. The same arrangement with an active stabilization of an operation point (1—low-frequency amplifier, 2—phase modulator). c. Adaptive interferometer employing a dynamic hologram (1—volume dynamic hologram, recorded here in a photorefractive crystal under dc voltage). d. Usage of an adaptive photodetector operating through the effect of non-steady-state photo-EMF (1— volume photoconductor with a dynamic distribution of a space-charge electric field $E_{sc}(x)$, J^{ω}—photo-EMF current, R_L—load resistor).

shortcomings, however. The first one is the need for precise adjustment (with an accuracy of about $\lambda/10$) of the wavefronts that interfere at the photodetector.

The second problem is the necessity to keep constant the average phase shift φ_0 between these two beams. In fact, the highest sensitivity of the device is achieved only for $\varphi_0 = \pi/2 + k\pi$ $(k = 0, \pm 1, \ldots)$. In this case, it operates in the linear regime, where the sinusoidal phase modulation of the input signal beam with a frequency ω is converted into an output electric signal of the same frequency. For $\varphi_0 = k\pi$, a quadratic regime of operation is observed when the first harmonic in the output signal disappears.

While the first problem can be solved by using simple Gaussian beams, single-mode optical fibers, and rigid adjustable beamsplitters, the second problem is a fundamental one. In laser vibrometers, the distance from the tested object usually is not fixed and can vary during the measurements. In fiber-optic sensors, minor changes in environmental pressure and temperature result in occasional phase shifts up to 10^3 rad and even higher [2]. The spectrum of the output electric signal proves to be broadened dramatically under these conditions, thus lowering the sensitivity of the whole measuring system as compared with the theoretical limit.

The problem is solved nowadays by the appreciable sophistication of the electronic and optical parts of the system. The proposed techniques include heterodyne detection [3] and active stabilization of the operation point [2]. The second method, which is based on the principles of adaptive optics, employs homodyne detection. The required average phase shift is kept constant here by means of a low-frequency electronic feedback loop and a phase modulator in the reference arm of the interferometer (Fig. 1b).

The necessity to develop a novel, simple tool for compensation of these unwanted phase shifts is obvious. One efficient way is the use of adaptive interferometers based on the principles of dynamic holography. This restores the initial simplicity of the interferometric system (Fig. 1a) and makes it really foolproof. A detailed discussion of the new technique is the main subject of this chapter.

2. Principles of Adaptive Interferometers

The main purpose of the feedback loop shown in Fig. 1b is to keep an average phase shift constant between the reflecting surface of the beamsplitter and the interference fringes. When a flat interface between two transparent media with different refractive indices is used as a beamsplitter, this shift will be $\Lambda/4$ to ensure the required optimal phase shift of $\varphi_0 = \pi/2$.

Matching of the beamsplitter and the pattern can be accomplished not only by the phase modulator, as shown in Fig. 1b, but also by means of the adaptive beamsplitter itself. The beamsplitter can be controlled not only electrically, but also optically, i.e., based on the principles of nonlinear optics. In addition, it can be multilayered. Such a multilayered adaptive beamsplitter is nothing else but a volume dynamic hologram recorded by the signal and reference beams.

Every dynamic hologram is characterized by its recording-erasure time τ_H. This time is needed for the hologram to change its position in accordance with a movement of the recording interference pattern. The adaptive beamsplitter under discussion also can be characterized by a transfer function, i.e., dependence of the amplitude modulation index in the output light beams on the modulation frequency ω. It is easy to show that the transfer function in question proves to be similar to that of a conventional differentiating RC electronic circuit with time constant τ_H [4]. Indeed, the dynamic hologram follows slow displacements of the recorded pattern, thus keeping the phase shift between the grating and the pattern nearly constant.

It is important to point out the main difference between this adaptive interferometer and a conventional, equivalent differentiating RC circuit at the output of the photodetector. In both cases, slowly varying signals with frequencies $\omega \ll \tau_H^{-1}$ are suppressed efficiently. However, in the former case, this is fulfilled in the process of transformation of the phase-modulated light beam into an amplitude-modulated one. As a result, high-amplitude, low-frequency jamming modulation does not affect transformation of a high-frequency signal of small amplitude.

There is another important feature of the dynamic holographic beamsplitters under consideration. It is their ability to compensate for the complicated transverse structure of the recording light beams. It relies on the basic principles of holography and obviates the need for adjustment. This allows, in particular, the use of multimode optical fibers in the arms of the interferometric fiber-optic sensors [5].

In fact, such adaptive holographic beamsplitters are employed to obtain precisely collinear interference of two mutually coherent light beams with slowly varying, complicated transverse structures. That is why we refer to them later as adaptive interferometers.

What are the main requirements for the dynamic holographic media in this application area? The first one is a continuous mode of operation that excludes the use of photothermoplastics, photochromics and all other materials that need cycling. The second requirement is a high steady-state diffraction efficiency ($\eta \gtrsim 10\%$) that is required for observation of high modulation indices in the output beams ($\Delta I/I_0 = 2\sqrt{\eta}$). The third requirement is a short recording time τ_H for the low intensities of the light beams used typically in the interferometric measuring systems. Taking into account these factors, photorefractive crystals now seem to be the main candidates for applications in this field [6].

3. Adaptive Interferometers Using PRCs

The main mechanism of holographic recording in PRCs is well understood now [6]. It consists of spatially non-uniform photoexcitation of mobile carriers by the recording interference pattern, their spatial redistribution via thermal diffusion or drift in an external (or efficient internal) electric field, and consequent trapping by deep trapping centers. The spatially periodic electric field $E_{sc}(x)$ of this trapped space charge is transformed into a phase relief in the crystal volume (i.e., volume phase hologram) via the linear electro-optic effect.

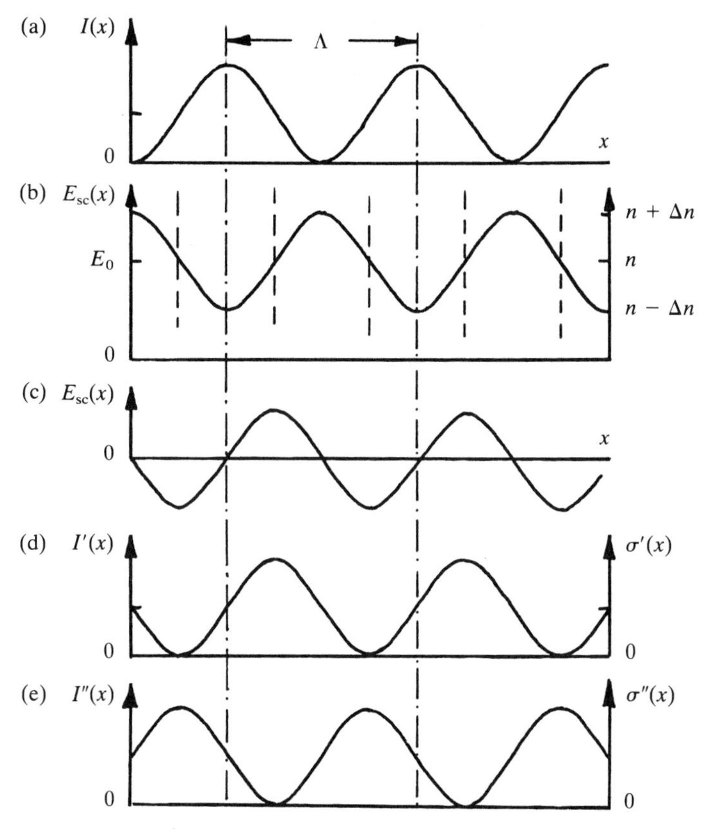

FIG. 2.a. Initial distribution of light intensity in the recording pattern (Λ—spatial period of the pattern). b. Electric field ($E_{sc}(x)$); and refractive index (n) (x)) Distributions in a PRC with the drift-recording mechanism—the so-called *unshifted grating* (effective reflecting boundaries are shown by dashed lines). c. Electric field distribution $E_{sc}(x)$ in a PRC with the diffusion recording mechanism—*shifted grating*. d. Light pattern shifted to the right by a quarter of a spatial period from the initial position. Efficient generation of photocarriers in positive semiperiods of the diffusion distribution of $E_{sc}(x)$ ($\bar{\bar{c}}$) results in a positive short-circuit photo-EMF current. e. A similar shift to the left results in efficient photocarrier generation in negative semiperiods of $E_{sc}(x)$ and, as a result, in a negative photo-EMF current.

Dynamic volume phase holograms recorded in PRCs afford, in principle, high steady-state diffraction efficiencies. There are two characteristic types of these phase gratings, the unshifted and the shifted (Fig. 2). The former grating (with the maxima of the refractive index distribution in phase or in antiphase with the interference pattern) usually is formed via the drift recording mechanism, when a dc voltage is applied to the crystal along the wave vector of the recorded grating. The latter grating (with the maxima shifted by a quarter of spatial period $\Lambda/4$ from the maxima of the pattern) is formed via the diffusion mechanism in the absence of a driving electric field. The question is what type of hologram will be chosen for our application.

Theoretical analysis predicts than an unshifted (drift) phase hologram is to be used when the linear regime of operation of an adaptive interferometer is needed [4]; and a conventional shifted (diffusion) phase hologram results in a quadratic regime of operation when only the second harmonic with a frequency 2ω appears at the output. One can understand this conclusion easily from the following qualitative speculations. For the sinusoidal phase grating under consideration, effective reflecting surfaces [between layers of high $(n + \Delta n)$ and low $(n - \Delta n)$ refractive index] are shifted by a quarter of spatial period with respect to the grating maxima (Fig. 2b). As shown in the previous section, these interface boundaries should be shifted in the same manner relative to the interference pattern maxima to observe linear operation. So, the maxima of the grating and the pattern are to be in phase or in antiphase, i.e., the case that is achieved for the drift recording mechanism.

However, as always, every coin has two sides. The result is good because the drift hologram usually shows the higher diffraction efficiency needed for higher output modulation indices. Yet, at the same time, it is bad because of the necessity of applying a high-voltage bias dc field to the PRC. Note that in the past several years, unusual geometries of holographic recording in cubic PRCs, which can afford linear regime of operation via self-diffraction from the shifted diffusion hologram, have been investigated [7, 8]. In both experiments mentioned earlier, anisotropic properties of the phase gratings formed in these PRCs were employed [9].

The characteristic time τ_{sc} for the hologram formation in a PRC [6] is determined by the characteristic Maxwell relaxation time

$$\tau_M = \frac{\varepsilon\varepsilon_0}{\sigma_0} = \varepsilon\varepsilon_0 \Big/ \frac{e\beta\alpha I_0 \mu\tau}{\hbar\omega_{ph}} \tag{1}$$

first of all. Here, $\varepsilon\varepsilon_0$ is the static dielectric constant of the crystal, σ is its average photoconductivity for the light intensity I_0, β is a quantum efficiency of photoconductivity, α is an optical absorption of the crystal, μ and $\bar{\tau}$ are mobility and average lifetime of photocarriers, and $\hbar\omega_{ph}$ is the recording photon energy.

One can see from Eq. 1 that there are quite natural trade-offs between the diffraction efficiency of the recorded grating and its characteristic response time in the PRCs. First, to reduce τ_M, one needs more absorbent cystals. However, this reduces the possible hologram thickness, which should be $d \gtrsim \alpha^{-1}$, and, as a result, the hologram diffraction efficiency

$\mu \propto d^2$. Second, the ratio between the linear electrooptic coefficient r and static dielectric constant ε proves to be nearly the same for all the PRCs used to date ($r/\varepsilon \sim 10^{-10}$ cm/V)[10]. That is why reduction of τ_M in crystals with low ε also is paid for by a proportional reduction in $\sqrt{\eta}$.

Third, if one tries to employ PRCs with photocarriers characterized by a high mobility-lifetime product, the maximum value of the dc electric field — and, as a result, the value of η, which is usually proportional to E_0^2 — also is found to be limited. This limitation is associated with a dramatic increase in the characteristic time for the drift hologram formation. In fact, beginning from $E_0 = \Lambda/2\pi\mu\tau$ it grows proportionally to E_0^2 from its lowest level of $\tau_{sc}^{min} = \tau_M$. Therefore, the spatial frequency of the pattern is chosen as low as possible to ensure application of the electric field $E_0 = 10$ kV/cm, limited typically by the surface electric breakdowns.

The typical characteristic times of the photorefractive hologram recording can be rather short, however. For cubic BSO ($Bi_{12}SiO_{20}$) crystals in the green region of the spectrum ($\alpha = 2.1$ cm^{-1}, $\beta = 0.7$, $\mu\tau = 1.4 \times 10^{-7}$ cm^2/V for $\lambda = 514$ nm, and $\varepsilon = 56$)[11], we obtain $\tau_M = 6 \times 10^{-4}$ sec for a moderate light intensity $I_0 = 1$ mW/mm^2. On the other hand, a theoretical estimate of the maximum diffraction efficiency

$$\eta = \left(\frac{\pi r n^3 E_0 d}{2\lambda}\right)^2 \tag{2}$$

of the drift grating with thickness $d = 2$ mm, recorded under external electric field $E_0 = 10$ kV/cm, proves to be equal to 14% for $r = 3.4 \times 10^{-10}$ cm/V and $n = 2.6$ [11]. To ensure application of this electric field without a drastic reduction in the speed of recording, one should use the interference pattern with spatial periods $\Lambda \gtrsim 2\pi\mu\tau E_0 = 90$ μm.

The maximum diffraction efficiency given earlier yields a rather high amplitude modulation index $\Delta I/I_0 = 2\sqrt{\eta} = 0.75$ in the output light beams. One should remember, however, that the effective value of E_0 inside the PRC volume usually proves to be lower because of the screening effect.

The main parameters of interest for every photodetector are sensitivity and frequency range of operation. As to the former parameter, there seems to be no remarkable difference between the adaptive system under discussion and a conventional scheme with a beamsplitter (Fig. 1a). The fundamental limit on the sensitivity is set in both cases by shot noise. Physically, it is associated with a discrete absorption of photons by the output photodetector used in the system. The practical sensitivity of a conventional geometry usually is determined by the amplitude noise of the laser employed. It is required that a special balanced scheme of detection be used to compensate for this amplitude noise [1]. This source of noise is not eliminated in the adaptive interferometer in question, but it can be overcome by the same method.

The frequency range of operation in our case is not associated with any temporal characteristic of the PRC used, and is determined by the output photodetector. Indeed, beginning from a frequency $\omega = \tau_{sc}^{-1}$, the dynamic hologram is as stable as a conventional beamsplitter. Its behavior is unaffected by a limited lifetime of the photoexcited electrons, which usually is much shorter than the Maxwell relaxation time at typical light intensities used in interferometers.

To our knowledge, T. J. Hall *et al.* were the first to put forward the idea of using PRCs in adaptive interferometers [5]. However, it is only in the past several years that there have appeared a number of papers that report on experimental studies of PRCs for similar purposes [4,7,12,13,14]. Specifically, in some cases, adaptive interferometers were used in laser vibrometers for detection of low-amplitude vibrations [7,12].

4. Adaptive Properties of an Interferometer with $Bi_{12}SiO_{20}$

In this section, we will give an example of a photorefractive adaptive interferometer—namely, an interferometer based on cubic photorefractive $Bi_{12}SiO_{20}$ (BSO). A complete set of the data obtained in these experiments is given in [4]. Here, we restrict ourselves to the adaptive properties of this device.

The BSO samples used in the experiments typically were $1 \times 10 \times 4$ mm^3 in size; their lateral (ground) faces (10×4 mm^2) were coated with silver paste electrodes for applying a dc-voltage U_0. The front and back (polished) faces of the sample (1×10 mm^2) were parallel to the (110) crystallographic plane with the $[1\bar{1}0]$ axis in the incidence plane. An ordinary HeNe laser ($\lambda = 633$ nm) with an average power $P_0 = 30 \sim 40$ mW was used as a source of coherent radiation.

Fig. 3 shows the experimental transfer function of the BSO adaptive interferometer. As expected, the characteristic cutoff frequency $\omega_0 = \tau_{sc}^{-1}$ and the amplitude of the output signal at high frequencies $\omega \gg \omega_0$ prove to be increasing linearly with the light intensity I_0. Because of the differentiating type of transfer function, only fronts of pulses are registered when a square-wave signal is applied at the input of the interferometer (Fig. 4).

The adaptive properties of the interferometer are most pronounced in the two-frequency excitation mode of operation, when two frequencies are applied simultaneously to the phase modulator in the signal arm. These are the signal frequency $\omega \gg \omega_0$ with a low amplitude $\Delta^{\omega} \ll 1$ rad and the low

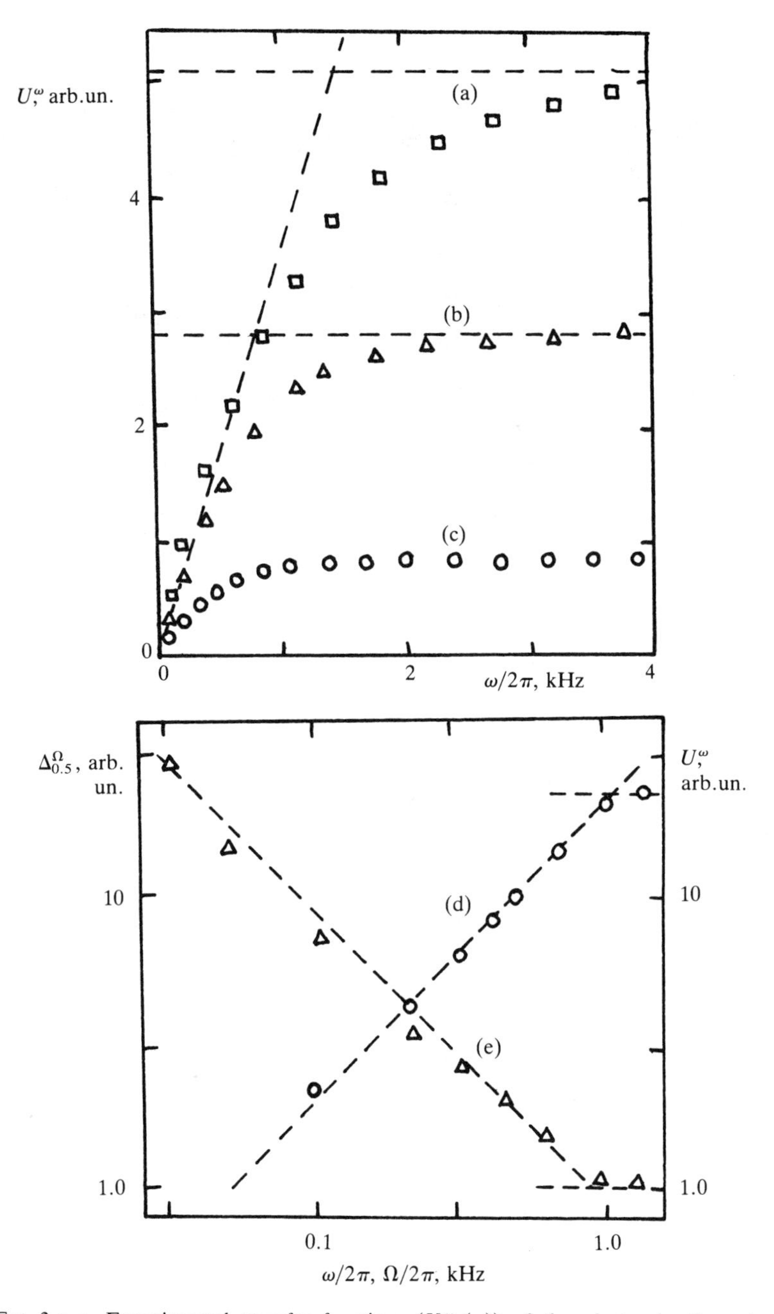

FIG. 3.a–c. Experimental transfer functions ($U^{\omega}(\omega)$) of the photorefractive adaptive interferometer using BSO for different light intensities I_0, mW/mm^2: a—160, b—80, c—40 (BSO, $\lambda = 633$ nm, $E_0 = 2$ kV/cm, $\Lambda = 20$ μm, $\Delta = 0.1$ rad) [4]. d–e. Transfer function (d) and frequency dependence of the characteristic amplitude of the jamming signal $\Delta_{0.5}^{\Omega}(\Omega)$ (e), resulting in a two-fold reduction of the intensity of the usable low-amplitude signai with a frequency $\omega/2\pi = 1$ kHz [4].

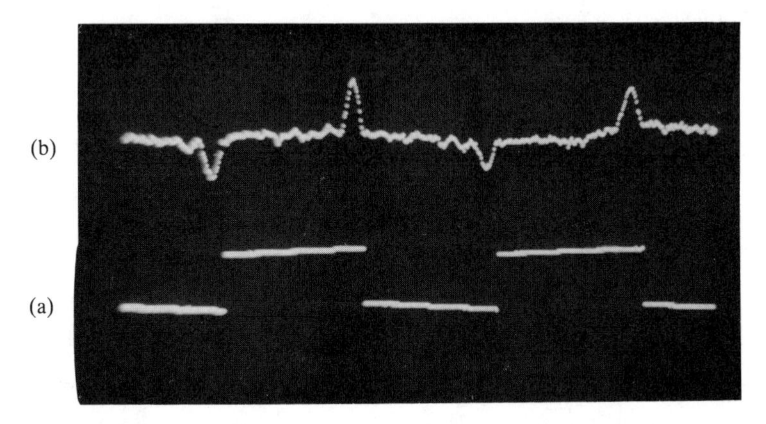

FIG. 4. Differentiation of the rectangular pulses at the output of the photorefractive BSO adaptive interferometer (a—input signal with a temporal period =3 ms, and b—output signal) [4].

frequency of a *jamming* signal $\Omega \lessdot \omega_0$ with a high modulation amplitude $\Delta^{\Omega} \gtrsim 1$ rad.

Because of nonlinear detection of the signal at high excitation levels, combinational harmonics with frequencies $\omega + 2l\Omega$ $(l = \pm 1, \pm 2 \ldots)$ arise in the output signal spectrum (Fig. 5). The amplitude of the signal with frequency ω becomes smaller, and this points to a reduction of the sensitivity of the interferometer in the presence of jamming.

The effect of suppression of the usable signal by jamming is frequency-dependent. At a specified Δ^{Ω}, it is more pronounced at a higher frequency Ω (Fig. 5). This is shown by a frequency dependence of the characteristic amplitude of jamming $\Delta^{\Omega}_{0.5}$ that leads to a two-fold reduction in the energy of the basic harmonic of the output signal. In the log-log scale, this curve looks like an inverted frequency transfer function of the interferometer measured under identical conditions (Fig. 3b).

To summarize, we note that a BSO crystal practically is ideal for applications in the blue/green region of the spectrum, which implies the use of an Ar-ion laser ($\lambda = 514$ nm) or the second harmonic of Nd:YAG ($\lambda = 530$ nm). Of course, these are not lasers that are promising for applications in interferometry. Semiconductor lasers ($\lambda = 0.85$, 1.3 μm) and even a conventional HeNe laser ($\lambda = 633$ nm) seem to be much more suitable here. As to the red spectral region, the question of an efficient PRC still is under discussion today. Semiconductor semi-insulating GaAs:Cr recently was proposed for usage in the near IR. The first results obtained show that this PRC can be as efficient here as BSO in the green light [15].

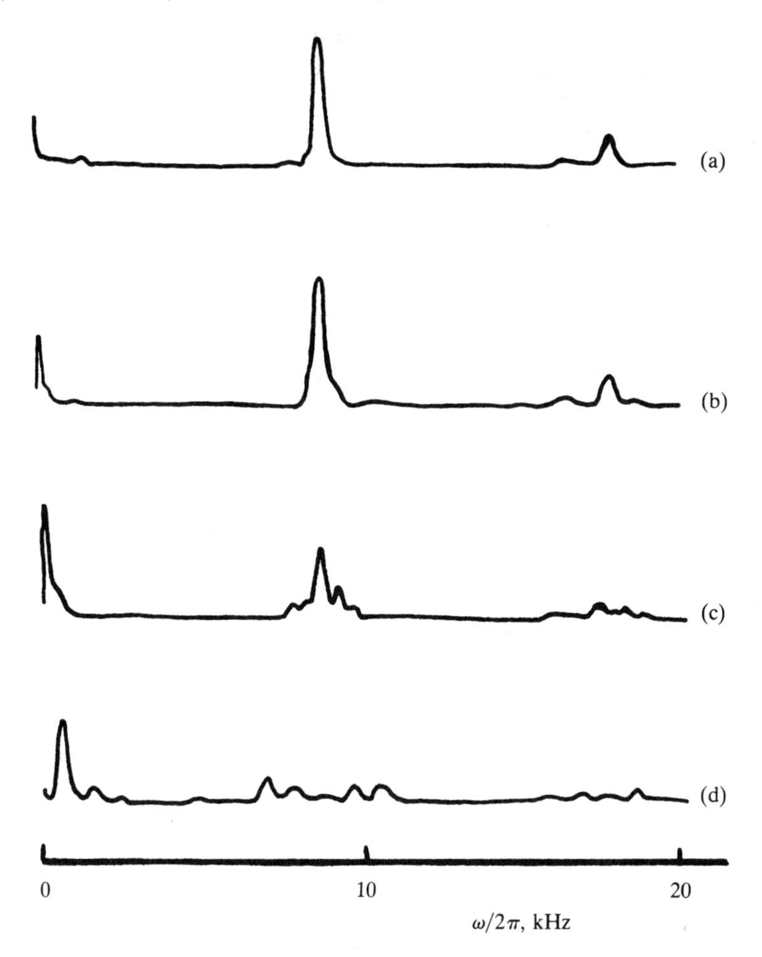

FIG. 5. Output spectra for two-frequency excitation of the BSO adaptive interferometer ($\omega/2\pi = 9$ kHz; $\Delta^{\omega} = 1$ rad; $\Omega/2\pi$, Hz: a—0, b—230, c—450, d—900; $\Delta = 2$ rad; $\omega_0/2\pi \simeq 1$ kHz) [4].

5. Adaptive Photodetectors Operating through Non-Steady-State Photo-EMF

Adaptive properties—i.e., efficient detection of a high-frequency phase-modulated light signal in the presence of a slowly varying phase shift—also are demonstrated by a novel photodetector based on non-steady-state photo-EMF [16]. This effect is associated closely with the formation of dynamic space-charge gratings within the crystal volume, identical with those observed during photorefractive holographic recording (Fig. 2)

[17, 18]. The photo-EMF manifests itself in an alternating current through a short-circuited sample illuminated by a vibrating interference pattern (Fig. 1d).

In this photodetector, we have a direct transformation of the phase-modulated optical signal into an output alternating electric signal J^{ω}. Light diffraction from the dynamic grating is eliminated in this case, and, therefore, a centrosymmetrical (nonphotorefractive) photoconductor can be used for these photodetectors as well.

Without going into detail—which can be found in original papers [17] and [18]—we note only that the photo-EMF signal in this case results from the periodic modulation of the spatial shift between spatial distributions of photoconductivity $\sigma(x, t)$ (caused by the interference pattern) and space-charge electric field $E_{sc}(x, t)$. For the diffusion mechanism of recording ($E_0 = 0$), the operation regime of this photodetector proves to be linear, with a short-circuit current amplitude given by

$$J^{\omega} = \Delta \frac{m^2}{2} \sigma_0 S K \frac{k_B T}{e} \frac{1}{1 + K^2 L_D^2} \frac{\omega \tau_{sc}}{\sqrt{1 + \omega^2 \tau_{sc}^2}}. \tag{3}$$

Here, σ_0 is the average photoconductivity of the sample, S is its cross-section, $K = 2\pi/\Lambda$ is the spatial frequency of the pattern, and L_D is the diffusion length of the photoinduced carriers. The amplitude of phase modulation and the modulation depth of the pattern were considered to be low ($\Delta, m \ll 1$) in the theoretical analysis.

As follows from Eq. 3, the photodetector demonstrates a differentiating transfer function similar to that observed for the photorefractive interferometer (Fig. 1c), with the same cutoff frequency $\omega_0 = \tau_{sc}^{-1}$. This means a linear growth of the output signal with the frequency of excitation up to ω_0 followed by a plateau. The adaptive properties of this photodetector in the geometry of the laser homodyne vibrometer are demonstrated clearly by Fig. 6.

The second advantage of the device based on holographic principles—low requirements of adjustment and the possibility of using complicated wavefronts—is characteristic of this photodetector as well. It can operate with speckle-like wavefronts, with only the condition that the carrier spatial frequency K is higher than the inverse characteristic size of the speckle spot.

In addition, the adaptive photodetector under consideration exhibits a unique property of suppressing the amplitude noise of the laser source. Unlike conventional photodetectors (photodiodes, photomultipliers, etc.), there is no average photocurrent J_0 through the device (Fig. 1d) when there is no biasing external dc voltage. As a result, amplitude laser noise

(a)

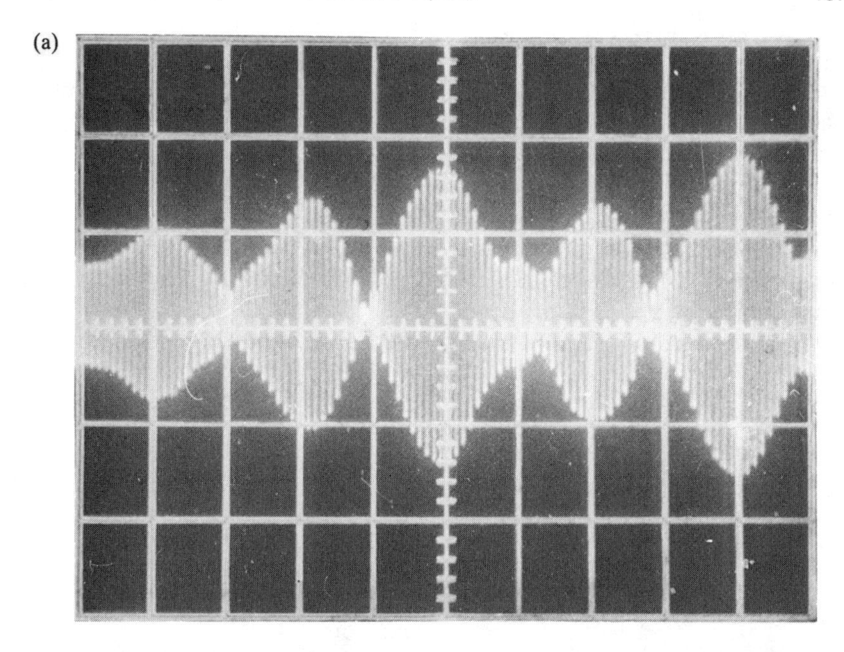

(b)

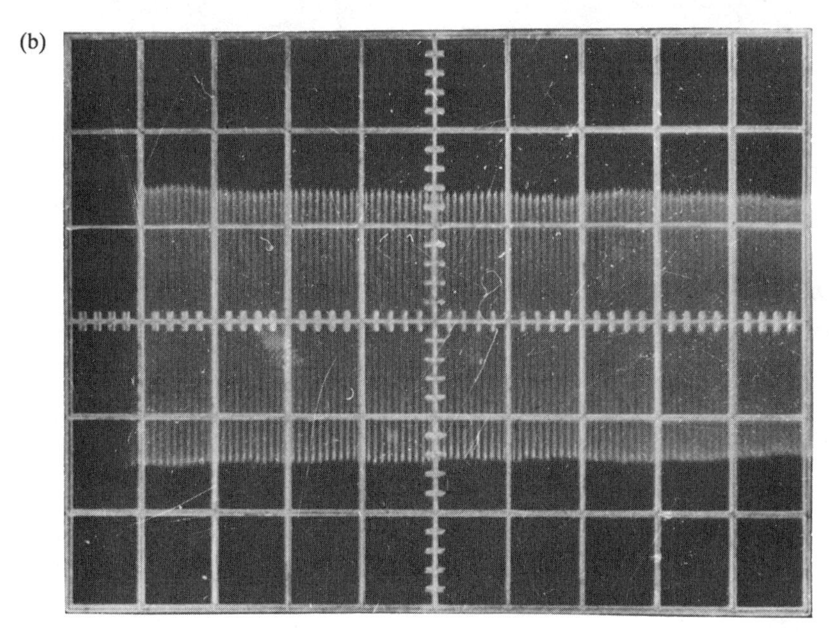

FIG. 6. Output signals from a homodyne laser vibrometer operating under laboratory conditions without special vibration insulation with a conventional photodiode (a) and GaAs:Cr adaptive photodetector with a cutoff frequency $\omega_0/2\pi \gtrsim 1$ kHz (b) (frequency of the signal $\omega/2\pi = 1$ kHz).

here proves to be multiplicative (i.e., proportional to the amplitude of a detected phase-modulated signal) rather than additive. So, suppression of the noise is performed automatically here and does not need usage of balance methods of detection used for conventional photodetectors [1].

What can we say about the operation frequency range and sensitivity of this photodetector? Its high cutoff frequency ω_0' is determined by the product of the sample capacity C and loading resistor R_L. In the photoconductors with a long average lifetime τ of photocarriers, ω_0' proves to be limited only by τ^{-1} [18].

In accordance with Eq. 3, the maximum of the output signal at $\omega \gtrsim \omega_0$,

$$J_{\max}^{\omega} = \frac{1}{4}\frac{1}{R}\frac{L_x}{L_D}\frac{k_B T}{e}, \tag{4}$$

is achieved for an optimal spatial frequency $K_{\text{opt}} = L_D^{-1}$. Here, L_x is the interelectrode spacing of the sample, R is its total resistance, and Δ, m are taken for simplicity to be equal to unity.

This means, in particular, that the maximum open-circuit voltage of the photo-EMF signal reached here for $\omega \simeq \omega_o$,

$$U_{\max}^{\omega} = \frac{1}{4\sqrt{2}}\frac{L_x}{L_D}\frac{k_B T}{e}, \tag{5}$$

is nearly equal to the characteristic voltage $k_B T/e$ (about 25 mV at room temperature) times the number of pattern stripes in the interelectrode spacing $(= L_x/2\pi L_D)$.

When only the thermal noise is taken into account, the maximum S/N ratio is observed for a matched loading $R_L \gtrsim R$. It equals

$$\left(\frac{S}{N}\right)_{\text{th}} = \frac{1}{4\sqrt{2}}\frac{L_x}{L_D}\frac{k_B T}{e}\frac{\sqrt{2}}{2\sqrt{k_B T R\,\Delta f}} = \frac{1}{8}\frac{\sqrt{V g_0}}{\sqrt{\Delta f}}, \tag{6}$$

where $V = L_x S$ is the volume of the sample and $g_0 = \sigma_0/e\mu\tau$ is an average generation rate of photocarriers. This estimate differs from the conventional equation for the S/N ratio determined by the shot noise

$$\left(\frac{S}{N}\right)_{\text{sh}} = \frac{1}{2}\frac{\sqrt{V g_0}}{\sqrt{\Delta f}} \tag{7}$$

by a numerical factor $\frac{1}{4}$ alone. This means that for properly matched loading, the sensitivity of the adaptive photodetector with no external biasing voltage proves to be only several times lower than its absolute minimum level limited by the shot noise.

6. Conclusions and Acknowledgments

In conclusion, we have shown that a new type of a photodetector for detection of phase-modulated light signals—the adaptive photodetectors—can be developed on the basis of dynamic holographic principles. It demonstrates the unique property of adaptivity, i.e., the possibility of efficient detection of usable signals in the presence of powerful low-frequency jamming, and it can be adjusted easily in the interferometer. In addition, adaptive photodetectors operating via nonsteady-state photo-EMF exhibit low sensitivity to the amplitude noise of the laser source.

The author acknowledges the contribution of Mr. I. A. Sokolov to experiments with adaptive interferometers, valuable discussions with Dr. G. S. Trofimov and Mr. V. V. Kulikov, and the help of Mrs. N. N. Nazina with translating the manuscript. He also would like to express his sincere gratitude to the head of the department of Quantum Electronics, where this work was performed, Prof. M P. Petrov, for his continuous support and interest in photorefractive investigations.

References

1. R. L. Forward, *Phys. Rev. D* **17**, 379–390 (1978).
2. D. A. Jackson, R. Priest, A. Dandridge, and A. B. Tveten, *Appl. Opt.* **19**, 2926–2929 (1980).
3. F. J. Eberhardt and F. A. Andrews, *JASA* **48**, 603–609 (1970).
4. S. I. Stepanov and I. A. Sokolov, *Proc. of the Second Intern. Conf. on Holographic Systems, Components and Applications* (Bath, UK), 95–100 (11–13 Sept., 1989).
5. T. J. Hall, M. A. Fiddy, and M. S. Ner, *Opt. Lett.* **5**, 485–487 (1980).
6. P. Gunter, *Phys. Rep.* **93**, 199–299 (1982); M. P. Petrov, S. I. Stepanov, and A. V. Khomenko, "Photorefractive Crystals in Coherent Optical Systems," to be publ. by Springer-Verlag, 1991.
7. A. A. Kamshilin and E. V. Mokrushina, *Sov. Tech. Phys. Lett.* **12**, 149–151 (1986).
8. G. Khitrova, D. Rouede, N. V. Kukhtarev, and H. M. Gibbs, *Phys. Rev. Lett.* **62**, 1110–1112 (1989).
9. M. P. Petrov, T. G. Pencheva, and S. I. Stepanov, *J. Optics* **12**, 287–292 (1979); A. Marrakchi, J. P. Huignard, and P. Gunter, *Appl. Phys.* **24**, 131–138 (1981).
10. S. H. Wemple and M. DiDomenico, Jr., in "Electro-optical and Nonlinear Optical Properties of Crystals" in Appl. Sol. St. Science (R. Wolf, ed.), pp. 264–383, Acad. Press, 1972.
11. J. P. Huignard and F. Micheron, *Appl. Phys. Lett.* **29**, 591–593 (1976).
12. A. V. Knyaz'kov, N. M. Kozhevnikov, Yu. S. Kuz'minov, V. V. Kulikov, N. M. Polozkov, and S. A. Serguschenko, *Sov. Phys.-Tech. Phys.* **29**, 801–802 (1984); Yu. O. Barmenkov, V. V. Zosimov, N. M. Koghevnikov, L. M. Lyamshev, and S. A. Serguschenko, *Sov. Phys.-Dokl.* **31**, 817–819 (1986).
13. A. A. Kamshilin, J. Frejlich, and L. H. D. Cescato, *Appl. Opt.* **25**, 2375–2381 (1986); P. A. M. Dos Santos, L. Cescato, and J. Frejlich, *Opt. Lett.* **13**, 1014–1016 (1988).

14. G. Hamel de Monchenault and J. P. Huignard, *J. Appl. Phys.* **63**, 624–627 (1988).
15. A. M. Glass, A. M. Johnson, D. N. Olson, W. Simpson, and A. A. Ballman, *Appl. Phys. Lett.* **44**, 948–950 (1984); M. B. Klein, *Opt. Lett.* **9**, 350–352 (1984).
16. S. I. Stepanov, I. A. Sokolov, G. S. Trofimov, V. I. Vlad, I. Apostol, and D. Popa, *Opt. Lett.* **15**, 1239–1241 (1990).
17. G. S. Trofimov and S. I. Stepanov, *Sov. Phys.-Sol. State* **28**, 1559–1562 (1986); M. P. Petrov, S. I. Stepanov, and G. S. Trofimov, *Sov. Tech. Phys. Lett.* **12**, 379–381 (1986).
18. M. P. Petrov, S. I. Stepanov, I. A. Sokolov, and G. S. Trofimov, *J. Appl. Phys.* **68**, 2216–2225 (1990).

CHAPTER 10

Water Wave Optics

Jakob J. Stamnes
Norwave Development, Oslo, Norway

1. Introduction

The best way to learn about water waves is to watch them. Thus, each time we throw a stone into a quiet pond, we readily can observe a diverging circular wave that spreads out from the center, where the stone hit the surface, and diminishes in wave height as the distance from the center gets larger.

Likewise, we readily can observe the focusing of water waves in a cup of coffee just by lifting the spoon and letting a small drop fall into the cup. If the drop hits the center, we first see a diverging circular wave, which subsequently is reflected from the wall and turned into a converging circular wave with the focus at the center. If the drop hits the surface outside the center, once again we see first a diverging and then a converging circular wave, but now the focus is displaced from the center to the opposite side, just as we would expect from elementary optical principles of imaging by means of a spherical mirror. Also, if we tilt the cup so that the water surface gets bounded by an ellipse, and let the drop hit the surface in one of its focal points, we will see the wave diverge towards the wall and, on reflection, converge at the other focal point.

From these considerations, it is clear that man's early knowledge about waves did not originate from acoustics or optics, but from observations

ISBN 0-12-289690-4

of waves on water. Thus, undoubtedly, the scientists involved in the early developments of the wave theories of sound and light not only have been aware of the analogy with water waves, but also have made use of it.

The time has come to pay back this debt. Therefore, I shall now consider the application of optical principles to study the propagation, diffraction, and focusing of water waves. Of course, the payback is not my real motivation. The real motivation is that in studies of non-optical waves, much can be gained by using the methods and principles developed in optics.

2. Linearised Water Waves

Perhaps it is not so clear that everyone will agree that the best way to learn about water waves is to watch them. In fact, when I expressed this point of view to a mathematician, his reaction was: "No, even better, *think* about them!" Of course, the conclusion to be drawn is that we should watch the waves and think about them at the same time.

Let us start the thinking process by asking a couple of questions: Are the displacements of the water surface really wave motions, and, if so, under what conditions does the familiar wave equation give a proper description of them?

2.1. Are the Displacements of the Water Surface Wave Motions?

The reason for asking this seemingly inappropriate question, after the evidence presented in the introduction, is that it is not obvious at first glance that the equations governing the motion of water *particles* have waves as solutions. In water of constant depth, the linearised equations [1,2] are

$$\nabla^2\phi(x,y,z)=0 \qquad -d<y<0. \tag{1}$$

$$-\omega^2\phi(x,y,z)+g\,\phi_y(x,y,z)=0 \qquad y=0. \tag{2}$$

$$\phi_y(x,y,z)=0 \qquad y=-d. \tag{3}$$

$$\eta(x,z)=\frac{i\omega}{g}\phi(x,0,z). \tag{4}$$

Here, ϕ, η, d, and g are the velocity potential, the free-surface displacement, the water depth, and the acceleration of gravity, respectively (Fig. 1). For simplicity, we have assumed that the water disturbance has been created by a time-harmonic source of angular frequency ω. For the

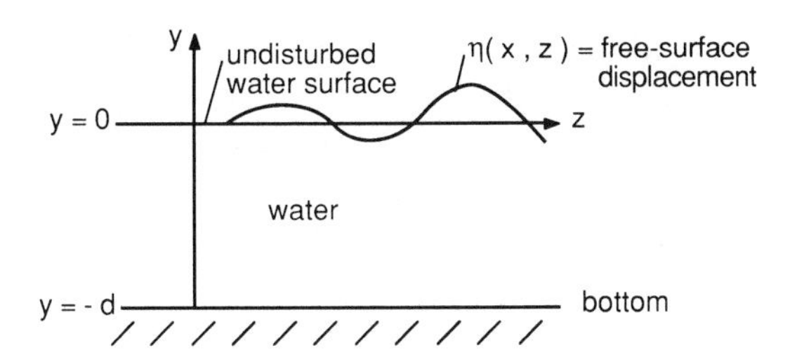

FIG. 1. Waves in water of constant depth.

sake of completeness, we mention that the motion of a water *particle* is given as the gradient of the velocity potential.

As noted above, Eqs. (1)–(3) are not recognised immediately as wave equations.

2.2. The Helmholtz Equation

To find the solutions of Eqs. (1)–(3), we follow Mehlum [3] and apply the standard technique of separation of variables. Thus, we write ϕ as a product of two factors, i.e., $\phi = A(y)B(x,z)$ and substitute this in Eq. (1) to obtain

$$\left(\frac{\partial^2}{\partial y^2} - k^2\right)A(y) = 0, \text{ and} \tag{5}$$

$$\left(\frac{\partial^2}{\partial x^2} + \frac{\partial^2}{\partial z^2} + k^2\right)B(x,z) = 0, \tag{6}$$

where k is a (generally complex) separation constant, yet to be determined.

The solutions of Eq. (5) are linear combinations of exp($-ky$) and exp(ky), but the only such combination that satisfies the boundary condition in Eq. (3) is $A = \cosh k(y+d)$. When substituting this in Eq. (2), we find the dispersion relation for water waves, i.e.,

$$\omega^2 = gk\tanh(kd), \tag{7}$$

and the phase velocity becomes

$$c(\omega) = \frac{\omega}{k} = \sqrt{\frac{g}{k}\tanh(kd)}. \tag{8}$$

Note that c depends on ω, since k is a function of ω according to Eq. (7). Thus, water waves are inherently dispersive, implying that waves at different frequencies travel at different speeds. Also, we note from Eq. (8) that the phase velocity depends strongly on the depth: The more shallow the water, the lower the wave speed.

At a given angular frequency ω and water depth d, the dispersion relation in Eq. (7) gives the following solutions for k [2, 3, 4]:

(1) A unique, real solution for the case in which $k^2 > 0$.

(2) An infinite number of imaginary solutions for the case in which $k^2 < 0$.

The imaginary solutions correspond to evanescent modes that decay in the xz plane, but oscillate in the vertical direction. These modes are peculiar to water waves and have no counterpart in optics.

For the case in which $k^2 > 0$, so that Eq. (7) gives a unique, real solution, we observe that Eq. (6) is nothing but the two-dimensional Helmholtz equation, with which we are familiar in optics.

Because of the peculiar evanescent modes for which $k^2 < 0$, it is *not* sufficient to know the free-surface displacement $\eta(x, z)$ on the line $z = 0$ to determine it at any point in the half-plane $z > 0$. In fact, we must know the velocity potential $\phi(x, y, z)$ in the plane $z = 0$ for all values of x and for all y values between the bottom at $y = -d$ and the undisturbed water surface at $y = 0$ [2].

On the other hand, if we can neglect the peculiar evanescent modes for which $k^2 < 0$, we are left with the familiar solutions of the Helmholtz equation, which means it is sufficient to know the free-surface displacement $\eta(x, z)$ on the line $z = 0$ to determine it at any point in the half-plane $z > 0$.

Still to be determined are the conditions under which we can neglect the peculiar evanescent modes. If we assume that the velocity potential is known in the plane $z = 0$, then these modes decay away from this plane exponentially [2]. Therefore, in situations where waves are observed at distances more than a couple of wavelengths from diffracting structures, there is good reason to believe that we may neglect the peculiar evanescent modes and work only with the familiar solutions of the Helmholtz equation.

From now on, we shall make this assumption, and then the connection between linear surface waves on water and linear two-dimensional optical waves is clear: They both are governed by the two-dimensional Helmholtz equation.

3. Huygens' Principle for Two-Dimensional Waves

For convenience, we denote the two-dimensional optical field or the free-surface displacement by $u(x, z)$ from now on. Further, we assume that the sources creating $u(x, z)$ are located to the left of the line $z = 0$, so that in the half-plane $z > 0$, we have waves travelling in the positive z direction.

If $u(x, z)$ is a solution of the Helmholtz equation and is known on the line $z = 0$, we can determine it everywhere in the half-plane $z > 0$. The solution can be expressed as a superposition of circular waves [2]:

$$u(x, z) = \int_{-\infty}^{\infty} u(x', z) \frac{-kz}{2iR} H_1^{(1)}(kR)\, dx', \tag{9}$$

where

$$R = \sqrt{(x - x')^2 + z^2}, \tag{10}$$

and where ${H_1}^{(1)}(kR)$ is the first-order Hankel function of the first kind.

Note that Eq. (9) is to be interpreted as Huygens' principle for two-dimensional waves.

4. Validity of Geometrical Optics

Ray tracing is an extremely important tool in optical design. It is based entirely on the laws of geometrical optics. Thus, wave theories or Huygens' principle rarely play(s) a role in lens designers' work. Since geometrical optics usually is considered to be valid at short wavelengths, and since an optical wavelength typically is 0.5 μm, while ocean swells can have wavelengths of 200–300 meters, there is good reason to ask the question: To what extent can we apply the techniques employed in the design of optical image-forming instruments to study the propagation and focusing of water waves?

Before answering this question, let us briefly review the basic principles of geometrical optics:

(1) In a homogeneous medium, light propagates along straight lines or rays.

(2) When light passes from one medium to another, the ray direction changes in accordance with Snell's law.

(3) In an inhomogeneous medium, the ray paths are curved and can be obtained by numerical integration of a differential equation.

(4) When light passes through an aperture, there is a sharp transition between the illuminated area and the dark area. These two areas are separated by the shadow boundary formed by the rays hitting the edge.

The advantage of the ray theory is that it is much simpler than the wave theory. Its main disadvantage is that it breaks down in certain areas, e.g., in focal regions and near shadow boundaries.

As noted earlier, it is common to define the range of validity of geometrical optics by saying that its laws are valid in the limit as the wavelength approaches zero. This definition is unfortunate, since even visible light has a wavelength of finite size, and since geometrical optics, as we shall see, can be used to describe wave phenomena at very long wavelengths, e.g., ocean swells.

With this background, it is worthwhile to consider whether it be would possible to redefine the range of validity of geometrical optics. To that end, we may start with Huygens' principle and consider the two-dimensional diffraction problem in Fig. 2, where light from a line source at P_1 passes through a slit aperture, or in the case of water waves, waves generated by a wave maker at P_1 pass through an opening in a breakwater. We assume that the observation point P_2 is far enough away from the aperture or breakwater that we can apply Huygens' principle and represent the Hankel function in Eq. (9) by the first term in its asymptotic series. Then we have, as shown in [2],

$$u(P_2) = \int_{-a}^{a} \left[\frac{\exp(ikR_1)}{\sqrt{R_1}}\right] \left[\frac{z_2}{R_2} \frac{\exp(ikR_2 - i\pi/4)}{\sqrt{\lambda R_2}}\right] dx, \tag{11a}$$

where

$$R_j = \sqrt{(x_j - x)^2 + z_j^2} \qquad j = 1, 2.$$

The expression inside the first bracket in Eq. (11a) is the incident wave, while the expression in the second bracket is Green's function or the secondary wavelet according to Huygens' principle. Thus, according to Huygens' principle, we add up all the secondary wavelets to obtain the field at P_2. However, by applying the method of stationary phase to the integral in Eq. (11a), we can show that

$$u(P_2) = u_G(P_2) + u_D(P_2), \tag{12}$$

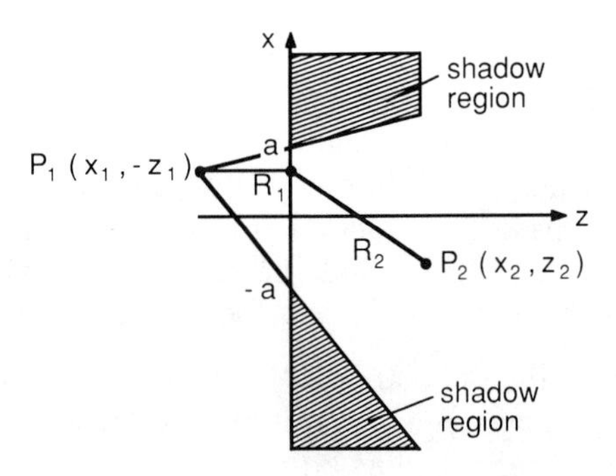

Fig. 2. Diffraction through a slit.

where $u_G(P_2)$ is the geometrical-optics contribution to the field at P_2, and $u_D(P_2)$ is the contribution due to diffraction at the two edges [2]. Thus, the only secondary wavelets included in Huygens' principle that contribute appreciably in this case to the field at P_2 are the geometrical-optics wave associated with the ray from P_1 to P_2 and the diffracted waves originating at the two edges. All the other secondary wavelets interfere destructively and contribute practically nothing to the field.

On the basis of Eq. (12), we may redefine the range of validity of geometrical optics as follows:

Geometrical optics is an approximation that is valid as long the edge-diffraction contribution u_D can be neglected in comparison with the geometrical-optics contribution u_G.

This new definition of the range of validity of geometrical optics has the advantage that it gives a clear meaning to the statement that "geometrical optics is valid in the limit as $\lambda \to 0$." In other words, the geometrical-optics approximation is good as long as $u_G \gg u_D$. Also, it now is clear how the geometrical-optics approximation can be corrected: by adding u_D to u_G. Since the ratio between $|u_D|$ and $|u_G|$ is proportional to $\sqrt{\lambda}$ [2], the geometrical-optics approximation steadily gets better as the wavelength decreases. Finally, we note that u_G drops abruptly to zero at the geometrical shadow boundary, so that the sharp transition between illuminated and dark areas is maintained when u_D is neglected.

5. Ray Tracing in Water of Variable Depth

In water of constant depth, the phase speed c is given by Eq. (8). In deep water ($d \to \infty$), Eq. (8) gives the phase speed $c_0 = \sqrt{g/k_0}$, so that on writing $k = k_0 n$, the refractive index in water of constant depth becomes

$$n = \frac{c_0}{c} = \coth(n\, k_0\, d), \tag{13}$$

where $k_0 = \omega^2/g$ is the wave number in deep water.

For a given angular frequency ω, the transcendental Eq. (13) for n can be solved by an iterative Newton-Raphson technique, which gives rapid and stable convergence for all depth values needed in practice [5].

If the depth is varying slowly, we assume that n is given by Eq. (13) locally, and the ray path then can be found by using the familiar differential equation for an inhomogeneous medium [6],

$$\frac{d}{ds}\left(n\frac{d\mathbf{r}}{ds}\right) = \nabla n. \tag{14}$$

By a transformation of variables, Eq. (14) can be solved by a Runge-Kutta method [7].

For the sake of completeness, we mention that it also is possible to include the presence of a current in the ray tracing procedure [5, 8]. Then, the wave propagation problem becomes anisotropic, since the phase speed along the current is different from that transverse to it. As in optics, the anisotropy implies that the propagation of energy no longer is along the phase rays.

6. Combined Method of Ray Tracing and Diffraction

When analysing waves in focal regions, we must combine geometrical optics and diffraction theory. The reason for this is that the geometrical-optics method breaks down in the focal region, and the use of diffraction theory alone could be extremely inefficient [2].

In the combined method, ray tracing is applied to determine how the wave propagates all the way from the source and up to some reference plane in front of the focal region. Thereby, one finds the amplitude and phase of the converging wave in the reference plane, and this knowledge is sufficient to carry out diffraction calculations from there on through the focal region.

Note that edge diffraction automatically is included in this method, since the integration only is over that finite region of the reference plane that is covered by geometrical rays. (For a detailed discussion, see Section 6.3.2 in [2]. This explains why the combined method also works well at long wavelengths, cf. Section 4.

7. Caustics Created by Devil's Hole

To explain how the combined method of ray tracing and diffraction can be applied to ocean waves, let us consider a stretch of ocean around the Ekofisk oil installation in the North Sea (Fig. 3). Waves from the Northwest will pass over a depression in the bottom called Devil's Hole. This hole acts as a negative lens spreading the energy out to the sides. Here, the waves will interact with the slightly focused waves coming from the more shallow region to the North, and we get a caustic of intersecting rays as shown in Fig. 4. Around this caustic, the energy is concentrated, as around a focus, and we would like to know the intensity here of course, since the area is dangerously close to the Ekofisk installation.

To solve this problem, we draw a line on the sea surface between Devil's Hole and the place where the caustic starts, as shown in Fig. 5. First, we use ray tracing to find the phase and amplitude of the wave disturbance

FIG. 3. Contour map of the ocean floor near Ekofisk and Devil's Hole in the North Sea.

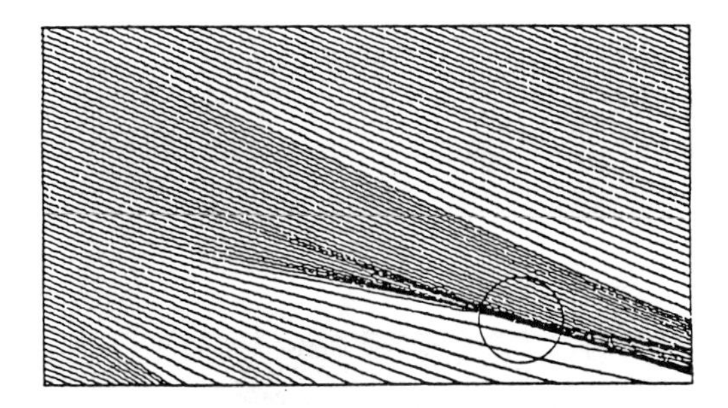

FIG. 4. Ray paths in the Ekofisk region showing a strong focusing effect.

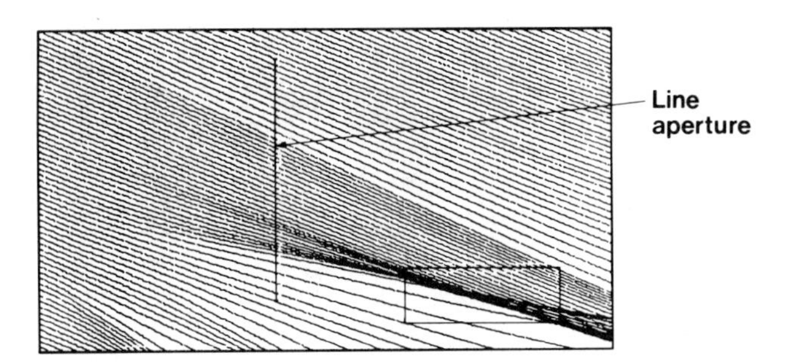

FIG. 5. Line aperture used for diffraction calculations. The diffracted field is computed inside the rectangle.

along this line, and then diffraction theory to compute the wave field east of this line, regarding the line as the *aperture* where the field is known. The resulting diffraction pattern is shown in Fig. 6 as a contour map of the wave height. The position of the diffraction pattern is well predicted by the ray theory, as can be seen by superimposing the magnified ray map in Fig. 7 on to the diffraction pattern in Fig. 6; but, of course, the intensity can *not* be obtained from the ray map.

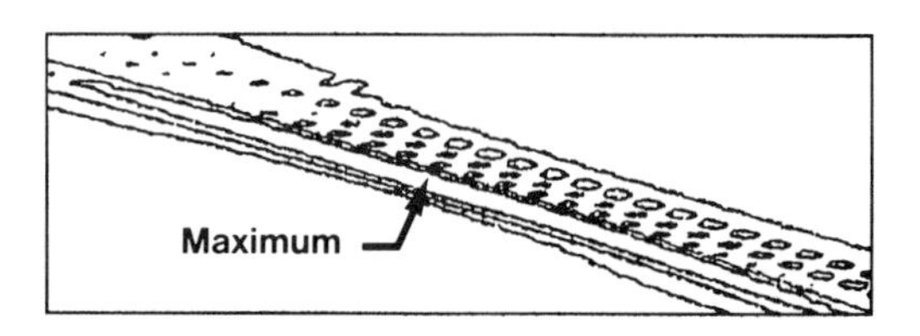

FIG. 6. Diffraction pattern near Ekofisk, i.e., inside the rectangle in Fig. 5.

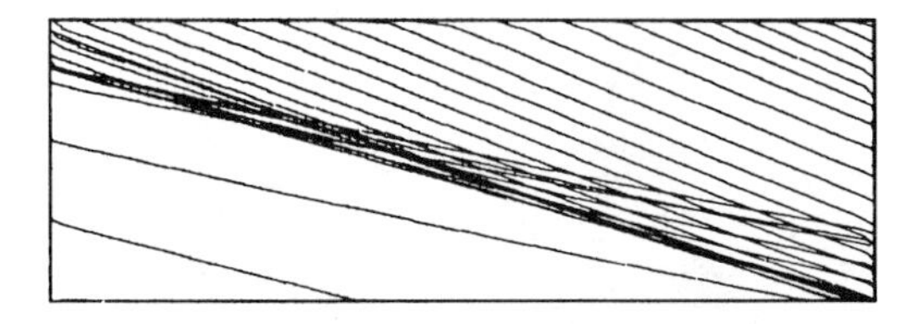

FIG. 7. Magnified ray plot for the same region as in Fig. 6.

FIG. 8. Focusing experiment with water waves using a zone-plate lens.

8. Zone-Plate Lens for Focusing of Ocean Swells

Zone-plate lenses can be used to focus water waves, as is demonstrated in the focusing experiment shown in Fig. 8. At the far end of the basin in the picture, a wave generator is emitting *diverging* circular waves. The zone-plate lens in the middle part of the picture transforms them into *converging* circular waves that propagate through the focal area. After passing through the focus, they are converted again into diverging circular waves, which pass on to the near end of the basin.

A lens for water waves works in a similar manner to an optical lens: As the wave passes over a submerged body (e.g., a lens element), it is retarded because it moves slower in shallow water than in deep water. By adjusting the shape and depth of the various lens elements, one obtains the intentional transformation from a diverging to a converging wave. For further details concerning the construction and use of such lenses, see [2, 4, 9, 10, 11, 12].

9. Speculations

In the preceding discussion of applying optics to water waves, we have assumed that we can ignore the inherent nonlinear character of water waves and also the peculiar evanescent modes associated with the linear waves. In addition, we have assumed that the Kirchhoff approximation is satisfied. Therefore, let us now consider some of the implications of these assumptions.

9.1. Nonlinear Focusing of Water Waves

Even for small-amplitude water waves, the nonlinear effects can not be ignored for propagation over long distances. Thus, a typical nonlinear phenomenon is the Benjamin-Feir instability, according to which a sinusoidal wave train of small amplitude will develop sidebands, causing a modulation of the wave. This effect has been observed experimentally in a focusing experiment similar to the one depicted in Fig. 8 [2, 13].

As the amplitude of the wave incident upon the lens in Fig. 8 increases, nonlinear effects will start to play a role in the focal region. To study such effects, we need a theory of nonlinear wave propagation that applies to focusing. A general theory of this kind has not been developed yet, but for water waves of small convergence angles, it is possible to apply a nonlinear theory, based on the cubic Schrödinger equation [14]. This was done by Stamnes *et al.* [13], who found consistent agreement between the results of the nonlinear theory and those of experiments. (See also Chapter 19 in [2].)

To my knowledge, the analogous theoretical and experimental work concerned with the nonlinear focusing of optical waves remains to be done. Thus, in this case, the water wave investigation is ahead of the corresponding optical work, and it seems plausible that the theoretical techniques applied to the water wave problem also may be applied to the corresponding optical problem.

9.2. Wave Shielding Based on Destructive Interference

The study described in Section 7 was motivated by the subsidence of the sea floor around the Ekofisk oil installation, caused by the removal of the hydrocarbons from the underground. This led to a serious concern that the air gap between the sea and the platform deck could be reduced to a dangerously low level. To compensate for this, our research group was asked to investigate whether it would be possible to shield the platform from the waves by using the principle of *destructive* interference, as opposed to the *constructive* interference applied in focusing. Our theoretical investigation showed that it is possible, indeed, to obtain shielding from waves of different frequencies coming from different directions by placing submarine structures around the platform in a carefully pre-designed pattern. These theoretical results also were confirmed experimentally.

Unfortunately, before our study was completed, the operating oil company decided to raise the whole Ekofisk installation! Nevertheless, our study did show that, in general, it may be possible to use the principle of destructive interference to construct submarine breakwaters for the protection of offshore installations, harbours, and eroding coastlines [15].

9.3. Near-Field Interactions

One thing we learned from the Ekofisk wave-shielding study in the previous subsection is that the Kirchhoff approximation is not reliable when the dimensions of the diffracting structures are comparable to those of the wavelength. Also, in the case of water waves, it may be important to include the peculiar evanescent modes to account correctly for the interaction between nearby submerged bodies.

We encounter an analogous situation in optics when dealing with the rigorous theory of diffraction gratings [16]. Also, in this case, the Kirchhoff or physical optics approximation is of limited value. Although the details are different in the two cases, I believe that much can be gained by taking the mathematical methods and techniques developed in one of these disciplines and using them in the other.

Another example in which near-field interactions play a dominant role is in the design of fiber-optic or integrated-optics couplers. There is a completely analogous coupling device in water waves: the Tapchan wave power converter [17]. Thus, in the Norwave wave power plant, a tapered channel (TAPCHAN) is used to guide the waves into a reservoir. In this manner, the kinetic energy of the incoming waves is converted into potential energy of water lifted to a higher level, and this water subsequently is fed through a turbine. Again, the mathematical techniques developed in water waves could be of potential use in optics, and vice versa.

Recently, E. Mehlum and B. Spjelkavik made a theoretical study of strongly nonlinear water waves in a tapered channel [18], a problem analogous to that of nonlinear light propagation in an optical fiber. In this case, the results obtained for water waves show a striking resemblance to the phenomena observed at Norwave's prototype power plant, which has been in operation since 1986 on the western coast of Norway.

9.4. Inverse Problems

From optics, we are familiar with *inverse problems*, which, in general, may be defined as being the task of determining a scatterer from knowledge of its scattered field. Thus, if we want to determine the index of refraction distribution of a material—for example, a non-uniform optical fiber—then we could illuminate the fiber from a certain direction and measure the scattered field, and possibly repeat this process, using several different directions of illumination. The inverse problem then consists of determining the index distribution of the fiber from the measured data, using, for example, holography or diffraction tomography [19, 20].

In the case of water waves, a similar inverse problem exists: From the observed scattered wave pattern due to a submerged body (e.g., a submarine), can we determine its position and also its shape, size, or other features of interest?

In the hope that I have triggered your imagination, let me end with this open question.

References

1. G. B. Whitham, "Linear and Nonlinear Waves," p. 431, Wiley, New York, 1974.
2. J. J. Stamnes, "Waves in Focal Regions," Adam Hilger, Bristol and Boston (1986).
3. E. Mehlum, "Large-Scale Optics: Ocean Waves," *Proc. 15th ICO Conf.* Garmisch-Partenkirchen, Federal Republic of Germany (1990).
4. E. Mehlum, "Bodies in Water," *Center for Industrial Research, Report* (1980).
5. E. Mehlum, H. Heier, K. I. Berg, S. Ljunggren, K. Løvaas, B. Spjelkavik, and J. J. Stamnes, "Computation of Water Wave Propagation by Means of Ray Tracing and

Diffraction Theory: WAVETRACK Computer Program Package," Norwave A. S., Oslo, Norway, 1985.
6. M. Born and E. Wolf, "Principles of Optics," 4th edition, Pergamon, New York, 1970.
7. A. D. Sharma, V. Kumar and A. K. Ghatak, *Appl. Opt.* **21**, 984–986 (1982).
8. K. I. Berg, "Propagation of Ocean Waves on Varying Depth and Current," *Center for Industrial Research, Report* (1984).
9. E. Mehlum and J. J. Stamnes, "On the Focusing of Ocean Swells and Its Significance in Power Production," *Center for Industrial Research, Report* (1978).
10. E. Mehlum and J. J. Stamnes, "Power Production Based on Focusing Ocean Swells," *Proc. 1st Symp. on Wave Energy Utilization*, Chalmers Institute of Technology, Sweden, 29–35 (1979).
11. E. Mehlum, "A Circular Cylinder in Water Waves," *Appl. Ocean. Res.* **2**, 171–177 (1980).
12. H. Heier and J. J. Stamnes, "Waves in Focus," *Fra Fysikkens Verden* **1**, 4–10 (in Norwegian) (1982).
13. J. J. Stamnes, O. Løvhaugen, B. Spjelkavik, C. C. Mei, E. Lo, and D. K. P. Yue, "Nonlinear Focusing of Surface Waves by a Lens—Theory and Experiment," *J. Fluid Mech.* **135**, 71–94 (1983).
14. D. K. P. Yue and C. C. Mei, "Forward scattering of Stokes waves by a thin wedge," *J. Fluid Mech.* **99**, 33–52 (1980).
15. E. Mehlum, G. Olsen, T. Simonstad, and J. J. Stamnes, "Submarine Breakwaters: A Novel Concept in Wave Shielding," *Proc. Offshore Technology Conference*, Houston, Texas (May 5–8, 1986).
16. R. Petit (ed.), "Electromagnetic Theory of Gratings," Springer, Berlin, 1980.
17. E. Mehlum, T. Hysing, J. J. Stamnes, O. Eriksen, and F. Serck-Hanssen, "TAPCHAN: The Norwave Power Converter," *Proc. Am. Soc. Mech. Eng.*, New Orleans (February, 1986).
18. E. Mehlum and B. Spjelkavik, "Strongly Nonlinear Waves in a Tapered Channel," *Proc. 5th Int. Workshop on Water Waves and Floating Bodies*, Manchester, U.K. (1990).
19. E. Wolf, "Three-Dimensional Structure Determination of Semi-Transparent Objects from Holographic Data," *Opt. Commun.* **1**, 153–156 (1969).
20. A. J. Devaney, "Reconstructive Tomography with Diffracting Wavefields, *Inverse Problems* **2**, 161-183 (1986).

CHAPTER 11

About the Philosophies of Diffraction

Adolf W. Lohmann*
Physikalisches Institut der Universität
Erlangen, Germany

1. Introduction

Philosophies? Not simply *philosophy*? It is not a typing error in the title, but implies that there is more than one philosophy of diffraction, more than I can mention in this short chapter.

So why an article about the philosophies of diffractions? Is it not enough if we know the facts of diffraction? Others might want to know the truth about diffraction, but not a wishy-washy philosophy.

This latter complaint can be illustrated by pointing out how Maxwell wrote about electromagnetic theory. He had the correct equations, of course. However, the verbal interpretation of those equations—in other words, his philosophy of electromagnetism—was such nonsense that Maxwell undoubtedly would have failed if he were examined today. Maxwell talked about "the mythical ether," and its "elastic module." His comments contradicted implicitly the special theory of relativity. That is funny, since others found out later that Maxwell's equations are, in fact, Lorentz-invariant—in other words, compatible with Einstein's theory, which did not exist when Maxwell presented his own theory.

Maxwell was a genius, no doubt. When developing his theories, he probably was guided by a philosophy. The fact that his philosophy now appears to be questionable—to say the least—does not diminish the value

* Present address: NEC Research Institute, 4 Independence Way, Princeton, NJ 08540.

ISBN 0-12-289690-4

of his theory. It also does not degrade the usage of philosophies as spiritual crutches while exploring something new.

Maxwell was not the only one to make progress based on a questionable philosophy. The same can be said about both Huygens and Young. More specifically, those followers of Huygens and Young, who perceived a contradiction between the viewpoints of the two, were comparing two mutually exclusive philosophies. Today, we might say both were right, in a sense, and each produced a fruitful, stimulating philosophy. By the way, Huygens published his principle 300 years ago. The Dutch Association for Mathematical Physics will celebrate this anniversary on November 19–22, 1990, in The Hague.

Philosophies often stimulate progress, more so than a set of coupled differential equations can do. This statement is probably subjective in nature. When I contemplate wave propagation, I use a philosophy, maybe the one by Huygens or another proposed by Ernst Abbe; but when I try to verify my guesses, I make use of mathematical tools.

Why all this fuss about philosophies? It has to do with the part in me that is a teacher. I try to teach when I write; but the generally accepted style for books such as this one, does not really allow me to do that, at least not as much as I would like. To be specific, I am not allowed to write as I would talk in front of a class. Some may say it is impossible to write as lively as one may be able to speak. Maybe that is due only to our lack of practice. After all, well-written novels can be more exciting than everyday life. In this chapter, I have tried to communicate quasi-orally. When I checked the manuscript before sending it to the editor, I was not really satisfied; but I learned a bit. On the next occasion, I hope to be more effective in quasi-oral communication.

Is this attempt to communicate in quasi-oral fashion just a private hobby? Maybe not. The way scientists are communicating today in printed form is changing rapidly. Not only our needs, but also commercial interests influence what happens. I dislike some of the trends that I perceive. Others are partially dissatisfied, too. I assume. Can things improve? Probably yes, but not unless we make constructive suggestions, possibly tried out through experimental publishing. More about publishing will be discussed at the end of this chapter.

2. A New Philosophy of Diffraction, Corner Diffraction—The Context

The motivation for developing this new philosophy is practical. As you probably know, most computer-generated holograms (CGHs) consist of rectangular micro-structures. The rectangle may be a micro-aperture or a

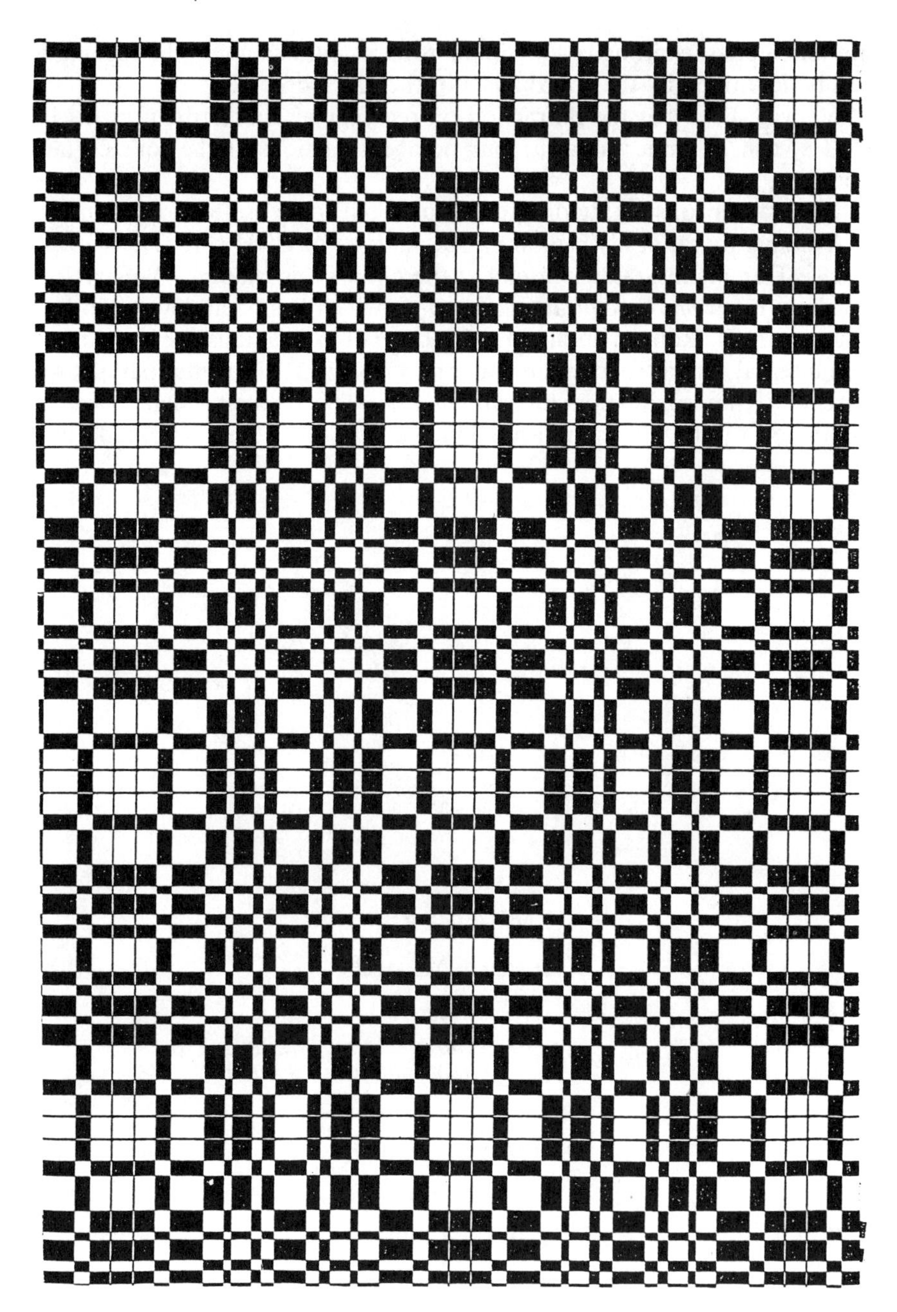

FIG. 1. Plot of a Damann grating.

mesa of a phase-shifting layer—in photoresist, for example. The rectangle can be identified by the coordinates of the four corners.

We will show that the diffraction on rectangular microstructures (as in CGHs) can be understood as four point-interactions at the four corner points.

This way of considering diffraction turns out to be useful in the context of Dammann gratings, which are a special type of CGH (Fig. 1). We will discuss Dammann gratings briefly at the end of this chapter. We start by touching upon *neo-positivistic philosophy*, which has been developed by, among others, Robert Kirchhoff, who otherwise is famous for his diffraction integral, as sketched in the upper part of Fig. 2. The center part shows Kirchhoff's approximation for *thin objects*. The incoming light wave is multiplied by the object transmission, yielding the object amplitude im-

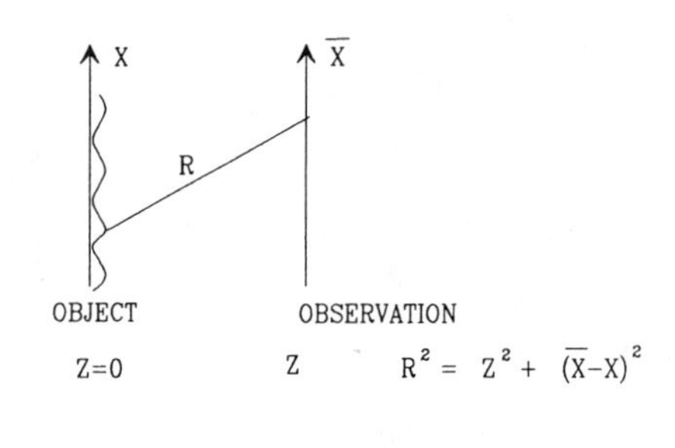

$$\int u_0(x)\ SPH(\bar{x}-x,R)\ dx = u(\bar{x},z)$$

<u>OBJECT APPROXIMATION</u>

$u(x,z=-0)$	$u_0(x)$	=	$u(x,z=0)$
ILLUMINATION	*OBJECT TRANSMISSION*		*OBJECT SOURCE*

POSITIVISTIC PHILOSOPHY:

THE JOB OF SCIENCE IS

TO DESCRIBE

IN A <u>CONVENIENT</u> AND <u>CONSISTENT</u> WAY.

EXAMPLE FOR CONVENIENT: COPERNICUS > PTOLEMAEUS

FIG. 2. 3x Kirchoff (1824–1887): Diffraction integral.

mediately behind the object. This approximation, which here is only of marginal importance, is another example of a very fruitful, yet inaccurate, philosophy. At the bottom of Fig. 2, we find the credo of the positivistic philosophy. Note: The aim of science is not the truth, but only the description of nature. The description should be convenient. Convenience is a matter of perspective, a matter of taste. Consider, for example, the geocentric system of Ptolemaeus and the heliocentric system of Copernicus. They differ only in the choice of the coordinate system. For astronomical computations, Copernicus is more convenient; but in everyday life, we say: "The sun is rising above the horizon." That statement uses—as Ptolemaeus did—the surface of the earth as coordinate system.

The difference between Ptolemaeus and Copernicus only is a matter of philosophy, not of truth. We now will mention several *philosophies of*

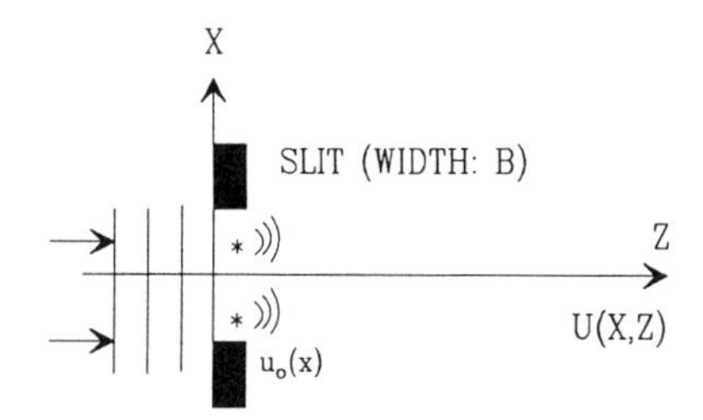

SLIT AS OBJECT

$$u_0(x) = \begin{cases} 1 & in\ |x| < B/2 \\ 0 & otherwise \end{cases}$$

MORE GENERAL:

$$u_o(x) = \underbrace{\int u_0(x')\delta(x-x')\, dx'}$$

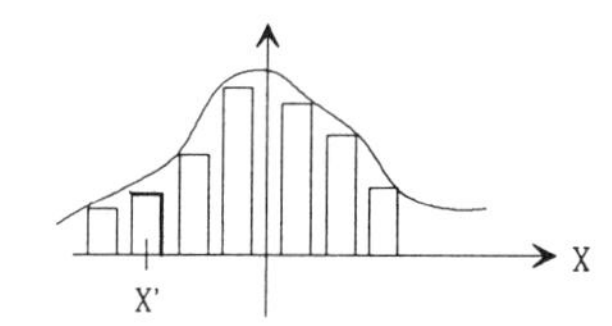

WAVEFIELD:

$$u(x,z) = \int u_o(x')Sph.(x-x',z)\, dx'$$

FIG. 3. Philosophy of Huygens: Secondary point sources.

diffraction, starting with Huygens and closing with a new philosophy, called *corner diffraction*, or *quadrant diffraction*.

3. The Philosophy of Huygens

Huygens considered every object as an assembly of point sources, as sketched in Fig. 3, mathematically equivalent to a DIRAC decomposition. Every Dirac pulse of the object creates a spherical wave.

4. The Philosophy of Young

Young argued that slit diffraction essentially is caused at the edges of the slit (Fig. 4). The rectangular transmission $U_o(x)$ can be composed of edges, as shown in Fig. 4, in three different ways: The version on the left has two

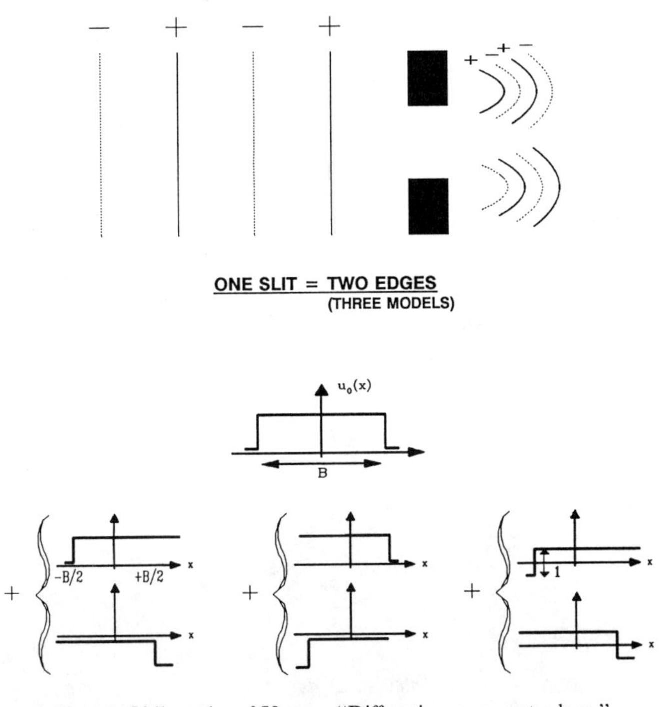

FIG. 4. Philosophy of Young: "Diffraction occurs at edges."

right-hand steps, one of them positive, the other one negative, located where the edges of the slit are. In the center version, two *left-hand steps* are used for the composition of the slit. The third version, on the right side, is the arithmetic mean of the two other step models.

The Young philosophy can be extended to arbitrary 1-D objects. The corresponding mathematics in Fig. 5 can be based on *integration by parts*. Apparently, the object derivative now is crucial. Hence, the steepness of the slit edges is significant for the observed diffraction pattern.

$$u_o(x) = \int_{-\infty}^{x} \frac{du_0(x')}{dx'} dx' = \int_{-\infty}^{+\infty} \frac{du_o(x')}{dx'} \cdot H(x-x')dx'$$

$$H(x-x') = \begin{cases} 1 & \text{if } x>x' \\ 0 & \text{if } x<x' \end{cases}$$

$$u_o(x) = -\int_{x}^{+\infty} \frac{du_0(x')}{dx'} = -\int_{-\infty}^{+\infty} \frac{du_o(x')}{dx'} H(x'-x)dx'$$

$$u_0(x) = \int_{-\infty}^{+\infty} \frac{du_0(x')}{dx'} E(x-x')dx'$$

$$u(x,z) = \int_{-\infty}^{+\infty} \frac{du_o(x')}{dx'} Cyl.(x-x',z)dx'$$

CONSEQUENCE: EDGE QUALITY CRITICAL

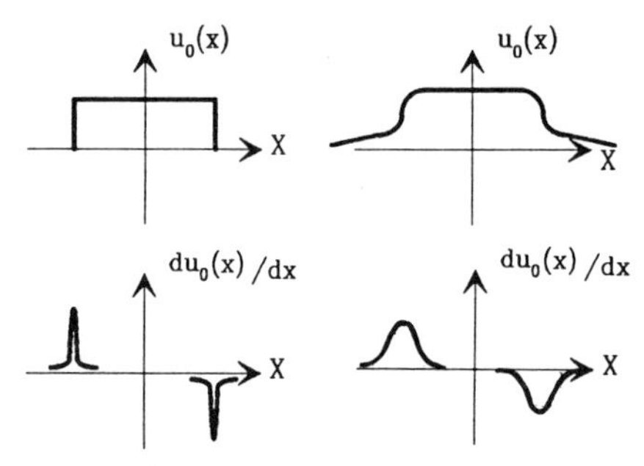

FIG. 5. Young theory (3 models).

5. Diffraction Philosophy Beyond Young

The theory associated with Young's point of view paid attention to the first derivative of the one-dimensional object $U_o(x)$. Now we extend this approach to two-dimensional objects, where the cross-derivative becomes the focus of our attention (Fig. 6). The elementary object now consists of two crossed edges. In Fig. 6, we used the $(+\frac{1}{2}; -\frac{1}{2})$ version. The point where the two edges cross is called a *corner* Q_o. Every corner gives rise to a corner wave or quadrant wave. In Fig. 7, we used two crossed right-hand steps, which generate a corner in the more common sense. Four such corners, properly located, compose a rectangle.

FROM YOUNG:

$$u_0(x) = \int_{-\infty}^{+\infty} \frac{du_0(x')}{dx'} \underbrace{E(x-x')}_{} \, dx'$$

$$\text{edge} = \begin{cases} +\frac{1}{2} & \text{if } x \geq x' \\ -\frac{1}{2} & \text{if } x < x' \end{cases}$$

NOW IN 2D:

$$u_0(x,y) = \iint \frac{\partial^2 u_0(x',y')}{\partial x' \, \partial y'} E(x-x')E(y-y') \; dx' \; dy'$$

$$\Downarrow$$

$$\text{Quadrant } Q_0(x-x',y-y') = \pm \frac{1}{4}$$

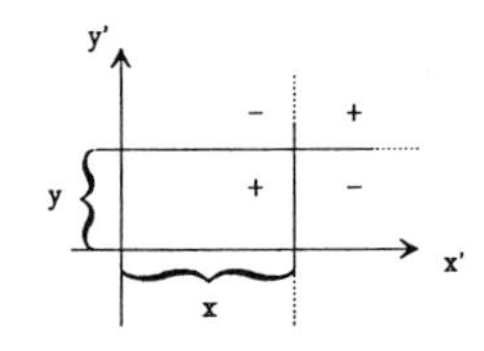

PROPAGATION TO Z ≥ 0

$$u(x,y,z) = \iint \frac{\partial^2 u_0(x',y')}{\partial x' \partial y'} Q(x-x',y-y',z) \; dx' dy'$$

$$\Downarrow$$

QUADRANT WAVE

starting in $z = 0$ *at* (x',y')

FIG. 6. Beyond Young: Quadrant diffraction.

$$u(x,y) = \int_{-\infty}^{x}\int_{-\infty}^{y} \frac{\partial^2 u(x',y')}{\partial x'\partial y'} dx'dy'$$

$$= \iint_{-\infty}^{+\infty} \frac{\partial^2 (x',y')}{\partial x'\partial y'} \underbrace{E(x-x')E(y-y')}_{Q(x-x',y-y')} dx'd$$

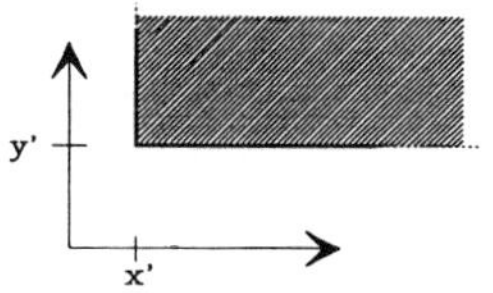

ESPECIALLY USEFUL (BUT NOT EXCLUSIVELY)
IF U(X,Y) PIECE-WISE CONSTANT, WITH CARTESIAN BOUNDARIES

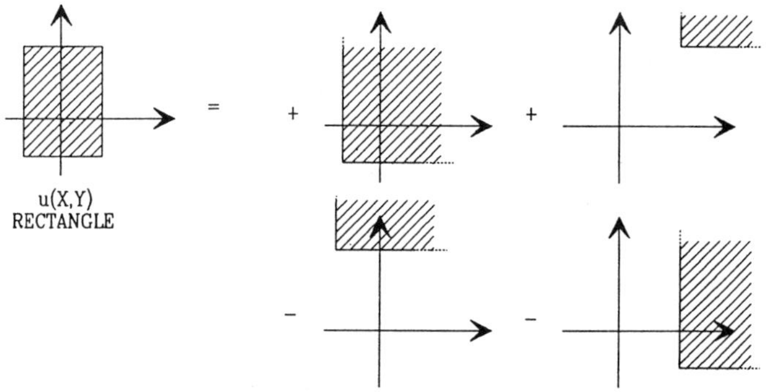

u(X,Y)
RECTANGLE

FIG. 7. And now: Quadrant decomposition.

6. Corner Diffraction on 2-D Dammann Gratings

A *Dammann grating* is a phase grating with phase values *O* or π in the simplest version. A grating cell consists of many π-phase mesas, often 100 or even 1,000 per elementary grating cell (Fig. 8). The cross derivative of such a phase-step grating consists of a periodic array of Dirac peaks with alternating $(+/-)$ signs. Each peak emits its own quadrant wave. If the peak is not located exactly where it should be, the *detour phase effect* comes into play. In other words, the Dirac peak will contribute to the overall diffraction with a wrong phase. The philosophy of quadrant diffraction, or corner diffraction, leads to error tolerance criteria that are easy to

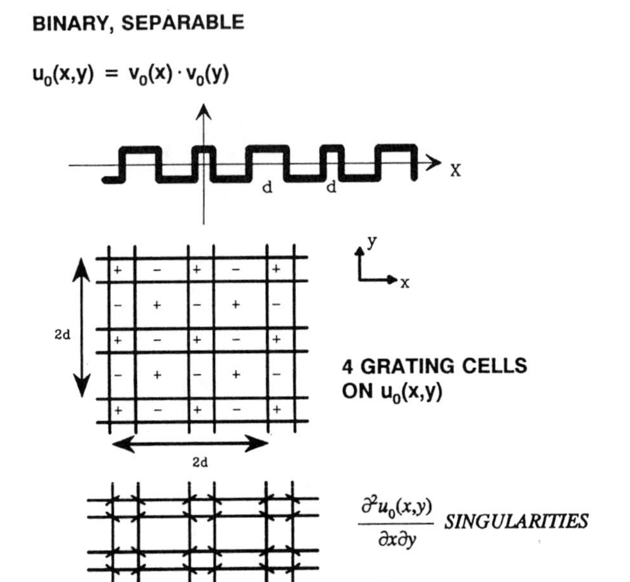

THE $\partial^2 u_0/\partial x \partial y$ OF DAMMANN GRATINGS AND OF THE MOST COMPUTER HOLOGRAMS CONSISTS OF ISOLATED δ-PEAKS AT (x_m, y_m). THE WAVEFIELD $u(x,y,z)$ CONSISTS OF A DISCRETE SET OF QUADRANT WAVES $q(X-X_m, y-y_m, z)$

FIG. 8. Quadrant diffraction on Dammann gratings.

understand. In other words, this new diffraction philosophy sometimes can be convenient. A detailed study of corner diffraction and its application to problems with Dammann gratings will be published elsewhere.

7. Summary

In the technical part of this chapter, we decomposed a diffraction object either into points (Huygens), edges (Young), or corners or quadrants. As Fig. 9 shows, there are at least two more prominent and useful ways available for decomposing an object. Which one of them is "best?" That is a matter of convenience. It depends on the class of objects and the problem under consideration.

For example, Ernst Abbe found it convenient to decompose a general

OBJECTS MAY BE DECOMPOSED IN MANY WAYS, DEPENDING ON THE PROBLEM

ABBE:

$$u(x) = \int A(\nu)\cos\{2\pi\nu x + \varphi(\nu)\}d\nu$$

COSINE GRATING *RESOLUTION*

(FOURIER):

$$u(x) = \int \tilde{u}(\nu)\cdot e^{2\pi i\nu x}\, d\nu$$

PRISM *SPATIAL FILTERING*

HUYGENS:

$$u(x) = \int u(x')\delta(x-x')dx'$$

POINT

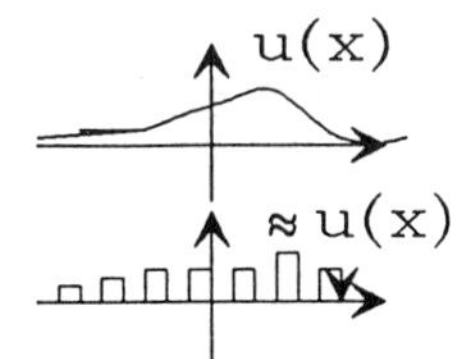

YOUNG:

$$u(x) = \int_{-\infty}^{x} \frac{du(x')}{dx'}dx' - \int_{-\infty}^{+\infty} \frac{du(x')}{dx'}E(x-x')dx'$$

EDGE

S1 X1 X2 X3 X

$$\approx S_1\cdot E(x-x_1) + S_2\cdot E(x-x_2) + S_3\cdot E(x-x_3) \quad \| \; (S_3<0)$$

$$u(+\infty) = S_1 + S_2 + S_3 = 0$$

FIG. 9. Summary.

object into cosine gratings. That is sensible if one wants to understand the limit of resolution. Abbe used the cosine-Fourier decomposition.

The complex Fourier decomposition can be interpreted as a superposition of many prisms. That viewpoint has been helpful in studies on spatial filtering logic and on interconnection networks such as the perfect shuffle.

It would require many, many more pages to show all those examples for which diffraction philosophies have been (or could have been) useful. Nevertheless, we hope the selection of examples was a convincing proof of convenience.

Reflections on Publishing

Several aspects of publishing deserve our attention: the style, the format and—last not least—the institutional and commercial aspects.

Style has been discussed already in the introduction. The acceptance of quasi-oral formulations would be helpful for conveying concepts and philosophies, not merely facts. Facts belong in data banks predominantly. Only a minimum number of facts is needed for those journal readers who want to be "informed in general." In our Annual Report 1987, I condensed such thoughts when I defined.

The Ideal Optics Journal

(1) One-page limit per article, on loose sheets. Back side for scribbling notes.

(2) On sturdy paper, with holes in it for collecting by the subscriber.

(3) The authors offer to send a detailed laboratory record on request.

(4) Annual index provided by the editor.

(5) Large wastebasket for discarding most of the sheets immediately.

(6) Liberal acceptance of manuscripts by the editor.

Not everyone will like this concept; maybe no one will. However, some kind of action is needed to stop the ever increasing flood of publications.

ICO would be an appropriate participant in the reform movement on scientific publications in optics. ICO is appropriate because it is genuinely non profit, with no permanent employees, yet with considerable insight into the philosophies of publishing in optics.

Dietmar Fey produced the figures and Norbert Streibl provided the Dammann plot. Thanks.

CHAPTER 12

The Essential Journals of Optics

John N. Howard
Newton Highlands, MA

1. Introduction

A dozen years ago, we made a few remarks concerning the essential standard journals serving the field of optics, and their costs to libraries and subscribers. Much of that data now is outdated, and perhaps a fresh examination of the journals of optics would be useful [1, 2].

What do we mean by *optics* these days? To some, it means the old classical optics of Isaac Newton and similar pioneers—the manipulation of light by lenses, mirrors, prisms, diffraction gratings, and interferometers. Others broaden this to include optical radiation of all wavelengths, from the most energetic ultraviolet radiation to the long-wavelength infrared. Still others add x-rays and even gamma radiation on the short wavelength side, and also include sub-millimeter and millimeter radiation on the long wavelength side. Some people feel that optics includes spectroscopy, but many others feel that spectroscopy—especially as applied to the chemical analysis of substances—now is a separate field of its own. Perhaps a rough working definition of optics is that it includes the study of generation, detection, and manipulation of electromagnetic radiation of all wavelengths. Then those who work in a specific subfield might say they work on infrared optics, or ultraviolet optics, or x-ray optics, and so on. Some of these workers consider themselves physicists; others call themselves optical engineers, photochemists, or perhaps even astrophysicists.

ISBN 0-12-289690-4

2. The Science Citation Index

One way to try to identify that group of scientists and engineers who work in optics is to examine the technical journals in which they publish their research; and for this purpose we can utilize some of the information presented in the *Science Citation Index (SCI)*, a voluminous set of indexes published each year (and available in many major libraries) in which all of the citations—the references at the ends of articles in the technical literature—are listed. (Citations are to the source items in a journal; the articles, letters, communications, and so on. They do not include book reviews, patents, abstracts, errata, and nontechnical matter.) In 1988, the most recent year of *SCI* available, about 3,200 standard journals were examined and the references in each article stored and sorted in a computer and then printed out in 18 large volumes of very fine print. In addition to the alphabetical index of all references by author—and cross-indexed by subject—the *SCI* also published two additional volumes called the *Journal Citation Index*, in which the number of citations to a given journal are compiled (an index of cited journal), and the distribution of other journals cited in the references of a given journal also are given (an index by citing journal).

The editors of the *SCI* also sort the journals they analyze by technical category, and one of these categories is optics. They list the 27 journals shown in Table I as optics. (They presumably assign a category to each journal by examining the scope of technical content printed in most journals.) How can we decide whether the SCI editors have made a complete listing? Perhaps most optics papers in India, say, are published in the *Indian Journal of Physics*. Perhaps there is a journal called *Photochemical Reactions* that the *SCI* editors simply listed as chemistry. One way to examine this list for self-consistency is to examine the ranked ordering of citations that occur within each of those 27 journals, as well as the distribution of other journals that cite each of the 27. Optical researchers tend to have favorite journals in which they publish their papers, or use regularly for reference material. So, if one examines the references in a given journal, or determines which other journals frequently cite that given journal, one finds that the handful of journals most frequently cited tend to represent the regular journals of the optics community.

In this chapter, we list each of the 27 journals called *optics* by the *SCI* editors, and also show for each one the total number of citations made in 1988 to that journal in all papers published in the 3,200 technical journals analyzed by the *SCI*. Thus, for example, in 1988 the most frequently cited journal, *Applied Optics*, was cited 9,858 times in all journals, and *Fiber and Integrated Optics* was the least cited, with only 26 citations. We have then included, for each journal, the ranking of the 10 journals most frequently

Table I. Times Cited in 1988 by All Journals

Appl Opt	**9858**	**IEEE JQE**	**7026**	**Electr Lett**	**6350**	**Opt Commun**	**4912**	**Opt Lett**	**4427**
• Appl Opt	2450	• IEEE JQE	1051	• Electr Lett	1210	• Opt Commun	608	• Opt Lett	571
• JOSA A	358	Appl Phys Lett	491	• J Lw Tech	824	Phys Rev A	355	• JOSA B	457
• Opt Commun	315	• J Lw Tech	480	Appl Phys Lett	435	• JOSA B	309	• Appl Opt	352
• Opt Lett	311	J Appl Phys	388	J Appl Phys	316	• Appl Opt	234	• IEEE JQE	290
• Opt Eng	295	• Electr Lett	295	• IEEE JQE	269	Appl Phys B	215	• Opt Commun	250
• J Lw Tech	227	• JOSA B	274	• IEE Proc J	201	• IEEE JQE	205	Phys Rev A	184
J Appl Phys	206	• Opt Lett	270	• Opt Lett	182	• Opt Lett	191	Appl Phys Lett	173
• J Mod Opt	181	• Kvant. Elek	249	• Appl Opt	141	• J Mod Opt	129	• J Lw Tech	142
• IEEE JQE	173	• Appl Opt	236	IEEE T Mic Th	123	• JOSA A	115	Appl Phys B	122
• JOSA B	153	Appl Phys B	225	J Crys Growth	110	• Kvant Elek	114		
49%		56%		60%		52%		57%	

Opt Spektrosk	**2855**	**Kvant. Elekt**	**2836**	**JOSA B**	**2432**	**JOSA A**	**1425**	**J Lw Tech**	**1421**
• Opt Spektrosk	802	• Kvant. Elekt	936	• JOSA B	420	• JOSA A	364	• J Lw Tech	439
J Chem Phys	108	• Opt Spektrosk	180	Phys Rev A	280	• Appl Opt	173	• Electr. Lett	239
Izv An SSR Fiz	91	Izv An SSR Fiz	178	• Opt Commun	120	Vision Res.	98	• IEEE JQE	85
• Kvant. Elekt	87	• IEEE JQE	165	• Opt Lett	114	• Opt Lett	54	• IEE Proc J	49
Usp. Fiz Nauk	79	Pisma Zh Tekh Fiz	102	• IEEE JQE	108	• Opt Commun	51	IEEE J Sel Area C	49
Phys Rev A	75	Zh Tekh Fiz	89	Phys Rev Lett	79	• JOSA B	40	• Appl Opt	44
Phys Rev	73	• JOSA B	78	J Phys Paris	76	Biol Cybern	26	• Opt Lett	43
Zh Eksp Teor Fiz	66	Appl Phys B	67	Appl Phys Lett	74	• Opt Eng	25	Appl Phys Lett	37
J Phys B	49	J Appl Phys	53	Phys Rev B	66	Ann Rev Psych	23	• Opt Quan Elec	33
Chem Phys	47	• Opt Commun	53	• Appl Opt	64	• J Mod Opt	23	• Kvant Elekt	28
51%		67%		57%		62%		74%	

Table I. (*Continued*)

Opt Eng	**1288**	**Opt Acta**	**1036**	**Optik**	**660**	**IR Phys**	**507**	**Int J IR Mil**	**330**
• Appl Opt	214	• Appl Opt	114	• Optik	167	• IR Phys	87	• Int J IR Millim	56
• Opt Eng	148	• JOSA A	110	Scanning Micros	40	Prog Quan Elec	67	Int J Electr	27
• Opt Commun	45	• J Mod Opt	103	• JOSA A	33	J Appl Phys	20	IEEE T Plasma	23
• JOSA B	44	• Opt Commun	66	• Appl Opt	30	J Cryst Growth	20	• IR Phys	21
• Opt Lett	43	Phys Rev A	54	J Vae Soc Tech	29	• Appl Opt	16	IEEE T Microw T	20
J Appl Phys	37	• Optik	34	Ultramicros	26	• Int J IR Millim	16	Rev Sci Instr	14
• JOSA A	36	• JOSA B	25	J Appl Phys	20	Mikrochim Acta	16	J Appl Phys	11
Appl Phys Lett	28	Phys Rep	23	Rev Sci Instr	16	Ferroelectr	15	JGR-Atmos	11
• J Mod Opt	22	Phys Rev B	19	J Elect. Micros	11	Phys Rev B	11	Appl Phys B	10
• J Lw Tech	19	• Opt Laser Eng	18	J Microsc Soc	11	Rev Sci Instr	9	J Mol Spec	9
49%		54%		58%		55%		61%	

Opt Quan El	**328**	**Prog Opt**	**284**	**IEE Proc H**	**267**	**Laser Focus**	**265**	**IEE Proc J**	**193**
• J Lw Tech	33	Phys Rev A	29	• IEE Proc H	65	• Laser Focus Elec	46	• IEE Proc J	63
Appl Phys B	23	• Appl Opt	26	IEEE T Microw T	56	Anal Chem	19	• IEEE JQE	22
• Opt Quan El	20	• JOSA A	20	IEEE T Ant Prop	24	• Appl Opt	14	• Electr Lett	20
• Opt Commun	19	• JOSA B	15	Microw J	8	• IEEE JQE	10	• J Lw T	12
• IEEE JQE	17	• Opt Commun	10	Ant Telecomm	5	Photosynthetica	10	J Phys Chem Solid	6
• Appl Opt	16	• IEEE JQE	8	• IEE Proc J	5	Proq Quan Opt	10	J Phys Paris	6
• JOSA B	15	• J Mod Opt	6	IEE Proc F	4	• Kvant Elekt	8	Ann Telecom	5
• Opt Lett	14	• Opt Lett	6	IEEE T Magn	4	• Opt Commun	8	• Appl Opt	5
• Electr Lett	11	Thin Solid Films	6	J Appl Phys	4	Appl Phys B	6	Ferroelectr	5
• IEE Proc J	11	J Appl Phys	5	• J Lw T	4	J Appl Phys	6	J Appl Phys	5
• Kvant Elekt	11								
58%		46%		67%		79%		77%	

Opt Laser Tech	134	J Opt	123	Image Vision Com	118	Sov J Opt Tech	102	J Mod Opt	97
• Appl Opt	14	• Appl Opt	16	• Comput Vision Graph	28	Izv An SSSR Fiz	19	• J Mod Opt	17
Appl Phys Lett	7	• JOSA A	12	• Image Vis Comput	23	Instrum Exp Tech	8	Phys Rev A	17
• Opt Eng	7	Phys Rev	10	IEEE T Pattern An	8	• Kvant Elekt	8	• Opt Commun	10
• Opt Laser Eng	7	• Opt Eng	9	Parallel Comput	8	J Am Ceram Soc	6	Phys Rev Lett	9
J Phys E	6	Astron Astrophys	8	IEE Proc E	6	Zh Narch Prikl Fot	6	• JOSA A	7
• Opt Commun	6	• J Opt	8	Let Notes Comp Sci	6	• Appl Opt	5	Phys Lett A	6
Appl Phys B	5	Opt Commun	8	Pattern Recoq Lett	6	Ferro electr	3	• JOSA B	3
• Optik	5	• J Mod Opt	5	Phil Trans Roy Soc	4	Meas Tech	3	• Opt Lett	3
Anal Chem	4	• Opt Laser Eng	4	P IEEE	4	• Opt Eng	3	Phys Scripta	3
• J Mod Opt	4	Phys Rev B	4	Pattern Recoq	4			Appl Phys B	2
JOSA A	4								
Opt Lett	4								
54%		68%		82%		60%		79%	

Opt Appl	52	Opt Laser Eng	48	Photon Spectr	39	Fiber Integ Opt	26		
• Opt Eng	16	• Opt Laser Eng	16	Comp Vision Gr	4	• J Lw Tech	3	The journal J Mod Opt is new name of Optica Acta. Entries for both should be combined.	
• Optik	4	Ann Rev Fl Mech	3	Nucl Instr Meth	3	• Appl Opt	2		
• Appl Opt	3	NDT Int	3	SMPTE J	3	IEEE Commun	2		
• J Mod Opt	3	• Appl Opt	2	• Appl Opt	2	• IEEE JQE	2		
Phys Status Solidi	3	Experientia	2	Appl Phys Lett	2	J Appl Phys	2		
Prog Quant Elect	3	• Opt Eng	2	• IEEE JQE	2	• Kvantor. Elekt	2		
• IR Phys	2	Phys Status Solidi	2	J Appl Phys	2	• Opt Eng	2		
J Appl Phys	2	Acta Oto Laryngol	1	• Opt Eng	2	• Opt Quan Elec	2		
• JOSA A	2	Anal Lett	1	Talanta	2				
Thin Solid Films	2	Comp Inter Manuf	1	Am Lab	1				
79%		69%		59%		65%			

• Optics Journals

Called "Optics" by SCI, but Deleted here

J Luminescence	**1519**
J Lumin	236
Phys Rev B	141
Phys Status Solids	102
Chem Phys Lett	68
J Chem Phys	51
J Phys C, Solid St	40
J Appl Phys	38
J Crystal Growth	32
Prog Solid State	31
J Phys Chem	28
50%	

No Optics in List of citing journals

General Interest Journals

J Appl Phys	**33787**	**Appl Phys Lett**	**28589**	**Phys Rev A**	**24388**
Appl Phys Lett	1749	Appl Phys Lett	4542	Phys Rev A	6537
J Appl Phys	6205	J Appl Phys	3696	J Chem Phys	1400
Phys Rev B	1725	Phys Rev B	1569	J Phys B	1349
Jpn J Appl Phys	968	Jpn J Appl Phys	1041	Phys Rev Lett	987
J Cryst Growth	680	J Cryst Growth	1006	Phys Rev B	733
Phys Status Solidi	672	• IEEE JQE	827	JOSA B	710
Thin Solid Films	554	J Vac Soc Tech	672	Phys Lett A	537
J Vac Soc Tech	539	• Electr Lett	497	Nuc Instr Meth	478
J Electrochem Soc	520	J Vac Soc Tech A	497	Phys Scripta	465
IEEE T Magn	516	J Electrochem Soc	400	Phys Rep	462
42%		45%		56%	
No Optics in upper half of citations		5% of citations to JQE and Elect Lett		3% of citations to JOSA B	

In 1988, in the 3200 journals covered by the Science Citation Index, there were 9858 citations of Applied Optics. The top 10 citing journals were *Appl Opt*, 2,450 citations; JOSA A, 358 citations; etc. These top 10 citing journals accounted for 49% of all citations to Appl. Opt.

citing that journal. It is interesting to note that the first 10 citing journals generally account for half of all citations to a given journal: For *Applied Optics*, the first 10 citing journals accounted for 49 percent of all 9,858 citations; for *Fiber and Integrated Optics*, the first eight accounted for 65 percent of the 26 citations (the nine remaining journals all were tied at one citation each). Those few journals that make up the first half of all citations to another journal we shall call the *Related* journals. As we mentioned, the first 10 or 12 journals in general account for about half of all citations to a given journal; as one examines the residual citations, one finds a broad scattering of citing journals bearing little or no predictable relationship to the topic of the journal cited. Even the detailed *SCI* listing cuts off at about the 80 percent point and lists the remaining stragglers as simply a residue. The tables here show the distribution of the citing journals accounting for the first half of all citations for each of the 27 optics journals. One of these journals, *J Luminescence*, was classified by *SCI* as optics. However, there is not a single optics journal in the top 50 percent of all 1,519 citations to that journal; and an inspection of the journals citing *J Luminescence* indicates that nearly all are journals of the solid-state community. (This is shown in Table I.) Furthermore, in the references of *J. Luminescence*, there is no optics journal in the top dozen journals that account for half of all the references in that journal. Therefore, we have corrected our list be removing this journal. It is not really related to optics.

An inspection of the distribution of citing journals also shows that *J Appl Phys*, *Appl Phys Lett*, and *Phys Rev A* frequently appear as citing journals for most of the optics journals. These three publications all are very large and vigorous journals of applied physics and electronics, and they actually represent a much broader area of physics than simply optics. *Phys Rev A* is a standard journal for atomic spectroscopy. Each of these journals is three or four times as large—in number of papers published, or in citations—as any of the standard optics journals. In Table I, we show the distribution of related journals in which these three very popular journals are cited. Only about five percent of all of the citations of *J Appl Phys*, *Appl Phys Lett*, and *Phys Rev A* are from the other journals we are considering as optics. These three journals represent the broader community of applied physics and electronics of which optics is a sub-discipline, and, therefore, we have not considered them as optics journals for the purpose of our present analysis.

The journal *Electronics Lett* is a somewhat different case. It serves as a rapid publication medium for a predominantly European optics and electronics community, in somewhat the same way as *Optics Communications* or *Opt Lett*, and furthermore, it is free of page charges, a mechanism that discourages many authors from publishing directly in *Opt Lett*. About half of the total citations to *Electron Lett* are from the other optics journals on our list, and it also appears as a citing journal for five of the optics journals on our list. Therefore, we have retained this journal in our list of optics journals.

Furthermore, we have added two journals, *IEEE J Quan Electr* and *Kvantovaya Elektron*, that were not included as optics in the *SCI* list, but which appear as citing journals for several of the other optics journals; they also are frequently cited by the other optics journals. They meet our criteria of relatedness, therefore, as being proper journals of the optics community.

So, to summarize to this point, we have started with the 27 journals that *SCI* identified as optics; deleted *J Lumin* as more properly a solid-state journal; and added *IEEE JQE*, *Kvantovaya Elektron*, and *Electron Lett* as journals related to the others. We are left with the journals shown in the tables. These journals are ranked in the order of the number of citations in 1988 in all 3,200 technical journals surveyed by *SCI*, from *Applied Optics*, with 9,858 citations from all journals, to *Fiber and Integ Opt*, with 26 citations. We also show for each of these journals the top 10 *related* journals that account for about 50 percent of the total citations. (At the bottom of each list, we show the actual percentage of citations in the top 10 related journals—it varies from 47 percent to 82 percent of all citations, and averages 62 percent.) The important point to be made here is that simply by examining the list of 10 principal journals that cite a given journal, and then examining whether that given journal in turn cites other journals on our tentative list, we can determine if any two journals seem to relate to the same audience. We have constructed a cluster of journals here for optics; if we had begun with astrophysics journals or remote sensing journals, we could have assembled a similar list of prinicpal journals in those fields. (*Applied Optics*, for example, even might appear on some of those lists for other fields, but it would be only a minor journal relating to another field, as *Astrophys J* is not high on the list of journals citing *Appl Opt*, or vice versa.)

We should also mention the obvious fact that it is to be expected that perhaps 15 or 20 percent of all citations of well-established journals will be self-citations, that is, to previous papers in the same journal. We should also remark that letters journals at the fast-breaking cutting edge of a field, or journals specializing in basic or fundamental physics, are more likely to be cited by applied or engineering journals than vice-versa; thus, *Opt Eng* is more likely to cite *JOSA A*, *JOSA B*, or *Opt Lett* than the reverse.

3. Establishing a Set of Closely Related Journals

How would one go about assembling a set of essential optics journals if *SCI* had not already identified a cluster of journals it called optics? (*SCI* did this, presumably, by reading the scope of each journal; a statement of intended technical coverage that is included in most journals.) One should start with any journal name that seems to suggest that optics is part of its content. Suppose one's first thought was *J Mod Optics*. By looking in *SCI* one would see that *J Mod Optics* was cited 97 times by all journals in 1988. One would also see the names of the 10 journals that most frequently cited *J Mod Optics*. Now examine these 10 citing journals, omitting titles that seem obviously non-optical (*Phys Rev A*, *Phys Rev Lett*, *Phys Lett A*, *Appl Phys B*) and determine for each the total number of citations: *Opt Commun*, 4,912; *JOSA A*, 1,425; *JOSA B*, 2,432; *Opt Lett*, 4,427. Another examination of the 10 highest citing journals of each of these four will turn up *Appl Opt*, 9,858; and *IEEE JQE*, 7,026. (The listings we just mentioned are given conveniently in Table I.)

Appl Opt, with 9,858 citations, has the highest citation number, and further iteration will not yield a higher-ranked optics journal. Now, starting from this most cited journal, construct a set of listings of the 10 most frequently cited journals, as we have done already in Table I. Examine each listing for journals that seem obviously related—say, with optics or electro-optics or photonics in the title—but omit those broad, umbrella-type general interest journals that attempt to serve entire areas of science, such as *Science*, *Nature*, and *J Appl Phys*. (What we are searching for here is the family of journals specifically serving one or two closely related subfields.) After two or three such iterations, one will find by inspection that the citation lists of the various peripheral journals (beyond our tentative list) do not contain any of our related journals on their lists of highly citing journals. For example, on our listings shown in Table I, we should now examine the citations to *J Quant Spectros RA*, *IEEE T Microw Th*, *J Crystal Growth*, *J Phys Paris*, *Vision Res*, *J Vac Soc*, *Ferroelectr*, *JGR-Atmos*, *J Mol Spec*, *Thin Solid Films*, *Anal Chem*, *J Phys Chem Solids*, *Astron Astrophys*, and *Phys Scripta*. We have done this for these journals, and none of our tentative family of optics journals figures substantially on the listings of these journals. One could even discard without looking further any journals whose citation ranking is hopelessly low compared with the sum of the citations to the first five or six tentative journals. In doubtful or marginal cases, one also could check the references of the doubtful journal in the citing index. One also could check related subfields, such as photography, astronomy, spectroscopy, or remote sensing. We have checked the principal journals in those fields and did not uncover any further hidden members of our optics family. Of course, if someone such as *SCI* already has compiled a tentative family of journals, it is less trouble to start with that list.

4. Comments on Individual Journals

We might make a few remarks about some of the individual journals on our list of 29. The letters journals *Opt Commun* and *Opt Lett* are cited about the same number of times and by very nearly the same mix of related journals. The letters journal *Electron Lett* is cited somewhat less by optics journals and somewhat more by applied physics and electronics journals. This indicates that perhaps there are other papers in that journal that are not principally optics.

The journal *Opt Spektrosk* (which exists also in an English translation version as *Optics and Spectroscopy*) has a respectable number of citations (2,855) but half of the principal journals citing *Opt Spektrosk* are Russian language journals not particularly read by the optics community, and the other principal citing journals are general-interest journals in physics and chemical physics. *Opt Spektrosk* also is not a principal citing journal of any of the other optics journals. Therefore, it is not a particularly essential journal in optics and probably should be considered simply a spectroscopy journal. It is included in the current listing partly because *SCI* classified it as optics and also because *Opt Soc Am* is the principal sponsor and administrator of the English translation version.

Optik is an ancient and respectable journal, but its content nowadays appears to be predominantly in microscopy and electron microscopy, rather than optics *per se*.

The *International Journal of Infrared and Millimeter Waves* largely serves as the proceedings of an annual meeting held by the community of workers in the far infrared and the sub-millimeter and millimeter region. The only other optics journal that contains a substantial number of citations (21) to *Int J IR Millim* is the journal *Infrared Physics*.

Progress in Optics is listed by *SCI* because it was cited 284 times by the 3,200 technical journals analyzed by *SCI*. It is not really considered by *SCI* to be a journal, as it is an annual survey of selected topics in optics, and *SCI* does not analyze annuals or semi-annuals. It is a scholarly annual review and survey of topics in optics, written by distinguished authors, but it is not a journal of original technical content.

Opt Laser Technol, *Photon Spectra*, and *Fiber Integ Opt* are trade magazines containing occasional survey and review articles, but are not journals of original technical content.

Sov J Opt Tech is a Soviet journal that also is translated by Opt Soc Am.

In a sense, it is the in-house publication of the Vavilov Institute (the State Optics Institute) in Leningrad, as all of the authors represented in *SJOT* are employed at the Vavilov Institute. This journal is not analyzed by *SCI*. (In the United States, we would regard this as an in-house publication somewhat like, say, the RCA Review.)

The journal *Optica Acta*, which was cited 1,036 times in 1988, recently has changed its name to *J Mod Opt*, which received an additional 97 citations. The citations to these two names, therefore, should be combined as 1,133 in the calculation of Table III.

5. Index by Citing Journal

The *SCI* also indexes all of the references in each journal and sorts them into a ranked ordering according to the journal being referenced. This produces a set of listings very similar to the cited journal listings already discussed. In Table II, we present the Index by Citing Journal for the same 29 journals relating to optics. We also rank the top 10 journals referenced in each journal. In Table II, we see that in 1988, *Appl Opt* contained 11,466 references. The distribution of the top 10 journals referenced in *Appl Opt* also is shown, with 2,450 self-references to *Appl Opt*, 493 to *JOSA*, 451 to *P SPIE*, etc. The top 10 journals referenced account for 46 percent of all 11,466 references. One can see that the set of principal journals referenced by *Appl Opt* is very similar to the set of principal journals that cited *Appl Opt* in Table I.

As was observed in Table I, we find also in Table II that the 10 most frequently cited journals in the references make up just about half of the total journals referenced. We have marked with a dot those journals on our list of 29 optics journals. For most of the optics journals, the frequently cited journals consist largely of other optics journals, plus three or four applied physics or electronics journals, such as *Phys Rev Lett*, or *J Appl Phys*.

6. Marginal and Peripheral Journals of Optics

The discipline we call optics is served by a variety of journals, some mostly optics, and others partially overlapping adjacent disciplines. Thus, if one mentally pictures the discipline served by a journal as a circle more or less centered on optics, on the edges or outside of optics there are other domains called spectroscopy, or electronics, or physiological optics, or radiative transfer, or atmospheric optics, and so on. Some of the journals we consider as optics may overlap into some of these adjacent areas, just as some of the journals serving electronics or spectroscopy or other areas may

Table II. References in 1988 Journals: (Citing)

Appl Opt	**11466**	**JOSA B**	**7237**	**IEEE JQE**	**7188**	**Elec Lett**	**5689**	**Opt Spektr**	**5624**
Appl Opt	2450	Phys Rev A	710	IEEE JQE	1051	Electron Lett	1210	Opt Spektr	802
JOSA	493	Opt Lett	457	Appl Phys Lett	827	Appl Phys Lett	497	Zh Eksp Teor Fiz	208
P Soc Photoopt Inst	451	JOSA B	420	J Appl Phys	327	IEEE JQE	295	J Chem Phys	203
Opt Lett	352	Phys Rev Lett	395	Phys Rev Lett	299	J LωT	239	Kvant. Elekt	180
Appl Phys Lett	283	Appl Phys Lett	379	Opt Lett	290	IEEE T Microw Th	122	Phys Rev A	144
IEEE JQE	236	Opt Commun	309	Electr Lett	269	IEEE T Elec Dev	112	Chem Phys Lett	102
Opt Commun	234	IEEE JQE	271	Phys Rev B	230	J Appl Phys	110	Zh Prikl Spektr	102
J Appl Phys	215	JOSA	173	Opt Commun	205	Opt Lett	109	Uzp Fiz Nauk	100
Opt Eng	214	Appl Opt	153	Appl Opt	173	Jpn J Appl Phys	91	Fiz Tverd Tela	95
JOSA A	173	Phys Scripta	153	Phys Rev A	172	IEEE T Commun	90	J Phys B	89
Electron Lett	141	J Appl Phys	144	Kvant Elekt	165	IEEE T Ant Prop	76	Phys Rev	85
		J Phys B	141			Appl Opt	75		
46%		51%		56%					
Kvant Elekt	**5397**	**JOSA A**	**5069**	**Opt Commun**	**4729**	**J LW Tech**	**4540**	**Opt Lett**	**4055**
Kvant Elekt	936	JOSA	521	Opt Commun	608	Electron Lett	824	Opt Lett	571
IEEE JQE	249	JOSA A	364	Appl Opt	315	IEEE JQE	480	Appl Phys Lett	329
Appl Phys Lett	210	Appl Opt	358	Phys Rev Lett	251	J Lω Tech	439	Appl Opt	311
Zh Eksp Teor Fiz	175	Vision Res	335	Opt Lett	250	Appl Phys Lett	398	IEEE JQE	270
JET P Lett	135	Opt Commun	115	Phys Rev A	233	Appl Opt	227	Phys Rev Lett	197
Prisma Zh Tech Fiz	124	Invest Ophth Vis	112	IEEE JQE	222	Opt Lett	142	Optics Commun	191
Appl Opt	121	Optica Acta	110	Appl Phys Lett	175	J Appl Phys	108	Electr Lett	182
J Appl Phys	115	P SPIE	79	JOSA	124	AT&T Tech J	95	JOSA B	114
Opt Commun	114	J Physiol Lon	68	JOSA B	120	JOSA	68	Phys Rev A	111
Opt Lett	110	Opt Lett	68	J Appl Phys	88	P SPIE	66	J Appl Phys	103
		IEEE T Ant Prop	50	J Chem Phys	74	IEEE T Microw	61	P SPIE	70
Opt Eng	**2283**	**J Mod Opt**	**2234**	**IEE Proc J**	**1299**	**Opt Quan Elec**	**1048**	**Optik**	**1041**
Appl Opt	295	Appl Opt	181	Electr Lett	201	Appl Phys Lett	135	Optik	167
P SPIE	225	Opt Commun	129	IEEE JQE	144	IEEE JQE	125	Appl Opt	68
Opt Eng	148	Phys Rev A	124	Appl Phys Lett	100	Electr Lett	93	JOSA	51
JOSA	58	Phys Rev Lett	119	IEE Proc J	63	Appl Opt	66	Opt Acta	34

Appl Phys Lett	54	JOSA	104	IEEE Elect Dev	62	Opt Lett	53	Opt Commun	31
Opt Lett	54	Optica Acta	103	J Appl Phys	58	Opt Commun	46	Ultrasucroscopy	19
IEEE JQE	39	Appl Phys Lett	88	Appl Opt	57	J Lw Tech	33	P SPIE	16
P IEEE	36	IEEE JQE	81	J Lω Tech	49	Opt Quan Electr	20	Opt Eng	15
Electr Lett	32	Opt Lett	60	IEEE J Lω T	33	J Appl Phys	19	Acta Phys Sinica	13
Opt Commun	28	J Appl Phys	42	IEEE T Microw T	33	Phys Rev Lett	19	J Appl Phys	13
J Appl Phys	26								
IEE Proc H	**995**	**IR Phys**	**847**	**Int JIR Millim**	**760**	**Opt Laser Eng**	**561**	**Image Vis Comp**	**472**
IEEE T Ant Prop	185	IR Phys	87	IEEE T Micro T	95	Appl Opt	146	Comp Vis Graph	47
IEEE T Microw Th	133	J Appl Phys	74	Int J IR Millim	56	P SPIE	23	Image Vis Comp	23
IEE Proc H	65	Appl Phys Lett	36	IEEE JQE	24	Exp Mech	21	IEEE T Pattern A	44
Electron Lett	60	Appl Opt	23	IR Millim	24	Opt Commun	20	Pattern Recoq	16
AT&T Tech J	18	Int J IR Millim	21	IEEE T Ant Prop	20	Opt Acta	18	Artif Intell	12
P IEEE	15	J Vac Soc Tech A	16	Int J Electr	18	Opt Eng	17	IEEE T Syst Man	10
Radio Sci	15	IEEE JQE	15	Appl Opt	17	Opt Laser Eng	16	Cylo	
J Appl Phys	13	Phys Rev	15	Astrophys J	17	JOSA A	13	Digital Image Proc	9
P I Elec Eng	13	IR Millim	13	Appl Phys Lett	16	Opt Lett	13	IEEE Comp Graph	9
P IRE	12	Opt Commun	13	IR Phys	16	JOSA	12	IEEE T Comput	9
								Int J Robot Res	9
				Not listed in		**No Optics, top**			
Laser Focus El	**299**	**J Opt**	**108**	**Sci Citation Index**		**10 cited Journals**		**Appl Phys Let**	**19937**
• Laser Focus Elect	46	• JOSA	15	Fiber Integ Opt		J. Appl Phys		• Appl Phys Lett	4542
• Appl Opt	16	• Appl Opt	10	Opt Appl		J Lumin		J Appl Phys	1749
P SPIE	16	• J Opt	8	Opt Laser Tech				Phys Rev B	1092
Appl Phys Lett	10	Nature	4	Photon Spectra				Phys Rev Lett	999
• Opt Commun	8	• Opt Commun	4	Prog Opt				Jpn J Appl Phys	715
• Opt Lett	7	• Optik	4	Sov J Opt Tech				• IEEE JQE	491
• Electr Lett	6	Functions Spheriq	3					Unpub	448
								• Electr Lett	435
• Laur Focus E/O	5	• Optica Acta	3					J Crys Growth	408
Nucl Instr Meth	5	Rev Sci Instr	3					J Electrochem Soc	357
IEDM Tech Dig	4	J Across S Sci	2					Less than 5% optics	

Table II. (*Continued*)

Borderline and Marginal Journals Cited 1988 by all Journals

Vision Res	**4572**	**IEEE T Microw Th**	**2897**	**J Quant Spectros Rad Tra**	**1794**	**JGR–Atmos**	**760**
Vis Res	891	⋮		J Q Spectr Rad Tr	389	JGR Atmos	363
• JOSA A	335	• Elect. Lett	122	J Chem Phys	126	Atmos Env	36
Inv Ophth	180	• Int J IR Millim	95	• Appl Opt	115	Geophys Res Lett	32
Clinic Vis	153	⋮		Phys Rev A	101	J Atmos Sci	29
Percept Psych	150	• J Lω Tech	60	J Mol Spec	93	J Atmos Chem	23
Exp Brain Res	140	Total Optics	277	Spectrochim Acta	53	• Appl Opt	14
J Comp Physiol	129			Astrophys J	46	J Phys Chem US	14
J Neurosci	126			Chem Phys Lett	43	Nature	14
J Neurophys	124			Int J Heat Namstr	43	JGR-Oceans	13
J Physiol	108	9% of citations optics		Astron Astrophys	38	Env Sci Tech	12
7% of citations JOSA A				6% of citations Appl Opt		2% of citations Appl Opt	

No Optics Journals in Top 50% of Citations

Astron Astrophys	Phys Lett A	J Mol Spec
Astrophysical J	J Phys Paris	Non Destr Testing
Applied Spectroscopy	J Vac Soc	Remote Sens Env
Anal Chem	J Crystal Growth	Photogram Eng & Rem Seus
Ferroelectrics	Thin Solid Films	Physicascripta
IEE Proc H	J Phys B At Mol Opt	AT&T Tech J
	J Phys Chem Solid	

partially overlap optics. A big busy journal such as *Applied Optics* may partially overlap several adjacent areas. Such overlap, where journals have fuzzy boundaries, makes it more elusive to define just what we mean by an optics journal; we have used the criterion here of examining only the first 10 or 12 journals that account for half of all cross-references to a given journal. This helps to eliminate the fuzzy overlap of a journal with its neighboring journals. Many of the journals have individual personalities concerning their technical content—specific characteristics that reflect their individual history or the topical interests of their audience. Thus, *JOSA A* overlaps the physiological journals relating to vision and visibility; *Opt Spektrosk* interfaces strongly with spectroscopy; *JOSA B* interfaces with quantum optics; *IEEE JQE* and *J Lightwave Tech* overlap with other areas of electronics, and *Appl Opt* has some overlap with atmospheric optics.

How one chooses a cutoff point—in deciding when a journal ceases being an optics journal and is simply a neighboring journal overlapping into optics—is a somewhat arbitrary decision. In this study, we have excluded those journals that have very few or none of our optics journals among the 10 journals accounting for half of all citations, and even when an optics journal is cited, its fraction of the total citations is small (say, less than 10 percent). Thus, *Vision Research*, with 4,572 citations, was cited 335 times by *JOSA A*, but all of the other frequent citations are physiological journals. The journal *Phys Rev A*, with 24,388 citations, was cited 710 times by *JOSA B*, but all of its other citing journals are applied physics. *J Quant Spectros* and *Radiation Transfer*, with 1,794 citations, was cited 115 times by *Appl Opt* (but conversely, *J Quant Spectr RA* was only a very minor citation in the references of *Appl Opt*). *JGR Atmos*, with 760 citations, was cited 14 times by *Applied Optics*. *Phys Scripta*, with 3,422 citations, was cited 153 times by *JOSA B*. *AT&T Tech J*, with 2,315 citations, was cited 95 times by *J Lightwave Tech* and 44 times by *Appl Opt*, for a total of 139 citations from other optics journals. *IEEE Transactions on Microwave Theory*, with 2,897 citations, was cited 60 times by *J Lightwave Tech*, 95 times by *Int J IR and Millimeter Waves*, and 122 times by *Electronics Lett* (for a total of 277 citations from our optics family) but all three of these journals already are strongly overlapped into electronics, so that it is only a very marginal optics journal at best.

All of the other journals we have checked have none of our optics family among their frequent citing journals. These include *Phys Lett A*, *J Phys Paris*, *J Vac* Soc, J Crystal Growth, Astron Astrophys, Thin Solid Films, Ferroelectr, J Phys B At *Mol Opt*, *Anal Chem*, *Appl Spectros*, *IEEE Proc H*, *J Phys Chem Solid*, and *J Mol Spec*. It is our feeling that even if we somehow have overlooked a possible journal serving optics, that neglected journal would have been only a very minor contributor.

7. Summary

We have used the information in the *Science Citation Index* to survey the citations of 29 journals in which a large part of the papers relate to optics. (See Table III.) Articles in these 29 journals were cited 49, 439 times in 1988 by the 3,200 journals analyzed by *SCI*, but about half of these 49, 439 citations were made to papers that appeared in this cluster of 29 journals, which, therefore, constitute a family of related journals serving the field of optics. Further analysis of citation data indicates that some of these journals are much more important than others; in fact, 95 percent of the citations to this family of journals are represented by the top 13 or 14 journals; or, to say this another way, if one completely neglected or ignored the lower-ranked half of this family of related journals, one presumably would lose only five percent of the articles considered worth citing. One obtains substantially the same mix of important journals whether one analyzes the citations to that family of optics journals made by all journals, or conversely, one analyzes the references in the articles contained in this family of 29 journals. (There were 70,380 references in all 29 journals in 1988, but 95 percent of these references occurred in the same cluster of 14 higher-ranked journals.) This smaller cluster of 14 journals, therefore, represents the *sine qua non*, the essential set of important journals that form the solid core and by far the largest portion of the standard periodical literature serving optics.

Not very long ago, in the 1940s and 1950s, the Journal of the Optical Society of America was the leading journal in optics by far. By some estimates nearly 70 percent of all papers on what most of us would call optics appeared in JOSA. But nowadays optics is a much bigger and more diverse field, and there are many new journals and even other societies active in optics. JOSA now contains a much smaller fraction of published optics. Even so, the five journals of the Optical Society (Appl Opt, JOSA A, JOSA B, Opt Lett, and J Lightwave Tech) still together account for 40 percent of all citations——by all 3200 journals——to the journals representing optics.

We should say again that we have been dealing here with statistics and rough quantitative relationships that do not really measure quality. One should also reflect that it is quite possible for excellent papers to appear in journals of lesser prestige, just as it sometimes happens that a below-average paper may manage to appear in a journal of very high prestige. Furthermore, even our cluster of 29 important journals does not include all optics papers—a very significant optics paper may appear with a highly unlikely title in some unexpected, highly unlikely journal. The only way to be certain of seeing them all would be to subscribe to all 3,200 standard

Table III. Ranking of Optics Journals

Cited range	1988 Citations by all Journals	Cited: Name of Journal	1988 Source Items			Citing range	Number References in Journal	Citing: Name of Journal	
	9858	Appl Opt	744				11466	Appl Opt	
	7026	IEEE JQE	278				7188	IEEE JQE	
	6350	Electr Lett	1051				4540	J Lightwave Tech	
	4912	Opt Commun	394				4055	Opt Lett	
	4927	Opt Lett	371				5689	Electr Lett	
	2855	Opt Spektrosk	425				7237	JOSA B	
	2836	Kvantovaya Elekt	475				4729	Opt Commun	
	2432	JOSA B	315				5069	JOSA A	
	1421	J Lightwave Tech	233				5624	Opt Spektrosk	
Upper 95%	1425	JOSA A	259				2283	Opt Eng	
Upper 95%	1288	Opt Eng	174			Upper 95%	2836	Kvantovaya Elekt	
Upper 95%	1036 / 97	Optica Acta			Now Named J Mod Opt	Upper 95%	2234	J Mod Opt	
Upper 95%	660	Optik	96			Upper 95%	1299	IEE Proc J	
Upper 95%	507	Infrared Phys	64			Upper 95%	1048	Opt Quant Elec	
Lower 5%	330	Int J IR Millim	87	16% residue		Upper 95%	1041	Optik	
Lower 5%	328	Opt Quant Electr	38	16% residue		Lower 5%	995	IEE Proc H	
Lower 5%	267	IEE Proc H	83	16% residue		Lower 5%	847	Infrared Phys	
Lower 5%	284	Proq Optics		16% residue		Lower 5%	760	Int J IR Millim	
Lower 5%	265	Laser Focus Elect	92	16% residue		Lower 5%	561	Opt Laser Eng	
Lower 5%	193	IEE Proc J	65	16% residue		Lower 5%	472	Image Vis Comp	
Lower 5%	134	Opt Laser Tech	37	16% residue		Lower 5%	299	Laser Focus Elec	
Lower 5%	123	J Optics	10	16% residue		Lower 5%	108	J Opt	
	118	Image Vision Comput	29			Lower 5%		Optica Acta	Now Named J Mod Opt
	102	Soc J Opt Tech	155					Sov J Opt Tech	Not in SCI
	(97)	J Mod Opt	155		formerly Optica Acta				
	52	Opt Appl	14				70380	Total	
	48	Opt Laser Eng	36						
	39	Photon Spectra	114						
	26	Fiber Integr Opt	30						
	49439	Totals	5824						

journals and look carefully at every paper. However, if your time and resources are limited, our cluster of 10 or 15 principal journals will expose you at least to the mainstream of optics papers—a very respectable cross-section of the whole.

References

1. John N. Howard, *Applied Optics* **16**, A173 (1977).
2. John N. Howard, *Applied Optics* **17**, A41 (1978).

CHAPTER 13

Optics in China: Ancient and Modern Accomplishments

Zhi-Ming Zhang
Lab of Laser Physics & Optics, Fudan University
Shanghai 200433, China

1. Introduction

The long, recorded history of China shows great scientific and technological achievements. Unfortunately, the individual intellectuals responsible for such contributions received only sorrow. Great inventions known worldwide—such as the armillary sphere, the compass, and gunpowder—were made during the period from several decades before the birth of Christ to the 11th century A.D. During this time, discoveries also occurred in the area of optical technology, which generated interesting results. Due to the long ruling feudal monarchy, however, creativity was suppressed until the mid-20th century. With relatively recent developments, Chinese intellectuals finally can learn and be recognized.

Within the past decade, Chinese physicists and opticists have had opportunities to interact frequently with foreign scientists through visits and international conferences. The Chinese Optical Society (COS), with its membership of approximately 5,000, organizes conferences for information exchange among the scientific and industrial communities.

Currently, several Chinese Academy of Science research institutes are very active in optics and optical engineering. Much work is being done on modern optics, laser physics, laser and optoelectronics, and the development of optical technology for applications. In addition, there are active university research groups focusing on basic problems in optics as well as optical and laser physics.

ISBN 0-12-289690-4

With regard to education, China's universities and institutes admit around 200 master's degree candidates annually. Of these, the top-level students can continue as candidates for Doctor of Philosophy in either optics or optical engineering, but only a limited number of Chinese universities and institutes are authorized to grant doctorates.

This chapter will discuss both the ancient and modern accomplishments in optics and optical technology. Also, it is the author's intention to introduce recent Chinese achievements, in light of the aforementioned oppression. This discourse has been prepared to remind the Chinese optics community that past history is no excuse for complacency. The reader may find much detail in some areas, but only partial detail in others. An apology is tendered beforehand for this flaw.

2. Accomplishments in Optics in Ancient China

Ancient achievements in optical science and technology by the Chinese people can be found in much literature, both classic and modern. Representative of the classic literature was the work of Shen Kuo (1033–1097 A.D.) of the Northern Song Dynasty [1]. In his book, Shen Kuo not only collected quite comprehensive details of the achievements prior to and during his own time, but also provided very valuable personal comments on these accomplishments. This book, originally written in classical Chinese literary style, now has been rewritten in modern Chinese style, and most people familiar with modern Chinese can understand it. Another monumental contribution was that of Dr. Joseph Needham of Cambridge, England, "Science and Civilization in China" [2]. It introduced the brilliant accomplishments and scientific ideology of ancient China. Needham's book surely promotes people's understanding of China.

Ancient Chinese achievements in optics and optical technology were very exciting and interesting. Though all of them were in the regime of classical geometrical optics, the respective authors' discussions of the results were quite logical and successful. The following subsections describe several examples.

2.1. Discovery of the Pinhole Camera Principle

The discovery of the *pinhole camera* phenomenon [1] was recorded in the literature of Mohism, founded by ancient Chinese Philosopher Mo Ti in the period of the Warring States of China (770–221 B.C.). Mohism said that a hawk or wild goose flew over a cottage and had its moving image cast on the floor through a light-leaking hole in the roof. Mohism explained that the hole in the cottage roof segregated the light. The light inclined from the East above the roof could pass only through the hole and be directed to the West; thus, the object and its image would move in opposite directions to each other.

One finds that this is the basic principle of imaging of a pinhole camera. The explanation for the image moving opposite to that of the object comprised the implicit idea of the rectangular propagation of light rays. It is marvelous that the ancient Chinese people of 2,500 years ago were so careful in observing the optical phenomena and so successful with their explanation by logical deduction.

2.2. On the Fabrication of Concave and Convex Mirrors

A contemporary of the philosopher Mo Ti was Huai Nan Zi, who described in his work an implement called the *sun-flint*. This device could make a fire when faced towards the sun. In the book of Shen Kuo (item 44) [1], he stated that the surface of this flint was concave and the image formed by the flint always was inverted when facing some object at a distance, and that all the light from the flint would aggregate to a tiny spot. Actually, this sun-flint was an ancient reflecting concave mirror, the tiny aggregated spot was its focus, and when this flint faced the sun, ignition would happen if some object was placed at that spot.

In another passage of his book (item 327) [1], Shen described the fabrication of a bronze concave mirror for ancient women's personal adornments. The book stated that when large-sized mirrors were cast, the surfaces of the mirrors remained flat as they gradually cooled down, but for small-sized mirrors, the surfaces usually became convex. It also stated that a concave mirror would give an enlarged image and a convex mirror would form a contracted one. The conclusion was that small mirrors used for adornments should be made with a little bit of convexity, so that people might use them to observe the whole image of a person's face. In his book, Shen claimed that ancient Chinese technicians were skillful enough to fabricate mirrors with proper size and with enough convexity to meet the requirements of good image and convenience.

The preceding descriptions were written during the 11th century. At that time, according to Chinese historical records, bronze mirrors were commonly used by royal and wealthy families. The accomplishments in fabricating these mirrors attained such maturity that one can conclude that the basic principle inherent in imaging with mirrors then was understood at least empirically.

2.3. The Miraculous Light-Penetrating Mirror

The so-called *light-penetrating mirror* [3] was an ingenious and fantastic achievement in optical and metallurgical techniques in ancient China.

The rear side of this light-penetrating mirror has a pattern with 16 or 20 classical Chinese epigraphs engraved around its periphery. The front side is an ordinary smooth reflecting surface with a little bit of convexity. However, if one faces this mirror to the sun in a dim room and casts the reflected beam on the wall, the rear-side pattern will be displayed clearly in the resulting spotlight on the wall. This really is a very curious optical phenomenon. It seems that a part of the light beam first penetrates through the front metal surface into the substrate of the mirror and then gets reflected from the rear side. The reflected light carries the information of the rear-side pattern, forming the image on the wall. This light-penetrating mirror really is a very mysterious and miraculous antique because it seems to violate the ordinary laws known in optics.

Only a few of these antique mirrors were found in China and some had been taken to Japan. Now, the Shanghai Museum keeps at least three intact and undamaged light-penetrating mirrors and several pieces of broken parts. According to the archaeologists' investigation, these mirrors were cast in the years of the Western Han Dynasty (207–9 B.C.), but the technique has been lost since. Though many people in the following decades tried to recover this sophisticated accomplishment, no one suc-

FIG. 1a. The *light-penetrating mirror* (duplicated by Fudan University). The left part shows the smooth reflecting surface of the mirror, and the right shows the pattern on the rear side of the mirror.

ceeded. It is only in recent years that research groups at Fudan University and at Jiaotung University in Shanghai have duplicated this accomplishment successfully. In addition, a group in Japan also has succeeded [4]. Figure 1a shows the front side and the rear side of a light-penetrating mirror duplicated by the Fudan group in 1975. Figure 1b gives the demonstrated display of the image from this duplicated mirror.

The inherent physical principle for this light-penetrating mirror, however, has not been explored completely yet. Different groups give different explanations according to their own experiences. Several modern techniques have been applied in investigation, including surface morphography,

FIG. 1b. The pattern on the rear-side of the mirror is displayed on a wall by holding the mirror in a sunbeam.

holographic study, optical testing, and composition analysis including metallographic and x-ray fluorescence analysis. The Chinese and Japanese scientists all agree that the thinner areas of the metal substrate cool down faster than the thicker areas, causing the former to become slightly convex and the latter slightly concave. The reflecting light from the concave areas has the effect of concentrating the rays, while the convex parts will diffuse them. These irregularities inherent in the rear-side pattern are so slight, they are imperceptible with the naked eye. However, the detailed procedures and mechanism for producing such an irregularity do not lead to a unified concept. One group claimed that heat treatment, followed by a bath of cold water for swift cooling with further polishing, would produce this "penetrating" effect, whereas the other group claimed the imaging property was created during their careful and sophisticated casting procedure. Nevertheless, the duplication of a light-penetrating mirror has been successful, regardless of the fact that the mechanism for producing such an ingenious optical phenomenon might need further detailed study.

3. Recent Progress of Optical Science in China

In recent years, interesting progress on optical science in China has occurred. This information can be found in Chinese journals and proceedings, most of which include in English summary and English captions for figures and tables. However, in recent years, more and more papers are being published in foreign English journals. This is a very good trend in that it enables more people outside China to understand the progress of Chinese science and technology directly.

3.1. Achievements in Astronomical Optics

In the 1950s, Chinese astronomers only had telescopes with small-sized apertures. One was the 40 cm telescope located at the Shanghai Observatory, another a 60 cm one at Nanjing Observatory. Now, Chinese opticists and optical engineers can design and fabricate large-aperture telescopes for astronomical study even up to an aperture size of two meters.

In 1989, a 1.55 m telescope for celestial measurements was set up at the Shanghai Observatory. This is one of the largest telescopes in the world for celestial measurements; the other is the 1.55 m telescope at the U.S. Naval Observatory. The relative aperture is 1:10 and the coma-free field of view is about $\varnothing 30'$ in angular diameter without a corrector. With the corrector, the coma-free field of view is $\varnothing 1°$.

A 2.16 m Cassagrain telescope is operating now at the Beijing Observatory, having been designed and fabricated by the optical engineers at the Nanjing Astronomic Instrument Factory. This telescope is of the Ritchey-

Chretien (R-C) type. With its relative aperture of 1:9, it also can be used alternately with the axis-folded configuration. The effect of such a folded system is achieved by inserting a relay mirror in the optical system. This is quite a successful achievement in that the auxiliary mirror of the Cassagrain system is kept in position, but still keeps the R-C system of the telescope and the folded-axis system free from spherical and coma aberrations. This telescope has been in operation at Beijing Observatory since 1988.

Another valuable accomplishment in astronomical optics is the design and fabrication of a birefringent filter for measuring the solar vector magnetic field and the sightline velocity field. This sophisticated filter actually is a kind of monchromator with a wide field of view that is suitable for telescopic observations. The filter has 151 plates 37×37 mm^2 in size, composed of materials like calcite, quartz, KD*P, polaroids, and wave plates. It consists of three Lyot elements and two Solc elements. The filter can receive many signals at the same time for real-time observation under polarization-free conditions. The accuracy for measuring the solar longitudinal magnetic field is about +10G. For the solar transverse magnetic field, the accuracy is +100G, and for solar sightline velocity field, the accuracy is +30m/sec.

3.2. Achievements in Lasers and Nonlinear Optical Crystals

There are several groups in China now doing research work on laser and nonlinear optical crystals. A research group at the Beijing Institute of Physics has grown the optorefractive crystal $BaTiO_3$ since 1987, and also studied its optical properties and performance in nonlinear optical phenomena [5]. They obtained an efficiency of 50% in self-pumped phase-conjugated reflectivity and an enhancement of 20,000-fold amplification for two-laser-beam coupling. The group from Shandong University at Jinan studies the growth and properties of crystals composed of yttrium aluminum borate (YAB) and either neodymium aluminum borate (NAB) or erbium aluminum borate (EAB). This work investigates crystals for multifunctions such as lasing and self-frequency-doubling in that same crystal. This group also works on high-quality KDP and KD*P crystals for nonlinear optical purposes.

The most active group in the study and growth of laser and nonlinear optical crystals is from the Fujian Institute of Research on Structure of Matter. They succeeded in growing the BBO crystal in 1979 and analyzed it with a theoretical model called the anionic group theory in hopes of obtaining higher nonlinear susceptibility of the crystals. They have succeeded again in growing a new crystal in the borate series: the lithium triborate (LBO) crystal. This crystal has a second-harmonic-generation coefficient comparable to that of BBO but with more advantages, such as high damage threshold up to 25GW (at 1.06u, 0.1ns), insensitivity in angular matching, and low absorption in the ultraviolet region. They recently succeeded in third-harmonic generation of Nd:YAG laser radiation with this crystal to obtain 355 nm radiation with an energy conversion efficiency of about 60% [6].

3.3. Research on Soft X-Ray Lasers

Early in the 1980s, a group at the Shanghai Institute of Optics & Fine Mechanics (SIOM) started to study x-ray lasers from highly ionized plasmas as the active media produced by a high-power laser. They obtained population inversion of the $n = 3$ and $n = 4$ levels of Mg^{10+} ions (154.8A) due to recombining pumping by using a high-power Nd:glass laser (100ps, 10^{11}W) with a given pre-pulse irradiating a magnesium slab target at the Six-Beam Laser Facility of SIOM. They observed population inversion not only in the plasma plume normally far from the target surface, but also, for the first time, in a new region of higher electron density near the target surface. This group also studied experimentally the formation and time evolution of the various plasma-jet structures in line-focused laser-produced plasma. It has been indicated that these inhomogeneous structures in plasma may initiate the turbulence and other nonlinear effects that seriously will influence propagation of x-ray laser radiation and the measurement of gain in the line-shaped plasma. They had obtained gain experiments of the Li-like Al ions and also obtained evidence of amplification of spontaneous emission for 5f-3d (10.57 nm) and 4f-3d (15.46 nm) transitions. In 1989, they performed soft x-ray laser experiments on Li-like Si ions at the LF12 Laser Facility at SIOM and obtained 5f-3d (8.89 nm) and 5d-3p (8.73 nm) transitions, and also at 6d-3d (7.58 nm) as well as at 6d-3p (7.46 nm) transitions [7]. These results, especially for the new x-ray laser transition of $n = 6$ levels, will renew greatly the understanding of recombining pumping x-ray lasers of the Li-like ions. In addition, their experimental results, obtained at very low driving laser intensity (10^{12}W/cm^2), have shown that the potentiality of the Li-like ion recombining pumping scheme, in driving the soft x-ray laser, might be very promising.

3.4. Research on Laser Physics

Most of the basic studies on laser physics are carried on by research groups in universities [5]. Short laser pulses are generated with the colliding pulse mode-lock (CPM) technique and around a 70-femtosecond duration has been achieved both at Fudan University at Shanghai and Tienjin University at Tienjin. The Fudan group also uses a copper ion metal vapor laser to amplify short pulses in a study of the transient phenomenon of the interaction of a laser with condensed matter [8]. Nonlinear optical studies of surfaces also are subjects of interest that have been carried on at Fudan University, where adsorption and desorption mechanisms on metal surfaces were studied [9]. A research group on laser physics in the Institute of Physics at Beijing also is very active in the field of degenerate four-wave mixing, and of chaos and optical bistability [10, 11]. Studies of laser physics and nonlinear optics are spread widely both theoretically and experimentally in the research groups at universities and research institutes. Many papers contributed by such groups have been published in journals and proceedings.

4. Conclusion and Acknowledgments

It is quite interesting to review the achievements in optics and optical science in ancient China, in that these could encourage the Chinese people to work harder. However, it also is very promising to see that in the past decade, Chinese scientists have worked hard to learn modern techniques. In addition, they have strived to train more students to make up for past oppression. It is hoped that in the coming decade, China will be more aggressive and interactive with the international optics and optical science communities.

The author would like to express his gratitude to Professor Pan Du-wu of Fudan University for his discussion of ancient accomplishments in optics in China, and to professors Yang Shi-jie of Nanjing Observatory and Xu Zhi-zhan of SIOM for sharing their recent results from astronomical optics and x-ray laser studies. Finally, the author would like to express his deep gratitude to the editor of this book, "International Trends in Optics," for inviting this manuscript. This provides an opportunity to bring more information to scientists worldwide in their efforts to understand Chinese scientists and their progress.

References

1. K. Shen, "Selected Work on Shen Kou's 'Meng Xi Bi Tan' (Notes on the Stream of Dreams)," Shanghai Ancient Chinese Literatures Publishing House, 1978.
2. Jixiang Pan (ed.), "Collected Papers of Joseph Needham," Liaoning Science and Technology Publishing House, Shenyang, 1986.
3. Q. Lin, "Uncovering the Secret of the 'Magic Mirror.'" *China's Reconstructs* (in English), 27, (Feb.-March, 1977).

4. Hideya Gamo, "Magic Mirrors: Optics, Technology and History," Demonstration and Poster Session WXl, OSA Annual Meetings, New Orleans Oct. 17–20, 1983.
5. Z. J. Wang and Z. M. Zhang (eds.), *Proceeding of Topical Meeting on Laser Materials and Laser Spectroscopy*, Shanghai, China (published by World Scientific Publishing Co. Pte. Ltd., Singapore) (July 23–25, 1988).
6. B. Wu, N. Chen, C. Chen, D. Deng, and Z. Yu, "Highly Efficient Ultraviolet Generation at 355nm in LiB_3O_5," *Optics Letters*, **14**, 1080 (1989).
7. Xu Zhizhan, Zhang Zhengquan, Fan Pinzhong, Chen Shisheng, Liu Lihuang, Lu Peixiang, Wang Xiaofang, Qian Aidi, Yu Jiajin, and Feng Xianping, "Experimental Study on Gains of Soft X-ray Lasers," *Scientia Sinica* (Series A) (in Chinese) no. 1, 27 (1990).
8. H. M. Ma and F. M. Li, "Picosecond Pulse Amplification by Wave Mixing in GaAs," *Appl. Phys. Lett.* **54**, 1953 (1989).
9. Li Le, Liu Yanghua, Yu Gongda, Wang Wencheng, and Zhang Zhiming, "Optical SHG study of Oxygen Adsorption on a Polycrystalline Ag Surface," *Phys. Rev.* **B40**, 10100, (1989).
10. Xin Mi, Haitian Zhou, Ruihua Zhang, and Peixian Ye, "Multilevel Theory of time-Delay Four-Wave Mixing with Incoherent Light in Absorption Bands" JOSA **B6**, 18, (1989).
11. Guang Xu, Jian-Hua Dai, and Hong-Jun Zhang, "Transition to Chaos from Quasiperiodicity in Optical Bistability with Competing Interaction," *Phys. Rev.* **A39**, 3508 (1989).

CHAPTER 14

Unusual Optics: Optical Interconnects as Learned from the Eyes of Nocturnal Insects, Crayfish, Shellfish, and Similar Creatures

Pál Greguss

Department of Non-ionizing Radiation, Frederic Joliot-Curie, National Research Institute for Radiobiology and Radiohygiene, Budapest, Hungary

1. Introduction

In general, biologists most often use analogies from physics and engineering to explain natural phenomena, while it is very uncommon for engineers to borrow from nature. When I was an eight- or nine-year-old boy—60 years ago—my father, a botanist, wrote a fascinating book entitled, "The Wonderful Life of Plants" [1], teaching that not only architects and bridge designers, but even those involved in aeronautics could learn a lot about how to master nature's laws by studying and trying to understand the physics (and chemistry) behind the observed biological phenomenon. I, however, became more interested in the life and *behavior* of the animal world.

Thus, I found myself confronted for the first time with problems associated with the fact that we are living and moving around easily in a 3-D environment, although all of our receptors for the signal-carrying waves—operating independently whether they are of electromagnetic or mechanical nature—are square-law detectors; i.e., depth-related phase-bound information should be lost (but it is not!). I then remembered my father's saying: Look for cues in the laws of physics governing the orientation strategies of life.

ISBN 0-12-289690-4

This way of thinking helped me through studying the behavior of echolocating animals to create the first acoustic hologram ever made [2] and to understand the signal-processing role of the pecten in the eyes of birds. [3]

2. Signal-Collecting Optics

Some years ago, I asked myself how human behavior would have developed if our retina as a sensor processed all the signal-carrying light waves from the 3-D space surrounding the living body and not only those restricted by the *see through a window* (STW) strategy (Fig. 1). Naturally, this means that we would need another type of *signal-collecting optic* (SCO), somewhat similar to the so-called fish-eye optics. However, present fish-eye optics used in photography are not adequate for such a psychological study, since by using fish-eye optics, we do not quit the STW strategy; as a consequence, when turning around to obtain a 360-degree panoramic view, the result is a distorted perspective.

According to the STW concept, there are several vanishing points, all lying on one line called the *horizon* (Fig. 2), and if a fish-eye-type SCO is used, the main disturbing distortions result from the fact that the horizon does not remain a straight line; instead, it becomes curved, which is the consequence of the fact that to map the sphere of vision, we would need an infinitely large plane, and no sensor surface can be considered as such.

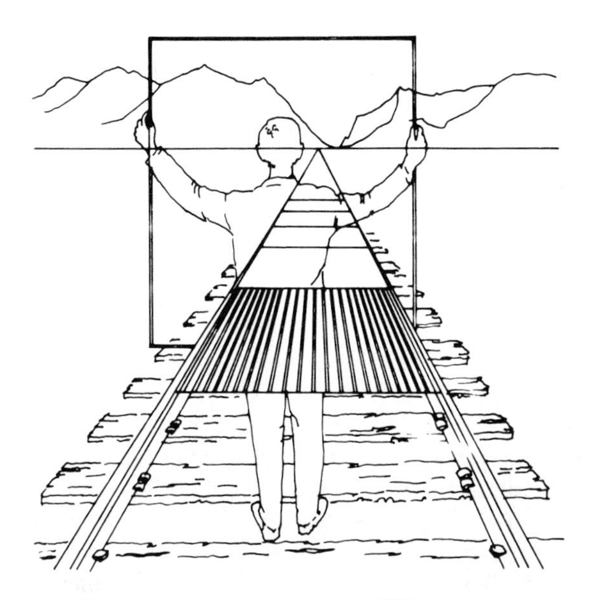

FIG. 1. *See through a window* (STW) concept.

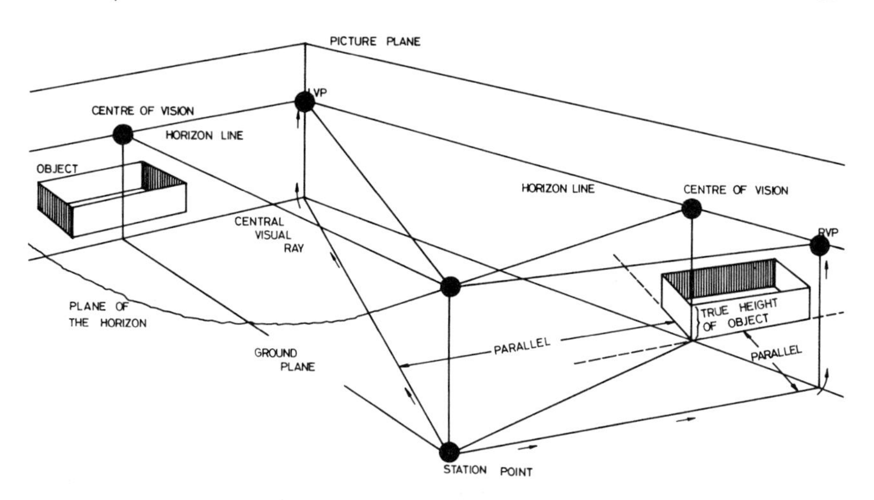

FIG. 2. Vanishing points are lying on a line called the *horizon*.

This problem perhaps could be solved by using some appropriate scanning-mapping technique, but in this case, a time factor also would be involved in the signal-processing procedure. However, time factors may influence psychophysical effects considerably; therefore, I decided to take a look at some of the really unusual SCOs of the animal world, which perhaps render a panoramic view without distortion by using a single vanishing point.

3. Strange Biological SCOs

There is a temptation to think of the difference between biological and technical optics as just a matter of material composition. The superficial resemblance between the human eye and a camera obscures the fact that there are other SCOs of design aspects that are basically different in how they convey the signal pattern originating from the 3-D environment. Thus, depending upon the *how*, the creature has to use different signal-processing schemes—starting from the retina as the sensor surface—and various *interconnect* strategies. I am convinced that there is much here the engineers may learn yet from zoologists.

Perhaps one of the most fascinating SCO designs used in the animal world is that of the scallop (Pecten and related genera). (Fig. 3) It forms the image by a concave mirror rather than a lens; i.e., it is the biological analog of the Newtonian telescope. In this case, the reflected signal pattern lies on the retina consisting of two layers; according to Hartline [4], the first

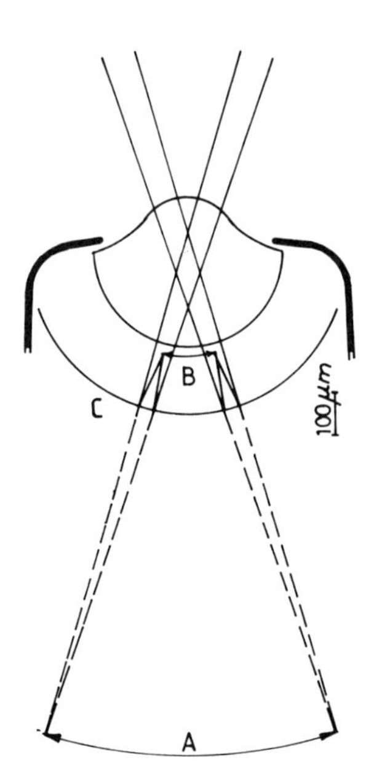

FIG. 3. Optical diagram of a scallop eye: A = retina, B = theoretical image plane, C = reflecting concave surface.

one is an *off* responding layer and lies immediately behind the lens, whereas the other, an *on* responding layer, is in contact with the mirror. While the role of the off layer may be interpreted as a feature that enables the scallop to see trouble coming and shut its shell, the function of the on layer still is not understood. We suggest that it may serve—at least partially—as an *interconnect*; further, it also may be involved in that type of signal processing that is described in Dennis Gabor's "Theory of Communication," and is characterized by the phrase: "It is necessary to know not only the past, but also the future."

With the preceding remark, I wish to refer to Gabor's elementary signal theory, the concept of *logon*, which perhaps may provide explanation for the strange optical architecture of the crab's vision system. The *biologon* signal-processing strategy probably allows the simultaneous *time analysis* and *frequency analysis* of the signal [5], which may be of vital importance in a really 3-D environment, such as water. Without going into detail, we only recall that a logon is the modulation product of any frequency with a pulse of the form of a probability function. To generate it with sufficient accuracy, the transmission of the signal has to be delayed by a small amount, which can be described, indeed, as "knowing not only the past, but also the future." The transmission of the unfocused light through the retina may serve this purpose well.

The advantage of this SCO strategy is that it allows a more panoramic view than observed by lens-based SCOs; however, it suffers from the fact that the resulting image has to have rather poor contrast, since focused light reaching the retina already has passed through it once unfocused.

Thus, the question may be raised whether the two SCO strategies, the lens and mirror, could not be combined in such a way that the concave parabolic mirror inspects the partially focused beam of a short focal lens and recollimates it on the receptor field, the retina, so that, as a result, a high-contrast panoramic view is produced. It seems that Mother Nature's answer is yes, since Nilsson has found a crab whose SCO is functioning as an intermediate of the lens-type and mirror-type SCO [6].

Our investigations showed that this architecture considers its surroundings—at least in first approximation—as cylindrical rather than spherical, as the fish eye does. As a result, the visual field is not conceived immediately as a flat surface, but as an image projected first onto a cylindrical surface and then transformed onto a flat surface, as shown in Fig. 4. The resulting annular image provides a two-dimensional skeleton of the geometric relations of the three-dimensional environment; i.e., the vanishing phenomenon of the three-dimensional visual space is abstracted and represented as a two-dimensional phenomenon of convergence: It has only one single vanishing point. Therefore, it is called a *flat cylinder perspective* (FCP). As a consequence of FCP, distortions of fish-eye-type SCO architectures are avoided. Thus, why not design such crab-eye-type optics?

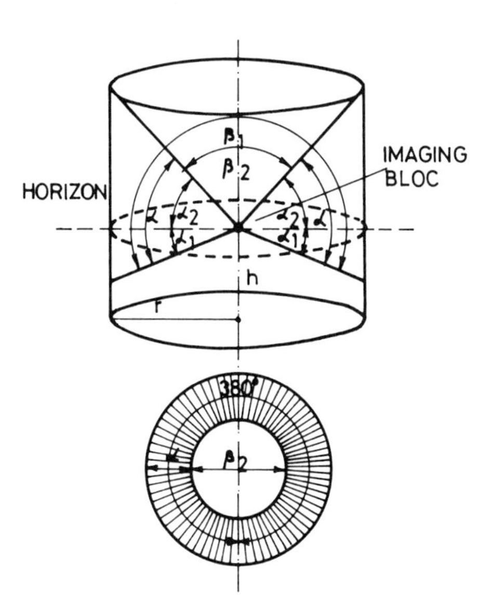

Fig. 4. Creating a 2-D skeleton of the geometric relations of the 3-D environment.

4. The PAL Optics

A 1:1 copying of the crab-eye architecture previously described would lead to various problems in evaluating the resulting panoramic image: According to this design, a real image is formed on the receptor field (retina) that is located, however, *inside* the SCO architecture. What type of a sensor surface should be used in this case, how does one read out the resulting electrical signals, etc.?

To avoid the aforementioned problems, we decided to modify this architecture so that a virtual image of the surrounding 3-D environment is formed in the lens-mirror system, and then this is projected by optical means to any type of sensor surface. [7]

As shown in Fig. 5, this is accomplished by using a single spherical refracting surface (1)—which is responsible for the first step of the panoramic image formation—instead of two, which is the situation in the crab-eye architecture. By using two parabolic reflecting surfaces—one concave parabolic mirror (2) for creating in the second step the real panoramic image, and a convex parabolic mirror (3) to convert this real image into a virtual panoramic image—the panoramic annular virtual image is formed inside the PAL optic. This virtual panoramic image then can be projected by a lens to any sensor surface in any dimension.

The width of the resulting annular image of the three-dimensional environment (Fig. 6)—which is the reason this SCO architecture is called *Panoramic Annular Lens* (PAL)—corresponds to the viewing angle in the direction of the SCO's optical axis, and each concentric ring in the image

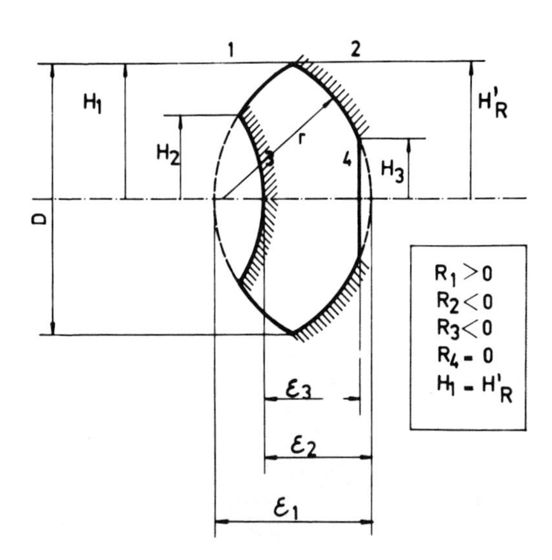

FIG. 5. The basic layout of a PAL SCO.

plane is the locus of points recorded at constant field angles perpendicularly to the optical axis.

The beauty of this architecture is that it kills two birds with one stone. The PAL can be made of a single block, since the reflecting surface—being opposite to each other—may be formed on the glass block itself; this means that the rays of light are changing their directions within a homogeneous medium rather than passing through media of various refractive indices, and this results in a more compact construction. Further, provided that the curvatures have the correct form and distances from each other, the PAL image practically will be free from spherical and chromatic aberration—as can be seen in Fig. 6—since the refracted rays are reflected twice before leaving the imaging block.

FIG. 6. PAL annular image showing flat-cylinder perspective (FCP).

5. From PAL Imaging to PAL Interconnects

That the PAL images with their single-point FCP really are the two-dimensional skeleton of geometric relations of the three-dimensional environment can be demonstrated well by reprojecting such a PAL annular image through a PAL optic onto a cylindrical surface. As can be seen from Fig. 7, the optical impression one may get really will be a panoramic one.

Since the demand for high-efficiency optical interconnects is increasing steadily, we thought: Why not use the unusual optical features of the PAL lens [7] as a SCO to do such a job, perhaps in a more convenient form than it is done today and also faster, allowing free-space interconnection for

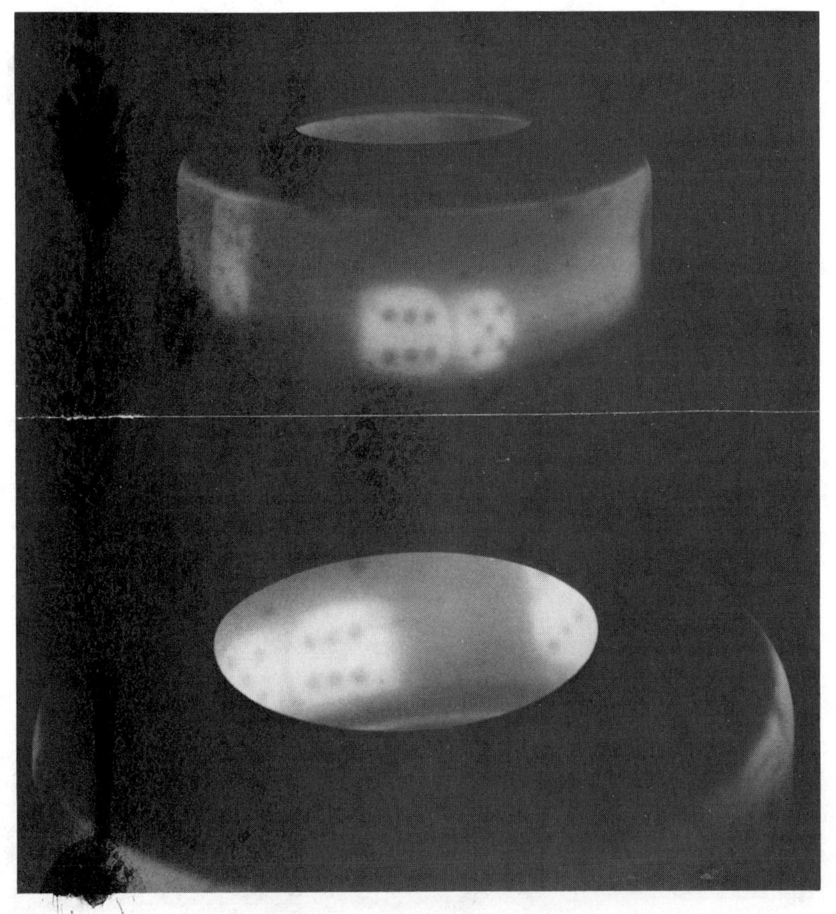

FIG. 7. Reprojected PAL images create a panoramic impression.

each of U users, giving independent access to any stored bit displayed on a small array of spatial light modulator (SLM) to overcome the limited capacity of a single SLM?

I wonder whether Tajima, Okada, and Tamura, when designing their multiprocessor system called DIALOG:H (Distributed computing Iconic programmable Asynchronous Logic-based Optical-basing General purpose: Hardware) [8], were aware of the fact that the principle they used can be found in nature, too. For example, their SCO architecture greatly resembles that used by some crayfish and shrimps. Although this interconnect system allows greater flexibility and speed over conventional bus design, the number of interrogating points is limited architecturally due to the restricted *viewing angle* of the cylindrical mirror. The introduction of PAL optics as SCO may overcome this limitation of DIALOG:H, since, as already indicated, it transforms the panoramic information inherent in a cylindrical volume onto a flat surface, creating a two-dimensional representation of the three-dimensional cylindrical volume surrounding the optics.

To illustrate the idea behind the PAL optical interconnect, let us suppose that we image a SLM in one direction and address it in another. If a vertical line is scanned across the SLM at a rate of P pixels per second, detectors at each of the row-condensed points then will read out the data stored in the corresponding row at bandwidth P. Electrical switches then can connect the selected row to the output line. If the scanning source has moderate or high coherence, holograms can be used in the SLM image plane to produce multiple readout lines.

However, instead of using a single flat SLM, we propose using a cylindrical SLM—or, what is equivalent to it, a set of flat SLMs arranged in a polygonal form—around a PAL [1], as shown in Fig. 8.

In this case, the three-dimensional panoramic arrangement of the SLM(s) can be imaged via a lens (3) onto a linear detector array (6). If this linear detector array coincides with the radius of the annular image of the SLMs, multiple output lines can be produced by a rotating dove prism (5) that is placed between the PAL and the detector array [9].

The advantage of this architecture is that:

(a) The interconnects do not require physical carriers (fibers, wires, etc.).

(b) Electromagnetic interference is precluded.

(c) Although time delay is scale-dependent, it is on the order of nanoseconds, and there is no capacitive slowing of propagation.

(d) Equal pathlengths between any two points lead to negligible (subpicosecond) clock skew.

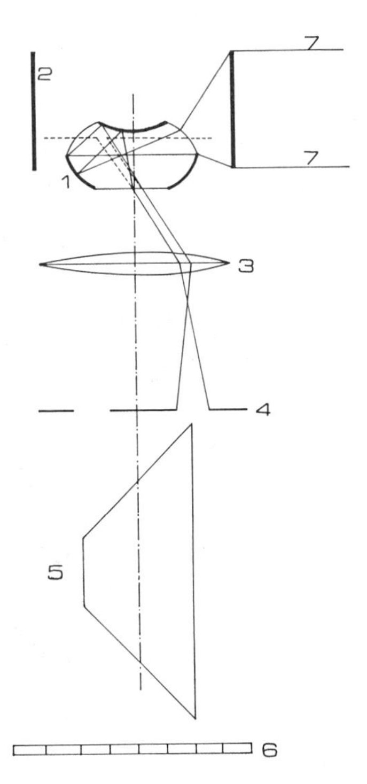

FIG. 8. Optical diagram of the PAL interconnect architecture.

(e) Depending upon the number of pixels of the linear detector array and the rotation speed of the dove prism, readout can occur at rates up to the GHz range, without losing parallel access by many users.

Naturally, one has to accept—at least at present—a trade-off between storage capacity and update speed. Considering magneto-optic SLMs, very fast operation can be expected; however, one has to take into account a limited capacity/SLM unit (512×512), while in the case of liquid crystal SLMs, huge capacity (10^8 pixels) but rather slow updating will be available [10].

6. Conclusions

We have demonstrated that studying the eyes of creatures living in a real three-dimensional environment, such as water, leads, not only to an optical system suitable for imaging the 3-D environment without scanning, but also to an element in the central part of an optical interconnect architecture on which optical neural networks can be built.

Acknowledgments

The author wishes to express his gratitude for the fruitful discussions he had with H. John Caulfield of CAO UAH, Jacques Ludman of Rome Air Development Center, and Ch. von Campenhausen of the University of Mainz.

References

1. P. Greguss, "A növények csodálatos élete," Franklin Társulat, Budapest, 1932.
2. P. Greguss, "Ultraschall Hologramme," *Research Film* **5**, 330–337 (1965).
3. P. Greguss, "Do Biological Systems Use Coded Aperture Technique?" *Proc. IEEE Int. Opt. Com. Conf.*, Zurich, 45–48 (1974).
4. I. K. Hartline, "The Discharge of Impulses in the Optic Nerve of Pecten, in Response to Illumination of the Eye," *J. Cell. Comp. Physiol.* **11**, 465–477, (1938).
5. D. Gabor, "Theory of Communication," *J. IEE* **93**, III, 429–457 (1946).
6. D. E. Nilsson, "Facets of Vision" (R. C. Haardie, and D. Stavange eds.), Springer Verlag, Berlin, 1989.
7. P. Greguss, USA Pat. No. 4.566.763.
8. S. Ishihara, "Recent Advances in Optical Computing in Japan," *SPIE Optical and Hybrid Computing* **634**, 31–50 (1986).
9. P. Greguss, Pat. pending.
10. H. J. Caulfield, P. Greguss, "Cylindrical Symmetry Lens Interconnect," *SPIE OE/Fibers '89: Optical Interconnects in the Computer Environment*, Boston (1989).

CHAPTER 15

The Opposition Effect in Volume and Surface Scattering

J. C. Dainty
Blackett Laboratory
Imperial College
London, UK

1. Introduction

In recent years, there has been renewed interest in the so-called *opposition effect*. This term refers to the enhanced intensity that is observed exactly in the backscatter or retro-reflection direction when certain volume and surface scatterers are illuminated with parallel light; it also is referred to sometimes as *enhanced backscattering*. There are many media that exhibit the opposition effect, such as paper, white paint, and even the surface of the moon. (The moon really *is* brighter when it is full [1, 2].) Materials used for diffuse reflectance standards, such as barium sulphate, also show the opposition effect. A similar phenomenon can be seen from scattering by clouds (the "glory" or "spectre of the Brocken" [3]), although the mechanism in this case is not the same as that for the examples discussed in this chapter. Figure 1 shows the effect for the case of scattering of a linearly polarised beam by a dense suspension (about 10% by volume) of latex spheres in water. The upper half shows a photograph taken in the backscatter direction and the lower half shows the co- and cross-polarised intensity as a function of scattering angle. The peak intensity in the backscatter direction in this case is up to twice that of the local background.

Observation of the effect is well-documented [4], but it is only recently that the mechanisms have begun to be understood. In this chapter, we

ISBN 0-12-289690-4

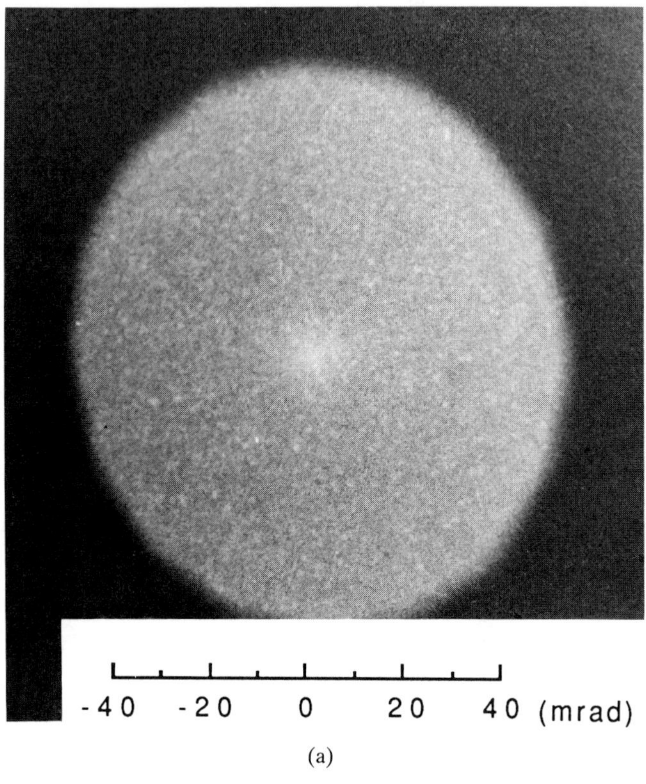

(a)

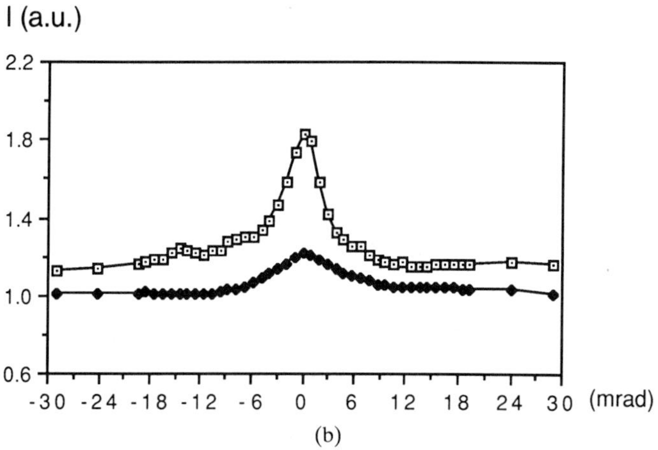

(b)

FIG. 1. Enhanced backscattering (opposition effect) by a dense suspension of latex spheres in water (Fig. 2 of *D N Qu* and J. C. Dainty, *Opt. Lett.* **13**, 1066–1068, 1988). The upper photograph is taken in the backscatter direction (co-polarised component) and the lower half shows the co- and cross-polarised intensities as a function of the scattering angle.

concentrate on one particular mechanism, the coherent co-operative effect of multiple scattering. This probably is the dominant mechanism for dense scatterers like white paint, but in other cases, there are geometrical optics effects, such as focussing, that also can give enhancement. To give some structure to the chapter, I shall discuss the cases of dense volume media, randomly rough surfaces, and double passage through a random screen in separate sections; the key unifying feature in all three cases is the occurrence of *multiple scattering*.

2. Scattering by Dense Volume Media

The scattering by many *surfaces* such as paper and paint is, in fact, predominantly a volume effect. To understand the scattering process, experiments often are carried out using controlled concentrations of monosized latex spheres suspended in water [5]. Although the opposition effect has been documented since early this century, in recent times the experiments of three groups [6–8] have resulted in a renaissance of the subject. (See [9] and [10] as a starting point for the literature.) The reason for this is not so much the importance of the opposition effect itself, but that the cause of it—multiple scattering— has other fascinating effects—in particular, the possibility of the localisation of photons in a disordered medium.

The basic mechanism for enhanced backscattering is illustrated in Fig. 2. If a linearly polarised plane wave is incident on a random medium, then a *ray* undergoing multiple scattering with particles $1, 2, 3 \ldots N$ has precisely the same complex amplitude as one interacting with $N \ldots 3, 2, 1$; the two *rays* are coherent and thus interfere constructively, giving an intensity up to twice that of the local background, where there are no such coherent effects. A more rigorous description defines each scattering event by a

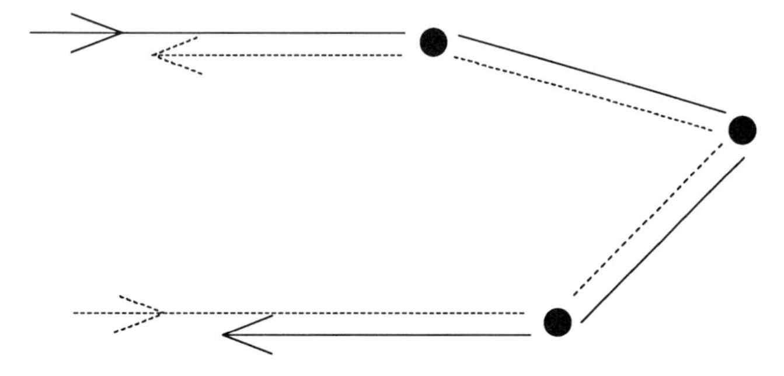

FIG. 2. Basic multiple-scattering explanation of the opposition effect: forward and reverse *rays* interfere constructively in the backscatter direction.

matrix equation, in which an incident polarisation vector is multiplied by the scattering matrix to yield the polarisation vector of the scattered light. Using this approach, one can show that the co-polarised multiple-scattered light is enhanced by a factor of two, but the cross-polarised forward and reverse paths only are partially coherent and give an enhancement factor less than two: in fact, as the degree of multiple scattering increases, the enhancement factor tends to unity for the cross-polarised case.

The width of the enhancement peak is related to the mean transport length of photons in the medium. Long paths give a contribution to the enhancement at very small angles, whereas shorter paths contribute over a wider range of angles. Thus, the shape of the enhancement peak gives information on the length of light paths in the medium. The path lengths also are contained in the polarisation behaviour, since, for the cross-polarised case, the degree of correlation between the forward and reverse paths decreases with increased order of scattering. The path lengths also can be probed directly using femtosecond pulses and observing their broadening after scattering.

The energy in the backscatter peak is very small indeed compared to the total scattered energy in all other directions, so, from an applied point of view, one could reasonably ask: Why is there all this fuss (over 100 publications in the past five years, symposia and workshops, etc.) over such a small effect? The reason probably is the following: The opposition effect is the result of scattering mechanisms that hitherto have been ignored almost completely and that could have profound influence on our understanding of the interaction of classical waves with disordered matter. Electrons can be *localised* in a medium if there is a certain degree of disorder and, in particular, if the mean free path of the electron is less than $\lambda/2\pi$, where λ is the electron wavelength. If one could make such a medium for visible electromagnetic waves, light inside would be *trapped*, and from the outside, such a medium would be a perfect reflector. It is still not clear whether, in fact, such a medium can exist for electromagnetic waves, but the first step, that of weak localisation, has been observed in the manifestation of the opposition effect.

3. Scattering by Randomly Rough Surfaces

There is an enormous amount of literature on the elastic scattering of electromagnetic radiation by randomly rough surfaces. Because an exact general solution of Maxwell's equations cannot be found for this problem, the theoretical treatments have tended to focus on approximate solutions, mainly using perturbation theory or the Kirchhoff boundary condition. For surfaces that are rough compared to the wavelength, virtually all theoretical work, until recently, has assumed single scattering. Experimental work on rough surface scattering has suffered from the fact that the detailed surface statistics frequently are unknown and thus, the measurements are only of empirical value; measurements made on poorly defined surfaces are of no value for checking the validity of approximate theories.

In 1987, E. R. Mendez and K. A. O'Donnell carried out some of the first optical scattering measurements on well-defined surfaces with relatively simple statistical properties [11, 12]. These surfaces were made by exposing photoresist to laser-produced speckle patterns whose statistics are well understood [13]; the surfaces can be coated with metal, typically gold, or replicated to form dielectric or metallised copies [14]. Using this technique, surfaces with a Gaussian probability density of surface height and a Gaussian auto-correlation function can be produced so that, for the first time, surface scattering experiments directly relevant to theory can be performed.

Some of the first experiments conducted by Mendez and O'Donnell were on surfaces of root-mean-square (rms) surface height equal to approximately two wavelengths and correlation length five wavelengths, i.e., surfaces with fairly large rms slope that should exhibit multiple scattering. Figure 3 shows one set of angular scattering curves for a one-dimensional version of such a surface [15]. Note the opposition-effect peak in the backscatter direction. This has the same origin precisely as the peak produced by scattering from dense random media; forward and reverse ray paths are coherent and thus interfere.

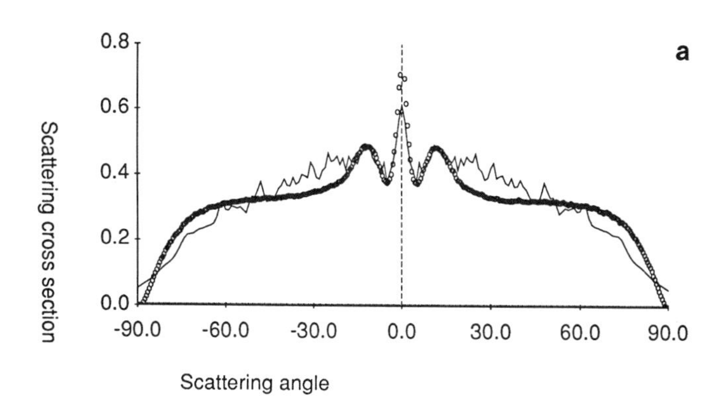

FIG. 3. Angular scattering of light from a one-dimensional gold-coated randomly rough surface of root-mean-square height fluctuation $\sigma = 1.2 \pm 0.1 \mu m$ and Gaussian correlation function of 1/e length $2.9 \pm 0.2 \mu m$ (Fig. 8 of [15]). The angles of incidence are (a) 0°, (b) -10°, (c) -20°, and (d) -40°, and the incident light is *s*-polarised (TE). The circles are experimental measurements and the jagged line the result of an *exact* calculation for a perfect conductor based on direct numerical solution of the Helmholtz equation. Note the enhanced backscatter peak (opposition effect) in the backscatter half-plane on the right hand side of each graph.

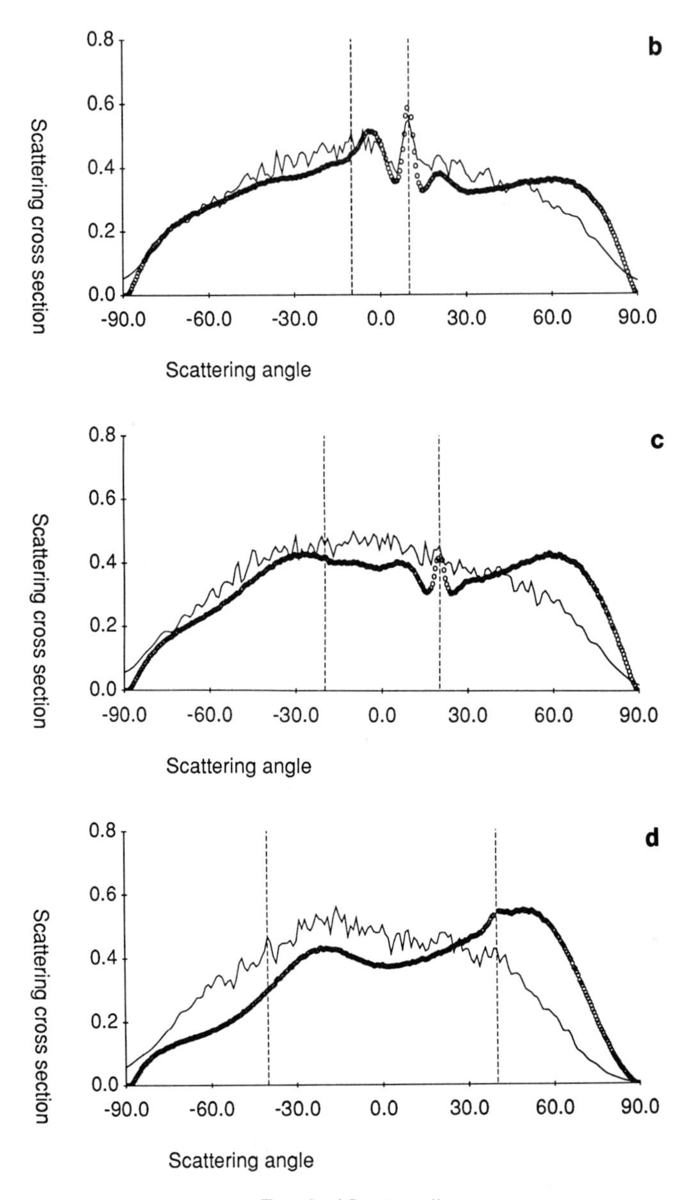

FIG. 3. (*Continued*)

Prior to the experimental work of Mendez and O'Donnell, the opposition effect was unknown in true surface scattering. (Here, we ignore the penetration depth of the wave in a metal, which is on the order of 80Å for visible light and a gold surface.) Their work has stimulated a large number of groups to search for a more "rigorous" explanation of the enhanced backscatter peak; electromagnetic theorists are not very fond of rays! Unfortunately, no analytical solution exists and thus one has to resort to — or exploit—numerical techniques. These give quite good agreement with the measurements [15–18], but otherwise give little physical insight to the nature of the scattering. Figure 3 shows the result of an "exact" numerical calculation for a perfect conductor based on the Helmholtz equation. The agreement is fair, although there is a significant discrepancy at higher angles. (This almost certainly is *not* due to the fact that the calculations are for a perfect conductor, whereas the experiments were done on gold-coated surfaces.)

An alternative approach is to adapt a theory based on approximate boundary conditions, such as the Kirchhoff or *physical optics* theory, to multiple scattering. Figure 4 shows the result of such a calculation carried

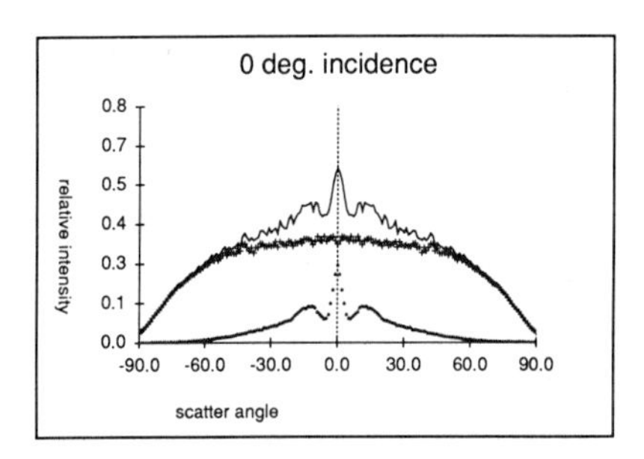

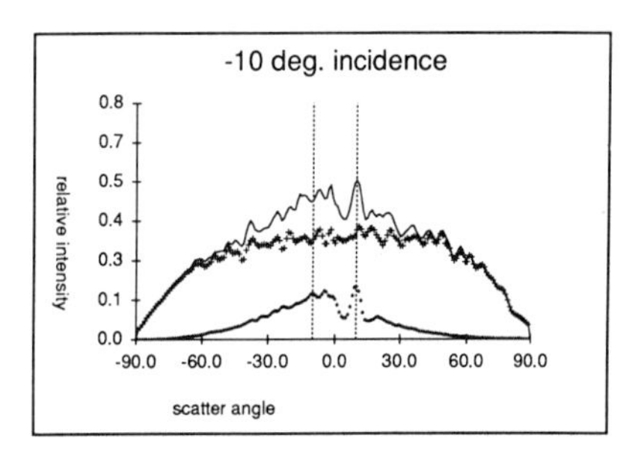

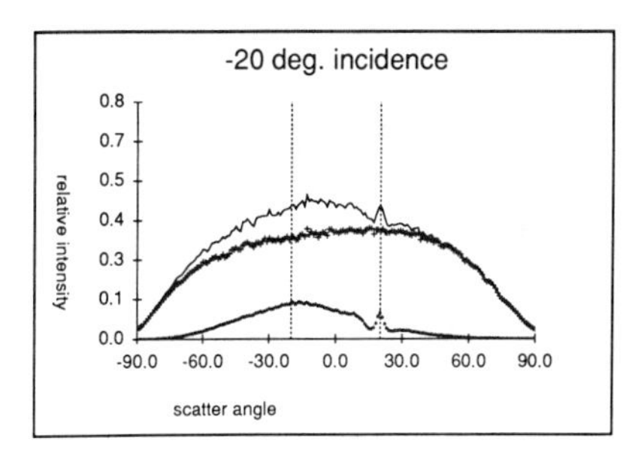

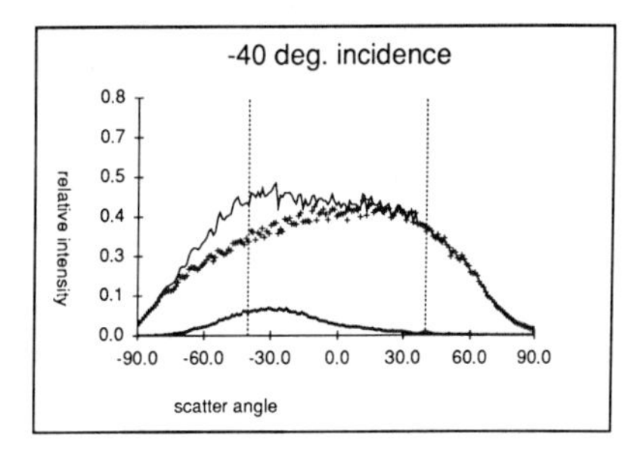

FIG. 4. The single (+++), double (***) and single+double (----) contributions to the angular distribution of scattered light for *s*-polarisation incident, for the same surface parameters as those in Fig. 3, calculated using a double-scattering version of the Kirchhoff approximation.

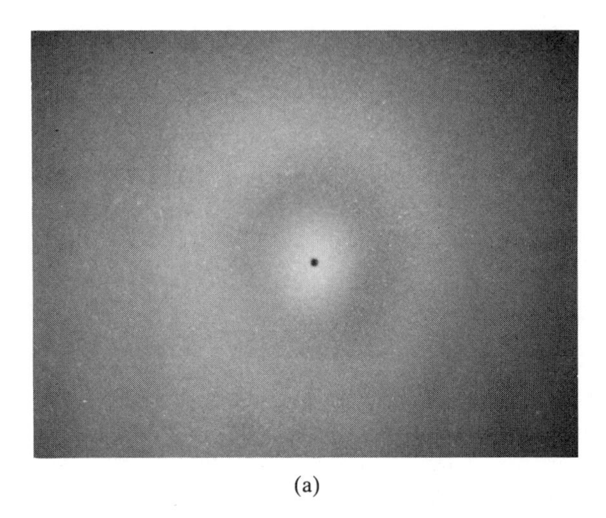

(a)

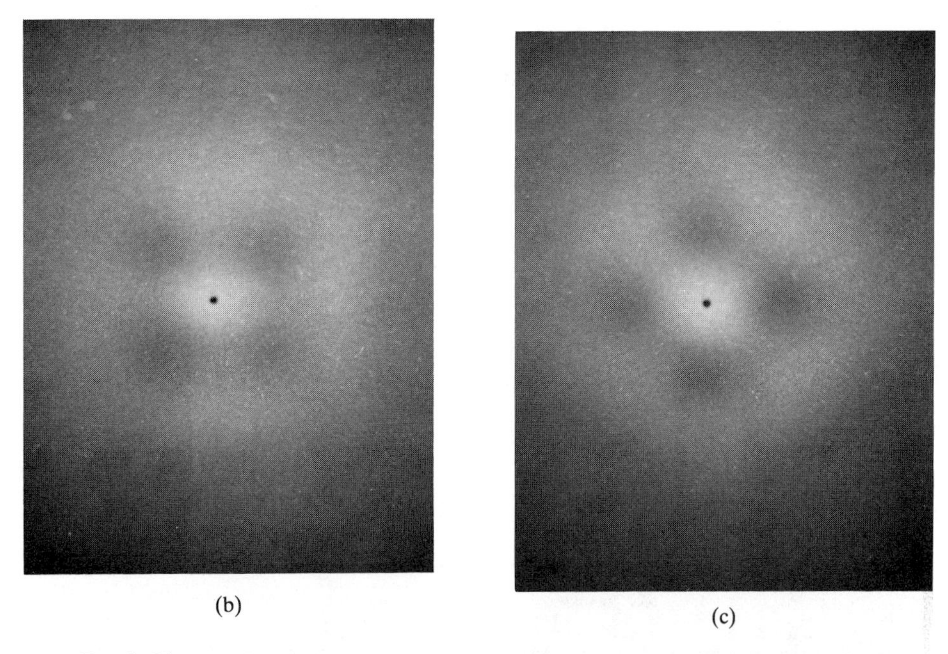

(b) (c)

FIG. 5. Photographs of the backscattered intensity from a two-dimensional randomly rough surface taken (a) with no analyser, (b) through an analyser parallel to the incident polarisation, and (c) with the analyser perpendicular to the incident polarisation (courtesy of E. R. Mendez and K. A. O'Donnell).

out to a double-scattering approximation (including shadowing) for a perfectly conducting surface of the same parameters as that in Fig. 3, illuminated by *s*-polarised (TE) light of wavelength 0.63 μm at normal incidence. The enhanced backscatter peak shows only in the double-scatter term, providing strong support of the ray theory that emphasises the crucial role of multiple scattering.

One important difference between the surface and volume scattering cases is that, in the surface case, only low-order multiple scattering occurs, mainly just second order. This means that the co- and cross-polarised contributions to the enhanced backscatter peak are equally large. For two-dimensional rough surfaces, there are strong polarisation effects. Figure 5 shows photographs of the backscattered light for a surface illuminated by linearly polarised light and viewed (a) with no analyser, (b) with the analyser parallel to the incident electric vector, and (c) with it perpendicular to the incident electric vector. The effect is very striking when observed: as the analyser is rotated through 90°, the scattering pattern appears to rotate 45°. Again, a simple ray argument seems to explain the observations [12].

As with the case of scattering by dense volume media, it is not the opposition effect itself that is particularly important— although it is a fascinating and unusual phenomenon—but, more significantly, the fact that it encourages us to question the *accepted* treatments of a subject. For example, one hears reference to Lambertian scattering surfaces, but it seems certain now that such a thing simply is a figment of the imagination and not realisable by any real surface, because of multiple scattering. Likewise, the angular scattering curves of Fig. 3 are quite remarkable, even ignoring the enhanced backscatter peak; for example, the existence of a large component of light scattered in the backscatter half-space at angles greater than the incident angle is unexpected and counterintuitive.

4. Double Passage through a Random Screen

There is an extensive amount of literature on the double passage of light through turbulence. (See [19] for a partial list.) By analogy with the above cases of surface and volume scattering, one expects to observe the opposition effect, and this has been reported [20, 21]. A practical consequence of this is that diffraction-limited information is present in the average image of a deterministic object that is illuminated and viewed through the same distorting medium. As an example of this, we consider the simplest case, that of a Michelson or two-slit interferometer illuminating and viewing an object through a random (phase) screen, as in Fig. 6.

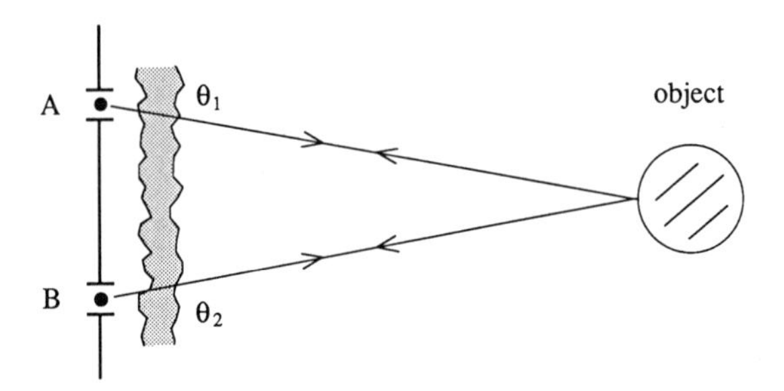

FIG. 6. Double-passage imaging with a Michelson interferometer.

Again, we consider a simple ray approach to explain why diffraction-limited information is present in the image of a point object. Consider light transmitted through slit A towards the object. It accumulates a random phase θ_1 on its passage through the random screen and then is scattered in all directions by the point object. That portion of the light that returns through slit A has a random phase $2\theta_1$ and does not contribute to the diffraction-limited process. However, another portion of the light travels to slit B, accumulating a total random phase of $\theta_1 + \theta_2$. Now consider light that is emitted from slit B. This also will accumulate a total random phase of $\theta_1 + \theta_2$ by the time it arrives at slit A and thus, constructive interference will occur between the forward (A to B) and reverse (B to A) waves. This constructive interference is superimposed on a background level caused by the other paths A to A and B to B.

Figure 7 shows a computer simulation of the fringes in a Michelson interferometer viewing a point source for (a) no random screen, (b) double passage through different random screens (2,000 realisations), and (c) double passage through the same random screen; (d), (e), and (f) show the Fourier transforms of (a), (b), and (c), respectively. Note the high contrast fringes formed in (c) and corresponding side-bands in (f), of magnitude one-half that of the no-random-screen case.

This technique can be generalised to non-redundant and filled imaging pupils [19, 22], with corresponding reductions in fringe visibility.

5. Discussion

The increased understanding of the opposition effect over the past five years has had an impact in at least three areas of physics. First, the importance of multiple scattering has been exposed: this has a profound importance for the behaviour of light in dense scattering media, and may lead to the phenomenon of photon localisation in disordered structures. Dense random media are a new type of scattering that may, in fact, be the

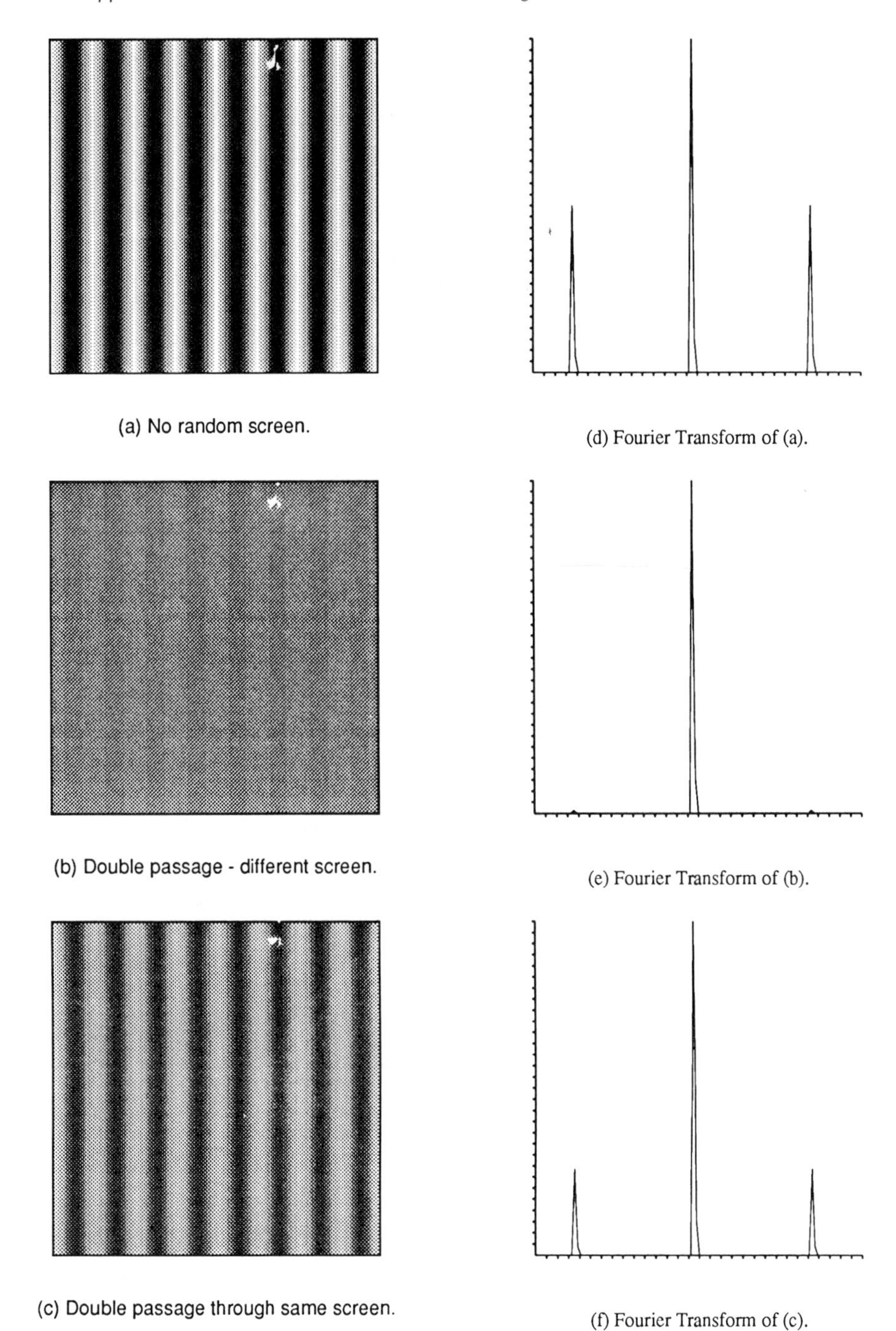

FIG. 7. Computer simulation of the fringes formed in a double-passage Michelson interferometer.

first *natural* random structure to show the unusual spectral shifts predicted by E. Wolf [23, 24].

Secondly, the opposition effect from well-defined surfaces has led to renewed theoretical studies, mainly using numerical techniques. Considering that this effect for true surface scattering was not predicted prior to its experimental discovery in 1987, it is quite amusing to observe that almost every study of light scattering is now considered incomplete without a discussion of it! Thirdly, it is becoming clear that the opposition effect—or more precisely, multiple scattering—has applications in areas other than scattering in volumes and surfaces. The example of the double passage of light through a random screen is one illustration of this, and doubtless there will be others in the future.

Acknowledgments

I am very grateful to my former and current research students, A. S. Harley, M. J. Kim, A. J. Sant, and N. Bruce, and international colleagues A. A. Maradudin, D. Maystre, E. R. Mendez, M. Nieto-Vesperinas, and K. A. O'Donnell for their contributions to this area of my research over the past few years.

References

1. N. Barabascheff, "Bestimmung der Erdalbedo und des Reflexionsgesetzes für die Oberfläche der Mondmeere: Theorie der Rillen," *Astron Nachr* **217**, 445–452 (1922).
2. P. Oetking, "Photometric Studies of Diffusely Reflecting Surfaces with applications to the Brightness of the Moon," *J Geophy Res* **71**, 2505–2513 (1966).
3. M. Minnaert, *The Nature of Light and Colour in the Open Air*, Dover Press, 1954.
4. W. W. Montgomery and R. H. Kohl, "Opposition-effect experimentation," *Opt Lett* **5**, 546–548 (1980).
5. Obtainable from, for example, Sigma Chemical Company, Fancy Road, Poole, Dorset BH17 7NH, UK as Product Number LB-5.
6. Y. Kuga and A. Ishimaru, "Retro-Reflectance from a Dense Distribution of Spherical Particles," *J Opt Soc Am A* **1**, 831–835 (1984).
7. M. P. van Albada and A. Lagendijk, "Observation of Weak Localisation of Light in a Random Medium," *Phys Rev Lett* **55**, 2692–2695 (1985).
8. P. E. Wolf and G. Maret, "Weak Localisation and Coherent Backscattering of Photons in Disordered Media," *Phys Rev Lett* **55**, 2696–2699 (1985).
9. P. E. Wolf, G. Maret, E. Akkermans, and R. Maynard, "Optical Coherent Backscattering by Random Media," *J Phys* (Paris) **49**, 63–75 (1988).
10. P. Sheng (editor), "Scattering and Localisation of Classical Waves in Random Media, World Scientific Press (Singapore), 1990.
11. E. R. Mendez and K. A. O'Donnell, "Observation of Depolarisation and Backscattering Enhancement in Light Scattering from Gaussian Rough Surfaces," *Opt Common* **61**, 91–95 (1987).
12. K. A. O'Donnell and E. R. Mendez, "Experimental Study of Scattering from Characterised Rough Surfaces," *J Opt Soc Am A* **4**, 1194–1205 (1987).
13. J. W. Goodman, *Statistical Optics,* Wiley, 1985."

14. A. J. Sant, M.-J. Kim, and J. C. Dainty, "Comparison of Surface Scattering between Identical, Randomly Rough Metal and Dielectric Diffusers," *Opt Lett* **14**, 1183–1185 (1989).
15. M.-J. Kim, J. C. Dainty, A. T. Friberg, and A. J. Sant, "Experimental Study of Enhanced Backscattering from One- and Two-Dimensional Random Rough Surfaces," *J Opt Soc Am A*, **7**, 569–577 (1990).
16. J. M. Soto-Crespo and M. Nieto- Vesperinas, "Electromagnetic Scattering from Very Rough Random Surfaces and Deep Reflection Gratings," *J Opt Soc Am A* **6**, 367–384 (1989).
17. A. A. Maradudin, T. Michel, A. R. McGurn, and E. R. Mendez, "Enhanced Backscattering by a Random Grating," *Ann Phys* **203**, 255–307 (1990).
18. M. Saillard and D. Maystre, "Scattering from Metallic and Dielectric Rough Surfaces," *J Opt Soc Am A* **7**, 982–990 (1990).
19. T. Mavroidis and J. C. Dainty, "Imaging after Double Passage Through a Random Screen," *Op Lett* **15**, 857–859 (1990).
20. Yu. A. Kratsov and A. I. Saichev, "Effects of Double Passage of Waves in Randomly Inhomogeneous Media," *Sov Phys Usp* **25**, 494–508 (1982).
21. P. R. Tapster, A. R. Weeks, and E. Jakeman, "Observation of Backscattering Enhancement through Atmospheric Phase Screens," *J Opt Soc Am A* **6**, 517–522 (1989).
22. T. Mavroidis, C. J. Solomon, and J. C. Dainty, "Imaging a Coherently Illuminated Object after Double Passage through a Phase Screen," *J. Opt. Soc. Am.* (in press).
23. A. Lagendijk (private communication).
24. J. T. Foley and E. Wolf, "Frequency Shifts of Spectral Lines Generated by Scattering from Space-Time Fluctuations," *Phys Rev A* **40**, 588–598 (1989). See also this volume, Chapter 16.

CHAPTER 16

Influence of Source Correlations on Spectra of Radiated Fields

Emil Wolf*

Department of Physics and Astronomy, University of Rochester, Rochester, New York

It generally is taken for granted that as light propagates through free space, its spectrum, appropriately normalized, remains unchanged. This assumption is implicit in all of spectroscopy and, until recently, has not been questioned, probably because with light from usual laboratory sources, one has never encountered any problem with it. Nevertheless, it was found in the last few years that this assumption is incorrect. More specifically, it was predicted theoretically [1] and demonstrated experimentally soon afterwards [2] that the spectrum of radiation depends not only on the spectrum of the source, but also on the coherence properties of the source; consequently, the normalized spectrum of light is, in general, not invariant on propagation. This discovery is likely to have an impact on several areas of optics and other fields, which utilize electromagnetic and other types of radiation. In this chapter, we give a brief account of this phenomenon and note some of its potential applications.

To bring out the essential features of this effect, we first consider the simplest radiating system that illustrates it: two small radiating scalar sources, located in the neighborhoods of points P_1 and P_2. For any realistic source, the source distribution (e.g., a Cartesian component of the current

* Also at The Institute of Optics, University of Rochester

ISBN 0-12-289690-4

density) will fluctuate randomly in time and, hence, must be described by an appropriate statistical ensemble. Instead of an ensemble of time-dependent realizations, one may characterize the fluctuations by an ensemble of frequency-dependent realizations [3], say $\{Q_1(\omega)\}$ and $\{Q_2(\omega)\}$. The corresponding realization of the statistical ensemble of the fluctuating field, which the two sources generate at a point P in space, then is given by

$$U(P, \omega) = Q_1(\omega)\frac{e^{ikR_1}}{R_1} + Q_2(\omega)\frac{e^{ikR_2}}{R_2}, \tag{1}$$

where ω is the frequency,

$$k = \frac{\omega}{c} \tag{2}$$

(c being the speed of light in vacuo), and R_1 and R_2 are the distances of the point P from the two sources (Fig. 1).

The spectrum of the field at P is given by the expression

$$S_U(P, \omega) = \langle U^*(P, \omega)U(P, \omega)\rangle, \tag{3}$$

where the asterisk denotes the complex conjugate and the angular brackets denote the ensemble average. On substituting from Eq. (1) into Eq. (3), we readily find that

$$S_U(P, \omega) = S_Q(\omega)\left[\frac{1}{R_1^2} + \frac{1}{R_2^2}\right] + \left[W_{12}(\omega)\frac{e^{ik(R_2-R_1)}}{R_1R_2} + \text{cc.}\right]. \tag{4}$$

In this expression, $S_Q(\omega)$ represents the spectrum, assumed for simplicity to be the same, of each of the two sources, i.e.

$$S_Q(\omega) = \langle Q_1^*(\omega)Q_1(\omega)\rangle = \langle Q_2^*(\omega)Q_2(\omega)\rangle, \tag{5}$$

and

$$W_{12}(\omega) = \langle Q_1^*(\omega)Q_2(\omega)\rangle \tag{6}$$

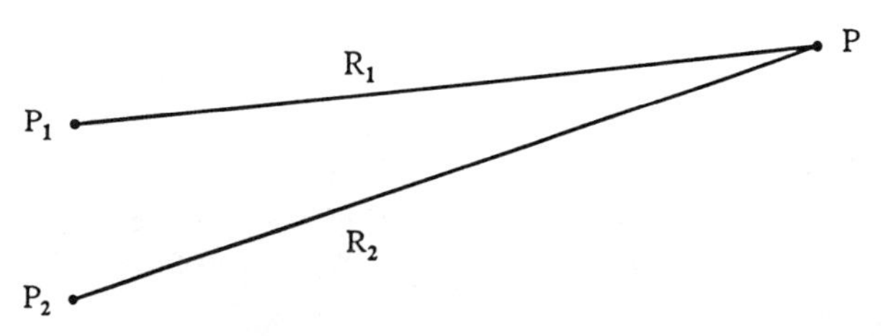

FIG. 1. Geometry and notation relating to the calculation of the spectrum $S_U(P, \omega)$ of the field at a point P, produced by two small fluctuating sources, with identical spectra $S_Q(\omega)$, located in the neighborhood of points P_1 and P_2.

is the so-called *cross-spectral density*, which characterizes the correlation (the spectral coherence) at frequency ω, between the fluctuations of the two sources. The abbreviation cc. on the right-hand side of Eq. (4) stands for the complex conjugate.

The formula (4) shows that the spectrum of the field at a point P depends not only on the spectra $S_Q(\omega)$ of the two sources, but also on the correlation between their fluctuations, characterized by the function $W_{12}(\omega)$. Only in the simplest case, when $W_{12}(\omega) \equiv 0$—i.e., when the sources are mutually uncorrelated (incoherent)—will the spectrum at every field point be proportional to the source spectrum, as is seen from Eq. (4).

Although Eq. (4) is reminiscent of the interference law of elementary wave theory, its significance is quite different. Interference, as commonly understood, gives rise to *spatial* variation of intensity throughout a region of space, whereas Eq. (4) describes its *spectral* variation at some fixed point [4].

It is clear from Eq. (4) that if the cross-spectral density varies appreciably with frequency over the effective spectral range of each source, the field spectrum may be very different from the source spectrum. The spectral changes for example, may, consist of narrowing, broadening, or shifting a spectral line, or generating several lines from a single line. Several examples of such spectral changes have been analyzed theoretically [5, 6] (Figs. 2 and 3) and demonstrated experimentally soon afterwards [7–9].

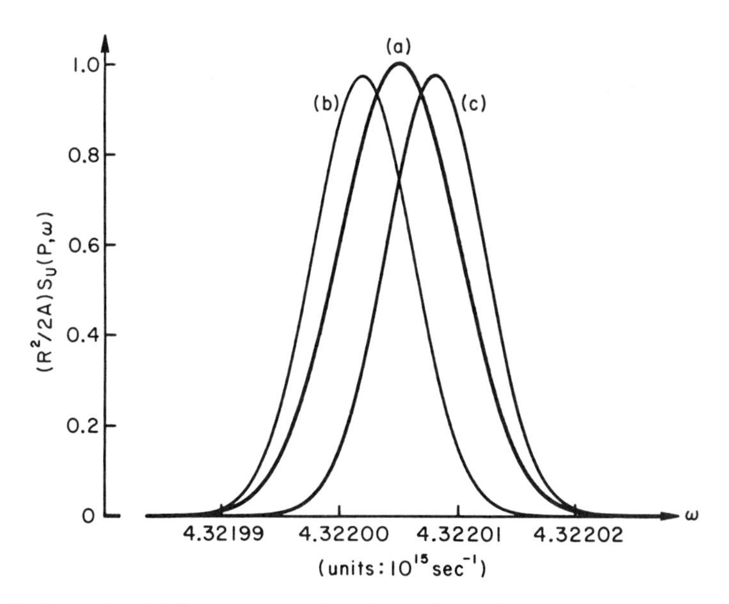

FIG. 2. Example of shifts of spectral lines induced by suitable correlation between two small sources: (a) source spectrum $S_Q(\omega)$; (b) and (c) field spectra $S_U(\omega)$ (After E. Wolf [5]).

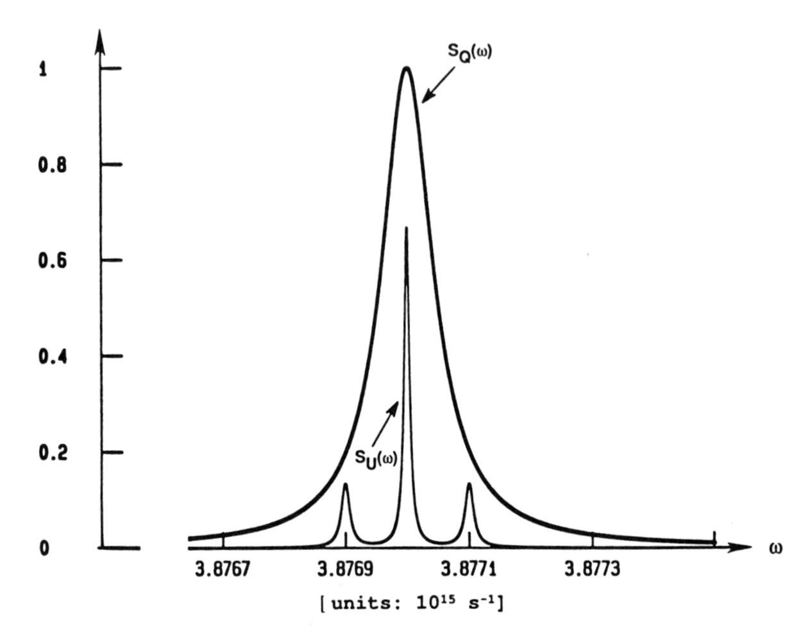

FIG. 3. Example of generation of three spectral lines from a single spectral line induced by suitable source-correlation. (After A. Gamliel and E. Wolf [6]).

The phenomenon that we just discussed is not restricted to radiation from the simple system consisting of two small sources, but applies generally to radiation from extended sources, both primary and secondary ones. For example, the normalized spectrum [10] $s^{(\infty)}(\mathbf{u}, \omega)$ of the field at a point P in the far zone, in a direction specified by a unit vector $\mathbf{u}$, produced by a planar, secondary, quasi-homogeneous source [11], is given by the expression [1]

$$s^{(\infty)}(\mathbf{u}, \omega) = \frac{k^2 S^{(0)}(\omega)\tilde{\mu}(k\mathbf{u}_\perp, \omega)}{\displaystyle\int_0^\infty k^2 S^{(0)}(\omega)\tilde{\mu}(k\mathbf{u}_\perp, \omega)\, d\omega}. \tag{7}$$

Here $S^{(0)}(\omega)$ is the spectrum of the field in the source plane, assumed for simplicity to be the same at every source point, specified by a position vector $\boldsymbol{\rho}'$;

$$\tilde{\mu}(f, \omega) = \frac{1}{(2\pi)^2}\int \mu^{(0)}(\boldsymbol{\rho}', \omega)e^{-if\boldsymbol{\rho}}\, d^2\rho' \tag{8}$$

is the two-dimensional spatial Fourier transform of the degree of spatial coherence $\mu(\boldsymbol{\rho}', \omega)$ of the field in the plane of the source [12]; and $\mathbf{u}_\perp$ is the projection, considered as a two-dimensional vector, of the unit vector $\mathbf{u}$.

We see, as an immediate consequence of Eq. (7), that the normalized spectrum $s^{(\infty)}(\mathbf{u}, \omega)$ of the light in the far zone in a direction specified by the unit vector $\mathbf{u}$ differs, in general, from the normalized source spectrum

$$s^{(0)}(\omega) = \frac{S^{(0)}(\omega)}{\int_0^\infty S^{(0)}(\omega)\, d\omega} \tag{9}$$

for two reasons: (a) because of the proportionality factor $k^2 = (\omega/c)^2$ and (b) because of the presence of the factor $\tilde{\mu}(k\mathbf{u}_\perp, \omega)$, which depends on the coherence properties of the light in the source plane. Examples of such spectral changes are given in Figs. 5b, 8, and 10.

Because the spectral changes we are discussing were neither predicted nor observed until recently, it is of interest to examine whether there are sources that do not give rise to such an effect. This problem has been discussed in [1] and led to the following result:

A sufficiency condition for the normalized spectrum of light produced by a planar, secondary, quasi-homogeneous source that which has the same normalized spectrum at every source point, to be the same throughout the far zone and across the source itself is that the degree of spatial coherence of the light in the source plane has the functional form

$$\mu^{(0)}(\boldsymbol{\rho}', \omega) = h(k\boldsymbol{\rho}'), \qquad (k = \omega/c). \tag{10}$$

When the degree of spatial coherence $\mu^{(0)}$ of the light at any two points $\boldsymbol{\rho}_1$ and $\boldsymbol{\rho}_2$ in the source plane has the form indicated by Eq. (10), it evidently is a function of the variable

$$\boldsymbol{\zeta} = k(\boldsymbol{\rho}_2 - \boldsymbol{\rho}_1) = 2\pi \frac{\boldsymbol{\rho}_2 - \boldsymbol{\rho}_1}{\lambda} \tag{11}$$

only, where λ is the wavelength associated with the frequency $\omega = kc = 2\pi c/\lambda$. A degree of spatial coherence that has the functional form (10) is said to obey the *scaling law*.

Examples of sources that satisfy the scaling law are planar, secondary, quasi-homogeneous, Lambertian sources, because all such sources are known to have a degree of spatial coherence given by the expression [12].

$$\mu^{(0)}(\boldsymbol{\rho}', \omega) = \frac{\sin k\rho'}{k\rho'}, \tag{12}$$

which, indeed, has the scaling-law form (10).

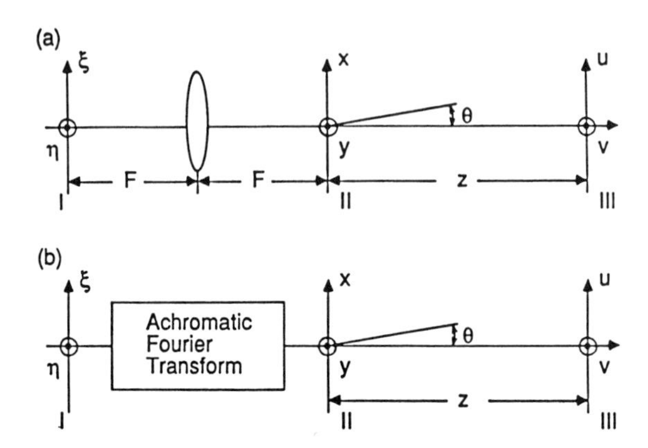

FIG. 4. The experimental configurations for realization of a planar, secondary, quasi-homogeneous source in plane II, which obeys the scaling law (a) and which does not obey it (b). The aperture in plane I was illuminated with broadband spatially incoherent sources, and the secondary sources were produced via a conventional lens (a) and via an achromatic Fourier transform system (b), respectively. In each case, the spectral intensity was measured in different θ—directions in the far field (plane III) of the secondary source and was compared with the spectral intensity in the plane of the source. Results are shown in Fig. 5 (After G. M. Morris and D. Faklis [2]).

Because Lambertian sources are encountered so commonly in the laboratory, and because the normalized spectrum of the light such sources generate throughout the far zone is the same as the normalized source spectrum, *spectral invariance* until recently was taken for granted—incorrectly as we now see—for light produced by all sources.

Some of these theoretical predictions regarding invariance and non-invariance of the normalized spectrum of light on propagation were confirmed experimentally by Morris and Faklis [2]. They measured the far-field spectra produced by two planar, secondary, quasi-homogeneous sources, one of which satisfied the scaling law (Fig. 4a) and the other which did not (Fig. 4b), and they compared them with the spectra measured in the source plane. The normalized spectra were found to be invariant on propagation in the first case (Fig. 5a), but not in the second case (Fig. 5b).

It is clear from Eq. (7) that by suitably controlling the degree of spatial coherence of the source, one can produce substantial modifications of spectra of radiated fields. In Fig. 6, an optical system for synthesizing sources of prescribed coherence properties, developed by G. Indebetouw, is shown [14]. An example of a pupil mask used in this synthesis is given in Fig. 7 and the spectral changes produced with it are shown in Fig. 8.

A potentially important application of the phenomenon, which is discussed in this chapter, was noted by H. C. Kandpal, J. S. Vaishya, and K. C. Joshi [16]. These authors pointed out that the large scatter that exists

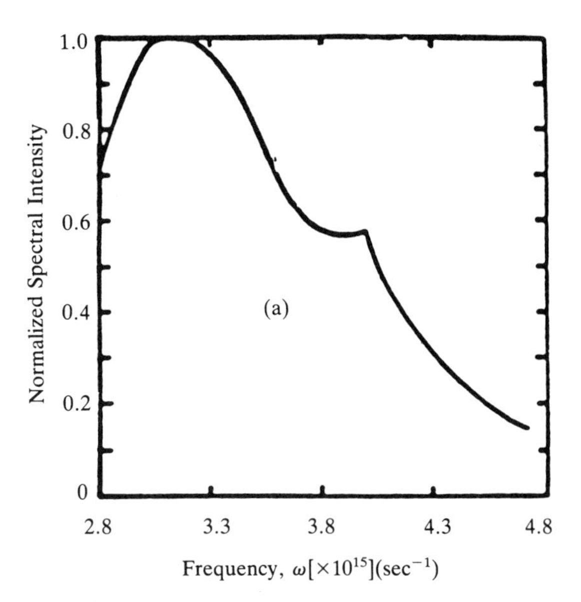

FIG. 5(a). Measured values of the normalized spectral intensity in plane II of the secondary source and in different directions in the far zone (plane III) produced by a source that obeys the scaling law. All the normalized spectral intensity distributions were found to be the same; hence, only one curve is shown (After G. M. Morris and D. Faklis [2]).

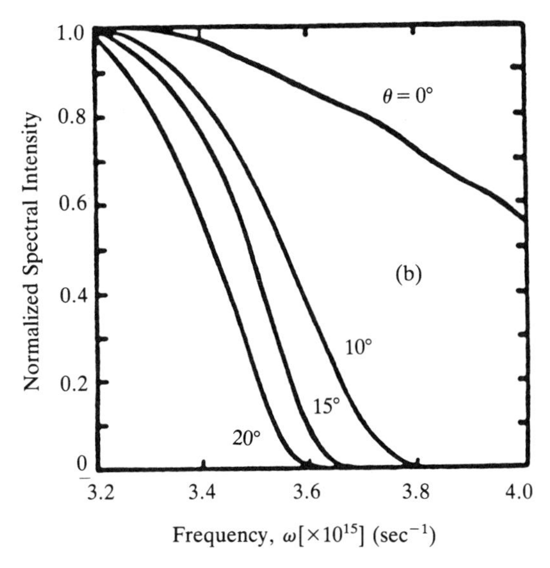

FIG. 5(b). Measured values of the normalized spectral intensity in different θ—directions in the far zone (plane III), produced by a source that does not satisfy the scaling law (After G. M. Morris and D. Faklis [2]).

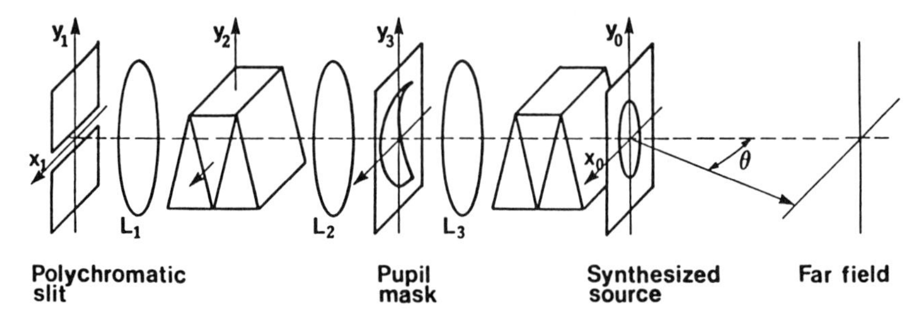

FIG. 6. An optical system used for synthesizing one-dimensional secondary sources of different degrees of spatial coherence. The degree of coherence of the synthesized source is proportional to the Fourier transform of the squared modulus of the transmission function of a pupil mask, modifying the light distribution in the dispersed image of a slit that is illuminated by incoherent polychromatic light of uniform spectrum (After G. Indebetouw [14]).

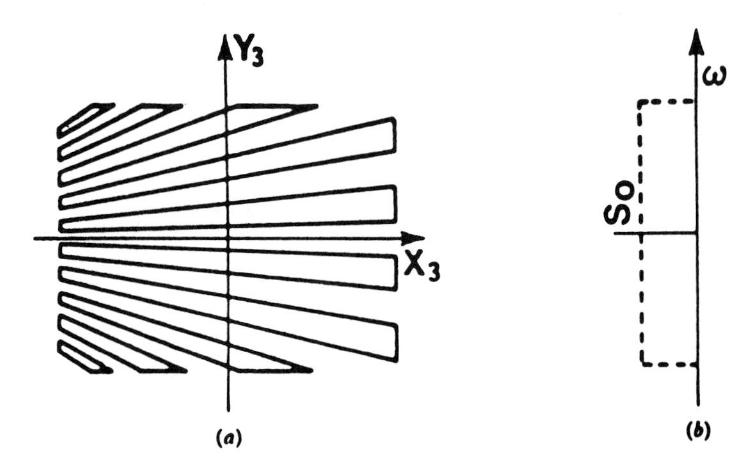

FIG. 7. A pupil mask (a) used to synthesize a source whose spectrum is uniform (b) and that produces the far-field spectra shown in Fig. 8 (After G. Indebetouw [14]).

in the spectroradiometry scales maintained by different national standards laboratories (*cf.* Fig. 9) has not up to now been explained satisfactorily. They noted that in setting up such scales, apertures are placed on the surface of an integrating sphere and within the optical system that focuses the radiation on a monochromator, and that the sizes of the apertures affect the coherence properties of the light transmitted by them. Controlled experiments by Kandpal *et al.* revealed that by changing the aperture diameters, the spectrum of the transmitted radiation undergoes frequency

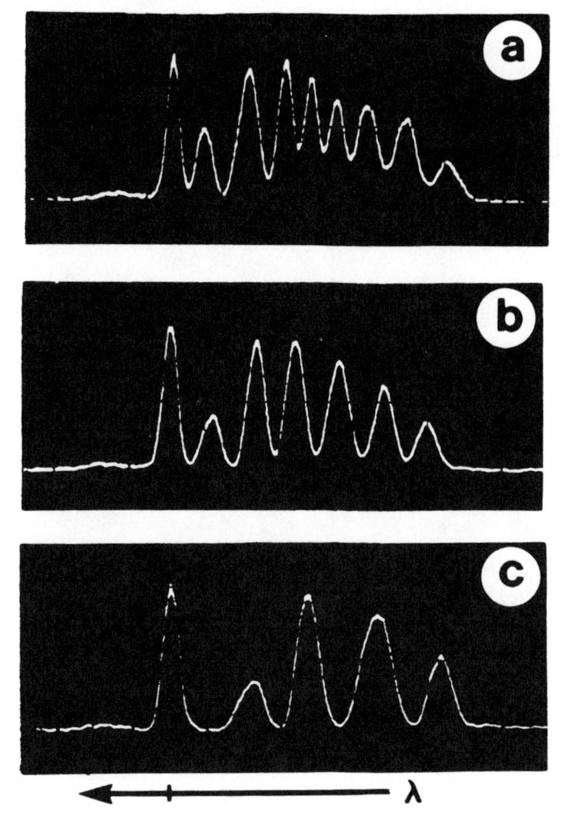

FIG. 8. Far-field spectra in directions of observations $\theta = 3°$ (a), $\theta = 0°$ (b) and $\theta = -3°$ (c), produced by the synthesized, spectrally uniform source that was generated by means of the pupil mask shown in Fig. 7. (After G. Indebetouw [14]).

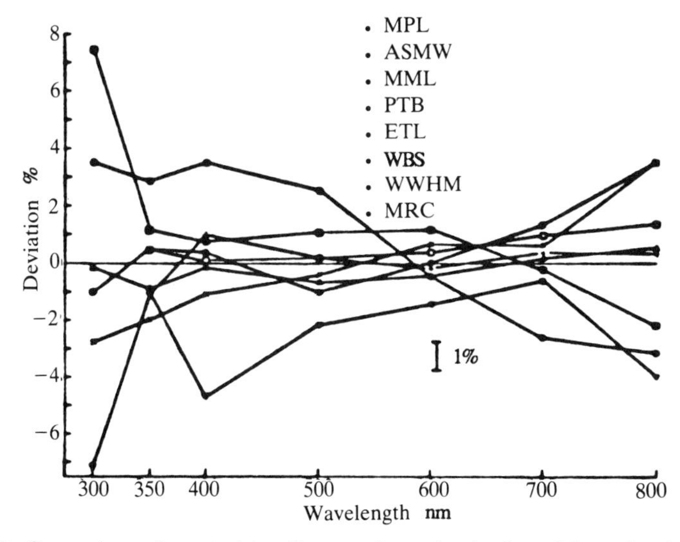

FIG. 9. Comparison of spectral irradiance scales maintained at eight national standards laboratories (After J. R. Moore [15]).

Table I. Observed changes in the location of the peak of the spectral distribution of light transmitted by circular apertures of two different sizes (After H. C. Kandpal, J. C. Vaishya, and K. C. Joshi [16]).

Incident Light		Transmitted Light	
		Peak (nm)	Peak (nm)
Peak (nm)	Half bandwidth (nm)	Aperture diameter 2.4 mm	Aperture diameter 10 mm
422.0	9	421.0	422.0
484.1	9	483.6	484.1
512.4	5	514.1	512.4
566.1	13	564.1	566.1
609.1	8	610.3	609.1
652.0	8	653.2	652.0

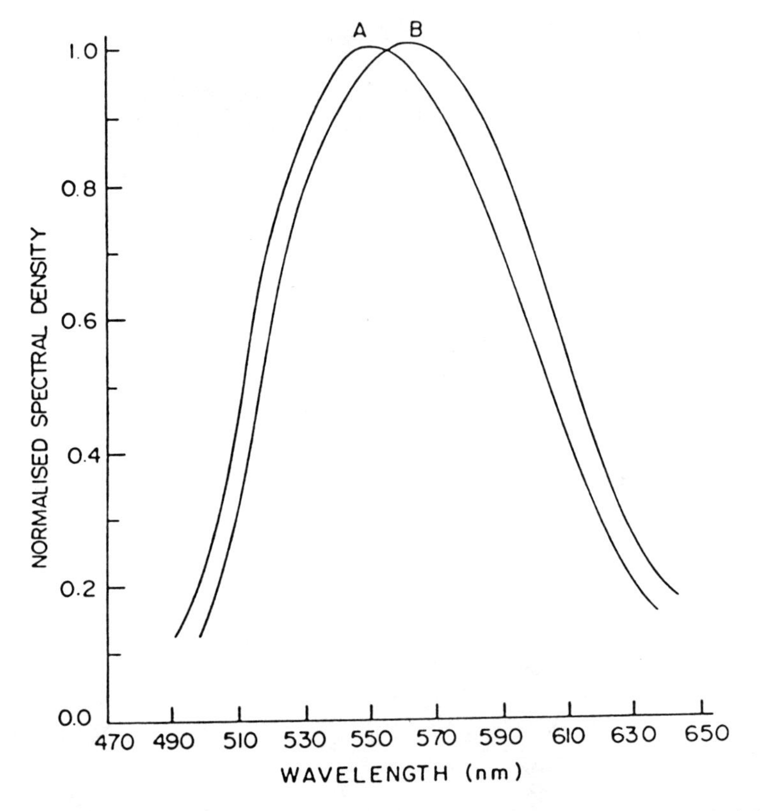

FIG. 10. Spectral distribution of a lamp—$V(\lambda)$ filter combination in the presence (A) and the absence (B) of a secondary aperture (After H. C. Kandpal, J. S. Vaishya, and K. C. Joshi [16]).

shifts, an effect that is to be expected according to the theory outlined in this chapter, but that has not been taken into account in specifying the spectroradiometric scales. The main results of Kandpal, Vaishya, and Joshi are reproduced in Table I and in Fig. 10. Similar conclusions based on detailed theoretical analysis were recently reached by J. T. Foley [17].

Throughout this chapter, we have restricted our discussion to correlation effects involving the simplest types of sources—either two small sources or quasi-homogeneous planar secondary sources. Influence of correlations between fluctuations in three-dimensional sources on the spectra of emitted radiation also has been considered [18, 19]. It also was shown that some interesting, previously unsuspected spectral changes may be introduced by an analogous mechanism involving scattering from suitably correlated random media [20–22, 24, 25]. These developments appear to be of particular interest for astronomy [23–25]. Other possible applications, which might utilize correlation-induced spectral changes and which are currently being explored, concern the development of new signal-modulation techniques and new techniques for testing theories of turbulence. The possible relevance of this phenomenon to precise interpretation of data from Doppler lidar satellite tracking and from enhanced back scattering are also being considered.

We conclude this brief chapter with the observation that the recent discovery that the state of coherence of a source affects the spectrum of the emitted radiation is an interesting phenomenon that not only tells us something new about light, but also promises some useful applications.

Acknowledgment

Some of the research described in this article was supported by the National Science Foundation and by the U.S. Army Research Office.

References

1. E. Wolf, "Invariance of Spectrum of Light on Propagation," *Phys. Rev. Lett.* **56**, 1370–1372 (1986).
2. G. M. Morris and D. Faklis, "Effects of Source Correlations on the Spectrum of Light," *Opt. Commun.* **62**, 5–11 (1987).
3. E. Wolf, "A New Theory of Partial Coherence in the Space-Frequency Domain, Part I: Spectra and Cross-Spectra of Steady-State Sources," *J. Opt. Soc. Amer.* **72**, 343–351 (1982).
4. The formula (4) is related to an expression for the normalized spectral density obtained by. L. Mandel a long time ago in his investigations regarding cross-spectral purity. [L. Mandel, "Concept of Cross-Spectral Purity in Coherence Theory," *J. Opt. Soc. Amer.* **51**, 1342–1350 (1961).] See also L. Mandel and E. Wolf, "Spectral Coherence and the Concept of Cross-Spectral purity," *J. Opt. Soc. Amer.* **66**, 529–535 (1976).

5. E. Wolf, "Redshifts and Blueshifts of Spectral Lines Emitted by Two Correlated Sources," *Phys. Rev. Lett.* **58**, 2646–2648 (1987).
6. A. Gamliel and E. Wolf, "Spectral Modulation by Control of Source Correlations," *Opt. Commun.* **65**, 91–96 (1988).
7. First such experiments were performed with sound waves rather than light waves by M. F. Bocko, D. H. Douglas, and R. S. Knox, "Observation of Frequency Shifts of Spectral Lines due to Source Correlations," *Phys. Rev. Lett.* **58**, 2649–2651 (1987).
8. D. Faklis and G. M. Morris, "Spectral Shifts Produced by Source Correlations," *Opt. Lett.* **13**, 4–6 (1988).
9. F. Gori, G. Guattari, C. Palma, and C. Padovani, "Observation of Optical Redshifts and Blueshifts Produced by Source Correlations," *Opt. Commun.* **67**, 1–4 (1988).
10. As is seen from Eq. (7), the normalization is chosen so that the integral of the spectrum over the whole frequency range is unity.
11. W. H. Carter and E. Wolf, "Coherence and Radiometry with Quasihomogeneous Planar Sources," *J. Opt. Soc. Amer.* **67**, 785–796 (1977).
12. L. Mandel and E. Wolf, cited in [4].
13. W. H. Carter and E. Wolf, "Coherence Properties of Lambertian and Non-Lambertian Sources," *J. Opt. Soc. Amer.* **65**, 1067–1071 (1975).
14. G. Indebetouw, "Synthesis of Polychromatic Light Sources with Arbitrary Degrees of Coherence: Some Experiments," *J. Mod. Opt.* **36**, 251–259 (1989).
15. J. R. Moore, "Sources of Error in Spectroradiometry," *Lighting Research and Technology* **12**, 213–220 (1980).
16. H. C. Kandpal, J. C. Vaishya, and K. C. Joshi, "Wolf Shift and its Applications in Spectroradiometry," *Opt. Commun.* **73**, 169–172 (1989).
17. J. T. Foley, "The effect of Aperture on the Spectrum of Partially Coherent Light," *Opt. Commun.* **75**, 347–352 (1990).
18. E. Wolf, "Non-Cosmological Redshifts of Spectral Lines", *Nature* **326**, 363–365 (1987).
19. E. Wolf, "Redshifts and Blueshifts of Spectral Lines Caused by Source Correlations," *Opt. Commun.* **62**, 2646–2648 (1987).
20. E. Wolf, J. T. Foley, and F. Gori, "Frequency Shifts of Spectral Lines Produced by Scattering from Spatially Random Media," *J. Opt. Soc. Amer.* A **6**, 1142–1149 (1989); *ibid.* **7**, 173 (1990).
21. J. T. Foley and E. Wolf, "Frequency Shifts of Spectral Lines Generated by Scattering from Space-Time Fluctuations," *Phys. Rev. A* **40**, 588–598 (1989).
22. E. Wolf, "Correlation-induced Doppler-like Frequency Shift of Spectral Lines," *Phys. Rev. Lett.* **63**, 2220–2223 (1989).
23. J. V. Narlikar, "Non-cosmological Redshifts", *Space Science Reviews* **50**, 523–614 (1989), Secs. 6.3 and 7.2.
24. D. F. V. James, M. P. Savedoff and E. Wolf, "Shifts of Spectral Lines Caused by Scattering from Fluctuating Random Media," *Astrophys. J.*, **359**, 67–71 (1990).
25. D. F. V. James and E. Wolf," Doppler-like Frequency Shifts Generated by Dynamic Scattering," *Phys. Letts. A.*, **146**, 167–171 (1990).

CHAPTER 17

Quantum Statistics and Coherence of Nonlinear Optical Processes

Jan Peřina
Joint Laboratory of Optics, Palacký University, Olomouc, Czechoslovakia

1. Introduction

Photon statistics and quantum coherence of light beams started to play an important role in physics after the discovery of strong-field nonequilibrium laser beams and the development of devices capable of detecting single photons. These circumstances led to a formulation of the theory of statistical and coherence properties of light beams in the framework of quantum electrodynamics. This was realized in 1963 by R. J. Glauber [1], who applied the so-called *coherent-state technique* to the description of quantum coherence as a boson cooperative phenomenon. (For a review, see [2].) The use of coherent states made it possible to treat radiation in a *classical* way, although all its quantum features were conserved. For this, quasi-distributions in a phase space had to be used to represent the density matrix of the system in which all information is included. However, some of these quasi-distributions need not exist as ordinary functions, but rather, they exist as generalized functions; they may take on negative values and be more singular than the Dirac function (representing a deterministic, or coherent, field), which is just the mathematical reflection of physical quantum properties of the system. Such purely quantum fields having no classical analogues may exhibit various quantum statistics and correlation properties, such as antibunching of photons (laser photons are unbunched,

ISBN 0-12-289690-4

natural source photons bunched), sub-Poisson photon statistics (laser photons obey Poisson statistics, natural source photons Bose-Einstein statistics, which are super-Poissonian), and squeezing of fluctuations of the real or imaginary part of the complex field amplitude below the level corresponding to isotropic (unsqueezed) fluctuations of the physical vacuum. There are other quantum features, such as violation of classical inequalities, oscillations in photon-number distribution, etc. These effects in light are purely quantum-mechanical and need a fully quantum approach for their description. They have fundamental meaning for the validity of quantum theory, and they offer a number of applications in high-precision measurements, optical communication, optical sensors, biology, etc.

To produce such nonclassical light beams, two methods can be adopted in principle—linear and nonlinear.

(a) When observing resonance-fluorescence light from one atom placed in an external pumping field of a tuned laser, photons emitted by the atom as a quantum system must exhibit antibunching and sub-Poisson photon statistics as a result of relaxation of the atom after the photon emission, which leads to a *dead-time effect* of the atom. A number of atoms can be used to generate sub-Poisson light, provided that sub-Poisson excitations are applied, such as by electrons or ions; also, various feedback arrangements can be used, particularly in semiconductor lasers [3]. This is the linear method.

(b) Interesting methods for generating nonclassical light beams—which are antibunched, sub-Poissonian and/or squeezed—are based on the use of nonlinear optical processes, particularly optical parametric processes (second-harmonic or sub-harmonic generation, frequency down- or up-conversion, etc.), four-wave mixing, multiphoton absorption, Raman or Brillouin scattering, free-electron lasing, soliton wave propagation, etc. Such nonlinear processes perform coincidence decimation of the incident light beams by means of the nonlinear consumation of *closed* photons in a coherent laser beam that is regularized in this way.

2. Quantum Coherence and Photocount Statistics

Denoting $\hat{A}^{(+)}(x)$ and $\hat{A}^{(-)}(x)$ (where $x = (\mathbf{x}, t)$, $\mathbf{x}$ is the position vector, and t is time) as the positive and negative frequency components of a component of the vector-potential operator of the electromagnetic field; and assuming detection of the field by means of a set of photodetectors, absorbing photons from the field, placed at n different space points, and operating at different times; then the probability of absorption of n photons from the field is proportional to the normal correlation function

$$\Gamma_{\mathcal{N}}^{(m,n)}(x_1, \ldots, x_{m+n}) = \mathrm{Tr}\{\hat{\rho}\hat{A}^{(-)}(x_1)\ldots\hat{A}^{(-)}(x_m)\hat{A}^{(+)}(x_{m+1})\ldots\hat{A}^{(+)}(x_{m+n})\}, \quad (1)$$

where $m = n$, $\hat{\rho}$ is the density matrix including all the field properties, Tr denotes the trace and $\hat{A}^{(+)}$ and $\hat{A}^{(-)}$ represent the annihilation and creation operators of a photon, respectively [1]. (See [2] for a review.) The normal order of the field operators in Eq. (1) will be changed to the antinormal order (with the order of the annihilation and creation operators changed) or to some alternating order if the field is detected by means of stimulated emission of photons to the field (photon counters) or by a scattering process. If a field possesses the correlation function (1) in the factorized form $V^*(x_1)\ldots V^*(x_m)V(x_{m+1})\ldots V(x_{m+n})$, where $V(x)$ is a classical field, then such a field is coherent in the $(m+n)$th order; it is fully coherent (to all orders) if such factorization is valid for all m, n. From the classical point of view, this means that the field is deterministic without any fluctuations. Quantum coherence is specified by the density matrix $\hat{\rho} = |\{\alpha_\lambda\}\rangle\langle\{\alpha_\lambda\}|$ using the coherent state $|\{\alpha_\lambda\}\rangle$ (α_λ are complex numbers, i.e., complex mode amplitudes), which is the eigenstate of the annihilation operator, $\hat{A}^{(+)}|\{\alpha_\lambda\}\rangle = V|\{\alpha_\lambda\}\rangle$, $\langle\{\alpha_\lambda\}|\hat{A}^{(-)} = V^*\langle\{\alpha_\lambda\}|$. Such states are as classical as they can be in quantum theory; i.e., they include only the isotropic fluctuations of the physical vacuum (which are represented by a circle in the α_λ -complex plane corresponding to the energy $\hbar\omega_\lambda/2$ of vacuum fluctuations, ω_λ being the frequency of the mode λ of the field). The quantum correlation functions (1) possess a number of interesting properties: They obey wave equations, they are analytical functions and fulfill various symmetry conditions and inequalities, etc. Although second-order coherence is related to the maximum visibility of interference fringes, as in the classical theory, higher-order quantum coherence is related to the statistical independence of counting rates of single photodetectors, which reflects no tendency toward any sort of correlation in the arrival times of photons; i.e., no correlation on intensity fluctuations in the Hanbury Brown-Twiss sense may occur (e.g. see [2]). Higher-order degrees of coherence defined by various normalizations of the correlation function (1) can be related to analyze photon correlation experiments, to specify higher-order partial coherence in terms of the n-fold intensities, etc. [2, 4].

Assume that light is incident on the cathode of a photodetector at time t for an interval T. Then the complete statistics specified by the set of quantum correlation functions are characterized by the photocount distribution giving the probability of n photoelectrons appearing in the time interval $(t, t+T)$,

$$\begin{aligned} p(n, T, t) &= \mathrm{Tr}\left\{\hat{\rho}\,\hat{\mathcal{N}}\,\frac{(\eta\hat{W})^n}{n!}\exp(-\eta\hat{W})\right\} \\ &= \int_0^\infty \frac{(\eta W)^n}{n!}\exp(-\eta W)P_{\mathcal{N}}(W)\,dW, \end{aligned} \tag{2}$$

where $\hat{W} = \int_t^{t+T} \hat{A}^{(-)}(t)\hat{A}^{(+)}(t)\,dt$ is the integrated number operator, $\hat{\mathcal{N}}$ denotes the normal ordering operator, η is the photoefficiency, and $P_{\mathcal{N}}(W)$ represents the pseudoprobability distribution of the integrated intensity W related to the normal ordering of field operators. We also have assumed a point and spectrally broadband detector. It is evident from the photodetection equation (2) that the photocount statistics of intensity-stabilized laser light for detection time intervals T shorter than the coherence time ($\hat{A}^{(-)}(t)\hat{A}^{(+)}(t)$ then is practically time-independent) are Poissonian. On the other hand, $P_{\mathcal{N}}(W)$ for natural sources is of the negative exponential form, and Bose-Einstein photocount statistics are appropriate in this equilibrium state. The inverse problem solution of Eq. (2) makes it possible to derive all the statistical properties of light from the photocount measurements.

3. Coherent-State Technique

In this section, we restrict our considerations to single-mode fields, for simplicity. The coherent states, including only incoherent fluctuations of the physical vacuum, are most convenient to describe boson cooperative phenomena, such as quantum coherence. However, they also can be used for representations of optical fields because they form an overcomplete system of states. Thus, the density matrix $\hat{\rho}$, containing full information about any quantum system, can be expressed in terms of the coherent states $|\alpha\rangle$ in the form of the Glauber-Sudarshan representation,

$$\hat{\rho} = \int \phi_{\mathcal{N}}(\alpha)|\alpha\rangle\langle\alpha|\,d^2\alpha, \tag{3}$$

where the integration is taken over the whole complex plane of the complex amplitude α, and $\phi_{\mathcal{N}}(\alpha)$ is a quasi-distribution function related to normal ordering of field operators $\hat{a}$ and $\hat{a}^\dagger$ (single-mode annihilation and creation operators of a photon, respectively). It may behave irregularly compared to classical probability distributions, which are nonnegative and regular; i.e., it may take on negative values and be much more singular than the Dirac δ-function corresponding to a deterministic field. These mathematical irregularities of $\phi_{\mathcal{N}}$ just reflect the quantum physical properties of optical fields. As a consequence of the wave-particle dualism and Bose-Einstein character of photons, we can obtain for fluctuations of the photon number n,

$$\langle(\Delta\hat{n})^2\rangle = \langle\hat{n}^2\rangle - \langle\hat{n}\rangle^2 = \langle n\rangle + \langle(\Delta W)^2\rangle_{\mathcal{N}}, \tag{4}$$

where $\hat{n} = \hat{a}^\dagger\hat{a}$ is the number operator, the commutation rule $[\hat{a}, \hat{a}^\dagger] = \hat{a}\hat{a}^\dagger - \hat{a}^\dagger\hat{a} = \hat{1}$ holds, $\langle n\rangle$ in Eq. (4) is the mean number of photons representing particle-like (Poisson) fluctuations, and

$$\begin{aligned}\langle(\Delta W)^2\rangle_{\mathcal{N}} &= \langle\hat{a}^{\dagger 2}\hat{a}^2\rangle - \langle\hat{a}^\dagger\hat{a}\rangle^2 \\ &= \int \phi_{\mathcal{N}}(\alpha)|\alpha|^4\, d^2\alpha - \left[\int \phi_{\mathcal{N}}(\alpha)|\alpha|^2\, d^2\alpha\right]^2\end{aligned} \tag{5}$$

represents generalized wave-like fluctuations. (The average is taken with the use of the quasi-distribution $\phi_{\mathcal{N}}$, so that $\langle(\Delta W)^2\rangle_{\mathcal{N}} < 0$ may occur!) Of course, classical fields always are Poissonian or super-Poissonian (e.g., fields from natural sources are Gaussian), since $\langle(\Delta W)^2\rangle_{\mathcal{N}} \geqq 0$ holds for them, whereas for quantum fields, $\langle(\Delta W)^2\rangle_{\mathcal{N}} < 0$ may occur, leading to sub-Poisson behavior with $\langle(\Delta\hat{n})^2\rangle < \langle n\rangle$ from Eq. (4). Hence, if we introduce the so-called Fano factor, $F_n = \langle(\Delta\hat{n})^2\rangle/\langle n\rangle$, we can distinguish the following classes of optical fields:

$$\begin{aligned}&F_n = 1\text{—coherent (Poisson) fields.}\\ &F_n > 1\text{—super-Poisson fields (e.g., } F_n = 1 + \langle n\rangle \text{ for natural Gaussian fields).}\\ &F_n < 1\text{—sub-Poisson (nonclassical) fields.}\end{aligned} \tag{6}$$

As a consequence of the Bose-Einstein statistics for photons and the maximum entropy principle for natural sources, the photons of their radiation possess bunching properties reflecting intensity fluctuations; i.e., such photons have a tendency to come to a photodetector in pairs. This effect is compensated completely by correlation inside the laser source, so that photons in a laser beam are unbunched. Such properties of photon bunching, unbunching, and antibunching are described by the normalized intensity correlation measured in the Hanbury Brown-Twiss arrangement involving two photodetectors and a correlation device for photosignals, which are proportional to the incident intensity,

$$\gamma_{\mathcal{N}}^{(2)}(\tau) = \langle I(t)I(t+\tau)\rangle_{\mathcal{N}}/\langle I(t)\rangle_{\mathcal{N}}^2, \tag{7}$$

expressing a probability of separation of two photons by τ seconds, $I(t)$ being the field intensity. For classical fields $\gamma_{\mathcal{N}}^{(2)}(\tau) \geqq 1$, and this quantity is maximal for $\tau = 0$. If $\phi_{\mathcal{N}}$ behaves irregularly, then $\gamma_{\mathcal{N}}^{(2)}(\tau) < 1$ is possible, which expresses antibunching of photons [3]. In this case, the well-known classical inequality $\langle I^2 \rangle_{\mathcal{N}} \geqq \langle I \rangle_{\mathcal{N}}^2$ is violated. In the first experiments on observing photon antibunching in resonance-fluorescence light by L. Mandel and his co-workers [5], a weaker definition of antibunching was used in terms of the derivative of Eq. (7), $\lim_{\tau \to +0} d\gamma_{\mathcal{N}}^{(2)}(\tau)/d\tau > 0$, i.e., $\gamma_{\mathcal{N}}^{(2)}(0) < \max_\tau \gamma_{\mathcal{N}}^{(2)}(\tau)$, which means that antibunching occurs naturally if $\gamma_{\mathcal{N}}^{(2)}(\tau)$ is increasing in the vicinity of $\tau = 0$. (However, the classical correlation function always is maximal at $\tau = 0$.)

One of the most important nonclassical effects observed in optical fields is the squeezing of fluctuations in the beam below the level corresponding to vacuum fluctuations, or fluctuations in the coherent state. If we define the hermitian quadrature operators $\hat{q}$ and $\hat{p}$, which are related to the generalized coordinate and momentum, or to real and imaginary parts of the complex field amplitude, respectively,

$$\hat{q} = \hat{a} + \hat{a}^\dagger, \hat{p} = \frac{1}{i}(\hat{a} - \hat{a}^\dagger), \tag{8}$$

then they must fulfill the Heisenberg uncertainty relation

$$\langle (\Delta\hat{q})^2 \rangle \langle (\Delta\hat{p})^2 \rangle \geqq 1. \tag{9}$$

The squeezed state is defined by the condition that $\langle (\Delta\hat{q})^2 \rangle$ or $\langle (\Delta\hat{p})^2 \rangle < 1$; i.e., the variance of one of the quadrature components is reduced compared to vacuum fluctuations $\langle (\Delta\hat{q})^2 \rangle$ or $\langle (\Delta\hat{p})^2 \rangle = 1$. Of cource, the reduction of fluctuations in one variable must lead to the increase of fluctuations in the other variable to fulfill Eq. (9). A squeezed state can be the *minimum uncertainty state* (squeezed coherent state; i.e., in Eq. (9), the equality sign holds) or it can be a *mixed state* (inequality sign holds in Eq. (9)). This reduction of the quadrature fluctuations usually is detected with the help of homodyne detection, where a coherent component of a local oscillator of the coherent complex amplitude γ is superimposed on the signal beam using a beamsplitter. Therefore, a detector placed beyond the splitter is detecting the variance $\langle (\Delta\hat{Q})^2 \rangle$, where $\hat{Q} = (\hat{a}^\dagger + \gamma^*)(\hat{a} + \gamma)/2 - (\hat{a}^\dagger - \gamma^*)(\hat{a} - \gamma)/2 = \gamma\hat{a}^\dagger + \gamma^*\hat{a}$, which can be maximized or minimized with respect to the phase of γ. This procedure provides the following rotational invariants for the large and small half-axes of the noise ellipse:

$$\begin{aligned} \lambda_{1,2} &= \frac{1}{2} \operatorname{Tr} \hat{M} \left[1 \pm \left(1 - \frac{4 \operatorname{Det} \hat{M}}{(\operatorname{Tr} \hat{M})^2} \right)^{1/2} \right] \\ &= \langle \{\Delta\hat{a}, \Delta\hat{a}^+\} \rangle \pm 2 |\langle (\Delta\hat{a})^2 \rangle|, \end{aligned} \tag{10}$$

where Det denotes the determinant, the matrix $\hat{M}$ equals the following:

$$\hat{M} = \begin{pmatrix} \langle(\Delta\hat{q})^2\rangle & \frac{1}{2}\langle\{\Delta\hat{q}, \Delta\hat{p}\}\rangle \\ \frac{1}{2}\langle\{\Delta\hat{q}, \Delta\hat{p}\}\rangle & \langle(\Delta\hat{p})^2\rangle \end{pmatrix}, \tag{11}$$

and $\{\Delta\hat{q}, \Delta\hat{p}\} = \Delta\hat{q}\ \Delta\hat{p} + \Delta\hat{p}\ \Delta\hat{q}$ represents the anticommutator [6]. One can see that this formulation is quite similar to that for partial polarization [7]. Instead of the Heisenberg inequality (9), now we have the Schrödinger-Robertson inequality

$$\text{Det}\ \hat{M} = \lambda_1\lambda_2 = \langle(\Delta\hat{q})^2\rangle\langle(\Delta\hat{p})^2\rangle - \frac{1}{4}\langle\{\Delta\hat{q}, \Delta\hat{p}\}\rangle^2 \geqq 1, \tag{12}$$

and the principal squeezing is defined by the following condition [6]:

$$\lambda_2 = 1 + 2[\langle\Delta\hat{a}^\dagger\Delta\hat{a}\rangle - |\langle(\Delta\hat{a})^2\rangle|] < 1.$$

The role of the anticommutator is particularly important in nonlinear interactions. If it is zero, the principal squeezing reduces to the standard squeezing. Some applications of principal squeezing can be performed [8]. Writing the anticommutator term in Eq. (12) on the right-hand side, one can introduce minimum-uncertainty coherent states with correlations [9].

It should be noted that the preceding quantum features of light generally are independent because they are related to different-order effects with respect to the complex field amplitude. Nevertheless, they may occur together in many cases.

4. Nonlinear Dynamics

To describe the quantum statistical properties of light, we can adopt two approaches in principle. First, there is (a) the *Heisenberg approach*, in which we solve the Heisenberg (Heisenberg-Langevin) equations for the field operators, substitute the operator solutions into a quantum characteristic function, and perform the average over the initial statistics; the Fourier transformation provides the corresponding quasi-distribution and the complete photocount statistics (photocount distribution, its factorial moments, characteristics of squeezing, etc.). Then there is (b) the *Schrödinger approach*, in which we start from the equation of motion for the density matrix and, using the coherent-state methods of quantum optics, we can derive the master equation and the generalized Fokker-Planck equation for a quasi-distribution. We can demonstrate these general methods using the simplest nonlinear process of second subharmonic generation with classical coherent pumping light, which proved to be very effective in the generation of squeezed light [10]. Such a process is described by the effective hamiltonian

$$\hat{H} = \hbar\omega\left(\hat{a}^{\dagger}\hat{a} + \frac{1}{2}\right) - \frac{1}{2}\hbar g(\hat{a}^2 \exp(i2\omega t - i\varepsilon) + \text{h.c.}), \tag{14}$$

where ω is the frequency of a single-mode field described by the annihilation and creation operators $\hat{a}$ and $\hat{a}^{\dagger}$, respectively, $g > 0$ is a real coupling constant, ε is a phase of pumping, and h.c. represents the hermitian conjugate term. The Heisenberg equation reads

$$\frac{d\hat{a}}{dt} = -i\omega\hat{a} + ig\hat{a}^{\dagger}\exp(-i2\omega t + i\varepsilon), \tag{15}$$

and can be solved, together with its hermitian conjugate equation, in the form

$$\begin{aligned} &\hat{a}(t) = u(t)\hat{a}(0) + v(t)\hat{a}^{\dagger}(0), \\ &u(t) = \cosh gt \exp(-i\omega t),\ v(t) = i \sinh gt \exp(-i\omega t + i\varepsilon). \end{aligned} \tag{16}$$

We see that the annihilation operator at time t is expressed in terms of annihilation, and creation operators at time $t = 0$. The solution (16) can be substituted into the normal quantum characteristic function with a complex parameter β,

$$\begin{aligned} C_{\mathcal{N}}(\beta, t) &= \operatorname{Tr}\{\hat{\rho}\exp(\beta\hat{a}^{\dagger}(t))\exp(-\beta^*\hat{a}(t))\} \\ &= \langle \exp[-B(t)|\beta|^2 + \left(\frac{C^*(t)}{2}\beta^2 + \text{c.c.}\right) \\ &\quad + (\beta\xi^*(t) - \text{c.c.})]\rangle_{\mathcal{N}}, \end{aligned} \tag{17}$$

where we have used the Baker-Hausdorff identity $\exp(\hat{A} + \hat{B}) = \exp\hat{A}$ $\exp\hat{B}\exp(-[\hat{A}, \hat{B}]/2)$, $[\hat{A}, \hat{B}] = \hat{A}\hat{B} - \hat{B}\hat{A}$, for any two operators $\hat{A}$, $\hat{B}$ fulfilling $[[\hat{A}, \hat{B}], \hat{A}] = [[\hat{A}, \hat{B}], \hat{B}] = \hat{0}$; we have performed the normal ordering in the initial operators; and we have used Eq. (3) to perform the average over the initial operator expressed by the angular brackets in Eq. (17). Here,

$$\begin{aligned} &B(t) = \langle \Delta\hat{a}^{\dagger}(t)\, \Delta\hat{a}(t)\rangle = |v(t)|^2 = |u(t)|^2 - 1 = \sinh^2 gt, \\ &C(t) = \langle(\Delta\hat{a}(t))^2\rangle = u(t)v(t) = \frac{i}{2}\sinh 2gt \exp(-i2\omega t + i\varepsilon), \\ &\xi(t) = u(t)\xi(0) + v(t)\xi^*(0), \end{aligned} \tag{18}$$

where $\xi(0)$ is the initial coherent complex amplitude. The Fourier transformation of the antinormal characteristic function $C_{\mathcal{A}}(\beta, t) = C_{\mathcal{N}}(\beta, t)$ $\exp(-|\beta|^2)$ with respect to β provides the quasi-distribution $\phi_{\mathcal{A}}(\alpha, t) =$

$\langle\alpha|\hat{p}(t)|\alpha\rangle/\pi$ related to antinormal ordering in the form of an averaged Gaussian distribution. The normal generating function determining the photon-number distribution and its factorial moments can be obtained as

$$\begin{aligned}\langle\exp(isW)\rangle_{\mathcal{N}} &= \frac{1}{\pi\lambda}\int\exp\left(-\frac{1}{\lambda}|\beta|^2\right)C_{\mathcal{N}}(\beta,t)\,d^2\beta\\ &= [1-is(E-1)]^{-1/2}\,[1-is(F-1)]^{-1/2}\\ &\quad\times\exp\left(\frac{isA_1}{1-is(E-1)}+\frac{isA_2}{1-is(F-1)}\right),\end{aligned}\tag{19}$$

where $\lambda=-is$ is a parameter of the generating function and

$$\begin{aligned}A_{1,2} &= \frac{1}{2}\left[|\xi(t)|^2 \mp \frac{1}{2|C|}(C^*\xi^2(t)+\text{c.c.})\right]\\ &= \frac{1}{2}|\xi(0)|^2\exp(\mp 2gt)\,[1\mp\sin(2\vartheta-\varepsilon)]\geqq 0,\end{aligned}\tag{20}$$

$$E-1 = B-|C| = \frac{1}{2}[\exp(-2gt)-1]\leqq 0,$$

$$F-1 = B+|C| = \frac{1}{2}[\exp(2gt)-1]\geqq 0;$$

here, ϑ is the phase of the initial amplitude $\xi(0)$. The quantities $A_{1,2}$ related to the initial coherent field are playing the role of the *coherent mean photon numbers*, whereas the quantities $E-1$ and $F-1$ related to quantum noise represent the *chaotic mean photon numbers* because Eq. (19) is the product of two generating functions for the superposition of coherent and chaotic fields $A_{1,2}$ and $E-1$, $F-1$, respectively (in classical terms). Therefore, the photon-number distribution $p(n,t) = d^n\langle\exp(isW)\rangle_{\mathcal{N}}/d(is)^n n!$ at $is=-1$ and its factorial moments $\langle W^k\rangle_{\mathcal{N}} = d^k\langle\exp(isW)\rangle_{\mathcal{N}}/d(is)^k$ at $is=0$ can be expressed in terms of the Laguerre polynomials [2]. For nonclassical fields, the necessary condition is $(E-1)<0$, which can lead to the narrowing of the initial Poisson distribution $(E-1=F-1=0)$ for a coherent field; i.e., sub-Poisson photon statistics can be reached. Under this condition, there always is principal squeezing, as seen from Eqs. (13) and (20), because $\lambda_2 = 1+2(B-|C|) = \exp(-gt)<1$. However, the principal squeezing condition is a necessary but not sufficient condition for sub-Poisson (antibunched) light. With respect to the product of generating functions for the superposition of coherent and chaotic fields in Eq. (19), the resulting photon-number distribution $p(n,t)$ is the convolution of partial distributions p_1 and p_2 for signals A_1 and A_2 and noise $E-1$ and $F-1$, respectively,

$$p(n,t) = \sum_{k=0}^{n} p_1(n-k,t)p_2(k,t). \tag{21}$$

Assume in the following the special initial phase condition $2\vartheta - \varepsilon = -\pi/2$. Then the signal $A_1 = |\xi(0)|^2 \exp(-2gt)$, which is attenuated, is dominant ($A_2 = 0$), quantum noise $F-1$ plays a negligible role at the beginning, and we have the superposition of signal A_1 and negative quantum noise $E-1$. This results in the sub-Poisson behavior of photon statistics. On the other hand, for $2\vartheta - \varepsilon = \pi/2$, the dominant behavior is determined by the signal $A_2 = |\xi(0)|^2 \exp(2gt)$, which is amplified, superimposed on the positive quantum noise $F-1$, and, consequently, the photon statistics are super-Poissonian. In the former case and for later time, the photon-number distributions $p_1(n,t) = p_1(n)$ (curve a), $p_2(n,t) = p_2(n)$ (curve b) and $p(n,t) = p(n)$ (curve c) are shown in Fig. 1 ($gt = 1$, $|\xi(0)| = 0.8$), demonstrating that the nonclassical behavior of light reflects itself in negative values of pseudoprobability p_2 and, consequently, in the oscillating behavior of the resulting distribution $p(n,t)$. Another explanation of the quantum origin of these oscillations was suggested using interference in phase space [11]. If dissipation (loss) is included with the mean number of dissipative particles $\bar{n}$ and with the damping coefficient γ, then $E-1$ and $F-1$ in Eq. (20) are replaced by $\exp(-\gamma t)$ $[\exp(-2gt)-1]/2 + \bar{n}(1-\exp(-\gamma t)$ and $\exp(-\gamma t)$ $[\exp(2gt)-1]/2 + \bar{n}(1-\exp(-\gamma t))$, respectively. Hence, with increasing $\bar{n}$, the reduced negative values of quantum

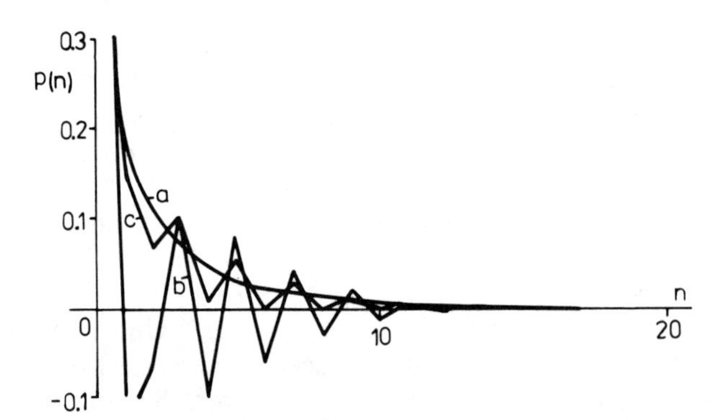

FIG. 1. Photon-number distributions $p_1(n)$ (curve a), $p_2(n)$ (curve b), and their discrete convolution giving the actual photon-number distribution $p(n) = p(n,t)$ (curve c); oscillations in $p(n)$ result from the negative values of pseudoprobability $p_2(n)$ as a consequence of negative quantum noise $E-1$; $|\xi(0)| = 0.8$, $gt = 1$, $2\vartheta - \varphi = -\pi/2$. (After J. Peřina and J. Bajer, *Phys. Rev. A* **41**, 516 (1990).)

noise $(E-1)\exp(-\gamma t)$ are compensated further by the positive values of the additional term $\bar{n}(1-\exp(-\gamma t))$ and the oscillations and nonclassical behavior are degraded.

Further nonlinear optical processes can be investigated in this way, such as second-harmonic generation, three- and four-wave mixing, nonlinear oscillations, and Raman scattering, including losses and nonclassical initial light [12, 13, 14].

5. Experiments with Nonclassical Light

Since 1977a new kind of light—*nonclassical (quiet) light*—is available. L. Mandel and his collaborators generated antibunched light in resonance fluorescence from Na atoms exposed to pump-laser light [5]. The atom behaves as a quantum system with a *dead time* related to its relaxation and therefore, the photons emitted by the atom are antibunched. Additional equipment was used to control the scattering volume so that photons were detected by the Hanbury Brown-Twiss intensity correlation technique only if one atom was present in the volume. (Otherwise, fluctuations of the number of atoms played an important role and nonclassical behavior was smoothed out.) This permitted observation of sub-Poisson light from one atom [15]. M. C. Teich and B. E. A. Saleh (see [3]) were able to generate stationary sub-Poisson light using the modified Franck-Hertz experiment, in which Poisson fluctuations of pumping electrons are reduced. Further improvements can be expected using superlattices or stimulated Franck-Hertz light [3]. In this method of generation of nonclassical light, the influence of statistics of the primary excitations and statistics of the individual emissions on the statistical and coherence properties of optical fields is fundamental. Sub-Poisson pumping in the laser also can lead to sub-Poisson light generation [16].

Effective arrangements for the generation of nonclassical light involve negative feedback [3, 17] controlling pump beams, the generation mechanism, or signal beams; e.g., one can use the pair of signal photons arising in down-frequency generation. As the sub-frequency photons are arising simultaneously from the pump photon, one of the sub-frequency beams can be used to influence the pumping beam or the other sub-frequency beam in producing nonclassical light; also, nonlinear dynamics inside the source can be modified. Yamamoto, Imoto, and Machida employed a semiconductor laser to produce sub-Poisson light with a Fano factor $F_n \approx 0.26$ in the negative feedback loop (Fig. 2), whereas the photocount distribution of the free-running laser was approaching the Poisson distribution [17]. Adopting a ring interferometer with a Kerr nonlinear medium, Machida, Yamamoto, and Itaya could realize a nondemolition measurement [16].

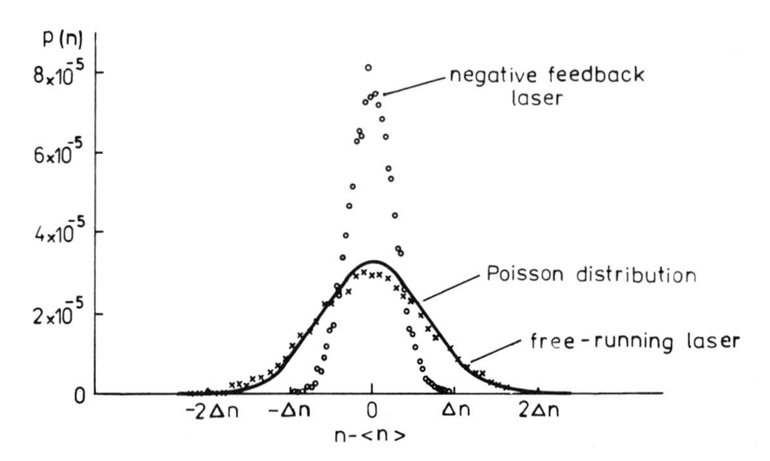

FIG. 2. Photocount distrbution of light from a semiconductor laser with negative feedback (circles) and in the free-running regime (crosses); the full curve represents the corresponding Poisson distribution for comparison. (After Y. Yamamoto, N. Imoto and S. Machida, *Phys. Rev. A* **33**, 3243 (1986).)

More substantial nonclassical effects were achieved in the generation of light with squeezed vacuum fluctuations. The first successful experiment was performed by Slusher *et al.*, based on four-wave mixing due to Na atoms in a cavity [18]. Two counterpropagating beams in a resonator served as pumping beams. Signal counterpropagating beams were generated from the vacuum in another resonator, both signal beams were mixed with a reference beam of a local oscillator, and the split beams were detected by two photodetectors whose photocurrents were subtracted and spectrally analyzed. When the four-wave mixing process was blocked, only a horizontal line of vacuum fluctuations was observed; when it was turned on, vacuum fluctuations were reduced for some phases of the local oscillator. Similar results for four-wave forward scattering on Na atoms were obtained by Maeda *et al.* [19]. Levenson *et al.* applied forward four-wave scattering to generate squeezed light in an optical fiber cooled to 2°K to avoid Brillouin scattering [20]. Effects of tens of percent were achieved.

The most pronounced effect of about 60% squeezing of vacuum fluctuations was observed by Wu *et al.* [10]. In a ring Nd:YAG laser operating at 1.06 μm with a nonlinear crystal, the second harmonic at 0.53 μm was generated and then used as a pump beam in an optical parametric oscillator in a cavity with the injected signal at 1.06 μm. Mixing this signal beam with a reference beam derived from the original laser beam at 1.06 μm, registering the sum and the difference of the beams, and performing spectral analysis of the difference of photocurrents, Wu *et al* were able to observe more than 1/2 of the squeezing of the circle of vacuum fluctuations. In fact, they estimated that fluctuations were squeezed to 1/10, but this high degree of squeezing was degraded by additional noise that was not under control in the measurement. They also proved experimentally that the generated squeezed light had minimal uncertainty.

Nonclassical behavior of light is a nice demonstration of quantum laws in nature, and it represents an excellent proof of the validity of quantum electrodynamics. Recent experiments on interference of nonclassical light exhibited wave-particle dualism for a single photon [21].

If squeezed light is injected into unused ports of optical interferometers, the sensitivity of the interferometer can be increased beyond the quantum limit, which offers the possibility of detecting gravitational waves and developing ultrahigh-precision nondemolition measuring techniques and optical sensors [22]. New perspectives have been opened in optical communications with nonclassical light, permitting an increase in channel capacity and a decrease in the number of photons per bit of information and error probability [3, 23]. Such a new kind of light represents a qualitatively new source for biological research, e.g., the research of human vision at threshold conditions [3].

References

1. R. J. Glauber, *Phys. Rev.* **130**, 2529 (1963); ibid. **131**, 2766 (1963).
2. J. Peřina, "Quantum Statistics of Linear and Nonlinear Optical Phenomena," D. Reidel, Dordrecht-Boston, 1984 (second enlarged edition in print); "Coherence of Light," D. Reidel, Dordrecht-Boston, 1985.
3. M. C. Teich and B. E. A. Saleh, "Progress in Optics," (E. Wolf, ed.), Vol. 26, p. 1, North-Holland, Amsterdam, 1988.
4. J. R. Klauder and E. C. G. Sudarshan, "Fundamental of Quantum Optics," W. A. Benjamin, New York, 1968.
5. H. J. Kimble, M. Dagenais, and L. Mandel, *Phys. Rev. Lett.* **39**, 691 (1977); *Phys. Rev. A* **18**, 201 (1978); M. Dagenais and L. Mandel, *Phys. Rev. A* **18**, 2271 (1978).
6. A. Lukš, V. Peřinová, and J. Peřina, *Opt. Comm.* **67**, 149 (1988); A. Lukš, V. Peřinová, and Z. Hradil, *Acta Phys. Pol. A* **74**, 713 (1988).
7. M. Born and E. Wolf, "Principles of Optics," Pergamon, Oxford, 1965.
8. R. Loudon, *Opt. Comm.* **70**, 109 (1989).
9. M. Kozierowski and V. I. Man′ko, *Opt. Comm.* **69**, 71 (1988) and references therein.
10. Ling-An Wu, H. J. Kimble, J. L. Hall, and Huifa Wu, *Phys. Rev. Lett.* **57**, 2520 (1986).
11. W. Schleich, D. F. Walls and J. A. Wheeler, *Phys. Rev. A.* **38**, 1177 (1988) and references therein.
12. V. Peřinová and M. Kárská, *Phys. Rev. A* **39**, 4056 (1989) and references therein.
13. V. Peřinová and A. Lukš, *Phys. Rev. A* **41**, 414 (1990) and references therein.
14. M. Kárská and J. Peřina, *J. Mod. Opt.* **37**, 195 (1990) and references therein.
15. R. Short and L. Mandel, *Phys. Rev. Lett.* **51**, 384 (1983).
16. S. Machida, Y. Yamamoto, and Y. Itaya, *Phys. Rev. Lett.* **58**, 1000 (1987); D. F. Smirnov and A. S. Troshin, *Opt. Spektr.* (USSR) **59**, 1 (1985); I. I. Katanayev and

A. S. Troshin, *J. Exper. Theor. Phys.* (USSR) **92**, 475 (1987); J. Bergou, L. Davidovich, M. Orszag, C. Benkert, M. Hillery, and M. O. Scully, *Opt. Comm.* **72**, 82 (1989); S. Machida and Y. Yamamoto, *Opt. Lett.* **14**, 1045 (1989).
17. J. G. Walker and E. Jakeman, *Opt. Acta* **32**, 1303 (1985); Y. Yamamoto, N. Imoto, and S. Machida, *Phys. Rev. A* **33**, 3243 (1986); J. G. Rarity, P. R. Tapster, and E. Jakeman, *Opt. Comm.* **62**, 201 (1987).
18. R. E. Slusher, L. W. Hollberg, B. Yurke, J. C. Mertz, and J. F. Valley, *Phys. Rev. Lett.* **55**, 2409 (1985).
19. M .W. Maeda, P. Kumar, and J. H. Shapiro, *Opt. Lett.* **12**, 161 (1987).
20. M. D. Levenson, R. M. Shelby, M. Reid, and D. F. Walls, *Phys. Rev. Lett.* **57**, 2473 (1986).
21. A. Aspect, P. Grangier, and G. Roger, *J. Optics (Paris)* **20**, 119 (1989); Z. Y. Ou, L. J. Wang, and L. Mandel, *Phys. Rev. A.* **40**, 1428 (1989).
22. C. M. Caves, *Phys. Rev. D* **23**, 1693 (1981); R. S. Bondurant and J. H. Shapiro, *Phys. Rev. D.* **30**, 2548 (1984).
23. Y. Yamamoto and H. A. Haus, *Rev. Mod. Phys.* **58**, 1001 (1986).

CHAPTER 18

One-Photon Light Pulses versus Attenuated Classical Light Pulses

Alain Aspect*

Collège de France and Laboratoire de Spectroscopie Hertzienne de l'Ecole Normale Supérieure
Paris, France

Philippe Grangier

Institut d'Optique Théorique et Appliquée
Université Paris XI
Orsay, France

1. An Old Question in a New Light: One-Photon Interferences

Almost as soon as Einstein suggested that light is made of grains of energy [1] (the term *photon* had not been invented yet), the question was raised whether a photon can interfere with itself. The first experiment on that question (to our knowledge) is the one by Taylor [2], where the diffraction pattern of a needle illuminated by an attenuated discharge lamp was recorded on a photographic plate [2]. The attenuated light was so feeble that the time of exposure was six months. The photon flux can be estimated to have been $10^6 s^{-1}$, so that "the average distance between two photons" (whatever this may mean) is 300 meters; one can conclude that "there never was more than one photon between the source and the photographic plate." With the progress of technology (especially the development of photodetectors and image intensifiers), experiments in this spirit have been repeated throughout this century, confirming that interferences do not disappear even for very attenuated light (see Table I).

* At the time of the experiments of Section 4, the author was with Institut d'Optique.

ISBN 0-12-289690-4

Table I. Feeble light interference experiments. All these experiments have been realized with attenuated light from a usual source (atomic discharge).

Author	Date	Experiment	Detector	Photon Flux (s^{-1})	Interferences
Taylor (a)	1909	Diffraction	Photography	10^6	Yes
Dempster *et al.* (b)	1927	(i) Grating	Photography	10^2	Yes
		(ii) Fabry Pérot	Photography	10^5	Yes
Janossy *et al.* (c)	1957	Michelson Interferometer	Photomultiplier	10^5	Yes
Griffiths (d)	1963	Young slits	Image Intensifier	2×10^3	Yes
Reynolds *et al.* (e)	1968	Young slits	Photomultiplier	2×10^4	Yes
Donstov *et al.* (f)	1967	Fabry Pérot	Image Intensifier	10^3	No
Reynolds *et al.* (g)	1969	Fabry Pérot	Image Intensifier	10^2	Yes
Bozec *et al.* (h)	1969	Fabry Pérot	Photography	10^2	Yes
Grishaev *et al.* (i)	1969	Jamin Interferometer	Image Intensifier	10^3	Yes

(a) G. I. Taylor, *Proc. Cambridge Philos. Soc.* **15**, 114 (1909).
(b) A. J. Dempster and H. F. Batho, *Phys. Rev.* **30**, 644 (1927).
(c) L. Janossy and Z. Naray, *Acta Phys. Hungaria* 7, 403 (1967).
(d) H. M. Griffiths, *Princeton University Senior Thesis* (1963).
(e) G. T. Reynolds *et al.*, "Advances in Electronics and Electron Physics," **28 B**, Academic Press, London, 1969.
(f) Y. P. Dontsov and A. I. Baz, *Sov. Phys. JETP* **25**, 1 (1967).
(g) G. T. Reynolds, K. Spartalian, and D. B. Scarl, *Nuovo Cim.* **B 61**, 355 (1969).
(h) P. Bozec, M. Cagnet and G. Roger, *C. R. Acad. Sci.* **269**, 883 (1969).
(i) A. Grishaev *et al.*, *Sov. Phys. JETP* **32**, 16 (1969).

However, it is clear that the image of a light beam as a stream of usual particles is too simplistic. For instance, we know from quantum optics that there is no position operator for the photon [3]. Thus, it is necessary to have a more clear-cut situation to be able to claim that only one photon is involved. A tempting solution is to produce a very short pulse of feeble light, so that the average energy per pulse is much less than the one-photon energy $\hbar\omega$ (where $\hbar$ is the Planck constant and ω the angular frequency of the monochromatic light). One could hope then to have either zero photons or one photon per pulse. In this chapter, we want to show that this view also is too naive. A very attenuated light pulse is not identical to a one-photon light pulse. We will demonstrate that this distinction is very clear in the language of quantum optics. Furthermore, it is possible to discriminate experimentally between these two different types of light, and we have performed the corresponding experiment. A true one-photon pulse has been produced, indeed, and characterized. We will show that this experiment also is relevant to the discussion about the necessity of quantum optics versus the possibility of embedding optics within the framework of semi-classical theories of light.

This chapter is organized as follows: In Section 2, we give the theoretical framework of the discussion; i.e., we show that a one-photon pulse and an attenuated light pulse are described differently in quantum optics. However, measurements corresponding to the first-order intensity correlation function are expected to yield similar behaviors. In Section 3, we present a more sophisticated experimental scheme—correlation measurements on both sides of a beam splitter-and we show that quantum optics predicts different behaviors here for the two situations. We also address the question of the semi-classical description of such a situation. In Section 4, we present an experiment performed with the scheme discussed in Section 3. In the conclusion, we will see how these results illustrate the question of wave-particle duality of light.

2. Theoretical Description

2.1. General Framework

We will describe light pulses in the standard formalism of quantum optics, in the Heisenberg representation [3, 4]. A given light field is represented by a state vector independent of the time $|\Psi\rangle$, or by a density matrix $\hat{\rho}$ representing a mixture of such states. The field observables then depend on time (and position). The electric field operator usually is decomposed into two adjoint operators, $\hat{\mathbf{E}}^{-}(\mathbf{r}, t)$ and $\hat{\mathbf{E}}^{+}(\mathbf{r}, t)$ corresponding respectively to positive and negative frequencies. These operators can be expanded on modes of polarized travelling waves:

$$\hat{\mathbf{E}}^{+}(\mathbf{r},t) = i\sum_{j} \mathscr{E}_{\omega_j} \boldsymbol{\varepsilon}_j \hat{a}_j \exp[i(\mathbf{k}_j \cdot \mathbf{r} - \omega_j t], \tag{1a}$$

$$\hat{\mathbf{E}}^{-}(\mathbf{r},t) = [\hat{E}^{+}(\mathbf{r},t)]^{\dagger}. \tag{1b}$$

The mode j is characterized by a wave-vector $\mathbf{k}_j$, an angular frequency $\omega_j = c|\mathbf{k}_j|$, and a polarization $\boldsymbol{\epsilon}_j$ orthogonal to $\mathbf{k}_j$. The factor

$$\mathscr{E}_{\omega_j} = \sqrt{\frac{\hbar\,\omega_j}{2\varepsilon_0 L^3}} \tag{1c}$$

is a normalization constant, depending on the volume of quantization L^3 that should not appear explicitly in the final results of the calculations.

The adjoint operators $\hat{a}_j$, and $\hat{a}_j^{\dagger}$ are the destruction and creation operators for photons of the mode j. They obey the fundamental commutation relations

$$[\hat{a}_j, \hat{a}_{j'}^{\dagger}] = \delta_{jj'} \tag{2}$$

with $\delta_{jj'}$ the Kronecker symbol. They allow the construction a complete basis $\{|n_j\rangle; n_j = 0, 1 \ldots\}$ of the state space associated with the mode j:

$$\hat{a}_j^{\dagger}\,|n_j\rangle = \sqrt{n_j + 1}|n_j + 1\rangle, \tag{3a}$$

$$\hat{a}_j|n_j\rangle = \sqrt{n_j + 1}|n_j - 1\rangle, \tag{3b}$$

$$a_j|0_j\rangle = 0. \tag{3c}$$

These so-called *number states* are eigenstates of the operator "number of photons in the mode j":

$$\hat{N}_j = \hat{a}_j^{\dagger}\hat{a}_j, \tag{4a}$$

the corresponding eigenvalue being precisely the number of photons n_j:

$$\hat{N}_j|n_j\rangle = n_j|n_j\rangle. \tag{4b}$$

One also defines the operator "total number of photons":

$$\hat{N} = \sum_{j} \hat{N}_j. \tag{4c}$$

We mentioned already that there is no position operator for the photon, so that one cannot define a density of probability of presence, as in quantum mechanics of massive particles. However, there is a very useful quantity which allows linking the theory to experiments—namely, the probability of a photodetection per unit of time at the point $\mathbf{r}$. For a field in the state $|\Psi\rangle$, the rate of photodetection around the point $\mathbf{r}$ at time t is written as:

$$W_I(\mathbf{r},t) = s\langle\Psi|\hat{\mathbf{E}}^{-}(\mathbf{r},t)\hat{\mathbf{E}}^{+}(\mathbf{r},t)|\Psi\rangle, \tag{5}$$

where s is the sensitivity of the detector. The rate of double photodetection at $(\mathbf{r}, t)$ and $(\mathbf{r}', t')$ is defined by

$$d^2\mathcal{P} = w_{II}(\mathbf{r}, t; \mathbf{r}', t')dt\, dt', \tag{6a}$$

where $d^2\mathcal{P}$ is the probability of a double photodetection at $\mathbf{r}$ during the time interval $[t, t + dt]$ and at $\mathbf{r}'$ during $[t', t' + dt']$. In addition,

$$\begin{aligned} W_{II}(\mathbf{r}, t; \mathbf{r}', t') = s^2\langle\Psi|\hat{\mathbf{E}}^-(\mathbf{r}, t)\hat{\mathbf{E}}^-(\mathbf{r}', t') \\ \hat{\mathbf{E}}^+(\mathbf{r}', t')\hat{\mathbf{E}}^+(\mathbf{r}, t)|\Psi\rangle. \end{aligned} \tag{6b}$$

This formalism will allow us to compare the properties of a one-photon pulse versus attenuated classical light pulses. Now we will introduce particular states of each of these types.

2.2. One-Photon Wave Packet

Any light state of the form

$$|1\rangle = \sum_j c_j|n_j = 1\rangle \tag{7}$$

is an eigenstate of $\hat{N}$ (see Eq. 4c) corresponding to the eigenvalue 1: It is a one-photon state. As a simple example, we can consider such a state with all the excited modes propagating along the same direction defined by the unit vector $\mathbf{u}$, so that

$$\mathbf{k}_j = \mathbf{u}\frac{\omega_j}{c} \tag{8a}$$

Equation (5) then gives the rate of photodetection:

$$W_I(\mathbf{r}, t) = s\left|\sum_j \epsilon_j \mathcal{E}_{\omega_j} c_j \exp\left[i\omega_j\left(\frac{\mathbf{r}\cdot\mathbf{u}}{c} - t\right)\right]\right|^2, \tag{8b}$$

which clearly suggests a propagation along $\mathbf{u}$ at velocity c.

We can be more specific yet and introduce a Lorentzian distribution for $|c_j|^2$, in the form:

$$c_j = \frac{K_1}{\omega_j - \omega_0 + i\Gamma/2}, \tag{8c}$$

with K_1 a real constant such that the state vector is normalized (See the remark in this section.). For simplification, we take all modes with the same polarization,

$$\epsilon_j = \epsilon. \tag{8d}$$

If Γ is small compared to ω_0, $\mathscr{E}_{\omega j}$ can be taken constant in the sum (8b). By transforming this sum into an integral, one gets

$$W_I(\mathbf{r},t) = \eta\Gamma H\left(t - \frac{\mathbf{r}\cdot\mathbf{u}}{c}\right)\exp\left[-\Gamma\left(t - \frac{\mathbf{r}\cdot\mathbf{u}}{c}\right)\right], \tag{9a}$$

where η is a positive constant, and $H(t)$ is the Heaviside step function. The rate of photodetection (9a) at point $\mathbf{r}$ is represented as a function of time in Fig. 1. We clearly see a wave packet propagating along $\mathbf{u}$ and exponentially damped with a time constant Γ^{-1}.

Note that Eq. 9a (or Fig. 1) must be understood in a statistical sense. One prepares a field in the form defined by Eqs. (8) at time $t = 0$, and one looks for photodetection by a detector at position $\mathbf{r}$. If there is photodetection, its time is recorded. The experiment then is repeated a great number of times, and the histogram of the results will look like Fig. 1.

Remark The modulus of the constant K_1 of Eq. (8c) is determined by the condition of normalization of the state vector $|\mathbf{1}\rangle$ corresponding to the coefficients (8c). It obeys

$$|K_1|^2 = \frac{\Gamma^2}{4\pi}\rho(\omega_0), \tag{9b}$$

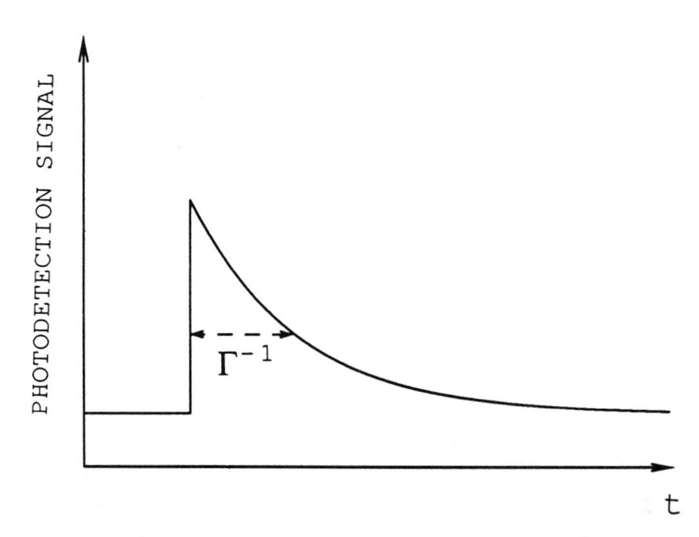

FIG. 1. Rate of photodetection at point $\mathbf{r}$ as a function of time, for the one-photon wave packet presented in this section. Since there cannot be more than one detection per pulse, this figure must be understood in a statistical sense.

where $\rho(\omega)$ is the density of modes, proportional to L^3. We then have

$$\eta = 4\pi^2 s|K_1|^2 \mathscr{E}_{\omega_0}^2. \tag{9c}$$

Note that η can be interpreted as a detection efficiency per photon (one just has to integrate Eq. (9a) over t).

If K_1 is taken to be imaginary, with a phase $\varphi = \omega t_0$, the result (9a) is changed to the value obtained by replacing t by $t - t_0$. In other words, the wave packet is translated in time by the quantity t_0.

2.3. Quasi-classical Wave Packet

When we speak about *classical light*, we have in mind a field that behaves according to classical electromagnetic theory, in which fields are C numbers obeying Maxwell's equations. We know that such fields exist in quantum optics: They are called *Glauber's coherent states* or *quasi-classical states*. A quasi-classical state $|\alpha_j\rangle$ of the mode j is an eigenstate of the destruction operator $\hat{a}_j$:

$$\hat{a}_j|\alpha_j\rangle = \alpha_j|\alpha_j\rangle, \tag{10a}$$

with α_j a complex number. A multimode quasi-classical state is

$$|\Psi_{\text{qc}}\rangle = |\alpha_{j=1}\rangle \otimes |\alpha_{j=2}\rangle \otimes \cdots \otimes |\alpha_j\rangle \otimes \cdots. \tag{10b}$$

This state vector is an eigenstate of the positive-frequencies electric field operator (1a):

$$\hat{\mathbf{E}}^+(\mathbf{r}, t)|\Psi_{\text{qc}}\rangle = \mathbf{E}_{\text{c}}^+(\mathbf{r}, t)|\Psi_{\text{qc}}\rangle \tag{10c}$$

with

$$\mathbf{E}_{\text{c}}^+(\mathbf{r}, t) = i \sum_j \mathscr{E}_{\omega_j} \alpha_j \boldsymbol{\epsilon}_j \exp\{i(\mathbf{k}_j \cdot \mathbf{r} - \omega_j t)\}, \tag{10d}$$

the positive frequencies being part of a classical field associated with $|\Psi_{\text{qc}}\rangle$.

Such a state—or, more generally, a statistical mixture of states of the form (10b)—is the quantum description of the light emitted by classical sources like a blackbody (or thermal sources), discharge lamps with many independently excited atoms, and a laser oscillating above threshold.

We consider again the case of a propagation along $\mathbf{u}$:

$$\mathbf{k}_j = \mathbf{u}\frac{\omega_j}{c}, \tag{11a}$$

and of a single polarization

$$\boldsymbol{\epsilon}_j = \boldsymbol{\epsilon}. \tag{11b}$$

If the α_j are taken with a distribution

$$\alpha_j = \frac{K_c}{\omega_j - \omega_0 + i\Gamma/2}, \tag{11c}$$

we find a rate of photodetection,

$$W_I(\mathbf{r}, t) = \eta' \Gamma H\left(t - \frac{\mathbf{r} \cdot \mathbf{u}}{c}\right) \exp\left[-\Gamma\left(t - \frac{\mathbf{r} \cdot \mathbf{u}}{c}\right)\right], \tag{12a}$$

which is similar to Eq. (9a); the positive constant η' is related to η (Eq. (9c)) by

$$\eta' = \eta \sum_j |\alpha_j|^2. \tag{12b}$$

Like (9a), Eq. (12a) suggests the propagation of a wave packet damped with a time-constant Γ^{-1}. However, the quasi-classical wave packet introduced here differs in many aspects from the one-photon wave packet of Section 2.2. The most striking difference will be seen in Section 3. Here, we can note already that the quasi-classical state $|\Psi_{qc}\rangle$ is not an eigenstate of the number of photons operator $\hat{N}$. It nevertheless is possible to calculate the average photon number:

$$\langle N \rangle_{qc} = \langle \Psi_{qc} | \hat{N} | \Psi_{qc} \rangle = \sum_j |\alpha_j|^2 = \frac{4\pi}{\Gamma^2} |K_c|^2 \tag{13a}$$

and the standard deviation

$$\langle \Delta N \rangle_{qc} = [\langle \hat{N}^2 \rangle - (\langle \hat{N} \rangle)^2]^{1/2} = [\langle N \rangle_{qc}]^{1/2}. \tag{13b}$$

It is important to realize that the constant K_c can be chosen arbitrarily (contrary to constant K_1 in the case of a one-photon wave packet). A quasi-classical wave packet thus can be built with any average photon number. In particular, K_c can be chosen small enough to get an average photon number smaller than one. Such a state is the theoretical quantum description of a quasi-classical pulse that has been attenuated by a neutral density. It is different from a single-photon state.

Remark By integration over time of Eq. (12a), one finds a total probability of detection per pulse equal to $\eta \langle N \rangle_{qc}$, which is consistent with the interpretation of η as the detection efficiency per photon.

Remark The calculation of $W_I(\mathbf{r}, t)$, starting from the quantum Eq. (5), yields in the case of a quasi-classical state $|\Psi_{qc}\rangle$

$$W_I(\mathbf{r}, t) = s|\mathbf{E}_c^+(\mathbf{r}, t)|^2, \tag{13c}$$

which is the same expression as the one that would be obtained for the classical field $\mathbf{E}_c^+(\mathbf{r}, t)$, provided that the probability of photodetection is taken proportionally to the light intensity. We thus find that for any quasi-classical state, there exists an associated classical field for which the classical theory of light predicts the same probability of photodetection as the quantum theory does.

3. Anticorrelation on a Beamsplitter

3.1. Physical Idea

We would like to find an experimentally realizable situation when the two different types of wave packets introduced in Section 2 behave differently. It is well-known that second-order intensity correlation functions are very sensitive to differences that do not appear in the first-order correlation function. This is the idea, for instance, underlying experiments on the *antibunching* effect [5a, 5b].

A very simple way to access to the second-order correlation function is to look for joint detections behind a beamsplitter. Before using the formalism of Section 2, we can try to guess the behavior of each type of wave packet. In the one-photon case, the most naive picture of a particle suggests that it either is transmitted or reflected by the beamsplitter, but it is not split, and no joint detection is expected. On the contrary, a quasi-classical wave packet is associated with the image of a classical light pulse, which is split into two parts, and there is some probability of joint detections.

To make the discussion fully quantitative, and to lead to an experimental test, we will consider a specific scheme taking into account the duration of the wave packet (Fig. 2). This scheme involves a pulsed source, launching light pulses towards a beamsplitter with two detectors in the transmitted

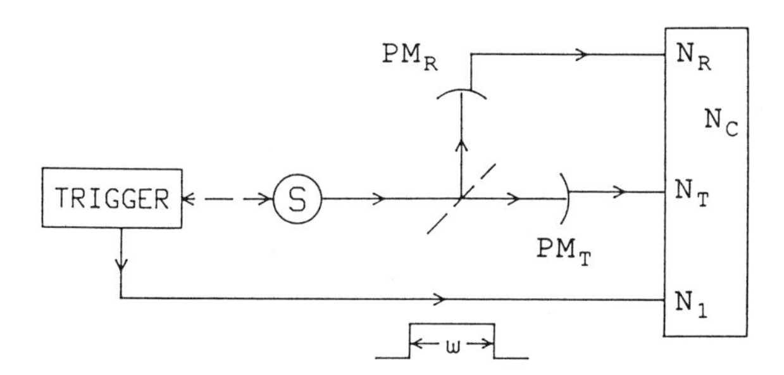

FIG. 2. Experiment to look for an anticorrelation on a beamsplitter. The source S emits light pulses that fall on a beamsplitter and can be detected in both channels (reflected, and transmitted) behind the beamsplitter. The detectors are enabled during a gate w synchronized with the light pulses. The rates of single detection (N_R and N_T) and coincidences (N_C) are monitored. If the light pulse is a one-photon wave packet, one detection per pulse is expected at most, and no coincidence is expected: This is the signature of the anticorrelation effect.

and reflected legs. Every time a light pulse is emitted, there is a triggering system that generates a gate of duration w, somewhat longer than the decay constant Γ^{-1} of the wave packet. This gate enables each detector, during the time interval when the light pulse is expected to fall onto the detectors. A counting system monitors the detection events. If both detectors are fired during the same gate, a coincidence is recorded.

A given experiment will consist of running the pulsed source for a given duration, and counting the total number of counts in the transmitted (N_T) or reflected (N_R) channels, the total number of coincidences (N_c), and the total number of gates N_1. We then can estimate the probabilities of single detections per gate,

$$P_R = \frac{N_R}{N_1} \quad \text{and} \quad P_T = \frac{N_T}{N_1}, \tag{14a}$$

and the probability of a coincidence per gate,

$$P_c = \frac{N_c}{N_1}. \tag{14b}$$

According to our intuitive reasoning, we expect P_c to be zero in the case of a single-photon wave packet, and to be non-zero in the case of the quasi-classical pulse. We now will see how large the expected difference is.

3.2. Predictions of Quantum Optics

To calculate the various detection rates in the experiment of Fig. 2, we need to take into account the beamsplitter. The simplest way to do it is to express the electric field operators at the detectors (after the beamsplitter) as a function of the electric field operator in the input space of the beamsplitter [6]. It turns out that the relation is exactly the same as the input/output relations on a beamsplitter for classical amplitudes in classical optics. Denoting R and T the reflection factor and transmission factor for the intensity, we have

$$W_I^{(T)}(\mathbf{r}, t) = TW_I(\mathbf{r}, t) \tag{15a}$$

and

$$W_I^{(R)}(\boldsymbol{\rho}, t) = RW_I(\mathbf{r}, t) \tag{15b}$$

for the single detection rates in the transmitted and reflected channels; $\mathbf{r}$ corresponds to the image of $\boldsymbol{\rho}$ by reflection in the beamsplitter; $W_I(\mathbf{r}, t)$ is the single detection rate that would be found at $\mathbf{r}$ with the beamsplitter removed. With similar notations, the joint detection rate for detectors at $\mathbf{r}$

and $\boldsymbol{\rho}'$ is

$$W_H^{(T,R)}(\mathbf{r},t;\boldsymbol{\rho}',t) = TRW_{II}(\mathbf{r},t;\mathbf{r}',t'). \tag{15c}$$

Let us apply these results to the case of a one-photon wave packet. Inserting the one-photon state (7) into the expression (6b) of the joint-detection rate, we get

$$W_{II}(\mathbf{r},t;\mathbf{r}',t') = 0 \tag{16a}$$

because

$$\hat{a}_j^2|1_j\rangle = \hat{a}_j|0_j\rangle = 0. \tag{16b}$$

We thus find a probability of joint detection

$$P_c = 0, \tag{16c}$$

as was guessed by intuitive arguments. The single-detection probabilities are

$$P_R = R\eta \quad \text{and} \quad P_T = T\eta. \tag{16d}$$

The most striking result for a one-photon wave packet is that the joint detection probability P_c is zero, although the single-detection probabilities are both different from zero.

If we now apply results (15) to the case of a quasi-classical wave packet $|\Psi\rangle$ (Section 2.3), we find single-detection probabilities

$$P_R = R\eta\langle N\rangle_{qc} \quad \text{and} \quad P_T = T\eta\langle N\rangle_{qc}, \tag{17a}$$

with $\langle N\rangle_{qc}$ the average photon number (Eq. (13a)). Similarly, the joint detection probability is

$$P_c = TR\eta^2\langle N\rangle_{qc}^2 = P_R P_T. \tag{17b}$$

As expected, in the case of a quasi-classical wave packet, the probability of a coincidence behind the beamsplitter does not go to zero even in the case of a very attenuated pulse.

If we now consider a statistical mixture of quasi-classical pulses $|\Psi_{qc}^{(i)}|$, described by a density operator

$$\hat{\rho} = \sum_i \rho_i |\Psi_{qc}^{(i)}\rangle\langle\Psi_{qc}^{(i)}| \tag{18.a}$$

the expressions for the simple probabilities of detection remain identical to (17a), provided that the average photon number is taken as the statistical average of the quantum average per pulse:

$$\langle N\rangle_{qc} = \sum \rho_i \langle N^{(i)}\rangle_{qc}, \tag{18b}$$

On the other hand, the probability of a coincidence is given by

$$P_c = RT\eta^2 \sum_i \rho_i \langle N^{(i)} \rangle_{qc}^2, \tag{18c}$$

and because of the trivial Cauchy-Schwartz inequality

$$\sum_i \rho_i \langle N^{(i)} \rangle_{qc}^2 \geq \left(\sum_i \rho_i \langle N^{(i)} \rangle_{qc} \right)^2, \tag{19a}$$

we finally get

$$P_c \geq P_R P_T. \tag{19b}$$

3.3. Discussions

Quantum optics thus predicts different results for the anticorrelation experiment described in Section 3.1, according to the kind of wave packet that is tested. For a one-photon wave packet, a perfect anticorrelation is predicted: There never is any coincidence between the transmitted and reflected channels. In the case of a quasi-classical wave packet, there is a finite probability of coincidence, which cannot be smaller than a minimum value given by Eq. (19b). If we consider again the experimental scheme of Fig. 2 allowing us to measure the relevant probabilities, the various counting numbers are expected to obey the inequality

$$\beta = \frac{N_c N_1}{N_R N_T} \geq 1 \tag{20}$$

in the case of a quasi-classical pulse.

We thus have a quantitative criterium, based on experimentally measurable quantities, allowing discrimination between a quasi-classical wave packet and a wave packet with a one-photon character.

This discussion can be rephrased in the context of a comparison between the quantum theory of light and the so-called *semi-classical* theories of light. These theories form a theoretical description where the light is treated classically; i.e., it is a classical electromagnetic field obeying Maxwell's equations. The photodetection then is described as an interaction between this classical field and a quantized detector [7], and the photodetection signal (for instance, the counting rate of pulses at the anode of a photomultiplier) is proportional to the intensity of the light. Such a theoretical model can be considered as a marginal case of quantum optics for a quasi-classical state of the light. Indeed, for a quasi-classical state that is an eigenstate of $\hat{\mathbf{E}}^+(\mathbf{r}, t)$ corresponding to the eigenvalue $\mathbf{E}_c(\mathbf{r}, t)$ (Eqs. (10)), quantum optics gives a photodetection rate

$$W_I(\mathbf{r}, t) = s\langle \psi_{qc} | \hat{\mathbf{E}}^-(\mathbf{r}, t) \hat{\mathbf{E}}^+(\mathbf{r}, t) | \psi_{qc} \rangle = s|\mathbf{E}_c^+(\mathbf{r}, t)|^2. \tag{21}$$

In this equation, $|\mathbf{E}_c^+(r,t)|^2$ is the light intensity of the classical field. More generally, any detection signal can be calculated in quantum optics as an average of a *normally ordered operator* [3, 4], i.e., a combination of electric field operators with $\hat{\mathbf{E}}^+$ acting on the right and $\hat{\mathbf{E}}^-$ acting on the left. It is possible then to generalize Eq. (21), i.e., to show that in the case of a quasi-classical state, the prediction of quantum optics is the same as the semi-classical formula applied to the classical field $\mathbf{E}_c(\mathbf{r}, t)$ associated with this quasi-classical state. In particular, the inequality (20) (or 19b), which has been proved in quantum optics for a quasi-classical state, also will hold for a semi-classical description of any wave packet. The semi-classical theories of light do not predict anticorrelation on a beamsplitter. As in the antibunching effect [5a, 5b], the observation of such an anticorrelation thus would be evidence against the semi-classical theories of light.

In conclusion, we have studied a situation where quantum optics predicts different behaviors for a one-photon wave packet and a quasi-classical wave packet (even strongly attenuated). We have found a criterion allowing us to discriminate between these two behaviors. This criterion (inequality 20, or 19b) also can be considered a way to test the semi-classical theories of light.

In the next section, we will present two experiments in which these two different types of light pulses have been produced and characterized, thanks to this criterion.

4. Anticorrelation on a Beamsplitter: Experimental Measurements

We have built an experimental setup corresponding to the scheme of Fig. 2, i.e., a setup allowing us to measure the single and coincidence rates on the two sides of a beamsplitter, during the opening of gates triggered by events synchronous with the light pulses. This system has been used to study light pulses emitted by a classical source (Section 4.1). Then, we have studied pulses from a source designed to emit one-photon wave packets, as will be explained in Section 4.2.

4.1. Attenuated Quasi-Classical Pulses

To confirm our arguments experimentally, and also to test the photon counting system, we first studied light from a pulsed photodiode. It produced light pulses with a rise time of 1.5 ns and a fall time about 6 ns. The gates, triggered by the electric pulses driving the photodiode, were 9 ns wide (as in 4.2) and had an almost complete overlap with the light pulses.

The source was attenuated to a level corresponding to one detection per 1,000 pulses emitted. With a detector quantum efficiency of about 10%,

Table II. Anticorrelation experiment for light pulses from an attenuated photodiode (0.01 photon/pulse). The last column corresponds to the expected number of coincidences for $\beta = 1$. All the measured coincidences are compatible with $\beta = 1$; there is no evidence of anticorrelation. Note that the singles rates are similar to the ones of Table III.

Trigger rates	Singles rates		Duration	Measured coincidences	Expected coincidences for $\beta = 1$
$N_1(s^{-1})$	$N_{2r}(s^{-1})$	$N_{2t}(s^{-1})$	$\theta(s)$	$N_c\theta$	$\frac{N_R N_T}{N_1}\theta$
4760	3.02	3.76	31 200	82	74.5
8880	5.58	7.28	31 200	153	143
12 130	7.90	10.2	25 200	157	167
20 400	14.1	20.0	25 200	341	349
35 750	26.4	33.1	12 800	329	313
50 800	44.3	48.6	18 800	840	798
67 600	69.6	72.5	12 800	925	955

the average energy per pulse can be estimated to be about 0.01 photon. In the context of Table I, this source certainly would have been considered a source of single photons.

Table II shows the results of the anticorrelation measurements. The quantity β (of inequality (20)), is consistently found equal to 1; i.e., no anticorrelation is observed. In fact, the coincidence rate is exactly in agreement with the limit of inequality (20).

This experiment thus supports the claim that light emitted by an attenuated classical source does not exhibit one-photon behavior on a beamsplitter. This has been found true, even with very attenuated light pulses with an average energy by pulse of the order of 10^{-2} photon.

4.2. One-Photon Pulses

When an atom is excited to a resonance-excited level, the fluorescent light emitted is a one-photon wave packet. (We must have energy conservation.) However, in usual sources, such as discharge lamps, many excited atoms are seen simultaneously by the detectors, and their time of excitation is random. The theoretical description of the light then is a mixture of one-photon wave packets of the form presented in Section 2.2, with random initial times (see remark). If one also takes into account the fluctuations of the number of excited atoms, one can show that the emitted light can be considered a mixture of quasi-classical states, and there is no hope in this situation to observe any non-classical effect.

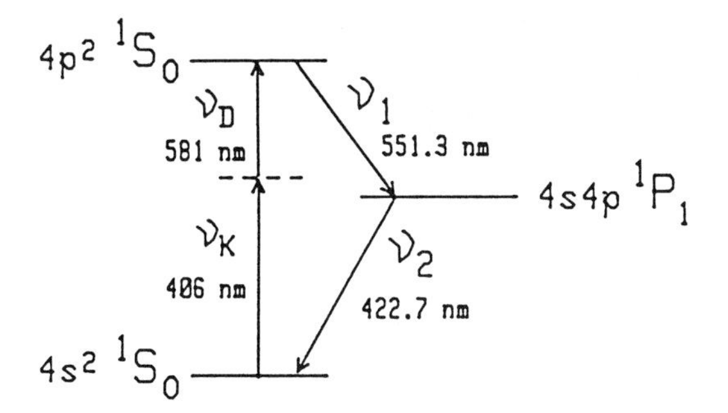

FIG. 3. Radiative cascade in calcium, used to produce the one-photon pulses. The atom is excited to its upper level by a two-photon excitation with two lasers. It then re-emits photons ν_1 and ν_2.

To observe non-classical properties in fluorescent light, it thus is necessary to isolate single-atom emission. This was realized in space, in the antibunching experiments of [5a, 5b, and 6]. In our experiment, we have been able to isolate single-atom emission, not in space, but in time. Our source is composed of atoms that we excite to the upper level of a two-photon radiative cascade (Fig. 3) [8]. The atom then decays by the emission of two photons at different frequencies ν_1 and ν_2. The time intervals between the detections of ν_1 and ν_2 are distributed according to an exponential law, corresponding to the decay of the intermediate state with a lifetime τ_s. By choosing the rate of excitation of the cascades much smaller than $(\tau_s)^{-1}$, we have cascades well separated in time. We can use the detection of ν_1 as a trigger for a gate of duration $w = 2\tau_s$, corresponding to the scheme of Fig. 2. During a gate opening, the probability of detecting a photon ν_2 coming from the same atom that emitted ν_1 is much bigger than the probability of detecting a photon ν_2 coming from any other atom in the source. We then are in a situation close to an ideal single-photon pulse [9], and we expect the corresponding anticorrelation behavior on the beamsplitter.

The expected values of the counting rates can be obtained by a straightforward quantum mechanical calculation. Denoting N the rate of excitation of the cascades, and η_1, η_T and η_R the detection efficiencies of photons ν_1 and ν_2 (including the collection solid angles, optics transmissions, and detector efficiencies) we obtain:

$$N_1 = \eta_1 N \tag{22a}$$

$$N_T = N_1 \eta_R [f(w) + Nw] \tag{22b}$$

$$N_R = N_1 \eta_R [f(w) + Nw] \tag{22c}$$

$$N_c = N_1 \eta_T \eta_R [2f(w)Nw + (Nw)^2] \tag{22d}$$

The quantity $f(w)$, very close to 1 in this experiment, is the product of the factor $[1-\exp(-w/\tau_s)]$ (overlap between the gate and the exponential decay) by a factor somewhat greater than 1, related to the angular correlation between ν_1 and ν_2 [10].

The quantum mechanical prediction for β thus is

$$\beta_{QM}=\frac{2f(w)Nw+(Nw)^2}{[f(w)+Nw]^2}, \tag{23}$$

which is smaller than 1, as expected. The anticorrelation effect will be stronger (β small compared to 1) if Nw can be chosen much smaller than $f(w)$. This condition actually corresponds to the intuitive requirement that N is smaller than $(\tau_s)^{-1}$ (where w is of the order of τ_s).

Counting electronics, including the gating system, was a critical part of this experiment. The gate w actually was realized by logical decisions based on the measurement of the time intervals between the detections at the various detectors. This allowed us to adjust the gates with an accuracy of 0.1 ns. The system also yields various time-delay spectra, useful for consistency checks.

Table III shows the measured counting rates for different values of the excitation rate of the cascade. The corresponding values of β have been plotted on Fig. 4 as a function of Nw. As expected, the violation of inequality (20) increases when Nw decreases, but the signal decreases simultaneously, and it becomes necessary to accumulate the data for long periods of time to achieve a reasonable statistical accuracy. A maximum violation of more than 13 standard deviations has been obtained for a

Table III. Anticorrelation experiment with single-photon pulses from the radiative cascade. The last column corresponds to the expected number of coincidences for $\beta=1$. The measured coincidences show a clear anticorrelation effect. These data can be compared to the ones of Table II.

Trigger rates	Singles rates		Duration	Measured coincidences	Expected coincidences for $\beta=1$
$N_1(s^{-1})$	$N_R(s^{-1})$	$N_T(s^{-1})$	$\theta(s)$	$N_c\theta$	$\frac{N_R\ N_T}{N_1}\theta$
4 720	2.45	3.23	1 200	6	25.5
8 870	4.55	5.75	17 200	9	50.8
12 100	6.21	8.44	14 800	23	64.1
20 400	12.6	17.0	19 200	86	204
36 500	31.0	40.6	13 200	273	456
50 300	47.6	61.9	8 400	314	492
67 100	71.5	95.8	3 600	291	367

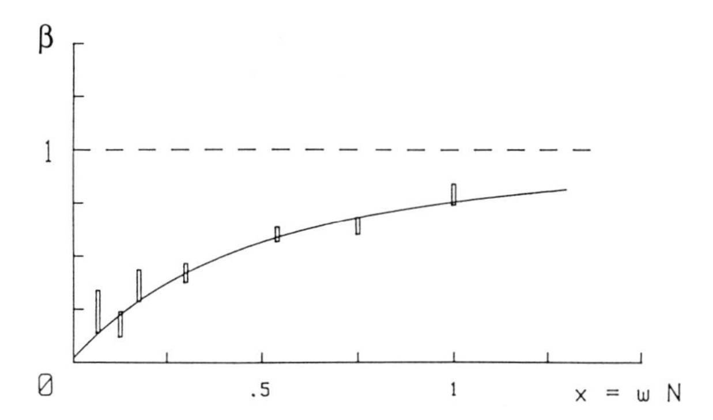

FIG. 4. Correlation parameter β as a function of the excitation rate of the cascade N. The value of β smaller than 1 is the signature of an anticorrelation, corresponding to the one-photon behaviour. The solid line is the prediction of qantum optics taking into account the possibility that several atoms are excited during the same gate w. For a pure one-photon pulse, β would be zero.

counting time of five hours. The value of β then is 0.18 ± 0.06, corresponding to a total number of coincidences of nine, instead of the minimum value of 50 expected for a quasi-classical pulse.

The experiments presented in this section confirm that anticorrelation on a beamsplitter is a very clear criterion for discriminating between a one-photon pulse and a quasi-classical wave packet. A pulse produced by a classical source, even attenuated to a level of 10^{-2} average photon number per pulse, has the behavior expected for a quasi-classical pulse: One observes coincidences in agreement with the inequality (20). However, we have been able to produce one-photon pulses for which the number of coincidences was so small that a violation of inequality (20) by more than 13 standard deviations has been observed. This last result also can be considered as experimental evidence against semi-classical theories of light, which never predict a violation of inequality (20).

5. Conclusion

There are other experiments in which a one-photon pulse is produced with information about the time of emission of the wave packet [11a, 11b]. In these experiments, pairs of photons are emitted by parametric down-conversion, and one of the two photons can be used as timing information about the second one. Non-classical statistics in the photoelectrons have been observed. The timing information also can be used to trigger a shutter placed on the path of the second photon, producing *sub-poissonian* light [12a, 12b]. In principle, parametric down-conversion in a nonlinear crystal is a better system than an atomic cascade, because the directions of emissions of the two photons are strongly correlated by the phase-matching condition. When all the technical problems are solved, this technique should lead to a better source of single-photon pulses than the one described in Section 4.2.

The understanding that an attenuated quasi-classical pulse is not equivalent to a one-photon pulse naturally has led us to revisit the question of one-photon interferences. We have performed an interference experiment, indeed, with the one-photon pulses emitted by our source, and we have observed interferences with a visibility close to 1 (Fig. 5) [9]. More recently, single photons produced in parametric down-conversion have

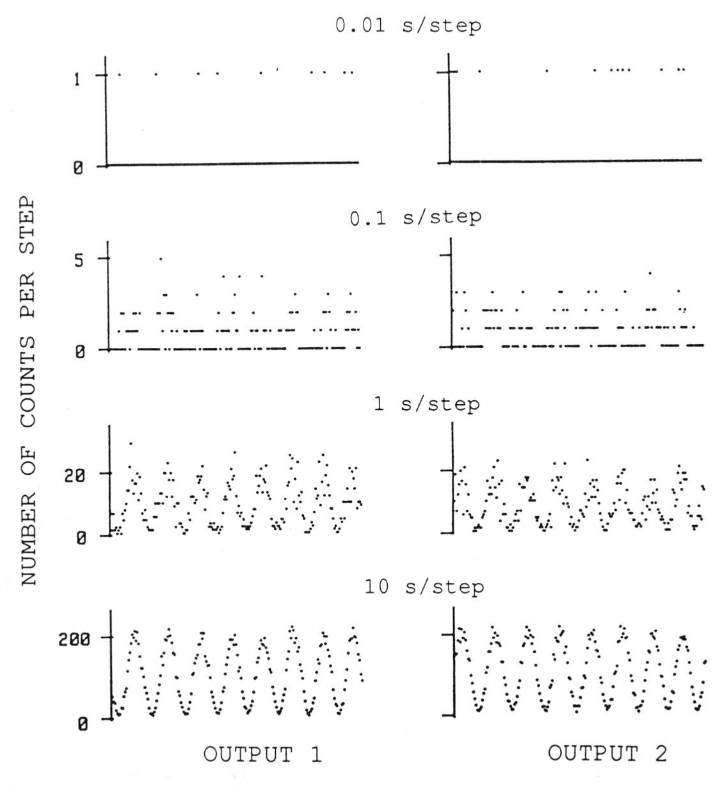

FIG. 5. One-photon interferences. This figure shows the number of detected counts in the two output channels of a Mach-Zehnder interferometer, as a function of the path difference in the interferometer, which is changed by steps. From top to bottom, the time spent at each path difference varies from 10^{-2}s to 10s. The interferometer was illuminated by the source of one-photon pulses described in Section 4.2, operated in the regime where $\beta = 0.2$ (strong anticorrelation), and the detectors were gated according to the procedure of Section 4.2. The visibility of the fringes is larger than 0.98.

been used in the context of *delay choice* interference experiments [13], i.e., a situation in which the wave-particle duality of light is especially amazing.

As a final conclusion, we would like to insist on the fact that more than 80 years after some properties of light suggested quantization, optics still is a domain in which the most specific features of quantum physics can be demonstrated. The non-exhaustive list—which includes wave-particle duality, non-classical statistics, violation of Bell's inequalities, squeezing, and quantum non-demolition measurements—certainly is not closed.

References

1. A. Einstein, *Annalen der Physik* **17**, 132 (1905).
2. G. I. Taylor, *Proc. Cambridge Philos. Soc.* **15**, 114 (1909).
3. See, for instance, C. Cohen-Tannoudji, J. Dupont-Roc, and G. Grynberg, "Photons and Atoms," Wiley, 1989.
4. R. Loudon, "The Quantum Theory of Light," 2nd edition, Oxford University Press, 1985.

5a. H. J. Kimble, M. Dagenais, and L. Mandel, *Phys. Rev. Lett* **39**, 691 (1977).

5b. J. D. Cresser, J. Hager, G. Leuchs, M. Rateike, and H. Walther "Dissipative Systems in Quantum Optics," *Topics in Current Physics* **27**, (R. Bonifacio ed.), Springer-Verlag (1982).

6. Z. Y. Ou, C. K. Hong, and L. Mandel, *Optics Commun* **63**, 118 (1977).
7. W. E. Lamb and M. O. Scully, in "Polarisation, Matière et Rayonnement," volume in honour of A. Kastler, Presses Universitaires de France, Paris (1969).
8. A. Aspect, P. Grangier, and G. Roger, *Phys. Rev. Lett.* **47**, 460 (1981).
9. P. Grangier, G. Roger, and A. Aspect, *Europhys. Lett.* **1**, 173 (1986).
10. E. S. Fry, *Phys. Rev.* **A8**, 1219 (1973).

11a. R. G. Brown, E. Jakeman, E. R. Pike, J. G. Rarity and P. Tapster, *Europhysics Lett.* **2**, 279 (1986).

11b. C. K. Hong, and L. Mandel, *Phys. Rev. Lett.* **56**, 58 (1986). In fact, these authors (Ref. 11a and 11b) did not analyse their experiment as producing one-photon pulses, but our analysis clearly applies to their experiment. Also, they were not looking for anticorrelation on a beamsplitter, but for non-classical statistics in photoelectron statistics.

12a. B. E. A. Saleh and M. C. Teich, *Optics Commun.* **52**, 429 (1985). This idea has been implemented by the authors in the following reference.

12b. J. G. Rarity, P. R. Tapster, and E. Jakeman, *Optics Commun.* **62**, 201 (1987).

13. J. Baldzuhn, E. Mohler, and W. Martienssen, *Zeitschrift Phys.* **B 77**, 77 (1989).

CHAPTER 19

Optical Propagation through the Atmosphere

Anna Consortini
Department of Physics, University of Florence
Florence, Italy

1. Introduction

Optical radiation that propagates through the atmosphere undergoes a number of effects that can be detected by the systems receiving it.

The first such system was—and still is—the human eye. It sees the wavelength-selective scattering of atomic and molecular gases when looking at the blue sky or a beautiful red sunset. It sees the effect of nonselective scattering due to fog and cloud particles, and their absorption in certain cases. Finally, it sees the effect of atmospheric turbulence, which makes stars twinkle at night and object images quiver when they are seen above a warm surface in summer.

The problem of the influence of the atmosphere on man-made systems has been of interest for more than a century in astronomical observations, and has become of great importance since the advent of the laser and the many systems based on it. A large number of systems are involved, ranging from astronomical telescopes to ground- or space-based laser systems with different applications, such as communications, remote sensing, and image formation.

Generally, the effects of the atmosphere on optical waves are divided into two main classes: propagation and scattering. For *propagation*, the atmosphere is considered as a continum with slow variations of the refractive index. *Slow* means that the changes in the refractive index take place

ISBN 0-12-289690-4

over scales much larger than the wavelength. Random fluctuations due to atmospheric turbulence and also continuous variations, such as the dependence of the refractive index on height, are among the most important slow variations. *Scattering* refers to discontinuities of the refractive index that are not slow, and includes all cases where there are particulates or corrugated boundaries. Although scattering and propagation are two different facets of an atmospheric scattering phenomenon due to refractive index variations, with propagation being a forward scattering, different analytical and experimental techniques are utilized to investigate them. Thus, they are generally kept separate.

Here, we will limit ourselves to coherent optical propagation through atmospheric turbulence by neglecting not only scattering, but also effects, like beam bending, related to the systematic variations of the refractive index. In addition, in spite of their present interest, we will also neglect the nonlinear effects that take place when powerful lasers are propagated.

Our field of discussion is still wide, however, despite considerable reduction.

In a few words, we can say that the effect of turbulence on optical waves is to produce random (in space and time) phase and amplitude fluctuations. The ultimate effect is generally a deterioration of the performance of an optical systems utilizing it. The problems of propagation study involve relating the fluctuations of amplitude and phase to those of the refractive index. Amplitude and phase second-order correlation functions, or their spectra, have been studied extensively for many different situations because they describe the *mean behavior* of systems, such as imaging systems. Great effort has also been devoted to the problem of the fourth-order correlation, which is of interest in deriving the *standard* deviation of system behavior. Big steps have been made toward the solution of the laser scintillation problem, as described later.

Recently, optical adaptive techniques and phase-conjugate methods have been introduced to correct the effect of phase deterioration. As is well known from optics, the main role in optical systems is played by the phase (think of the way the Airy pattern is obtained from an apeture), so that even a partial correction of phase fluctuations can give rise to big improvements. These techniques are not described here, being the subject of the chapter by F. Merkle in this volume.

Alternatively, the laser has become a powerful tool for probing turbulence by measuring effects, such as beam wandering, or scintillation produced by turbulence. All these things have required the development of models to describe turbulence, and inversion methods to derive atmospheric turbulence parameters from measured laser quantities.

An extensive amount of literature has been published in the last 30 years and many authors deserve mention. Here, however, we will limit ourselves to pointing out a number of books [1–5] that have already become classics. Recent results can be found in special issues of optics journals [6, 7].

There are many problems that have been solved, but still a number of unsolved ones and new subjects of interest for practical applications. Here, we will begin with a section on turbulence and its models. Then we will describe briefly some solved problems, problems still under study, and some new subjects.

2. Description of Atmospheric Turbulence and Models

Due to turbulence, the refractive index n of the atmosphere fluctuates at random in time and space around the mean value r_0 in the absence of turbulence. One can write

$$n = n_0 + \delta n. \tag{1}$$

According to the ergodic theorem, ensemble averages can be replaced by infinite time averages. One has

$$\langle n \rangle = n_0 \qquad \text{and} \qquad \langle \delta n \rangle = 0, \tag{2}$$

with brackets denoting time averages.

To have complete statistical information, one should know the probability density function of δn at one point, and also its joint probability density distributions at all possible points. A basic function is the correlation function of the refractive index fluctuations $B_n(P_1, P_2)$ at two points P_1, P_2, also called *covariance* for obvious reasons. We assume here that we are in conditions of homogeneity and isotropy so that B_n depends only on the distance r between any two points. One has

$$B_n(r) = \langle \delta n(P_1) \delta n(P_2) \rangle. \tag{3}$$

Alternatively, the spatial power spectrum of turbulence, defined as the tridimensional Fourier transform of B_n, is given by

$$\Phi_n(\kappa) = \frac{1}{2\pi^2 \kappa} \int_0^\infty r B_n(r) \sin(\kappa r)\, dr \tag{4}$$

and, inversely,

$$B_n(r) = \frac{4\pi}{r} \int_0^\infty \kappa \Phi(\kappa) \sin(\kappa r)\, d\kappa. \tag{5}$$

Another function that allows one to treat the problem of locally homogeneous turbulence is the structure function of the refractive index,

$$D_n(P_1, P_2) = \langle [\delta n(P_1) - \delta n(P_2)]^2 \rangle. \tag{6}$$

A slowly varying common part is removed, indeed, by difference. In our case,

$$D_n(r) = 2\,[B_n(0) - B_n(\mathrm{r})]. \tag{7}$$

An explicit expression for $B_n(r)$ or $D_n(r)$, or for the spectrum $\Phi_n(\kappa)$ is called a *turbulence model.*

From the several models found in the literature, we will consider some analytical forms derived from *Kolmogorov's theory of turbulence*. According to this theory, the shape of $D_n(r)$ is given by

$$D_n(r) = C_n^2 \ell_0^{2/3} \left(\frac{r}{\ell_0}\right)^2 \quad r \ll \ell_0, \tag{8}$$

$$D_n(r) = C_n^2 r^{2/3} \qquad \ell_0 \ll r \ll L_0, \tag{9}$$

where C_n^2 is called the structure constant, ℓ_0 the inner scale, and L_0 the outer scale. These equations are related to the physics of turbulence, according to which an initially laminar motion becomes turbulent when its Reynold's number increases to a limiting value. Turbulent eddies are produced of a large size, say L_0; these in turn are not stable and break up, transferring their energy to smaller and smaller eddies. In large-size eddies, energy dissipation is negligible, but it increases when the size decreases, until a stable dimension ℓ_0 is reached, where the energy is dissipated into heat. For these reasons, region $r \ll \ell_0$ where Eq 8 is valid is called the *dissipative region* and region $\ell_0 \ll r \ll L_0$ is referred to as the *inertial subrange*. Eq. 9 is very well known as the *two-thirds law*. In the region $r \gg L_0$, one can only affirm that the correlation of fluctuations tends to zero. For computational reasons, sometimes one assumes

$$D_n(P_1, P_2) = C_n^2 L_0^{2/3} \qquad r \gg \mathrm{L}_0. \tag{10}$$

A way to avoid the problem when L_0 is large, following Tatarskii, is to consider L_0 as infinite and to extend the range of validity of the 2/3 law up to infinity:

$$D_n(r) = C_n^2 r^{2/3} \qquad r \gg \ell_0. \tag{11}$$

This form can be referred to as the *Tatarskii-Kolmogorov model*. The corresponding spectrum is given by

$$\Phi_n(\kappa) = 0.033 C_n^2 \kappa^{-11/3} e^{-\kappa^2/\kappa_m^2}, \tag{12}$$

where $\kappa_m = 5.92/\ell_0$ and the exponential factor allows for the dissipation range.

Sometimes ℓ_0 is made to tend to zero so that the exponential factor is equal to 1. This corresponds to extending the validity of the 2/3 law over the entire r range.

An analytical form of the spectrum that allows one to reproduce $D_n(r)$ in both the dissipative and inertial subranges and lets D_n become constant for $r \gg L_0$ is the so-called *von Karman* (or simply *Karman*) *modified spectrum*:

$$\Phi_n(\kappa) = \frac{0.033 C_n^2 e^{-\kappa^2/\kappa_m^2}}{(\kappa^2 + \kappa_0^2)^{11/6}}, \tag{13}$$

where

$$\kappa_0 = 1/L_0. \tag{14}$$

A model suitable to describe the spectrum better in the inner-scale region was given by Hill. The model closely follows Eq. 12 apart from the region $\kappa \approx \kappa_0$, where it has a "bump"[8].

Another model sometimes used for its mathematical simplicity is a Gaussian model given by

$$B_n(r) = \langle \delta n^2 \rangle e^{-r^2/r_0^2}, \tag{15}$$

where $\langle \delta n^2 \rangle$ denotes the mean square refractive index fluctuation and r_0 the scale of turbulence. Note that r_0 plays the role of both inner and outer scale. This model reproduces the correct behavior of $D_n(r)$ in the dissipative range ($r \ll r_0$), but it does not include the inertial subrange. It seems suitable at near ground levels where Reynold's number is not too high and, consequently, the inertial subrange is small. In this case, it seems reasonable to consider r_0 as an inner scale.

3. Second-Order Coherence

Amplitude and phase fluctuations produced by turbulence on coherent optical radiation give rise to effects—such as loss of coherence, beam wandering, fluctuations of angle of arrival, loss of isoplanatism, etc.—that deteriorate the performance of systems. One of the first effects investigated after the invention of the laser was the limitation of the useful receiving aperture imposed by turbulence. Interest in the utilization of laser radiation for communication through the atmosphere soon declined. Many other applications, however, requiring atmospheric propagation and including astronomical systems, solicited theoretical investigation. Subsequently, methods of overcoming the limitations imposed by turbulence were introduced. Stellar speckle interferometry and adaptive optical systems belong to these classes.

The average effect of turbulence on systems—say, on an imaging system—was described by introducing a mean optical transfer function of a system, which for homogeneous and isotropic turbulence is given by

$$\langle T(\mu) \rangle = T_0(\mu) T_A(\mu), \tag{16}$$

where μ denotes spatial frequency, $T_0(\mu)$ is the transfer function in the absence of turbulence, and $T_A(\mu)$ is called the transfer function of atmospheric turbulence.

Owing to the fact that turbulence does not introduce depolarization, as recently reaffirmed, only one component of the optical field is necessary. As usual, an analytic signal is utilized. The complex amplitude is written as

$$\nu(P) = e^{i\Phi(P) + \ell n A(P)}, \tag{17}$$

where Φ and A are phase and amplitude. They fluctuate randomly around mean values Φ_0 and A_0, respectively, and also depend on the shape of the wave. By introducing the quantity

$$\chi = \ell n \frac{A}{A_0} \tag{18}$$

correlation and structure functions of Φ and χ were defined. One has

$$T_A(\mu) = \Gamma_2(\lambda f u), \tag{19}$$

where λ denotes wavelength, f the focal length of the receiving optical system, and Γ_2 the mutual correlation function (or cross correlation, or simply correlation or covariance) of the signal, sometimes also called the second moment, second-order correlation, or two point coherence function,

$$\Gamma_2(r) = \langle \nu(P_1) \nu^*(P_2) \rangle = \langle e^{i(\Phi_1 - \Phi_2) + (\chi_1 + \chi_2)} \rangle. \tag{20}$$

The problem for evaluating Γ_2 has been solved, and it has been shown that

$$T_A(\mu) = \exp\left[-\frac{1}{2} D_w(\lambda f \mu) \right]. \tag{21}$$

Quantity

$$D_w(r) = D_\Phi(r) + D_\chi(r) \tag{22}$$

is called the wave structure function and

$$D_\Phi(r) = \langle (\Phi_1 - \Phi_2)^2 \rangle \quad \text{and} \quad D_\chi(r) = \langle (\chi_1 - \chi_2)^2 \rangle \tag{23}$$

are the structure functions of phase and log amplitude, respectively. The main problem was to express $D_w(r)$ or the corresponding spectrum in terms of the analogous quantities of the refractive index.

Initially, $D_w(r)$ was evaluated theoretically in the case of small fluctuations using the so-called Ritov approximation, and, subsequently, also for large fluctuations and in the so-called saturation region that will be described later. The main role is played by the phase structure function because, in the case of strong fluctuations, $D_w(r)$ coincides with $D_\Phi(r)$, obtained by using a geometrical optics approximation. This means that the average behavior of a system in (locally) homogeneous and isotropic turbulence is now completely understood. One can find more detail on this subject, including the case of noncoherent fields, in a recent book on statistical optics [9].

Up to now, we have considered homogeneous and isotropic turbulence. In the literature, the case of a dependence of C_n^2 on one coordinate, such as height, was also treated. Vertical profiles of C_n^2 with height were introduced and treated. A random variability of the turbulence parameters has also been investigated. This allows one to account for practical atmospheric situations.

Subsequent research has been devoted to the problem of the short-term coherence function. It has contributed to the solution of the problem of tilt correction, a first step towards adaptive optical systems.

The two-frequency mutual coherence function has also been studied in view of its application to pulse propagation.

4. Higher-Order Coherence and Scintillation

The third-order or three-point coherence function Γ_3 and the corresponding bispectrum have been studied in astronomy because they also give information on the phase of the field. Here, we will not treat this subject, but refer instead to the chapter by Weigelt in this volume.

The fourth-order or four-point correlation function (or moment) is required to describe intensity fluctuations (scintillation). The problem was raised long ago by the renowned experimental results of scintillation measurements by Gracheva and Gurvich. They found that the variance $\sigma_{\ell nI}^2$ of the log intensity (irradiance) fluctuations with increasing turbulence or path length, after an initial increase according to the theory of small fluctuations, tends to saturate. The dependence of $\sigma_{\ell nI}^2$ on turbulence and path length L, as obtained by the Ritov approximation in the case of smooth fluctuations and a plane wave, is

$$\sigma_{\ell nI}^2 = 1.24 C_n^2 k^{7/6} L^{11/6}, \tag{24}$$

where $k = 2\pi/\lambda$. This quantity has become an important reference parameter and is generally denoted as σ_I^2, since $\sigma_I^2 \simeq \sigma_{\ell nI}^2$ for small fluctuations. Sometimes, it is denoted as β_0^2, but it is better that β_0^2 maintains its original meaning of log amplitude variance $\beta_0^2 = \sigma_{\ell nI}^2/4$.

In the evaluation of fluctuations, four-point correlations are required, the so-called gamma four (sometimes called fourth moment),

$$\Gamma_4 = \langle \nu(P_1)\nu^*(P_2)\nu(P_3)\nu^*(P_4)\rangle. \tag{25}$$

When the points coincide, $\Gamma_4(P) = \langle I^2 \rangle$, with $\langle I \rangle$ the intensity at P. The problem of finding Γ_4 for four points in a plane has been treated extensively in the literature in the past 20 years, and has led to the formulation of equations for the correlation functions of any order by many authors with different approximations, such as those based on the Feymann path integrals, the Markov approximation, the extended Huygens-Fresnel principle, and the radiative transfer equation.

The final equations one arrives at are of the parabolic type. As already mentioned, the equation of second-order coherence has received a general solution for arbitrary models of refractive index and any beam wave.

No analytical solutions can be found for fourth-order coherence apart from the approximate one for small fluctuations. Thanks to a number of authors (Gochelashvili and Shishov, Uscinski, Beran, Gozani, Fante, Flatté), progress has been made and some understanding reached in the region of strong fluctuations. First, an approximate solution was found for the strong fluctuation region that reproduced the saturation and, finally, an approximate solution valid for all the regions was found. It was called a *two-scale solution* because it can take into account both the inner and outer scales of atmospheric turbulence. In practice, the outer scale is shown to make a negligible contribution to scintillation, and only the inner scale is taken into account. Recently, Furutsu has presented an exact version of the two-scale method and has shown that it is based on a variational principle.

A general trend appears from the approximate results from the different authors. For a fixed value of the inner scale, the normalized variance of the intensity σ_I^2 initially increases with C_n^2 and the distance, then reaches a maximum value (strong fluctuation region) and, finally, decreases (saturation region) to the limiting value of 1. Moreover, from [10], it appears that for a fixed distance, the value of the maximum in σ_I^2 increases for increasing ℓ_0, while the maximum is reached at larger values of C_n^2. Unfortunately, the theoretical results still underestimate the existing experimental variance, although some improvements in accuracy and turbulence models have been introduced. On the other hand, there are few data where all relevant quantities, including the inner scale, were measured simultaneously in good atmospheric conditions, so that comparison is difficult. At present, there is a strong need for more experimental results.

5. Probability Distribution of Intensity

The problem of intensity probability distribution (p.d.f) is still an unsolved one. There is general agreement that intensity follows a log-normal distribution for small fluctuations and a negative exponential one in the limit of very strong fluctuations. However, the latter condition has never been reached experimentally. For intermediate values, a number of distributions have been proposed—like the K, the lognormally modulated exponential, the IK, the log-normal Rician, and the Furutsu distribution—that have proved useful in different ranges.

However, the main problem is to obtain good experimental results to compare with the distributions. In principle, the probability density distribution can be determined by measuring intensity moments, but a number of factors affect the measured signal and the moments as well as the probability density distribution. They give rise to a deterioration of information and sometimes even to information loss. One point that has been studied thoroughly in the recent literature is the reduction of higher-order moments in the region of strong fluctuations due to saturation of the collecting system [11] and the finite sample size [12].

Additional reduction effects can also be introduced on the probability density function and moments by the averaging effect of the receiving aperture and bandwidth. Recently, criteria have been given for a suitable choice of the apparatus for making good measurements in the strong scintillation region as well [13].

Saturation is typical of an automatic real-time acquisition system that digitizes the signal with a given number of bits. It was shown that the effect of the saturation is described by quantity α, defined by

$$\alpha = \frac{I_{\text{sat}}}{\langle I \rangle}, \tag{26}$$

where I_{sat} denotes the saturation value of the system and $\langle I \rangle$ the mean intensity to be measured. The lowering of higher-order moments also depends on the distribution to be measured. Typical required values of α are of the order of 100. This means that values of intensity 100 times the average cannot be neglected in order to measure, say, a fourth moment. Depending on α, the effect of the saturation can be a simple deterioration of information. In this case, correction of the moments for given distributions has been shown to be possible. If information is lost, there is no way of recovering it.

The situation is even more intriguing as far as the finite sample size is concerned. The effect of moment reduction is due to the high asymmetry of the probability density functions in the strong fluctuation region. In principle, it is also possible to think of increasing the length of the sample required for measuring a given moment, with the required accuracy, by collecting together all the data of many experiments done under the same conditions. However this is neither practical nor easy and would require a big experimental effort. Maybe some kind of statistical correction could be considered in the future.

6. Other Turbulence Effects and Conclusions

A subject of present interest is *enhanced backscattering* (sometimes called *backscatter amplification*) produced by turbulence on laser light that has been retro-reflected by, say, a mirror. Enhanced backscattering is well-known in the field of discrete scatterers and surfaces [14]. The effect of the electromagnetic reflection and enhancement by an extended turbulent medium was initially investigated at radar wavelengths by de Wolf for evaluating the radar backscatter cross-section of atmospheric turbulence. There is a considerable amount of Soviet literature on the subject, the earlier part of which was surveyed in [15]. The enhancement by turbulence is due to the double pass of the light through the same turbulence. This enhancement has been shown to be related to the fourth moment of the field [5]. Sometimes, the effect is explained as a kind of *phase conjugation* experienced by the light in the direction of backscattering. It is clear that these enhancement effects present practical interest in imaging as well as lidar application [16].

Before concluding, we will briefly refer to some effects related to instantaneous situations that have been investigated recently and seem of interest for application. One example, already mentioned, concerns wavefront deformations that adaptive optical systems can correct instantaneously (say in ~ 1 μsec) and continuously. Another case is turbulence intermittency, in which instantaneous realizations can occur where improvement instead of deterioration takes place. The most interesting example is a super-resolution effect through a distant turbulent layer found in 1989 by Zavorotni *et al* [17]. In their experiment, resolution of an object (a grating) was obtained in the presence of turbulence by an aperture whose resolving power in the absence of turbulence was not adequate to resolve it. The explanation is that in certain conditions, the instantaneous energy spectrum of the image due to turbulence can correspond to that of the source. This is a recent achievement and certainly an interesting field for future investigation.

For the future, we expect a continuation of research in the forementioned directions toward the solution of unsolved problems—such as scintillation and probability density functions of intensity fluctuations—and the development of many applications that utilize the effects summarized in these sections. A development of techniques for practical applications is desirable. We also hope that an effort will be made to narrow the spectrum of function names and normalization constants. Another problem that we would like solved in the future is that of the speed of publication of research results translated from lesser known languages.

References

1. V. I. Tatarskii, "The Effect of the Turbulent Atmosphere on Wave Propagation," *NTIS,* U. S. Department of Commerce, Springfield, VA (1971).
2. J. W. Strohbehen (ed.), "Laser Beam Propagation in the Atmosphere," in *Topics in Applied Physics*, Vol. 35, Springer-Verlag, 1978.
3. A. Ishimaru, "Wave Propagation and Scattering in Random Media," 2 volumes, Academic Press, 1978.
4. V. E. Zuev, "Laser Beams in the Atmosphere," Consultants Bureau, New York, 1982.
5. S. M. Rytov, Yu A. Kravtsov, and V. I. Tatarskii, "Principles of Statistical Radiophysics," 4 volumes, in particular "Wave Propagation through Random Media" (Vol. 4), Springer-Verlag, 1989. (First Russian edition, Nauka, 1978.)
6. A. Ishimaru (ed.), "Wave Propagation and Scattering in Random Media," special issue of *J. Opt. Soc. Am. A* **2**, no. 12, 2066–2348 (1985).
7. A. Consortini and R. A. Elliott (eds.), "Propagation and Scattering in the Atmosphere," special issues of *Applied Optics* Atmosphere **27**, no. 11, 2109–2272 (1 June, 1988) ibid **27**, no. 12, 2375–2548 (15 June, 1988).
8. R. J. Hill and S. F. Clifford "Modified Spectrum of Atmospheric Temperature Fluctuations" J. Opt. Soc. Am. **68**, 892–897, 1978
9. J. W. Goodman, "Statistical Optics," Wiley-Interscience, New York, 1985.
10. A. M. Whitman and M. J. Beran, "Two-Scale Solution for Atmospheric Scintillation from a Point Source," *J. Opt. Soc. Am. A* **5**, 735–737 (1988).
11. A. Consortini, E. Briccolani, and G. Conforti, "Strong Scintillation-Statistics Deterioration due to Detector Saturation," *J. Opt. Soc. Am. A,* **3**, 101– 107 (1986).
12. N. Ben-Yosef and E. Goldner, "Sample-Size Influence on Optical Scintillation Analysis 1: Analytical Treatment of the Higher-Order Irradiance Moments," *Appl. Opt.* **27**, 2167–2171 (1988).
13. J. H. Churnside, R. J. Hill, G. Conforti, and A. Consortini, "Aperture Size and Bandwidth Requirements for Measuring Strong Scintillation in the Atmosphere," *Appl. Opt.* **28**, 4126–4132 (1989).
14. M. Nieto-Vesperinas and C. Dainty (eds.), "Scattering in Volumes and Surfaces," North-Holland Delta Series, North-Holland, 1990.
15. Yu. A. Kravtsov and A. I. Saichev, "Effects of Double Passage of Waves in Randomly Inhomogeneous Media," *Soviet. Phys. Usp.* **25** 494–508 1983.
16. V. A. Banakh and V. L. Mironov, "Lidar in a Turbulent Atmosphere," Artech House, 1987 (translation from Russian).
17. M. I. Charnotskii, V. A. Myakinin, and V. Zavorotny, "Observation of Superresolution in Nonisoplanatic Imaging through Turbulence," *J. Opt. Soc. Am. A* **7**, 1345–1350 (1990).

CHAPTER 20

Are the Fundamental Principles of Holography Sufficient for the Creation of New Types of 3-D Cinematography and Artificial Intelligence?

Yury N. Denisyuk
Ioffe Physical-Technical Institute,
Academy of Sciences,
Leningrad, USSR

1. Principles

Holography generally is associated by the public with a method by which images creating a full illusion of the reality of objects can be obtained. The principle according to which a hologram creates this illusion is simple. We see an object (a doll O in Fig. 1) due to the fact that we perceive light waves W_1, W_2, W_3 emitted by points Q_1, Q_2, Q_3 of the object. The eye's crystalline lens L focuses these waves onto retinal light-sensitive elements r_1, r_2, r_3, and then the signals of the light-sensitive elements are perceived by the neural network of the brain.

One need not be a physicist to understand a simple fact: If the image of the doll is created in our consciousness by the waves W_1, W_2, W_3, then an image of the doll identical to the original will appear if these waves are reproduced. In this case, the doll itself may be absent. The method permitting us to produce this operation—i.e., recording and reconstructing wave fields—was invented not long ago. In 1949, D. Gabor proposed recording arbitrary wave fields of objects by mixing them with a simple reference wave [1]. Unfortunately, using the so-called one-beam scheme proposed by D. Gabor, only wave fields of very simple objects could be recorded, e.g., a thin hair against an extended light background.

The starting point of present-day holography, a method that permits the recording and reproduction of arbitrary wave fields, can be traced to the

ISBN 0-12-289690-4

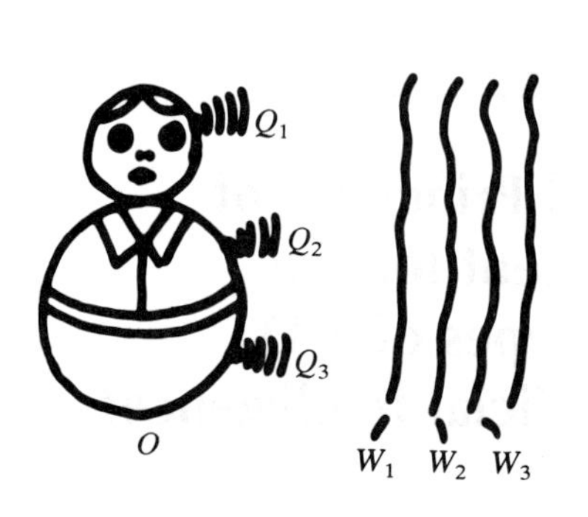

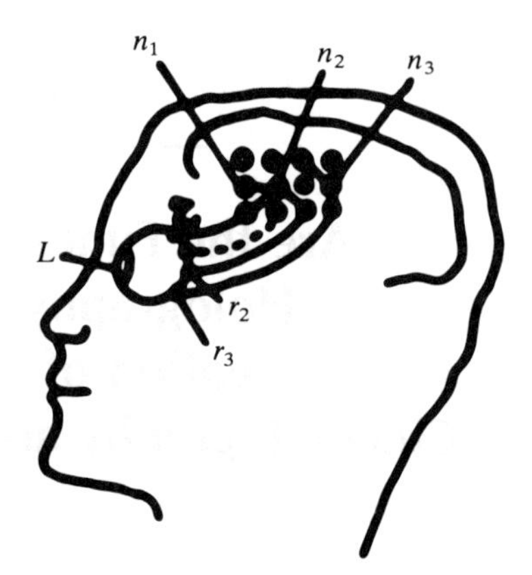

FIG. 1. Principle by which a hologram creates an illusion of the presence of the object. To make such an illusion for an observer, it is enough to reproduce the waves W_1, W_2, W_3 of the light scattered by the object O. The object itself may be absent here.

works of E. N. Leith and J. Upatnieks [2, 3]. According to their scheme, a photoplate F records the result of the interference between the wave W_O propagating from the object O, and the wave W_S produced by a special point source S (Fig. 2). At the stage of reconstruction, the radiation of the reference source S is directed onto a developed photoplate, i.e., a hologram H. This results in the reconstruction of the wave of the object and that of a so-called conjugated wave W_O^*. An observer perceiving these waves sees two images of the object, O^1 and O^*.

In the early 1960s, it also was found that the property of a photographic model of an interference picture to reproduce waves takes place not only in the case of a plane interference picture, but also extends to a spatial interference picture, a so-called *standing wave* [4, 5]. The recording scheme for such a 3-D hologram is shown in Fig. 3. In this case, the radiation of a monochromatic source S is directed onto an object O through a transparent light-sensitive material V. When interfering with a falling wave W_S, the wave W_O reflected by the object forms a system of standing waves inside the hologram volume, the antinode surfaces being denoted by d_1, d_2, d_3. After development, these surfaces transform into a system of peculiar curved mirrors.

The reconstruction is produced by a point source situated in the very same point as the source S during the recording. However, in this case, a laser is not necessary: The reconstruction may be accomplished by means of an ordinary source having a continuous spectrum, i.e., an incandescent lamp. As a result of the light interference at the system of proceeding

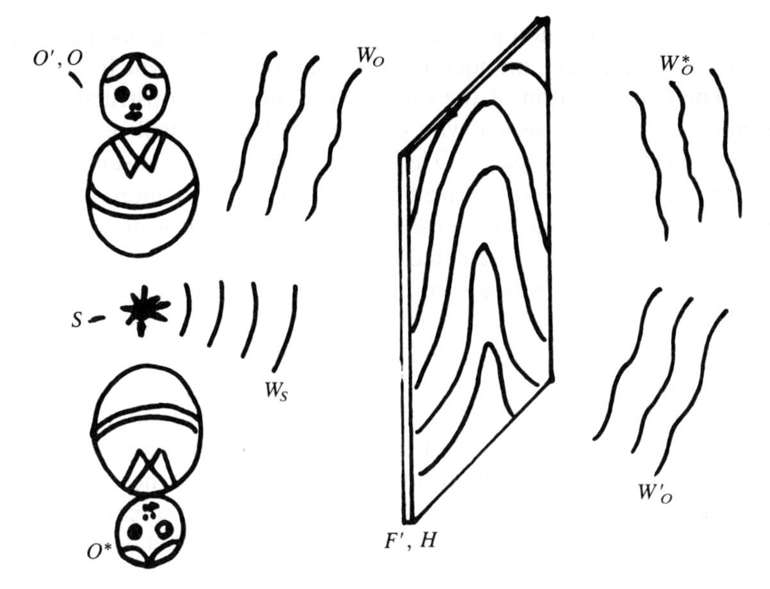

FIG. 2. Recording and reconstruction of a 2-D hologram: S, reference source; O, object; F, photoplate; W_O and W_S, object and reference waves; H, hologram; W'_O, reconstructed object wave; W_O^*; reconstructed conjugated wave; O', reconstructed image of the object; O^*, conjugated image of the object.

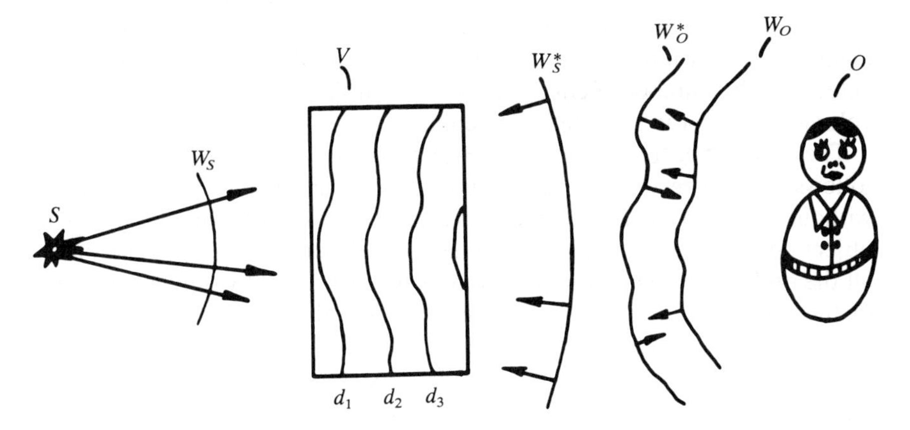

FIG. 3. Recording, reconstruction, and conjugate reconstruction of a 3-D hologram: S, radiation source; O, object; V, 3-D transparent light-sensitive material; W_S, W_O, W_O^*, W_S^*, waves of the source, object, and conjugated waves; d_1, d_2, d_3, antinode surfaces of the standing wave.

mirrors d_1, d_2, d_3, the hologram chooses out of the continuous spectrum—and reflects—that very monochromatic component by which it was exposed. The configuration of the object wave also is reconstructed by the 3-D hologram. The observer who perceives this wave will see a volume colored image of the object O. The conjugated image is absent here.

A 3-D hologram is characterized by a number of very interesting properties, reversibility being one of them. In particular, it was discovered that if a 3-D hologram is reconstructed by a wave W_S, which converges to a source S instead of diverging from it, then a so-called *reversed*, or *conjugated, wave* W_O^* will be reconstructed. This wave coincides in form with that of the object, W_O, but it converges to the object (see Fig. 3). Mathematically, the operation of complex conjugation corresponds to the transition to a reversed wave [5]. The associativity of a 3-D hologram and the possibility of recording many holograms onto one and the same photomaterial region are its specific properties, too [6]. Both these properties, revealed by P. I. Van Heerden in 1963, are based on the so-called selectivity of the 3-D recording. The selectivity properties of a 3-D hologram manifest themselves, in particular, in the fact that the hologram, on which the radiation of two points is registered, interacts only with the radiation of these very points and ignores that of all others.

During recording onto one and the same photomaterial region, objects—e.g., O_1, O_2... (Fig. 4)—are recorded sequentially in the form of holograms by using reference waves R_1, R_2... of different directions and wavelength. Van Heerden showed that in a 3-D hologram, the number of independent harmonics that can be used to record the information coincides with the number of cubes of $\lambda \times \lambda \times \lambda$ in size that are contained in its volume (where λ is the light wavelength).

The associative properties of a 3-D hologram manifest themselves as follows. At the recording stage, a hologram of an object O_1 is registered without the help of a reference wave in a light-sensitive medium. Each object point of such a hologram serves as a reference source in relation to all the other ones. Then the hologram is reconstructed by only a fragment of the object recorded on it (in Fig. 4, by the dashed part of the arrow point O_1). By interacting with a hologram, the radiation of this fragment of the object reconstructs the image of the object as a whole.

However, the most essential peculiarity of a 3-D hologram apparently is the very fact that the laws of holography extend into the third dimension. Indeed, practically all physical processes of light and material interaction take place in three dimensions; hence, the notions of holography can be applied to a wide range of physical phenomena. In particular, on this basis, as a result of the combination of holography and nonlinear optics ideas, a new promising field of optics arose, so-called *dynamic holography*. Unlike

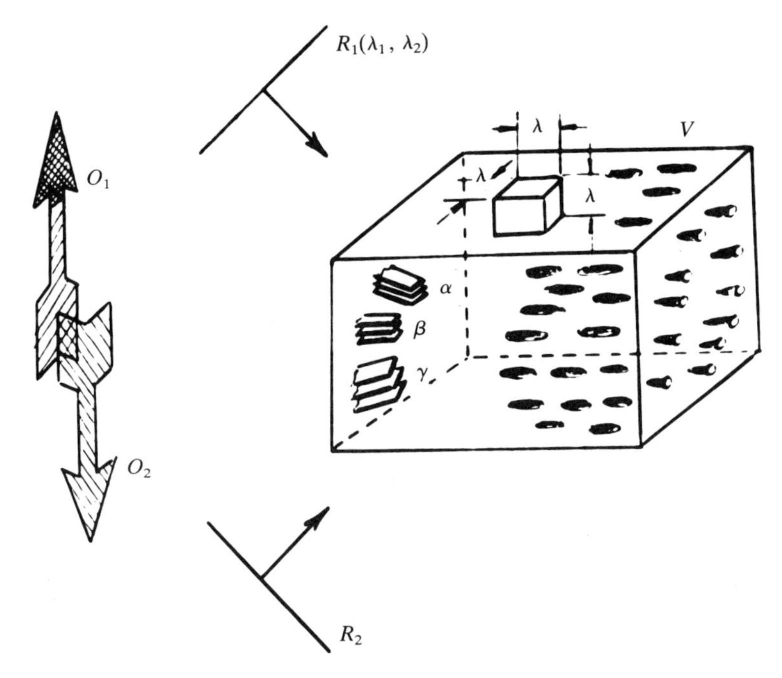

FIG. 4. Associative properties and information capacity of a 3-D hologram: O_1, O_2, objects recorded on the hologram; R_1, R_2, reference waves; α, β, γ, 3-D gratings connecting different points of the object recorded on the hologram.

common holograms, dynamic holograms are recorded in a nonlinear light-sensitive medium that responds to light immediately at the moment of its action. An evident advantage of dynamic holograms is their ability to transform wave fields in real time. However, there are a number of other specific laws of dynamic holograms that turned out to be very useful in some practical applications.

In particular, when studying the process of recording a 3-D hologram in lithium niobate, D. L. Staebler and J. J. Amodei revealed the effect of an energy transfer between the waves interfering in the crystal volume [7]. This effect was found to be caused by the shift of the picture of the intensity distribution in a standing wave relative to the corresponding grating of the refractive index being recorded in the crystal. The effect of an energy transfer from a strong wave to a weak one under a non-stationary condition or recording a 3-D dynamic hologram also is very interesting and useful [8]. This effect together with a number of other effects of dynamic holography is used widely now in the correction of wavefronts of laser emission, optical information processing, etc.

The advent of dynamic holography has opened up the possibility of recording and transforming wave fields formed by moving objects. In particular, it was found that not only standing waves, but also travelling intensity waves that arise upon the interference of waves of differing frequencies, possess reflectivity properties [9]. In particular, these travelling intensity waves appear around a moving object O on which the radiation of a source S falls (Fig. 5). Reflecting from the object, this radiation is shifted in frequency due to the Doppler effect and forms a travelling intensity wave d_1, d_2, d_3... when interfering with the falling radiation.

It would appear that if this moving picture is materialized in a nonlinear medium, then in addition to all other wave field parameters, the 3-D hologram obtained also will reconstruct the Doppler shift of the object wave. It is interesting to note that if this hologram is reconstructed by a wave converging into the source S (rays l_1, l_2, l_3 in Fig. 5), then a reversed wave will appear and it will be brought into focus in front of the object at a point P, with a lead depending on both the velocity of the object and that of the reversed wave propagating in the direction of the object [10].

An important feature of an electromagnetic wave field is its state of polarization. The method of recording and reconstructing this parameter by a so-called *polarization hologram* was suggested by a Georgian physi-

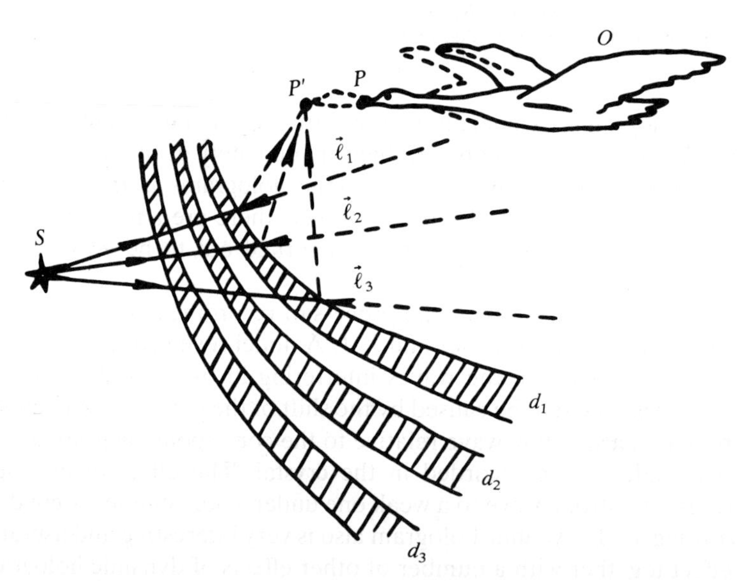

FIG. 5. Dynamic 3-D hologram on which travelling intensity waves d_1, d_2, d_3 formed as a result of the reflection of the radiation of the source S from the moving object O; $\mathbf{l}_1$, $\mathbf{l}_2$, $\mathbf{l}_3$, beam pass at the conjugate reconstruction of such a hologram.

cist, Sh. D. Kakichashvili in 1974 [11, 12]. The scheme of this method is given in Fig. 6. Let the radiation of the source S with the vertical orientation of the vector of electric field $\mathbf{E}_S$ fall onto a light-sensitive medium F and an object O. Let us suppose that the object rotates this vector by the angle $\pi/2$: As a result, two waves $\mathbf{E}_S$ and $\mathbf{E}_O$ characterized by mutually orthogonal polarization states meet at the photoplate F. In a usual sense, these waves do not produce an interference picture. Here, it is not the intensity that is modulated, but a state of polarization. Fringes of linear (δ_1) and circular (δ_2) polarization alternate.

The hologram recording is produced in a medium characterized by a so-called *Weigert effect*, i.e., the ability to acquire the anisotropy of the transmittance coefficient corresponding to the polarization of the impinging radiation. The transmittance coefficient of the hologram recorded in this way will be maximum in the place of linear polarization and remain constant in those of circular light polarization. It is easy to note that in this case, two holograms are formed—shifted by half a period—that polarize the light in mutually perpendicular directions. As a result of the addition of the radiation reconstructed by these holograms, the vector of electric intensity of the reconstructing wave will be rotated by $\pi/2$; thus, not only the object wave $\mathbf{E}_O$, but also its polarization state, will be reconstructed.

Let us go on to so-called *echo-holograms*, which are capable of recording simultaneously both spatial and temporal characteristics of wave fields. By

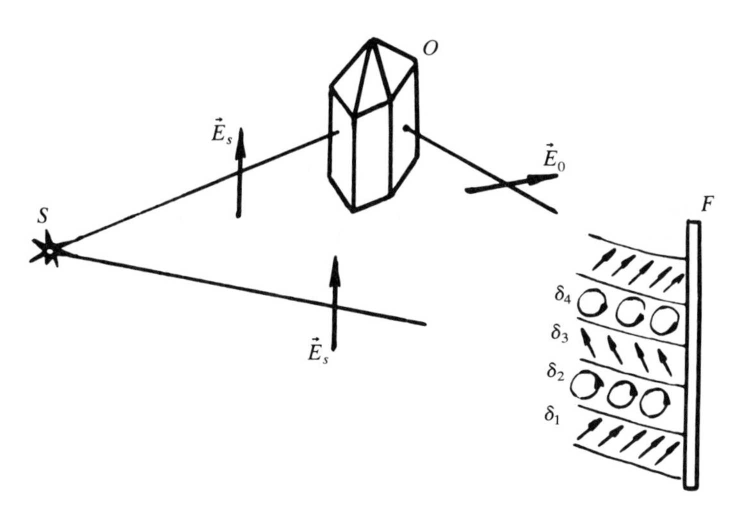

FIG. 6. Recording of a polarization hologram: S, radiation source; O, object; $\mathbf{E}_O$ and $\mathbf{E}_S$, vectors of electric intensity of the object and reference waves; F, polarization-sensitive medium; δ_1, δ_2, δ_3, fringes of different polarization of the polarization-interference picture.

generalizing light echo notions, E. I. Styrkov and V. V. Samartsev proposed echo-holograms, which are recorded by pulses of object and reference radiation not overlapping in time [13]. According to this method, at a certain instant in time, an object wave pulse is directed to a resonance medium, resulting in the excitation of the oscillations of atoms in the medium. Then, after some time interval τ, a reference wave pulse is directed on this medium. The wave field of the pulse interferes with the oscillations of excited atoms in the medium, thus forming a hologram. According to the light-echo mechanism, after the next time interval τ, the medium will emit a light-echo pulse, the wavefront of which is conjugated to the object wave.

The resonance echo-holography notions stimulated Estonian physicists P. Saari *et al.* to develop an original method of recording holograms capable of reconstructing simultaneously both spatial and temporal characteristics of wave fields [14, 15]. This method is based on the use of spectrum-selective light-sensitive media. When illuminated by monochromatic light of the wavelength λ, these media change the transmittance coefficient at this very wavelength. Figure 7 shows the process of hologram

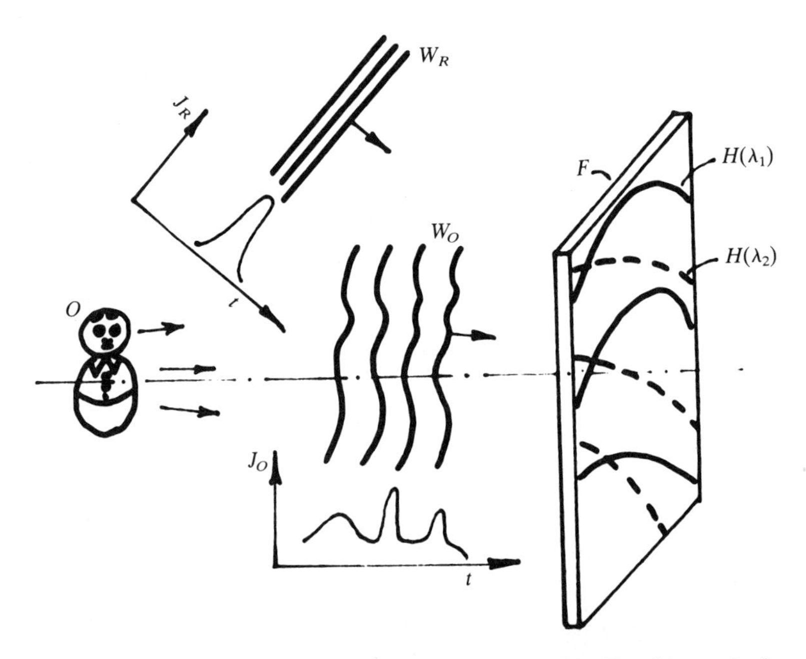

FIG. 7. Space-time holography: O, object changing in time; W_O, W_R, object and reference waves; F, spectrum-selective light-sensitive media; $H(\lambda_1)$, $H(\lambda_2)$, holograms recorded a different monochromatic components of the object wave.

recording for this case. An object wave W_O of the intensity J_O depending in a certain way on time t, and a plane reference wave W_R of the intensity depending on time as a δ-function (Fig. 7), are falling onto a spectrum-selective material F. Each molecule of a such light-sensitive material interacts only with a definite narrow monochromatic component of the incident light. As a result of the interference of each couple of corresponding monochromatic object and reference components, many independent monochromatic holograms are formed. For example, a hologram $H(\lambda_1)$ is formed at a wavelength λ_1 (a solid line in Fig. 7), a hologram $H(\lambda_2)$ is formed at a wavelength λ_2 (a dotted line), etc.

Under reconstruction, when a reference wave δ-pulse falls on the hologram recorded in this way, each monochromatic component of this pulse will reconstruct a corresponding object-wave monochromatic component by interacting with the given monochromatic hologram. When added together, these monochromatic components form an object wave W_O changing in time.

2. Perspectives

Considering applications of holography, it should be noted that the time when one could hope to review all the known holography applications in one review passed long ago. So, we shall restrict ourselves the consideration of the two greatest and "eternal" problems, so to speak: 3-D cinematography and artificial intelligence.

The history of holography demonstrates that the public is most interested in one thing: holograms with images reproducing a full illusion of the reality of the objects being recorded on them, and especially the prospect of reproducing the process of the motion of these images. Since 1962, when the use of holography was proposed as an imaging technique [4, 5], considerable success was achieved in the development of the technology of producing holographic images of non-moving objects.

In particular, based on the method of reflective 3-D holograms [4, 5], large-scale holograms of about 1m × 1m in size were obtained. They often are demonstrated at various expositions and museums [16]. These holograms create such a full illusion of the existence of the objects recorded on them that many unexperienced spectators assume them to be the objects.

Another advanced technique of recording and copying holograms, which also can be reconstructed by common white light sources, has been developed on the basis of a so-called method of *rainbow holograms* proposed by S. A. Benton in 1969 [17]. Rainbow holograms can be copied simply enough and thus are used widely nowdays as book illustrations, stamps, and even in the production of bank notes.

As to the problem of producing a holographic cinema, it appeared to be considerably more complicated as compared to that of images of non-moving objects. The most considerable attempt to produce holographic cinematography was made in the USSR by Professor V. G. Komar during the years 1976–1984 [18]. Figure 8 shows a schematic diagram of the setup. A hologram, H, on which a 3-D image of scene is recorded, is reconstructed by reference radiation R. A holographic screen E, which is, in fact, a multiplexing focusing element, projects the hologram H into a hall by multiplying its image according to the number of seats. For a spectator, each of these images (H^1, H^{11} ... in Fig. 8) is a window into the space of objects through which he can observe a 3-D image of the scene of action.

According to this scheme, shooting and projection equipment and special holographic screens have been produced. As a result, two holographic films of a duration equal to several minutes have been made and demonstrated. In one of them, Monochromatic, a 3-D image of a woman viewing her treasures was reproduced. This film was made by using a pulsed laser as

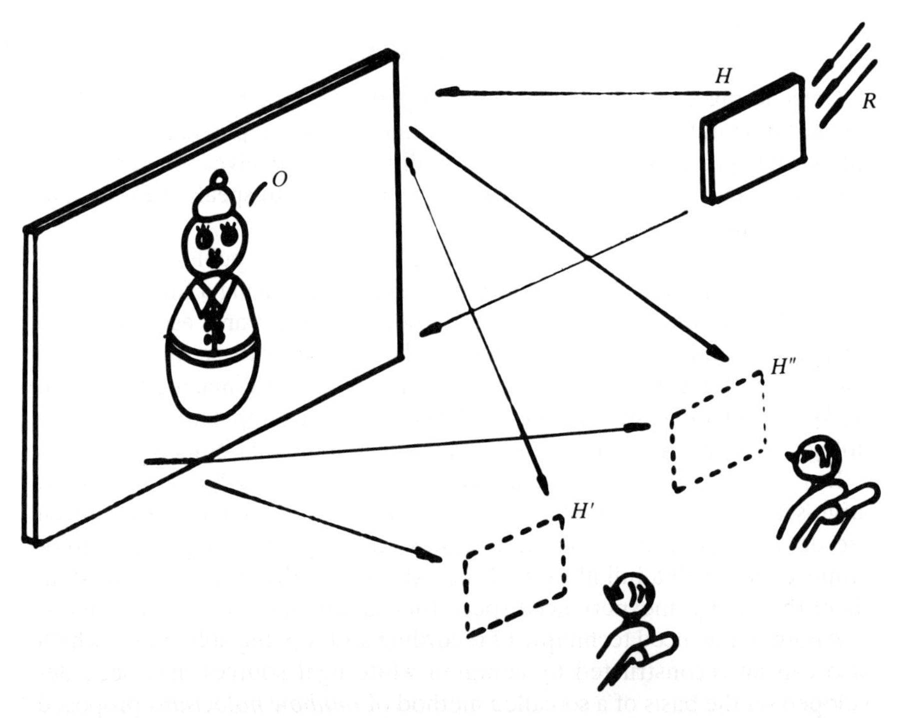

FIG. 8. Scheme of the experimental setup for holographic cinematography: H, hologram; R, reconstructing radiation; E, holographic screen; H^1, H^{11}, images of the hologram formed by the screen; O, image of the scene of action.

a source of the illumination of the scene. The second film, an animated color cartoon in which moving dolls were imaged, has been made by using continuous gas lasers. However, the difficulties that accompanied the creation of these films were so great that work on holographic cinematography was stopped.

The natural question arises: Why did the specialists try to create a 3-D form of cinematography based on holography? At first glance, more simple stereoscopic cinematography systems, which create a sufficiently full illusion of the volume of the scene being imaged, are known. The answer follows from the physiology of vision.

It appears that a human eye is highly adapted to perceive natural objects, and it uses the entire set of optical phenomena accompanying the appearance of an object, i.e., the stereo effect, accommodation, etc. The eye considers a plane image of the object on the screen as a separate object, but not as an imitation of the real oject. In such an image, all the components of vision mentioned here are matched mutually, and consequently, routine cinematography and TV allow a long-duration observation. As to spectacles that reproduce only part of the 3-D features of the object (e.g., the stereo effect only) while not reproducing all the others (e.g., the state of an eye accommodation), they are very harmful to the nervous system.

All the components of vision will be reproduced and matched mutually in holographic cinematography, since a hologram essentially creates a full illusion of the object presence. Being a safe and natural sight, such cinematography could give a completely new dimension concurrently to the imaging technique. In this case, a spectator would be a direct participant in the scene presented to him.

From a technical perspective, first of all, the problem of creating holographic cinematography and TV comes down to the development of methods for the removal of information redundancy, i.e., the elimination of excessive information from the holographic image. In particular, the observer does not use the information on the radiation-phase distribution, for example. Besides, as a rule, the shape of objects is relatively simple, and thus, the view of the scene changes slowly enough when one changes the viewing point. A so-called vertical parallax is redundant for the observer, too. Furthermore, a hologram is capable of reconstructing far smaller parts of an object than is necessary for the observer.

If all these particularities are taken into account, then the scene area on which the hologram is recorded or, equivalently, the frequency bandwidth of the TV channel, can be reduced considerably, with the size of the window through which the observer looks at the scene remaining constant. A certain part of the operations on reducing the redundancy of information can be done by optical methods of information transformation, as has been accomplished already, e.g., in [17, 19]. However, a decisive part in the redundancy reduction may be played by a computer that has to synthesize holograms relatively small in size that allow considerable magnification when projecting through a screen.

The part played by a computer apparently will be still more essential when synthesizing holograms by analyzing photos obtained in natural light from different viewpoints. In general, optical methods for the synthesis of these holograms are amply used (e.g., see [20]). However, the holograms obtained so far by these methods do not produce such perfect images as holograms recorded by lasers.

The development of holographic cinematography according to the scheme considered here is a very complicated problem, and it will require that much time be spent on the creation of fundamentally new systems and elements. However, even this relatively distant prospect is not final. Holographic cinematography using a moving tape and many other elements of the traditional cinematography scheme still do not conform well with the fundamental principles of holography.

Holography is a world of gratings and waves that is convenient to represent in a so-called *Fourier space*. In this space, the wave fields changing in time and space are considered as a sum of waves constant in time and space that are characterized by different frequencies, directions, and initial phases. When added up, they produce a picture of moving images of objects that are familiar to us.

Thus, a holographic system producing moving, "living" images of objects can be represented as one 3-D hologram composed of a multiplicity of independent monochromatic holograms, each of them producing a wave constant in time that is characterized by a certain initial phase. When added up, these monochromatic waves will produce time-changing images of objects.

A prototype of such a "wave supercinema" is the spectral holography considered in Fig. 7. Of course, the development of a wave supercinema is a problem of the very distant future, whereas in the near future, it will be developed by physicists and specialists in the field of optical communications and optical computers who need to reproduce the processes taking place in periods of time from 10^{-9} to 10^{-12} s.

The second important direction in the development of holography is the study of its use in artificial intelligence, which was begun by P. I. Van Heerden as long ago as 1963. Nowdays, almost 30 years later, the development of computers has achieved the level where it is possible to apply holography in the implementation of computing operations and, in particular, in the creation of interconnections between elements. The interconnections of the optical computer are very numerous: networks in the associative memory, adaptive networks in the neural computer, changing networks between processors. The part played by networks increases steadily, so much so that they have stopped being conductors only, but are becoming the very meaning of the operations being passed to them. In general, this tendency is easy to understand: It is a reflection of the well-known notion that the brain is a universal system of connections and comparisons of different objects and events rather than a store of truths.

In many known systems of intelligence—i.e., the brain, electronic computers—and in many types of optical computers being developed today, linear networks are used—i.e., nerves, wires, fiberglass light guides—along which particles (e.g., electrons, photons) propagate. The connections of this type are natural in our usual Euclidean 3-D space. The intelligence based on this type of connection can be called *Euclidean intelligence* or *intelligence on-particles*.

Holography suggests another alternative. In this type, signals are transferred by means of superimposed waves propagating in space, while the connection between these waves is accomplished by means of holographic gratings that also are superimposed. The intelligence based on this system can be called *Fourier intelligence*, or *wave intelligence*. The properties and capabilities of this intelligence must differ essentially from those of such systems as, e.g., the electronic computer. In particular, J. Caulfield and D. Shamir considered a similar system they called a *wave-particle* or *coherent processor*. They showed that, unlike a common processor in which much energy is spent on a performance of intermediate operations, the main part of the energy in such a processor will be spent only at the stage of the readout of the result [21]. Apparently, the wave computer has many other peculiarities. However, we shall not get involved in these special questions and shall go on to the consideration of various types of holographic connections.

A two-dimensional hologram like the one shown in Fig. 2 is the most simple in production and convenient in application. In this case, the elements connecting the pairs of waves, or equivalent, the pairs of point sources, are two-dimensional gratings being formed during the recording of the radiation of these points on the hologram. However, the connection by two-dimensional gratings is not unambiguous, since each of these gratings interacts not only with the points that have formed it, but with all other points as well. This limits considerably the potentialities of two-dimensional holograms.

An unambiguous system of connections between waves can be established by means of a three-dimensional hologram like that shown in Fig. 4. Such a hologram is a sum of three-dimensional gratings, each of which interacts, according to the Bragg condition, only with that very pair of waves that has formed it. This system of connections is very effective, and hence, it is being widely studied nowadays in connection with the neuro-computer problem [22, 23]. Its main disadvantage is that the 3-D gratings that form it are superimposed and thus, it is practically impossible to control them: to erase, to regenerate, etc.

Waveguide holograms occupy an intermediate position between 2-D and 3-D holograms [24, 25]. In this case, in a planar waveguide H, a hologram of a line object ab is recorded by radiation propagating along the wave-guide (Fig. 9). Like a 3-D hologram, this one has a depth and consequently, it couples the waves unambiguously. Unlike 3-D holograms, waveguide holograms give the possibility of separating spatially pages of information recorded on them. For example, the object ab can be recorded in a strip A by means of a reference source r, an object $a'b'$ can be recorded in a strip by a reference source r', and so on. Here, the extent of strips along the depth d (see Fig. 9) plays the part of the width of a 3-D hologram and defines its selectivity properties.

The fact that properties of a waveguide hologram can be controlled from the 3rd dimension also is very important. In particular, by using auxiliary sources S_1, S_2, S_3, for example, gratings connecting arbitrary points of the object ab can be recorded into the strip A' of the hologram. However, for all the advantages, waveguide holograms have not yet found an appropriate application in information processing devices due to the technology difficulties in their production.

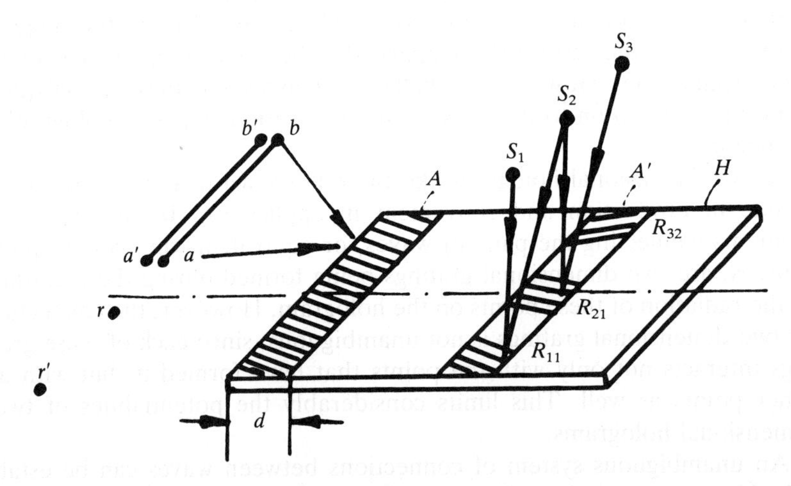

FIG. 9. Information recording on a waveguide hologram: H, waveguide hologram; ab, line object; r, r^1, reference sources; A, A^1, strips with the record of pages of information; S_1, S_2, S_3, auxiliary sources of radiation recording the gratings R_{11}, R_{21}, R_{32}.

To use the advantages of waveguide holograms and to avoid the aforementioned difficulties, a so-called method of *pseudodeep inclined holograms* was proposed [26, 27]. The properties of such holograms are similar to those of 2-D, 3-D, and waveguide holograms. The recording scheme of a pseudodeep hologram is presented in Fig. 10. A light-sensitive layer H, on which a fan of beams propagating from a line object ab is recorded, is inclined at a small angle β relative to the plane of these beams' propagation. A slit D is placed behind the hologram. At the stage of reconstruction, the slit chooses only the beams situated in the object ab plane and cuts off all the others.

In general, a pseudodeep inclined hologram acts as a common 2-D one. Unlike waveguide holograms, in this case, the radiation intersects the layer, rather than propagating along it. Unlike a 3-D hologram, the physical width of a light-sensitive layer of a pseudodeep hologram is not important. However, the combination of an inclined hologram and a specially-chosen slit results in properties close to those of a waveguide and a 3-D hologram. In particular, the pages of information on such holograms, as in the case of a wave-guide hologram, can be recorded in the form of strips, the selectivity properties of the recording on strips being defined by their extent along the reading beam (d in Fig. 10). Like a waveguide hologram, the pseudodeep one permits the use of the third dimension to control the

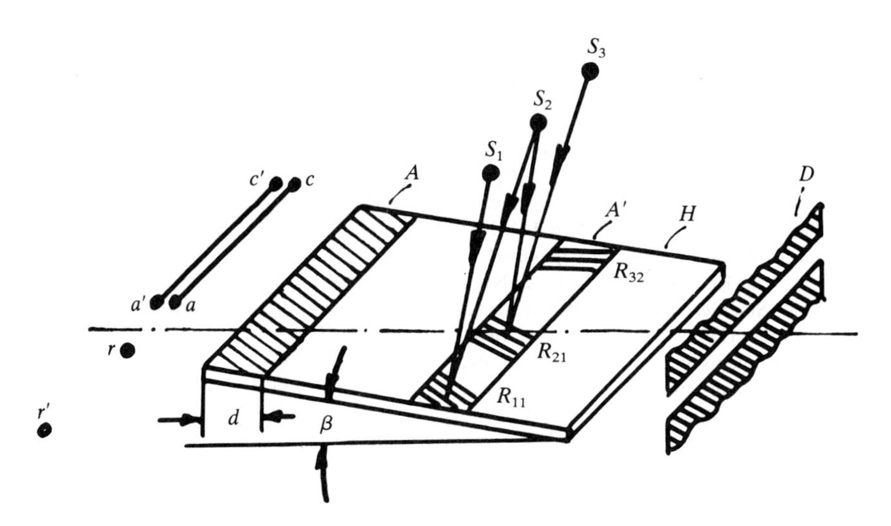

FIG. 10. Information recording on a pseudodeep inclined hologram: H, 2-D thin hologram; β, angle of the hologram inclination relative to the optical axis of the system; ab, line object; r, r^1, reference sources; A, A^1, strips with the record of the pages of information; S_1, S_2, S_3, auxiliary radiation sources recording gratings R_{11}, R_{21}, R_{32}; D, slit that selects the beams lying in the object plane out of the reconstructed radiation.

structure of the recording, e.g., by recording gratings R_{11}, R_{21}, R_{32} with the help of auxiliary sources S_1, S_2, S_3. At the same time, in the case of the pseudodeep hologram, many technology difficulties and limitations associated with special requirements for light-sensitive material are absent.

A review of effects on which holography and its applications are based cannot be complete without mentioning a decisive part played by light-sensitive media. In fact, the optical laws of holography essentially are only a sort of "framework." It is the light-sensitive medium that draws the pictures inside this framework, the variety of them being infinite, like those of properties manifested by matter.

The properties of a computer based on the principles of holography also will depend essentially on those of the light-sensitive media applied in it. In particular, by applying various nonlinear media for hologram recording, one can produce very fast associative memory systems and systems of-adaptive interconnections changing with a great rate. The use of polarization-sensitive media opens up the possibility of controlling the state of polarization of reconstructed wave fields and, consequently the possibility of producing different logical operations [28]. Resonance media can be used in producing interconnections changing during unimaginably small periods of time according to a prescribed law. The possibility in principle of producing all these operations is evident already. There is no doubt that the range of these possibilities will be increasing permanently.

References

1. D. Gabor, *Proc. Roy. Soc.* (London) **A197**, 454–465 (1949).
2. E. N. Leith and J. Upatnieks, *J. Opt. Soc. Am.* **52**, 1124–1129 (1962).
3. E. N. Leith and J. Upatnieks *J. Opt. Soc. Am.* **54**, 1295–1302 (1964).
4. Yu. N. Denisyuk, *DAN SSSR* **144**, No. 6, 1275–1278 (1962).
5. Yu. N. Denisyuk, *Optika i Spektroskopija* **15**, 522–532 (1963).
6. P. I. Van Heerden, *Appl. Opt.* **2**, No. 4, 393–400 (1963).
7. D. L. Staebler and J. J. Amodei, *J. Appl. Phys.* **43**, No. 3, 1043–1046 (1972).
8. F. Gires, *Comt. Rend. Acad. Sci.* **266B**, No. 10, 596–598.
9. Yu. N. Denisyuk, *Zh. Tekh. Fiziki.* **44**, No. 1, 131–136 (1974).
10. Yu. N. Denisyuk, *Zh. Tekh. Fiziki.* **52**, No. 7, 1338–1345 (1982).
11. Sh. D. Kakichashvili, *Kvantovaja Elektronika* **1**, No. 6, 1435–1441 (1974).
12. T. Todorov, L. Nikolova, and N. Tomova, *Appl. Opt.* **23**, No. 24, 4588–4591 (1984).
13. E. I. Shtyrkov and V. V. Samartsev, *Optika i Spektroskopija* **40**, 392–393 (1976).
14. A. K. Rebane, R. K. Kaarli, and P. M. Saari, *Optika i Spektroskopija* **55**, No. 3, 405–407 (1983).
15. P. M. Saari, R. K. Kaarli, and A. K. Rebane, *J. Opt. Soc. Am.* **3**, No. 4, 527–533 (1986).
16. Yu. N. Denisyuk, *Science and Mankind* (international annual in Russian), Znanie, Moscow, 299–314, (1982).
17. S. A. Benton, *J. Opt. Soc. Am.* **59**, 1545–1550 (1969).

18. V. G. Komar and O. B. Serov, "Imaging Holography and Holographic Cinematography" (in Russian), Iskusstvo, Moscow, 1987.
19. K. A. Haines and D. B. Brum, *Proc. IEEE.* **55**, 1512–1515 (1967).
20. R. V. Pole, *Appl. Phys. Letters* **10**, 20–22 (1967).
21. H. J. Caulfield and J. Shamir *Appl. Opt.* **28**, 2184–2186 (1989).
22. D. Psaltis, D. Brady, and K. Wagner *Appl. Opt.* **27**, No. 9, 1752–1759 (1988).
23. Y. Owechko, G. L. Dunning, E. Marom, and B. N. Soffer, *Appl. Opt.* **26**, No. 10, 1900–1910 (1987).
24. W. Lukosz, *Wüthrich Opt. Comm.* **19**, No. 2, 232–235 (1976).
25. T. Sihara, H. Nishihara, and J. Kogama, *Opt. Comm.* **19**, No. 3, 353–358 (1976).
26. Yu. N. Denisyuk, *Pis'ma v Zh. Tekh. Fiziki.* **15**, No. 8, ("Zh. T. F. Letters"), 84–89 (1989).
27. Yu. N. Denisyuk and N. M. Ganzherli, *Pis'ma v Zh. Tekh. Fiziki.* **15**, No. 14 (Zh. T. F. Letters), 4–7 (1989).
28. A. W. Lohmann, *Appl. Opt.* **25**, No. 10, 1594–1597 (1986).

CHAPTER 21

Medical Applications of Holographic 3-D Display

Jumpei Tsujiuchi
Faculty of Engineering, Chiba University
Chiba, Japan

1. Introduction

Among many 3-D-image display techniques, holography provides the most natural 3-D images to the eye. With the development of white-light reconstruction holograms and synthesized holograms, holographic 3-D image display has become an excellent 3-D image display technique.

On the other hand, many kinds of medical images have been in practical use for medical diagnosis, and their display in a 3-D format strongly. In this area, holographic 3-D-image display has become one of the most promising techniques for that purpose. Our research group—consisting of the Tokyo Institute of Technology, Chiba University, Fuji Photo Optical Co., and Toppan Printing Co.—has been developing holographic 3-D image display techniques for medical purposes for the past 10 years, and two kinds of holograms have been studied.

One is a *multiple-exposure hologram* of sectional images obtained by computer-assisted tomography (CT), nuclear-magnetic resonance imaging (MRI), and ultrasonic imaging; the other is a *holographic stereogram* synthesized from a series of x-ray images taken from different directions. As the recording methods and the properties are different for these two types of holograms, we will discuss each case separately in the following sections.

ISBN 0-12-289690-4

2. Multiple-Exposure Holograms for Medical Use

CT provides sectional images ordinarily perpendicular to the body axis of a patient, and if a series of CT images is taken with a sufficiently small interval between exposures, the 3-D structure of internal organs and bones of the patient can be recorded in these CT images. To observe the 3-D structure of the patient's body, these tomograms are arranged in regular order and observed from the direction of the body axis.

Holography is an excellent way to record and display these images, and an optical system as shown in Fig. 1 was proposed by M. Suzuki *et al.* [1]. This is an ordinary Fresnel hologram recording setup with laser light reconstruction, and a series of tomograms with a diffuser are recorded successively in a photographic plate with a fixed reference beam. The position of tomograms D should be chosen at every additional distance of md, where m is the magnification of tomogram and d is the interval of tomograms when taken. After recording all the tomograms, the hologram can be reconstructed by illuminating it with the identical laser beam with the reference beam.

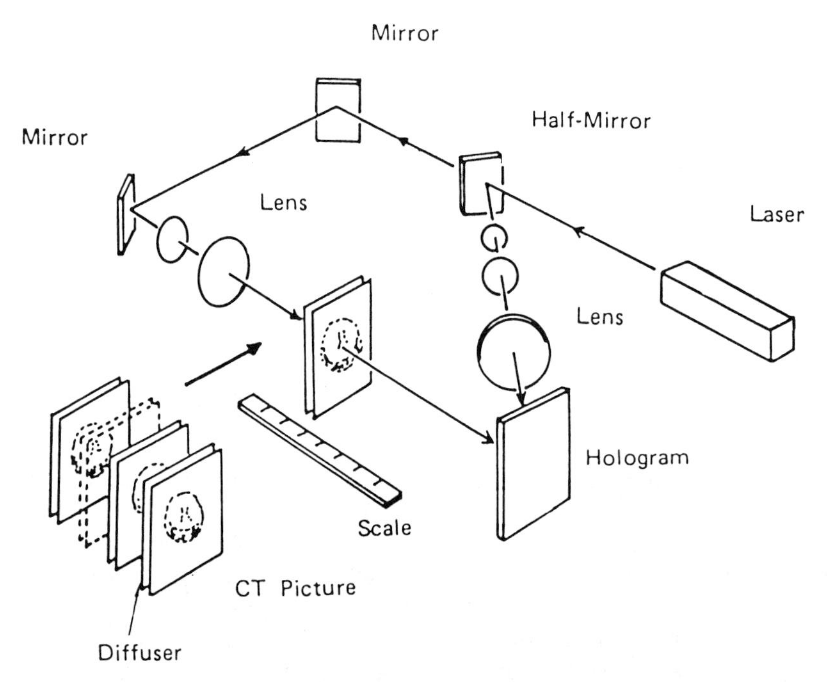

FIG. 1. Optical system for making a multiple-exposure hologram for CT images [1].

The reconstructed image of this hologram is not satisfactory because only laser light is available for reconstruction, and speckles due to the diffuser accompanying the tomograms generate much noise. So, the hologram often is transformed to a rainbow hologram [2] or a Lippmann hologram [3] for easy reconstruction with white light, and this will make less speckle noise as well.

However, since the hologram is recorded with a technique of incoherent superposition, the brightness of the reconstructed image decreases as the number of tomograms increases, and this fact will limit the number of tomograms to a certain value, probably about 10.

An improved technique for multiple-exposure holograms was proposed by Okada *et al.* [4]. Figure 2 shows principles of the method. As shown in (a), a series of tomograms $T_1 \ldots T_n$, with a diffuser are recorded on a photographic plate H_1 similarly to the previous method, but the plate is set with an angle of α, and the recording of each tomogram is made through a narrow horizontal slit S, which limits the recording area to avoid the incoherent superposition and moves along the plate by its width as the tomogram changes. This hologram H_1 becomes a master hologram, and is

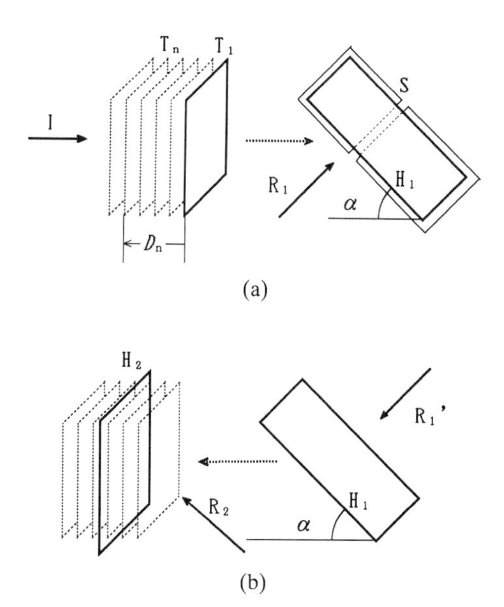

FIG. 2. Improved method of a multiple-exposure hologram with coherent superposition [4]: (a) Making a master hologram by spatially separated recording. (b) Recording a final hologram for practical use.

illuminated by a laser beam R_1' that is conjugate with the reference beam R_1; then the real image is reconstructed at the original position of the tomograms, as shown in (b). The second hologram is recorded with a reference beam R_2 by setting another photographic plate H_2 at an appropriate position, where the real image of the tomograms is reconstructed. This hologram is the final hologram, recording many rainbow holograms with coherent superposition, and can be reconstructed by a white-light beam conjugate with the reference beam R_2. In this case, if the angle α is chosen in such a way that the plate is inclined by the achromatic angle of the rainbow hologram, images of the tomograms are distortion-free but not in the direction of the depth of image, i.e., the body axis. This can be compensated for by changing the position of the tomograms D_n in making the master hologram. A computer-controlled positioning of tomograms by calculating D_n is employed in practice [4].

This hologram can record many tomograms without any decrease of brightness, and the reconstructed image shows a color change according to the position of tomograms in depth, i.e., the depth of object. Such an image is a little bit unusual, but seems convenient to estimate the depth of object by observing colors. If an incandescent lamp with a fine vertical filament of a certain length is used for reconstruction, each image of the tomogram becomes whitish, and the reconstructed image seems more natural.

This hologram is expected to be used as one of the excellent techniques for demonstrating 3-D imaging of tomograms. Also, observation of the reconstructed image of this hologram needs an accommodation of the eye, and this will give a very natural feeling to 3-D image display.

Reduction of the time required for making the hologram is always of great importance in medical applications. A spatial light modulator (SLM) connected to CT equipment will make the processing required to produce holograms easy and rapid. A prototype machine for such a purpose is under investigation in the author's group.

3. Holographic Stereograms

3.1. Multiplex Holograms

Another hologram for 3-D display of medical images is a holographic stereogram. The holographic stereogram is a hologram synthesized from a series of photographs taken from different directions around the vertical axis of the object. The most popular holographic stereogram is a cylindrical white-light reconstruction hologram called a *multiplex hologram* [5]. For medical purposes, x-ray images of a patient are taken from different directions around the body axis of the patient, and these images can be used

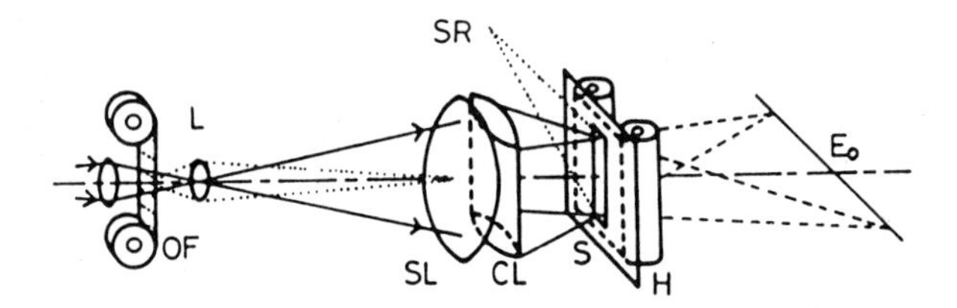

FIG. 3. Optical system for making a multiplex hologram [6].

as original images for synthesizing the hologram. Usually, these original images are recorded successively on movie film, often called the original film. This kind of hologram shows 3-D images by binocular stereoscopy, but with no accommodation of the eye.

Figure 3 shows equipment [6] used to synthesize multiplex holograms. A frame of the original film OF is projected onto the pupil of a combination of large-aperture cylindrical (CL) and spherical (SL) lenses. Such a lens system produces an astigmatic beam and has two foci; one is the vertical focus and the other is the horizontal one. A film for recording holograms is placed at the vertical focus, and a narrow vertical strip hologram of the projected image is recorded through a slit *S* by using a reference beam emitted from a point *SR* located above the pupil of the lens system. This strip hologram becomes an elementary hologram with a size of about 250mm in length and 0.5mm in width. After recording an elementary hologram, both the original film and the hologram film are fed one frame and repeat the same operation, and the entire hologram can be completed.

The multiplex hologram thus made is shaped into a cylinder by winding it around a transparent cylinder, and illuminating it from the bottom with a lamp having a vertical fine filament, preferably less than 1mm in diameter and 5mm in length to get high-resolution images [7], located on the axis of the cylinder.

The reconstructed image appears inside the cylinder, and if the illuminating lamp is located at the same position as the reference source, and the observer puts his eyes at the same position as the horizontal focus E_0 of the lens system (see Fig. 3) with regard to the hologram, the image has the same color as that of the laser in hologram synthesis and also has minimum distortion.

3.2. Distortions of Reconstructed Images

The formation of reconstructed images from multiplex holograms is done in an unusual way. The most important problem is how to make the distortions small. An *elementary hologram* is an image hologram in the vertical direction but a Fourier transform hologram in the horizontal direction. So, if the hologram is illuminated by a small white-light source, each elementary hologram produces the image of the corresponding frame of the original film. At the same time, the horizontal focus (E_0 in Fig. 3) also is reconstructed, and if the observer puts his eyes in that position, the pupil of an eye acts as the exit pupil of a monochromator, and he can observe the reconstructed images with monochromatic light. In this case, the entire image is composed of many narrow vertical segments, which were taken from different directions when the original film was made, and the reconstructed image can be observed as a continuous 3-D image by binocular stereoscopy.

The image thus obtained has two kinds of distortions; one is *static distortion* due to the change of line of sight in the reconstruction from taking the original images, and the other is *dynamic distortion* due to the movement of the object during the taking of the original image.

Figure 4 shows schematically why static distortions of multiplex holograms take place. Suppose that a simple object ABC is recorded and reconstructed, and the reconstructed image is shown in both the horizontal and vertical directions. Figure 4(b) shows the vertical section of the hologram, CRE_0 is a ray recording a point object in the hologram Q, and the image C projected in the pupil of the lens system is recorded around a point R in the hologram. In the reconstruction, if an observer put his eyes at E_0, with b_0 apart from the axis, the same ray CRE_0 as recorded is reconstructed and enters the eyes, and the reconstructed image appears at C without any distortion in the vertical direction. If the eyes move from E_0 to E , for example, the reconstructed image of C appears at C' by a ray diffracted at R, and the image shift from C to C' gives distortion in the vertical direction. In addition, the color of the reconstructed image C' changes due to the change of diffraction angle δ, and this change of δ depends on the height of C from the optical axis. This means that the

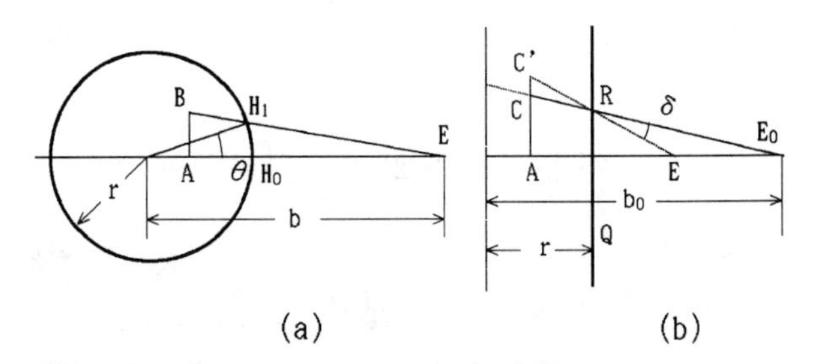

FIG. 4. Generation of static distortion of multiplex holograms [17]: (a) Horizontal section. (b) Vertical section.

reconstructed image no longer is monochromatic but has rainbow colors, and such a change of local colors brings a local change of magnification, i.e., a new type of distortion.

Figure 4(a) shows the horizontal section of the hologram, where the image is reconstructed as AB. If the original image of AB is taken from the direction perpendicular to AB, the corresponding elementary hologram is recorded in H_0. If an observer looks at the reconstructed image AB from a point E, with b apart from the axis, the ray conveying the information of B comes from an elementary hologram H_1, but H_1 is produced by the original image taken from an offset angle θ, and the image looks somewhat different than that for $\theta = 0$. So, this offset angle produces a distortion in the horizontal direction. Accordingly, a small angle θ seems effective for making the distortion small, and a distortion-free image can be obtained at $\theta = 0$; i.e., the observer puts his eyes close to H_0. In fact, a laser reconstruction multiplex hologram can be made by using a diffusing screen instead of the lens system CL + SL in Fig. 3, and such a hologram can satisfy this condition [8]. However, since the hologram under discussion is a sort of rainbow hologram in the vertical direcion, no image can be observed if the observer puts his eyes at H_0. This means that there will be no possibility of getting distortion-free images in the white-light reconstruction multiplex hologram.

Two methods of decreasing static distortions have been proposed: One is a suitable choice of geometrical configuration in taking the original film of objects and synthesizing holograms. This can be realized by taking the condition $ma = 2r$, where a is the distance between the object and the camera lens when the original film is taken, r is the radius of the cylindrical hologram, and m is the magnification of the reconstructed image relative to the original object [6, 8, 9]. However, images thus obtained are not distortion-free, even though the amount of distortion is small—negligible—in most cases.

Another approach to making the distortion small is computer processing applied to the original images [10]. As shown in Fig. 4 (a), if the observer at E looks in the direction $\theta = 0$ at an image point B, reconstruction is made by an elementary hologram H_1, whose offset angle is θ. Accordingly, if the vertical segment recording B in a frame at θ of the original film is replaced by that of $\theta = 0$ and a hologram is made by using this new original film, a distortion-free image can be obtained. This means, as shown in Fig. 5, that each frame of the original film is divided into many narrow vertical segments, and a frame of $\theta' = 0$ of the new original film is synthesized by rearranging these segments. Such processing is carried out easily by computer, and this new original film can produce a hologram reconstructing distortion-free images.

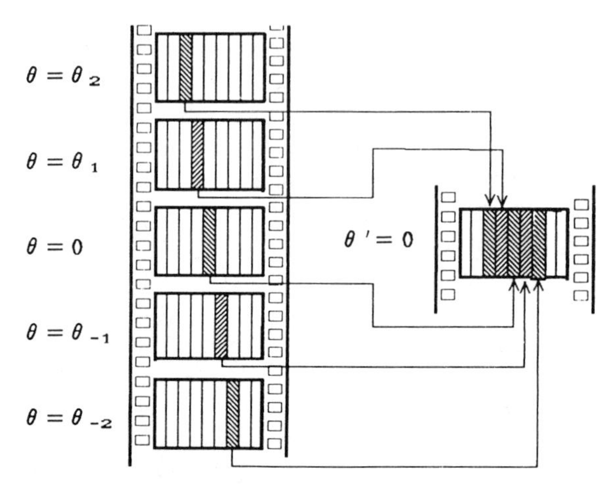

FIG. 5. Rearrangement of original image for compensating distortions [10].

The other kind of distortion, dynamic distortion, takes place if the object moves rapidly during the taking of the original film. In fact, the object for a multiplex hologram has to be static because the reconstructed image is the combination of a series of vertical segments of reconstructed images from the corresponding elementary holograms positioned on the line of sight of the observer. If the object moves, these segments show images of the object at different moments, and the reconstructed image becomes discontinuous. This phenomenon produces a new sort of distortion—dynamic distortion—which depends on the movement of the object. For sufficiently slow-moving objects, however, this distortion becomes negligible, and the reconstructed image moves as the hologram rotates around the axis.

It also is possible to compensate for dynamic distortion by using a similar technique to that used in the static [10], because the offset angle θ corresponds to a time difference in taking the original film. In addition, conditions to compensate for both static and dynamic distortions can be satisfied at the same time, and perfectly distortion-free images can be expected.

4. Medical Applications of Multiplex Holograms

4.1. Holograms of Medical Images

Many kinds of medical images are in use and their 3-D display is expected for easy understanding in medical diagnosis. Recently, the advancement of computer technology has made possible the display of these images using 3-D graphic techniques [11]. Holographic 3-D display

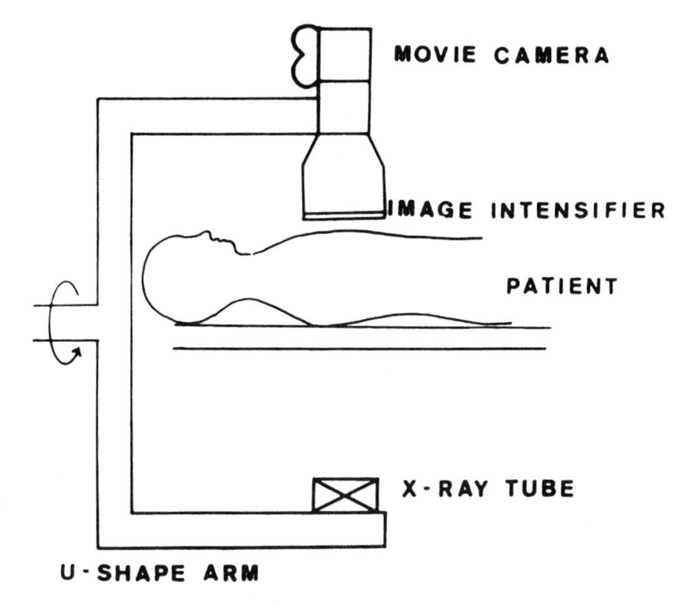

FIG. 6. Taking x-ray original film for making medical multiplex holograms [12].

seems more attractive in its natural visual feeling and easy presentation as a sort of hard copy, however, and a multiplex hologram is a suitable technique for that purpose. For example, a series of x-ray photographs taken around a patient are used as an original film. Figure 6 shows an example of equipments used to take such an original film [12]. This equipment has a U-shaped arm with a mechanism to turn around the patient, and the arm has a point x-ray source at one end and a 35mm movie camera ready to record the output screen of an image intensifier at the other end. The distance between the x-ray source and the rotating axis along which the patient lies corresponds to the distance a between the camera and the object in the ordinary case.

Multiplex holograms made from such an original film provide an advantage for gaining a better understanding of the 3-D structure of bones, blood vessels, and internal organs of the patient, and also the possibility of measuring 3-D coordinates of particular points of the objects [13].

However, since medical x-ray images in general are of low contrast, the images reconstructed from such original images cannot be expected to be of good quality. Computer processing of the original images, such as enhancement and distortion correction, is very effective in obtaining better images. For that purpose, a computer-aided hologram synthesizing system for medical applications is proposed as shown in Fig. 7 [14].

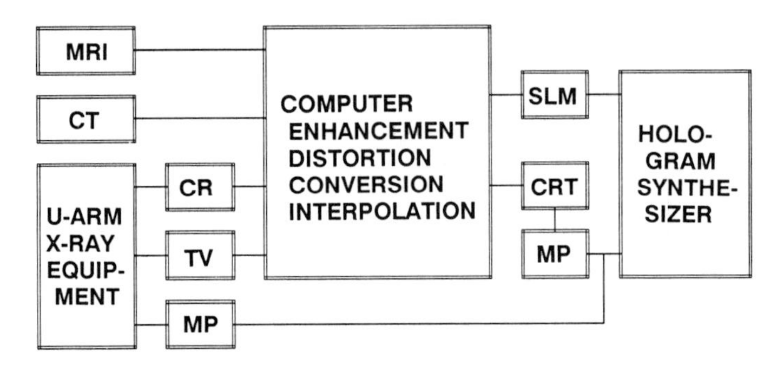

FIG. 7. Proposal of a computer-aided hologram synthesizing equipment [14].

This system provides the possibility of synthesizing holograms from various kinds of medical images. The simplest case is to use x-ray images recorded by a movie camera with the aid of the U-shaped arm equipment; the image recorded in a movie film is sent to the hologram synthesizer without any processing, so that reconstructed images may have some distortions and low image quality. To apply computer processing, the aforementioned images must be scanned by an appropriate scanner and the A–D converted data sent to a computer, but scanning many frames of the original film does not seem practical. Another possibility is to record the original x-ray images by a TV camera, preferably of the high-definition type, on a videotape by using the U-shaped arm equipment, and sending the digital data to a computer through a frame memory. A digital x-ray imaging system using computed radiography (CR) or a digital image intensifier also can be used for taking the original x-ray images, and the digital data is obtained directy from the equipment.

Image processing in the computer includes image enhancement, correction of both static and dynamic distortions, and interpolation to generate intermediate images if the total number of original images is limited [15]. After processing, two methods for sending processed images to the hologram synthesizer are to be considered: One is to display processed images on a CRT screen and record these images on film by using a camera controlled by the computer. The other is to use a computer-accessible SLM as an interface between the computer and hologram synthesizer, and processed images can be sent directly to the synthesizer.

This computer-aided hologram synthesizing system brings to mind the possibility of using other kinds of medical images, such as CT or MRI images. They are given as sectional or volume images of the patient, and the original film for synthesizing multiplex holograms can be made by computer.

4.2. Holographic Display of Surface-Reconstruction Images

Since CT images are sectional images in planes perpendicular to the body axis of the patient, a series of CT images taken with a small interval give enough information about the 3-D structure of the patient. These images are very well suited for making a hologram because they already are enhanced and available in digital form. However, these images are impossible to use directly as original images for synthesizing multiplex holograms, and the images should be transformed into the ordinary original images recording the patient around the body axis.

In this section, an experiment that made a hologram of a skull from a series of CT images [16] is described. Ninety-three CT images of the head of a patient are taken with an interval of 2mm by a CT system especially designed to make the x-ray dose to the patient small, and these images are used as the original data for synthesizing a hologram. Each tomogram contains 256 × 256 pixels, and the interval between adjacent pixels corresponds to about 1mm in the object space. Since the subject of medical interest is the anterior half of the skull, only the skull is extracted by level slicing from the front half of the tomograms.

To synthesize a multiplex hologram, an image of the object projected in a plane parallel with the body axis should be obtained. This is a technique in computer graphics, and is shown in Fig. 8 [16, 17]. The Nth tomogram is taken as an example, and is projected in a plane $V(N, \theta)$, where θ is an azimuthal angle of the plane from the front of the skull, $P(K,N,\theta)$ is a point with a discrete coordinate K, where a point M on the surface of the skull is projected. The projected image has an intensity of 256 levels (eight bits) due to the shadow by a parallel uniform beam illumination from the direction θ, where D, the distance between M and P, becomes an important parameter to calculate the shadow in the projected image of the skull

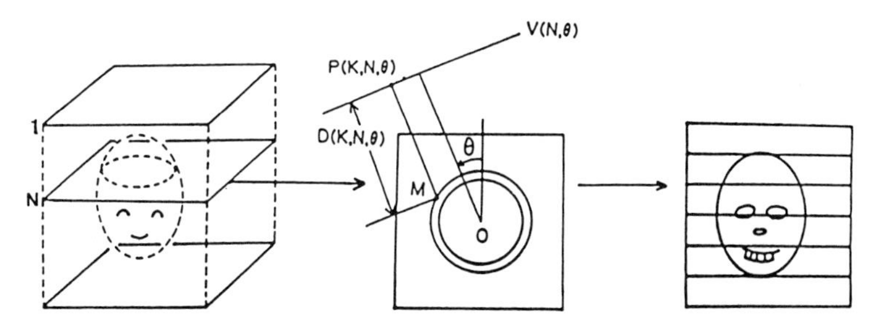

FIG. 8. Schematic diagram of surface reconstruction from CT images [17].

surface. Tomograms from $N = 1$ to $N = 93$ are projected in regular order, and a projected image of the entire surface of the skull in a direction θ is obtained by interpolating in the vertical direction. Such an image is called a *surface reconstructed CT image* [18], and it seems much better to display in holography than ordinary transmitted x-ray images.

The synthesized hologram has a diameter of 400mm and a height of 300mm, and is composed of 2,514 elementary holograms 0.5mm in width and 250mm in height. So, the total number of frames of the original film becomes 419 if six elementary holograms are recorded repeatedly for every frame of the original film. This means that the aforementioned operation has to be carried out for every 0.86° of θ, and 419 projected images in total have to be calculated. These images are displayed successively on a CRT screen and recorded on movie film, frame by frame, using a movie camera controlled by computer; the original film thus obtained is sent to the hologram synthesizer. Figure 9 shows an example of the reconstructed images, and a bright 3-D image of the skull of a living patient can be observed in the cylinder.

4.3. Further Developments

The multiplex hologram is very useful for medical diagnosis, the planning of surgical operations, the education of medical students, and the collection of samples of diseases. Another kind of medical imaging, such as MRI or ultrasonic imaging, also is possible to use, and a wider application can be expected. However, some problems still remain.

One is the lower resolution of the reconstructed image in the vertical direction. This is due to the difficulty of taking CT images with a smaller interval. One of the possibilities for overcoming this difficulty will be to use a cone beam reconstruction method from 2-D x-ray images. This is worth studying if a digital 2-D detector of x-ray images is available. According to a preliminary experiment, the point-spread function of the reconstructed tomographic image becomes space variant and the reconstruction algorithm is not analytically exact [19], but if we confine the processing to extracting only the bones, it can be achieved with high spatial resolution. This means that the resolution in the vertical direction is expected to increase, and the dose of x-rays to the patient becomes smaller.

The other problem is the time necessary to synthesize a hologram. To make the computation time of the computer short, the number of original images now is limited to 360; i.e., projected images are made for every 1°. This corresponds to the specification of the surface reconstruction facility of most CT equipment now available on the market, and seven elementary holograms are recorded repeatedly for every frame. In addition, a computer-

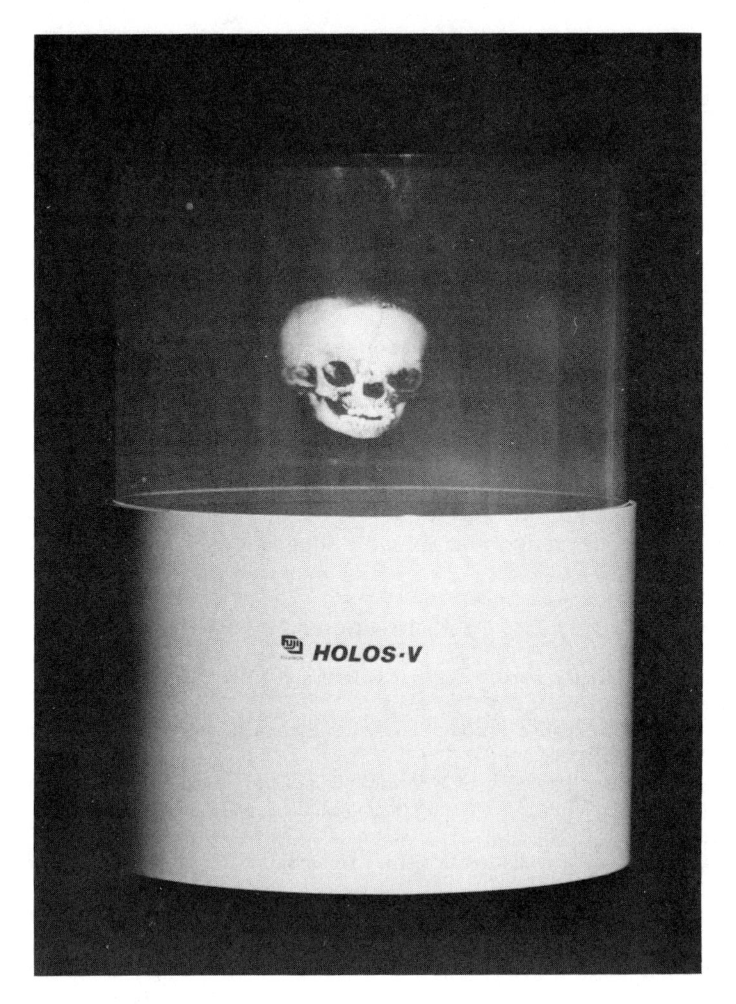

FIG. 9. Reconstructed image of a skull by a multiplex hologram made of CT images (courtesy of M. Suzuki).

controlled automatic hologram synthesizer is under construction, and a liquid-crystal TV screen is to be employed as an interface SLM between the computer and the hologram synthesizer [20]. This device will convert output images of the computer directly to input images for synthesizing holograms without taking any original film. Preliminary experiments show the feasibility of this system [21] and it will contribute to quick and easy operation of hologram synthesis.

5. Conclusion and Acknowledgments

Holographic 3-D display techniques in medicine are reviewed together with problems of practical application. Recent medical diagnosis and treatment take advantage of many kinds of imaging techniques to make understanding of the conditions of patients easy, and 3-D image display is one of the most useful techniques for that purpose. Holographic 3-D display techniques using multiple-exposure holograms and/or holographic stereograms will be very promising because of their natural 3-D appearance and easy observation.

The author would like to express his thanks to his co-workers, notably Profs. T. Honda, N. Ohyama, Tokyo Institute of Technology; Dr. K. Okada, Chiba University; Messrs. M. Suzuki and T. Saito, Fuji Photo Optical Co.; Mr. F. Iwata, Toppan Printing Co.; Prof. M. Fujioka, M. D., Dokkyo Medical University; Prof. S Hashimoto, M. D., Keio University School of Medicine; and Prof. S. Ikeda, M. D., National Cancer Center, for their kind cooperation and valuable suggestions.

References

1. M. Suzuki, M. Kanaya, and T. Saito, "3-dimensional Illustration by Multiply Exposed Hologram," *Proc. SPIE,* **523**, 38 (1985).
2. S. A. Benton, "Hologram Reconstruction with Extended Incoherent Light Sources," *J. Opt. Soc. Am.* **59**, 1545 (1969).
3. Yu. N. Denisyuk, "Photographic Reconstruction of the Optical Properties of an Object in its Own Radiation Field," *Soviet Physics-Doklady* **7**, 543 (1962); *Optics and Spectrosc.* **18**, 152 (1965).
4. K. Okada and T. Ose, "Holographic Three-Dimensional Display of X-Ray Tomogram," *Proc. SPIE,* **673**, 84 (1986).
5. First realized by Lloyd Cross, Multiplex Company, 454 Shotwell Street, San Francisco, CA 94110.
6. J. Tsujiuchi, T. Honda, K. Okada, M. Suzuki, T. Saito, and F. Iwata, "Conditions for Making and Reconstructing Multiplex Holograms," *AIP Conf. Proc.* **65**, 594 (1981).
7. K. Okada, T. Honda, and J. Tsujiuchi, "Image Blur of Multiplex Holograms," *Opt. Commun.* **41**, 397 (1982).
8. T. Honda, K. Okada, and J. Tsujiuchi, "3-D Distortion of Observed Images Reconstructed from a Cylindrical Holographic Stereogram: (1) Laser Light Reconstruction Type," *Opt. Commun.* **36**, 11 (1981).
9. K. Okada, T. Honda, and J. Tsujiuchi, "3-D Distortion of Observed Images Reconstructed from a Cylindrical Holographic Stereogram: (2) White Light Reconstruction Type," *Opt. Commun.* **36**, 17 (1981).
10. K. Okada, T. Honda, and J. Tsujiuchi "A Method of Distortion Compensation of Multiplex Holograms," *Opt. Commun.* **48**, 167 (1983).
11. G. M. Hunter "Three-Dimensional Frame Buffers for Interactive Analysis of Three-Dimensional Data," *Opt. Eng.* **25**, 292 (1986).
12. J. Tsujiuchi, T. Honda, M. Suzuki, T. Saito, and F. Iwata, "Synthesis of Multiplex Holograms and Their Application to Medical Objects," *Proc. SPIE* **523**, 33 (1985).

13. K. Okada, T. Honda, and J. Tsujiuchi, "3-D Measurement by Using a Multiplex Hologram," *Opt. Commun.* **45**, 320 (1983).
14. J. Tsujiuchi, "Multiplex Holograms and Their Applications in Medicine," *Proc. SPIE* **673**, 312 (1987).
15. K. Okada, T. Honda, and J. Tsujiuchi, "Multiplex Holograms Made of Computer Processed Images," *Proc. SPIE* **402**, 33 (1983).
16. N. Ohyama, Y. Minami, A. Watanabe, J. Tsujiuchi, and T. Honda, "Multiplex Holograms of a Skull Made of CT Images," *Opt. Commun.* **61**, 96 (1987).
17. J. Tsujiuchi, "3-D Image Display Using Holographic Stereograms," *Proc. SPIE*, **1033**, 410 (1988).
18. M. Fujioka, N. Ohyama, T. Honda, J. Tsujiuchi, M. Suzuki, S. Hashimoto, and S. Ikeda, "Holography of 3-D Surface Reconstructed CT Images," *J. Comput. Assist. Tomogr.* **12**, 175 (1988).
19. N. Ohyama, S. Inoue, H. Haneishi, J. Tsujiuchi, and T. Honda, "Three-Dimensional Reconstruction of a Bone Image from Cone Beam Projection," *Appl. Opt.* **28**, 5338 (1989).
20. M. Kato, S. Hotta, and K. Kanai, "Generation of High Quality Holograms with Liquid-Crystal SLM," *Proc. SPIE* **1212**, 93 (1990).
21. T. Honda, M. Yamaguchi, D. K. Kang, K. Shimura, J. Tsujiuchi, and N. Ohyama, "Printing of Holographic Stereogram Using Liquid-Crystal TV," *Proc. SPIE* **1051**, 186 (1989).

CHAPTER 22

Moiré Fringes and Their Applications

Olof Bryngdahl
Department of Physics, University of Essen
Essen, Federal Republic of Germany

1. Introduction

Moiré is a customary designation for watered silk. This luxurious textile was manufactured in ancient China. The structural appearance of this lustrous material seems to change when it moves, which indicates that the effect could be applied to examine small movements. It is common practice nowadays to call those configurations moiré that appear when periodic or quasiperiodic patterns are superposed.

In 1874, Lord Rayleigh pointed out about the moiré effect that "this phenomenon might perhaps be made useful as a test" [1]. He described in detail how line gratings can be tested: "The lines themselves are, of course, too close to be seen without a microscope; but their presence may be detected, and even the interval between them measured, without optical aid, by a method not depending on the production of spectra or requiring a knowledge of the wavelength of light. If two photographic copies containing the same number of lines to the inch be placed in contact, film to film, in such a manner that the lines are nearly parallel in the two gratings, a system of parallel bars develops itself, whose direction bisects the external angle between the directions of the original lines, and whose distance increases as the angle of inclination diminishes."

Moiré between superposed 2-D gratings were examined: circular and radial gratings by Righi (1887) [2] and zone plates by Schuster (1924) [3].

ISBN 0-12-289690-4

Ronchi (1925) [4] and Raman and Datta (1925) [5] investigated the phenomenon further. However, it seems that interest diminshed in the field, and it would take until the end of World War II before a gradually increasing number of published works started to appear. This is reflected in the fact that the effect was not even mentioned in technical and scientific textbooks. The situation then changed swiftly and even books on moiré became available [6].

2. Moiré—Physical Parameters

Interaction between periodic/quasiperiodic structures produces miscellaneous periodicities in addition to the original ones. These regularities, which in optics are called *beats* or *intermodulations*, constitute the moiré appearance.

Moiré as an optical phenomenon is tied to the properties of the light involved. It can occur under the most varied circumstances comprising the range from complete incoherent to coherent illumination. Moiré is formed by structural changes in intensity, color, polarization, and phase. Among these, we are able to visualize the first two, and the others have to be converted into one of these two.

The interacting patterns are formed by material structures—which influence intensity, color, polarization, and phase—or by interference and diffraction phenomena. The latter are predominant when the structural dimensions approach the order of the wavelength of the light.

Frequently in technical applications, we are faced with opposing occurrences of a physical phenomenon. With moiré, this also is the case. When we intend to use the effect, it is important that we are able to amplify and isolate it. In other situations, moiré fringes are disturbing, and we then try to eliminate them. In general, these extremes are difficult to realize. To become successful, we need to understand and master the phenomenon.

In optics, information is stored in general as a spatial and temporal variation of one or more of the parameters that characterize a wavefield, viz., amplitude, phase, frequency, and polarization. Periodic structures that influence one of these quantities are called absorption, phase, color, and polarization gratings. These can be used to generate moiré effects.

3. Superposition of Periodic Structures

Superposition of patterns can be accomplished in different ways. They can be added, subtracted, and multiplied easily by optical means. To enhance, and in some cases, to form moiré fringes, a nonlinear operation has to be applied, too.

To illustrate the situations, structures with cosinusoidal profiles are superposed [7]. The different situations are

(a) addition:

$$(1+\cos 2\pi u_1 x)+(1+\cos 2\pi u_2 x) \\ = 2+2\cos \pi(u_1+u_2)x \cos \pi(u_1-u_2)x; \tag{1}$$

(b) subtraction:

$$(1+\cos 2\pi u_1 x)-(1+\cos 2\pi u_2 x) \\ = -2\sin \pi(u_1+u_2)x \sin \pi(u_1-u_2)x; \tag{2}$$

(c) multiplication:

$$(1+\cos 2\pi u_1 x)(1+\cos 2\pi u_2 x) \\ = 4\cos^2 \pi u_1 x \cos^2 \pi u_2 x \\ = [\cos \pi(u_1+u_2)x+\cos \pi(u_1-u_2)x]^2. \tag{3}$$

The effects are different: In Eqs. (1) and (2), a cosinusoidal structure of the average frequency $(u_1+u_2)/2$ is modulated by a cosinusoid of half the difference frequency $(u_1-u_2)/2$, and in Eq. (3), the sum of the cosinusoidal structures is squared.

Frequently, the superposed structures are inclined to each other. Then moiré occur even when they have the same frequency. To enhance the moiré effect, a directional averaging or filtering process may be performed.

In Eqs. (1) and (2), a nonlinear operation has to be applied for moiré fringes to be formed. If one of the substructures does not have a strong bias in the subtractive superposition, negative values result. This cannot occur in reality where the rectified values appear. This nonlinearity is sufficient to generate moiré.

In general, many fringe systems appear. Besides the original structures, patterns occur whose frequencies are the sum and difference of the fundamentals and higher harmonics of the original ones.

In metrological situations, we want to single out one fringe system. This is possible by adjusting the relative inclination of the individual structures to obtain fringes with a difference frequency an order of magnitude smaller than the original frequencies. The lowpass character of the observing system then can filter out the desired pattern.

4. Flexible Fringe Variables

To analyze and predict properties of moiré fringes, several approaches have been applied. Among them are the indicial representation of sets of curves and the extraction of components in the Fourier expansion of the combined structure configuration.

The individual structures are defined by their spatial frequencies, orientations, and profiles within their periods. The transform of the first two is called the *interference function* and can be illustrated as a vector; the transform of a single period is called the *diffraction function.*

When two structures are combined, the formation of a low-frequency beat separated from other existing frequencies is limited to small angles between structures of comparable frequencies. This is shown in the reciprocal (Fourier) space representation of Fig. 1. In Fig. 1(a), a structure of frequency r_0 and another of r_1 form the low-frequency moiré R. Obviously, the spatial frequency and orientation of a two-structure moiré cannot be varied independently. A relative rotation between the structures only can cause combinations of frequency and orientation indicated by an arc.

Incorporation of additional structures increases the number of parameters. Already three superposed structures allow independent adjustment of the moiré orientation and frequency [8]. In Fig. 1(b), a further structure r_2 is added. This allows R to be adjusted to any low frequency, and the orientation and frequency of the moiré fringes can be influenced independently.

Even higher harmonics of the periodic structures can be combined in this way to form a lowest-frequency moiré. This is shown in Fig. 1(c), where the second harmonic $2r_0$ (or higher harmonics) is combined with r_1 and r_2.

The microconfiguration consisting of rows of diamonds typical for two Ronchi-type structures will change with superposed additional ones. The moiré fringes then will be built up from several rows of micropolygons that differ in shape even within the fringes. The contrast of the moiré decreases

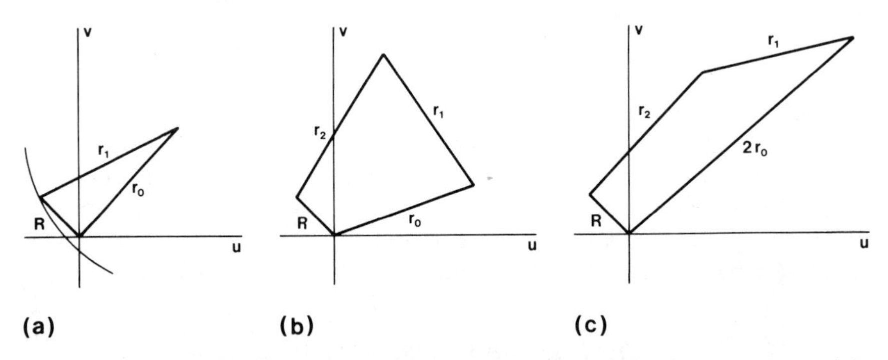

FIG. 1. Frequency space representation of superposed gratings: (a) two-grating moiré; (b) three-grating moiré; (c) by changing the relative orientations of the gratings, the second harmonic $2r_0$ participates to form the moiré R.

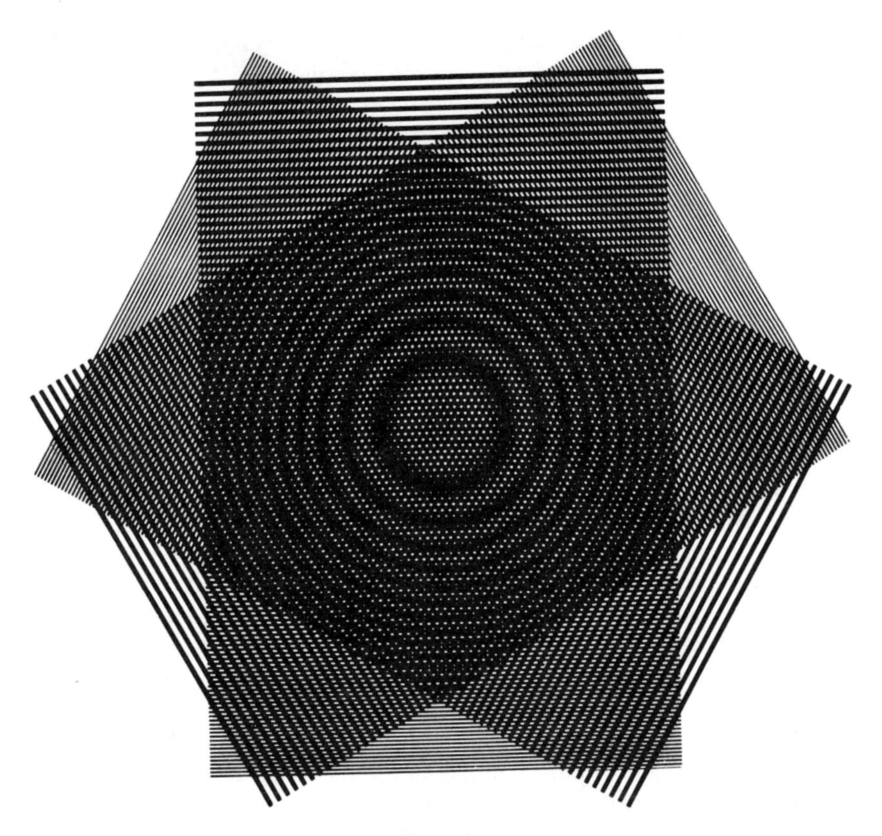

FIG. 2. Visualization of three-grating moiré from one-dimensional, one-sided zone-plate gratings.

with each superposed structure. However, the rather isotropic microconfiguration of the multiple-structure moiré introduces a smoother fringe profile and is advantagous in a local averaging (smearing) procedure. This is illustrated in Fig. 2.

5. Parameters for Fringe Selection

One way to expand the possibilities of variation and presentation in moiré fringe formation is to include more than two structures [8]. Another way to introduce additional parameters is to incorporate polarization [9] and color [10] effects. These parameters constitute physical states with orthogonal character that allow for interesting and valuable selective operations. When the periods of the structures consist of a specific sequence of different states, these structures contain defined spatial directions and allow spatially shifted (phase-shifted) components to be influenced separately.

A straightforward superposition of two states is the multiplicative one, which is analog to the superposition of transmission filters. These filters are orthogonal to each other if the state of the light, after passing through a filter, is completely blocked by the state of the next one. For polarization, two orthogonal states exist—i.e., two crossed linear or left and right circular polarizations—whereas for color, as many orthogonal states exist as nonoverlapping spectral bands can be formed.

There are some differences between the properties of these physical states. Polarization is a two-dimensional phenomenon in the space domain. Its states can be changed; e.g., one orthogonal state may be transformed into another. Color, on the other hand, is a multi-dimensional phenomenon in the frequency domain, and it is not possible to transform one of its states into another. Polarization and color states function independently of one another.

When structures of orthogonal state sequences are superposed, the different states of the structures are multiplied separately. These types of structures allow optional operations like selective addition and multiplication. They can be used for spatial-frequency filtering purposes in the spatial domain, where the process is independent of the actual value and the local variation in spatial frequency. After the operation, it is possible to remove the distinction between the different states in the moiré by a state blind detector (recording media, TV cameras and monitors, etc.). Figure. 3 shows an example to demonstrate this.

Superposition of polarization and color-coded structures lead to unique results. For example, the intensity of the moiré fringes, resulting from superposed aligned transparencies in which the state of linear polarization is rotated with the lateral position along one direction, is

$$\cos^2[2\pi(u_1 \mp u_2)x]. \tag{4}$$

The minus sign is valid when the sense of rotation of polarization is the same in both structures, and the plus sign when they are opposite. This is a way to choose beat frequency in a moiré display. When using polarization for this purpose, two orthogonal states will suffice. When using color, three states are necessary to indicate a certain direction. The corresponding moiré pattern for subcoded color structures is

$$1 + \cos[2\pi(u_1 \mp u_2)x], \tag{5}$$

where the minus sign is valid for identical color state sequences within the periods, and the plus sign when they are different. A 180° rotation of the

FIG. 3. Demonstration of filtering in the spatial domain. Superposition of two circular gratings: (a) conventional Ronchi rulings: (b) gratings with identical three-color sequence; (c) gratings with different three-color sequence.

polarization structure changes the sense of rotation among the states, and of the color structure the sequence among the states. Thus, a mere relative rotation of the structures will change the difference frequency moiré to the sum frequency.

6. Presentation and Processing of Optical Information

Moiré has become a simple and versatile technique to display optical information. Information in physics is stored in general as a distribution of energy. A spatial carrier usually is introduced to be able to manipulate and process spatial information that modulates the physical and geometrical properties of the carrier. The moiré, in the form of a superposition of a periodic structure onto one modified by the information, constitutes a demodulation process that serves to display the information in a convenient way.

To process the information, it is possible to introduce new degrees of freedom that are independent of the energy distribution. These information-independent parameters (polarization and color) for example, can be applied to mark (subcode) the carrier directionally, which in turn may serve to determine the sign of the variation of the information and to perform object-dependent spatial filtering.

After first being recommended as a method to test the quality of optical gratings, moiré has become the standard technique to determine periodic errors of gratings and rasters. Some new applications had been suggested and a handful of papers had appeared when the procedure suddenly started to attract attention around 1960. The papers by Oster *et al.* summed up the potential and showed the simplicity of the technique [11]. By imaging a grating structure through or via an optically varying medium or surface, a deformed image of the structure is obtained. The deformation reflects the optical state of the medium or surface, which can be displayed using the moiré technique.

To indicate the flexibility and adaptability of the moiré technique, it is possible to mention only a few of its diverse applications here: Besides gratings, other optical elements also can be tested and characterized by moiré, e.g., the examination of aberrations and focal power of lenses. The technique even can be used to obtain data of birefringent materials. In this respect, moiré can be compared with interference. However, in general, moiré is simpler to apply and can be made less sensitive so that measurements can be performed over a larger range. For example, moiré can be used to determine contours of the human body. The technique also is applicable for precision measurements of refractive index gradients formed by centrifugation, diffusion, heat conduction, and electrophoresis. Moiré is used in combination with holographic interferometry; moiré fringes can be formed between two interferometric or holographic recordings of different or the same states of the object. In the *dupligram method*, two copies of an interferogram are superposed and the moiré fringes represent the spatial derivative of the interferogram fringes [12]. The technique also has been applied successfully to investigate crystals in electron microscopy. Even though the crystal structure itself cannot be resolved, the moiré pattern between two superposed crystals can show information about dislocations.

Among the natural and important applications of moiré are precision measurements of small angles and translations. Furthermore, it is a sensitive method for optical alignment.

A moiré pattern can be regarded as the mathematical solution of the interaction between periodic functions. For example, the technique may be used as an analog computer to show the solution to a physical problem in graphical form. Of special value is its application to the study of complex physical fields and wave phenomena.

Simulation experiments of complicated practical situations can be performed. Moiré has been used to solve acoustical problems in buildings and to construct wave breakers in harbors.

7. Moiré Techniques in Metrology

Among established techniques in metrology, there are interferometric, holographic, and moiré methods that are based on the principle of superposition of optical fields. The position and frequency of the fringes formed contain information about the object under testing. When using a carrier, it is possible to display information about pure phase objects as a spatial variation of a fringe pattern.

In contrast to interferometry and holography, where the fringes are formed by interaction between waves, the occurrence of moiré fringes is a geometrical phenomenon of superposition of periodic structures. In moiré, the light is used for illumination and, in principle, can be produced by any source.

Several methods have been presented that are based on the principle that the object under testing is combined with a grating structure to form a modulated light distribution, which in turn falls onto a second grating to display moiré fringes. The methods differ mainly in the relative positions of the object and the two gratings. They may be divided into three groups:

(1) To examine object deformation, one of the gratings is attached to or etched onto the surface of the object. This is the procedure in strain analysis.

(2) For topographic investigations, the *shadow moiré* [13] and *grating projection* [14] methods are becoming standard techniques. The object is placed between the two gratings. In the shadow moiré, the same grating performs two functions. If it is placed close to the object and properly illuminated, its shadow onto the surface of the object can be viewed through the same grating. In the grating projection method, a grating is projected onto the object that, in turn, is imaged onto the second analyzing grating with a camera system.

(3) To investigate phase objects, the two gratings are placed a distance apart and the object is placed either between or in front of the gratings. The procedure has been called *Talbot interferometry* [15] and *moiré deflectometry* [16].

8. From Perception to Art

Moiré techniques are used to examine the visual system: not only the imaging system itself—i.e., the portion forming an image and the receptor geometry of the retina—but also the perceptual interaction of periodic spatial and temporal patterns.

Some unusual effects occur: The translation of the moiré fringes is magnified in comparison to the relative movement between the structures that causes them. Further, striking depth sensation can be perceived as exaggerated due to parallax and moiré motion.

Moiré can be used to achieve fascinating artistic effects. It is possible to perceive moiré fringe motions simultaneously in different directions by a slight translation of the head or the design pattern. For example, polarizing colored-film patterns on revolving glass doors have been applied for both decorative and warning purposes.

In the future, we may find that moiré effects will add new dimensions and forms to art in which the viewer takes an active part. Art and science can help us to understand better the perception of moving objects in our three-dimensional surroundings.

9. Trends and Expectations

In established measuring techniques like interferometric and holographic methods, a carrier is introduced. Moiré also belongs to this class. It has the advantage that manipulation of the information carrier is allowed in an easy, accessible way. The potentials of the carrier recording have not been utilized fully yet. Especially in the field of parallel processing, many applications and inventive ideas are to be expected. The moiré technique is a simple way to demodulate a signal contained in the form of a modulated carrier structure—not in a conventional way, but as an intermodulation (beat) pattern. In this respect, moiré is different from interferometry and holography, and it is not to be regarded as a pure alternative, but as a supplementary approach.

The moiré technique is well suited to be realized as a hybrid system: A video system with data memory can be used in the superposition step to form the moiré. A computer also may be added for additional processing and experimental control.

For illumination and display, a scanning mode frequently is used. This also is the case in electronic displays. Unwanted moiré effects then may turn up in spatial as well as temporal form. Further, sampling and quantization sometimes cause strange patterns, which may be disturbing especially in printing situations. Moiré suppressing algorithms will become available.

The unscientific term *moiré*, covering intermodulation patterns, seems to have been coined in the last century's Parisian fashion world. The circle now closes when moiré techniques are used in dressmaking, fitting, and tailoring work.

References

1. Lord Rayleigh (J. W. Strutt), *Phil. Mag.* **47**, 81, 193 (1874).
2. A. Righi, *Nuovo Cimento* **21**, 203 (1887).
3. A. Schuster, *Phil. Mag.* **48**, Ser. 6, 609 (1924).
4. V. Ronchi, "Attualita scientifiche," no. 37 Ch. 9, N. Zanichelli, Bologna (1925).
5. C. V. Raman and S. K. Datta, *Trans. Opt. Soc.* **27**, 51 (1925/26).
6. J. Guild, "*The Interference System of Crossed Diffraction Gratings*," Clarendon, Oxford, 1956; J. Guild, "*Diffraction Gratings as Measuring Scales*," Oxford University Press, London, 1960; P. S. Theocaris, "*Moiré Fringes in Strain Analysis*," Pergamon, London, 1969; A. J. Durelli and V. J. Parks, "*Moiré Analysis of Strain*," Prentice-Hall, Englewood Cliffs, N. J., 1970; B. Drerup, W. Frobin, and E. Hierholzer, "*Moiré Fringe Topography and Spinal Deformity*," Fischer Verlag, Stuttgart, 1983; O. Kafri and I. Glatt, "*The Physics of Moiré Metrology*," Wiley, New York, 1990.
7. O. Bryngdahl, *J. Opt. Soc. Am.* **66**, 87 (1976).
8. O. Bryngdahl, *J. Opt. Soc. Am.* **64**, 1287 (1974); *ibid.* **65**, 685 (1975).
9. O. Bryngdahl, *J. Opt. Soc. Am.* **62**, 839 (1972).
10. O. Bryngdahl, *Opt. Commun.* **39**, 127 (1981); *ibid.* **41**, 249 (1982).
11. G. Oster and Y. Nishijima, *Sci. Amer.* **208**, 54 (May, 1963); Y. Nishijima and G. Oster, *J. Opt. Soc. Am.* **54**, 1 (1964); G. Oster, M. Wasserman, and C. Zwerling, *J. Opt. Soc. Am.* **54**, 169 (1964).
12. E. Lau, *Optik* **12**, 23 (1955).
13. D. M. Meadow, *Appl. Opt.* **9**, 942 (1970); H. Takasaki, *Appl. Opt.* **9**, 1457 (1970).
14. M. Suzuki, M. Kanaya, and K. Suzuki, *Precision Instruments* **40,** no. 9, 36 (1974).
15. S. Yokozeki and T. Suzuki, *Appl. Opt.* **10**, 1575, 1690 (1971); A. W. Lohmann and D. E. Silva, *Opt. Commun.* **2**, 413 (1971).
16. O. Kafri, *Opt. Lett.* **5**, 555 (1980).

CHAPTER 23

Breaking the Boundaries of Optical System Design and Construction

Christiaan H. F. Velzel
Philips–CFT
Eindhoven, The Netherlands

1. Introduction

In this chapter, we discern three types of limits to optical system design and construction. We restrict the discussion to imaging systems, of which we define the performance as the quotient of field area and pixel area, or the number of degrees of freedom in the image.

This performance is limited, in the first place, by the laws of geometrical and physical optics. The pixel size cannot be smaller than the diffraction limit; we assume that the aberrations of the systems we deal with are small, so that we are in the neighborhood of diffraction-limited imaging. From the principles of geometrical optics, we can deduce to what extent systems without aberrations are possible. In Section 2, we discuss this problem, using Hamilton's method of characteristic functions.

The second limitation to the performance of an optical system is given by the mathematical design process. This process usually consists of (at least) two phases: initial design and optimization. In the initial design phase, one chooses the type of optical system that can be expected to fulfill the system specifications required. This includes analytical calculations, mostly limited to third-order aberrations.

In Section 3 of this chapter, we discuss the number of degrees of freedom of optical systems and their use in initial design. When an initial design has been made up, its parameters will be varied until a design is found that

ISBN 0-12-289690-4

agrees with the required specifications. This process of optimization, usually performed with the aid of standard computer programs, has its own limitations, which we also describe in Section 3.

An optical system that has been designed on paper is not necessarily fit for production. This depends on the material properties and the tolerances required for its realization. In Section 4, we discuss the sensitivity analysis of optical systems from which these requirements can be deduced, and the state-of-the-art technology with which the requirements can be met. The accuracy with which components of optical systems can be made and systems can be assembled from them forms the third type of limitation that we encounter in optical system design.

Finally, in Section 5, we give our opinion on the possibility of overcoming the limitations of the existing methods of design and construction of optical systems as described in this chapter.

2. The Perfect Optical Instrument

We define a *perfect optical instrument* as a configuration of finite dimensions that transforms a spherical wave coming from an arbitrary point on a closed surface, the object surface, into a spherical wave centered on a point of another closed surface, the image surface. The situation has been depicted in Fig. 1.

From this figure, it is clear immediately that, when a sphere of diameter d can be circumscribed around the instrument, an upper limit for the number of resolved pixels N must be given by

$$N \leq 4\pi(d/\lambda)^2 \tag{1}$$

because the angular resolution is given by λ/d, when λ is the wavelength of the radiation used. Here, we have assumed that the index of refraction in object and image space is equal to 1. With unequal indices, the argument is

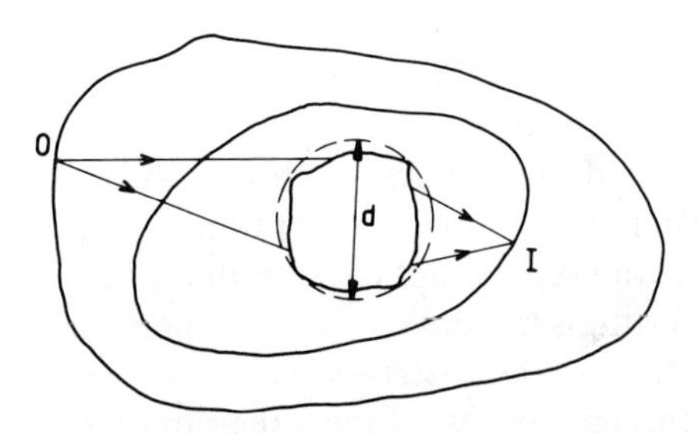

FIG. 1. A perfect optical instrument: O is a point on the object surface, I on the image surface. The instrument is circumscribed by a sphere of diameter d.

more complicated, but the result is the same; we will give an example of that case shortly.

We derive the point eikonal of a perfect instrument, in order to study its properties more closely. For that purpose, we need the relation between the coordinates $x', y', z'(x', y')$ of the image surface and the coordinates x, y, $z(x, y)$ of the object surface. (See Fig. 2.) Let this relation be given by

$$x' = x'(x, y), \qquad y' = y'(x, y), \tag{2}$$

where we assume that the imaging is reciprocal; that is, x, y can be solved from Eq. (2). The point eikonal between the object surface O and the exit pupil surface P, with coordinates $x_p, y_p, z_p(x_p, y_p)$ in image space can be found from the argument that the optical path from an object point (x, y) to its conjugate image point (x', y') given by Eq. (2) is independent from (x_p, y_p) when the imaging is perfect [1].

The point eikonal S, therefore, takes the form:

$$S = c(x, y) - ((x_p - x'(x, y))^2 + (y_p - y'(x, y))^2 + (z_p - z'(x, y))^2)^{1/2}, \tag{3}$$

where we take $n' = 1$ in the image space. Note that S is a function of x, y and x_p, y_p.

Physically meaningful values of the radiation amplitude in the image point are obtained by taking S real over that part of the pupil surface that can be seen from the image point, and imaginary over the rest of the pupil surface so that no radiation leaves that part.

We now consider the possibility of perfect imaging for more than one pair of surfaces. When we consider a pupil surface Q in the object space, the point eikonal between the surfaces P and O that gives a perfect image of P on Q must be of the form:

$$S = c'(x_p, y_p) - ((x - x_q(x_p, y_p))^2 + (y - y_q(x_p, y_p))^2 + (z - z_q(x_p, y_p))^2)^{1/2}, \tag{4}$$

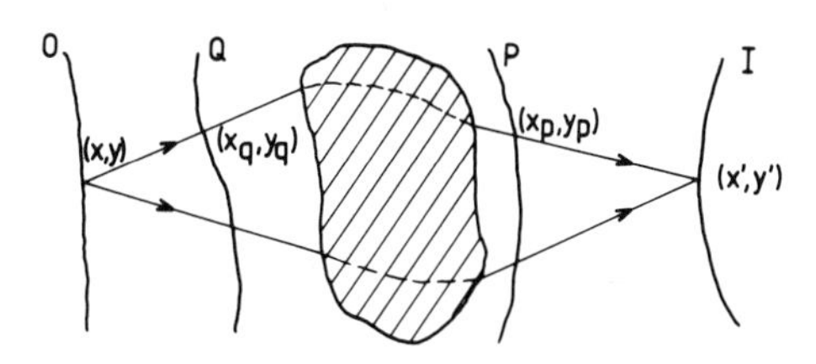

FIG. 2. The point eikonal is the optical path between a point (x,y) on the object surface o and a point (x_p, y_p) on the exit pupil surface P. The entrance pupil is denoted by Q, the image surface by I.

where we have to take $n = 1$ in the object space and $x_q(x_p, y_p)$, $y_q(x_p, y_p)$, $(z_q(x_p, y_p)$ are the coordinates of a point on Q conjugate to (x_p, y_p).

Inspection of the expressions (3) and (4) shows that the two eikonals can be made identical only when all the object space is imaged on all the image space with a magnification of ± 1. An example of this perfect volume imaging is the image formation by a plane mirror.

The surface P can be chosen arbitrarily; it should not be identical to I. It follows that, in general, it is not possible to image more than one pair of surfaces perfectly on each other.

We still have the function $c(x, y)$ in Eq. (3) available to enrich the specifications of our instrument. When we take, for instance,

$$\begin{aligned} c(x, y) = c_o - ((x - x_{qo})^2 + (y - y_{qo})^2 + (z - z_{qo})^2)^{1/2} \\ + ((x' - x_{po})^2 + (y' - y_{po})^2 + (z' - z_{po})^2)^{1/2}, \end{aligned} \tag{5}$$

where c_o is a constant, and where x', y' are given again by Eq. (2), we see that the optical path from the point x_{po}, y_{po}, z_{po} on P to the point x_{qo}, y_{qo}, z_{qo} on Q is equal to c_o and independent of x and y. This means that apart from the surfaces O and I, it is possible to image two arbitrary points stigmatically on each other. An example of this case was given by Th. Smith [2].

In his book, *Traité de la Lumière*, published 300 years ago, Huygens gives an example of a perfect instrument by his construction for the refraction of a ray at a spherical surface. (See Fig. 3). An incoming ray, focussed on a point O on a sphere with radius nr, where r is the radius of the refracting sphere and n its refractive index, is refracted to I, on a sphere with radius r/n and on the radius MO, where M is the center of the refracting sphere. We see that the two spheres on which I and O lie are imaged perfectly on one another; it is easy to show that the number of

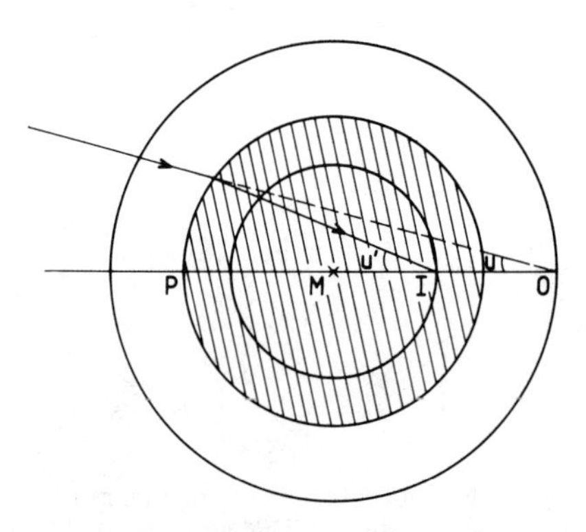

FIG. 3. Construction of Huygens: $MP = r$, $MI = r/n$, $MO = rn$.

pixels is given by Eq. (1), where $d = 2r$ is the diameter of the refracting sphere. Here, we have an example of a perfect instrument where the refractive index in the image space is different from that in the object space. The wavelength in the image space is smaller by a factor n than that in the object space; we also have

$$\sin u' = n \sin u \tag{6}$$

so that the pixel size in the image space is a factor $1/n^2$ times that in the object space. The magnification between the conjugate spheres also is equal to $1/n^2$, so that the number of pixels on both spheres is equal, as it should be.

The refracting sphere is an impractical instrument because its object is virtual and the image lies inside the sphere. When we want to use an optical instrument to project an image of a real object on a flat surface a more complex instrument must be designed.

Although we do not know the geometry of such an instrument, we can describe its eikonal. For distortion free imaging of a plane object in a plane, we insert in Eqs. (3) and (5)

$$x' = Gx, \qquad y' = Gy, \qquad z' - z_p = t', \qquad z - z_q = t, \tag{7}$$

where t and t' are constant distances, and G is a constant magnification factor. It is the task of optical design to approximate as well as possible (or necessary) the ideal eikonal over a specified area of the image plane and the pupil plane.

3. Initial Design and Optimization

The number of degrees of freedom that the designer can use to correct the aberrations of an optical instrument is approximately proportional to the chosen number of components of the instrument. With a symmetrical system consisting of lenses with spherical surfaces, the number of degrees of freedom per lens is five (two curvatures, thickness, axial position, and refractive index). In practice, the thickness of the lens often is neglected and the number of different glass types is limited. By aspherizing one or more surfaces, degrees of freedom can be added; we will not consider this topic further in this chapter.

In a triplet of thin lenses with two glass types, therefore, we have 12 degrees of freedom when we also take into account the stop position.

What can we do with these degrees of freedom? In the previous section, we derived the eikonal of a perfect instrument; for a nearly perfect instrument, the eikonal can be written as

$$S = S_o + \triangle S, \tag{8}$$

where S_o is the ideal eikonal in the sense of Section 2. We consider again the case of distorsion-free imaging of a plane object (with $n = 1$ in the object space) on a plane surface (with $n' = 1$ in the image space) with given magnification. To a good approximation (we assume that $\triangle S$ is of the order of the wavelength and many orders of magnitude smaller than the focal length of the system), the transverse aberrations in the image plane are given by

$$\begin{aligned} x' - Gx &= s' \frac{\delta \triangle S}{\delta x_p}, \\ y' - Gy &= s' \frac{\delta \triangle S}{\delta y_p}, \end{aligned} \tag{9}$$

where s' is the distance from a point x_p, y_p in the pupil plane to the point Gx, Gy in the image plane. We see that $\triangle S$, for all practical purposes, is equivalent to the wavefront error in the pupil plane. We want, for a good correction, that the variance of $\triangle S$ over the pupil be small compared with λ for all points in a specified image field. Because the instrument is axially symmetric, S can be written as a function of the three variables u, v, w, defined by

$$\begin{aligned} u &= (x^2 + y^2)/t'^2, \\ v &= (xx_p + yy_p)/t'^2, \\ w &= (x_p^2 + y_p^2)/t'^2. \end{aligned} \tag{10}$$

(See Eq. (7).) When S is written as a power series in u, v, w, the number of coefficients of order k is given by

$$n(k) = \tfrac{1}{2}(k+1)(k+2). \tag{11}$$

The coefficient with $k = 0$ determines the optical path along the axis. There are three coefficients of the first order, which determine the paraxial constants of the system. The six coefficients of second order determine the Seidel aberrations. The total number of coefficients up to $k = 2$ is 10. With a triplet as described earlier, we can control, therefore, all coefficients up to $k = 2$, plus two coefficients of color correction. Also controlling the monochromatic coefficients of order $k = 3$ would require at least 10 more degrees of freedom.

When the number of degrees of freedom has been chosen, it remains to be seen whether in the part of parameter space that can be reached in practice, a useful initial design can be obtained. For a thin-lens triplet, indeed, solutions can be found analytically where all the paraxial and Seidel coefficients have satisfactory values; see the work of Berek [3] and Cruikshank [4].

Departing from the initial design, we minimize the variance of $\triangle S$ over a given range of field coordinates x, y and pupil coordinates x_p, y_p using the available degrees of freedom. The minimization is subject to a number of constraints: Curvatures cannot be too high; axial surface positions must be compatible; and refractive indices have discrete values. A number of algorithms for optimization exist [5] and computer programs to execute these are available; nevertheless, some problems remain. When the algorithm has led to a minimum of Var ($\triangle S$), we are never certain whether this is a local minimum or a global minimum in parameter space. Even when the global minimum has been reached, residual errors of $\triangle S$ remain; we would like to have an estimate of these before optimization. When the residual errors are too large, we must start anew with a better initial design. That means that we must take different initial values of the design parameters or introduce more degrees of freedom, that is, more components.

As an example of a design with a great number of resolved pixels, ($N \approx 3 \times 10^8$) over a field of 20mm diameter, we discuss a microprojection objective designed by Braat [6]. In this instrument, diffraction-limited imaging is achieved over a numerical aperture of 0.38 at a wavelength of 405 nm.

The wish to obtain a flat field and freedom of distortion leads to the introduction of a number of negative components. (See Fig. 4.) By using these negative components at a small beam diameter and the positive components at a large diameter, so that bulges and constrictions occur in the layout of the system, the field curvature can be controlled accurately. With the negative elements at the beginning (2) and the end (16), distortion and astigmatism are suppressed. A low-power element (5) is introduced to correct fifth-order aberrations. The doublets (10), (11) assist in the fine correction of axial chromatism.

To reduce the residual higher-order aberrations, the number of elements has been increased. The gradual bending of the rays obtained in this way also leads to more friendly manufacturing tolerances, as we will see in the

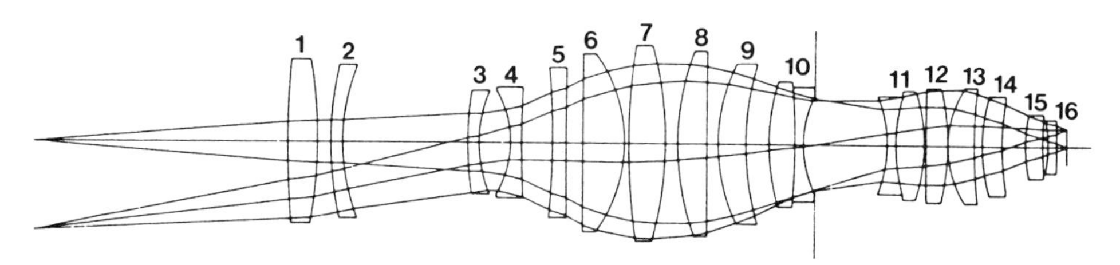

FIG. 4. Outline of a microprojection lens, designed by Braat [6].

next section. The addition of errors due to the individual components reduces this advantage.

The resulting system, with a distance between the object and image plane of 600mm and a diameter of 110mm for the largest component, has 16 components (among which are two doublets) and about 60 degrees of freedom. Obviously, the high number of resolved pixels and the requirements of manufacturing have led to a complex design; it is difficult to see whether an alternative solution is possible.

4. Sensitivity Analysis and Assembling

The manufacturing tolerances of an optical system can be derived from its design parameters. We describe briefly the analysis of the sensitivity of symmetrical systems to centering errors of their components. (See [7] for more details.)

From Fig. 5, we see that the optical path change dS along a ray caused by a vector shift dx, dy, dz of a component is given by

$$dS = (n'L' - nL)\,dx + (n'M' - nM)\,dy + (n'N' - nN)\,dz, \qquad (12)$$

where L, M, N and L', M', N' are the direction cosines of the ray before and after refraction, respectively. The component may be a single surface, a lens, or a group of lenses; the direction cosines can be expressed analytically in the field and aperture coordinates of the system—like u, v, w defined in Eq. (10)—or can be obtained by ray tracing [7].

Here, we content ourselves with a qualitative discussion. The sensitivity for transverse shifts, embodied in the factors $n'L' - nL$ and $n'M' - nM$, is roughly proportional to the power of the component and to the diameter at which it is used. In Fig. 4, we can see directly which elements are most sensitive to transverse shifts (6, 13). We remark in passing that the factors $n'L' - nL$ and $n'M' - nM$ also give an indication of the higher-order

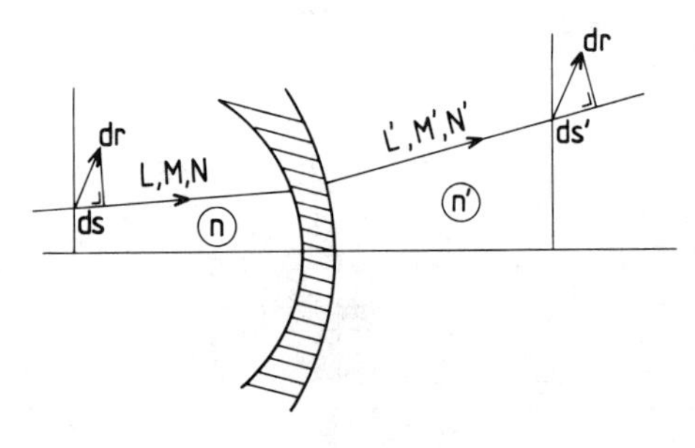

FIG. 5. Shift of a component: dr is the displacement vector; the optical path change is given by dS = $-nds + n'ds'$.

aberrations induced by the component in the rest of the system. See the discussion on *strain* in [7].

A spherical surface is insensitive to rotations about its center of curvature. A tilt about an axis through its vertex can be composed of a rotation about the center plus a shift. Therefore, the effect of tilts can be described by Eq. (12) when we insert

$$dx = \alpha r, \qquad dy = \beta r, \tag{13}$$

where α, β are the tilt angles and r is the radius of curvature. Note that this is valid for a single surface only. With a symmetric optical system, the factors $n'L' - nL$ and $n'M' - nM$ contain terms of uneven order only. The aberrations that follow from dS by differentiation (see Eq. (9)), therefore, are of even order. The linear terms in dS give image shift and tilt of the axis. The third-order terms give rise to axial coma, linear astigmatism, and second order distortion. These aberrations should not be confused with Seidel aberrations.

The sensitivity for axial shifts is given, according to Eq. (12), by $n'N' - nN$. This quantity contains terms of even order in the field and aperture coordinates only, for a symmetric system. That means that axial shifts change the paraxial constants and the third- and higher-order aberrations of a system.

The tolerances for shifts and tilts of components can be obtained by setting an upper limit on dS. Thickness errors of components can be treated as axial shifts; errors of curvature and refractive index must be treated differently.

In a system with many components, the effects of the different components must be added. In [6], an explicit discussion of the tolerances of a compound system is given. The author gives tolerances of 2 μm for shifts and 25 microrads for tilts.

Now let us discuss how these tolerances are realized in the construction of optical systems. We suppose that components (single lenses or doublets) have been made with the required curvatures, thickness, and refractive index. Small errors in these parameters can be corrected by axial shifts. The components must be positioned, and kept into position, along a common axis of symmetry. This constitutes the problem of optical assembling.

In most optical systems, the lenses are positioned by inserting them in a mounting ring or directly in a barrel that is made with the required precision. The lenses are kept in place usually by retainers in the form of a ring. The resulting construction is overconstrained and contains materials with different coefficients of expansion (e.g., glass and steel) and different stiffness. Due to external forces (mechanical or thermal), small random changes of position of the lens in the mounting will occur. This makes the conventional technology unsuitable for the more exacting designs, such as that of Fig. 4.

It has been proposed that an elastomer be used between the lens and the mounting ring, with a thickness and an expansion coefficient chosen to compensate for a difference in expansion of the lens and the mounting [8]. With this technique, the lens can be kept in position with a precision of the order of 1 μm; it still must be centered in its mounting.

Centering can be achieved by measuring the alignment between the optical axis of a lens and the mechanical axis of the mounting, and correcting it by machining the mounting. In the literature, one finds a great number of papers on this subject [9]. Here, we do not discuss the different methods, but we point out that by interferometric methods, it is possible to measure displacements with nanometer precision, so that it must be possible to measure centering errors with the required accuracy.

After their centering errors have been corrected, the components still must be mounted together to form the optical system that we want to produce. About this subject, there is practically no literature. It is clear that if the centered components are simply stacked in a barrel, even if it is accurate and stable enough, the addition of errors can give rise, with a complex system, to intolerable degradation of performance. Unless one is prepared to adjust some of the components afterwards, the mounting procedure then must be repeated until a satisfactory result is obtained.

5. Looking to the Future

In the preceding sections, we described the limitations of optical system design and construction. We saw that there are three types: physical limits, connected with the laws of wave propagation; mathematical limits, connected with the choice of the number and arrangement of the components and the finding of the optimal system; and technological limits, connected with the construction of optical systems.

It seems that the first type of limitation is the most difficult to overcome. In Section 2 we discussed the limitations of perfect imaging and the diffraction limit to resolution. The first way to increase the performance of a perfect instrument that comes in mind is increasing the scale of the instrument, and thus the parameter d in Eq. (1). Note, however, that, when the wavelength remains the same, the effect of centering errors and that of component imperfections increases, and also the rest of the aberrations, so that design and construction become more difficult. Also, in most systems, limits are set on volume and weight.

A second way to increase the number of pixels of a perfect instrument is to decrease the wavelength of the radiation used. Recently, excimer lasers, with a wavelength of 245 nm, have been used in microlithography [10]. Because all optical glasses become opaque at that wavelength, the projection lens should be built from fused quartz; an alternative is the use of reflective optics, and perhaps the best system could be found by combining the two [11]. Reflective optics certainly is necessary when the wavelength is reduced further. X-rays with a wavelength of 13 Å have been used recently for projection imaging (0.1 μm lines over a field of 25×50 μm) [12]. We point out that at such a short wavelength, the micro-roughness of the reflecting surfaces used for projection may become a problem.

Let us consider if it is possible to improve the existing methods of optical design. We begin with the initial design. Since Berek [3], little headway has been made in the development of analytical design methods. We think that headway can be made in two directions. In the first place it must be possible to find simpler, more elegant formulas with which to describe higher-order aberrations. In recent work, a start has been made in this direction, using Hamilton's method of characteristics [13] and, alternatively, using group-theoretical techniques [14].

In the second place, more use could be made of the symmetry properties of surfaces and systems. For instance, aplanatic surfaces like those described in Section 2, are insensitive to shift and tilt. This follows from Eqs. (6) and (12), if we take $\sin u = -L$ and $\sin u' = -L'$. A disadvantage of the use of aplanatic surfaces is that they do not contribute to the correction of aberrations. The consequence is that this method leads to designs with a great number of degrees of freedom, and thus of components. Perhaps equivalent designs can be found where the burden of correction and of centering errors is distributed more evenly among a smaller number of components.

The use of the properties of aplanatic surfaces in optical design has been advocated by Shafer [15]. Both approaches are combined in the work of Stavroudis on modular design [16]. We think that the idea to build up systems from modules (singlets, doublets, etc.) that have special correction properties is promising and should be pursued further.

Improving optimization programs—which is a different thing from expanding them by the addition of new features—is a matter of mathematical technique about which we have little to say. To solve the problems with optimization indicated in Section 3, it would seem worthwhile to have a theory predicting the number of minima of the merit function in a given region of parameter space. Such a theory exists [17], but its application to optical design is still pending.

A new approach to optimization can be made by calculating the characteristic function of an optical system rather than tracing a great number of rays, and using the squared difference of this function and a target function as a merit function. The advantage of this approach is that one obtains a direct insight into the correction and balancing of the aberrations of the system, and the influence of single components. An optimization program based on the angle eikonal has been written by Kruizinga [18].

The technologies that are used in the construction of optical systems are: mechanics (for positioning and stability), fixing technology, and optical measurement. We discuss the progress possible in these fields.

In positioning, there are potentialities that have not been applied yet to the construction of optical systems. In the mass production of optical recording systems, for instance, tolerances within 1 μm are achieved routinely [19]; it seems worthwhile to transfer the technologies used there and in other parts of the electronics industry to the production of more complex imaging systems like microscopes and projection lenses.

Accurate positioning must be accompanied by adequate stability of the construction. Many of the constructions used in optics are overconstrained; we advocate the use of kinematic constructions to obtain better stability.

We subdivide the fixing technologies into three categories: mechanical, adhesive, and thermal. A combination of the first two is used, as we saw in Section 4, in the construction of high-performance optical systems. The nonelastic properties of plastic materials used in lens mountings, together with the overdetermination of the mechanical construction, can cause small but significant displacements of components during product life, even if the initial positioning is accurate enough. Nevertheless, it is possible to obtain stable submicron precision by the use of adhesives when special care is taken in the construction of parts [19].

Thermal fixing technologies comprise soldering, thermocompression, and laser welding. These technologies have not been used often for the assembling of optical systems. Laser welding has been used recently in the electronics industry for the assembling of electron guns for T.V., for example; it holds promise with respect to stability and speed of operation, with its initial accuracy of the order of 1 μm [20].

The last technology we discuss is optical measurement. In Section 4, we have argued that by the use of interferometry, it is possible to measure centering errors with the required precision. This does not mean that interferometers available on the market are suitable for this purpose. In recent years, much progress has been made in the accurate processing of interference patterns [21].

The application of these new developments to optical assembling is obvious. Apart from interferometric methods, other approaches, such as autocollimation and laser scanning, also are used or could be used in optical assembling. It should not be thought that these techniques are less accurate than interferometric methods. With state-of-the-art position-sensitive devices, it is possible to measure beam displacements of 0.5 μm or less. At a distance of 500 mm, this means an angle deviation of 1 microrad. With a pupil diameter of 50 mm, this corresponds to wave-front errors of 25 nm (for distortion) or 6nm (for spherical aberration) at the rim of the pupil.

The conclusion of this chapter should be clear to the reader by now: The limits of optical system design and construction can be shifted in many ways. The most immediate progress can be made in the fields of assembling technology: accurate positioning, stable constructions, fixing technology, and optical measurement. The methods of optical design can be improved, and a better balance can be found between the correction of aberrations and the manufacturing tolerances. Whereas an increase in the dimensions of optical systems is not always possible, the use of shorter wavelengths can lead to the resolution of smaller details. It is impossible, however, to design a system that has better specifications than the perfect instrument defined in Section 2.

The treatment in this chapter follows, of necessity, a rather straight course from theory to practice. We have left aside several interesting alternatives; a delightful collection of these can be found in [22].

References

1. R. Luneburg, "Mathematical Theory of Optics," 23, U. of California Press, Berkeley, 1964.
2. Th. Smith "The Changes in Aberrations when the Object and Stop are Moved," *Trans. Opt. Soc.* (London) **23**, 311–322 (1922).
3. M. Berek, "Grundlagen der Praktischen Optik," W. de Gruyter, Berlin, 1970.
4. F. D. Cruikshank, "The design of Photographic Objectives of the Triplet Family (Parts I and II)," *Austr. J. Phys.* **11**, 41–54 (1958); ibid. **13**, 27–42 (1960).
5. T. H. Jamieson, "Optimization Techniques in Lens Design," Adam Hilger, London, 1971.
6. J. J. M. Braat, "Quality of Microlithographic Projection Lenses," *SPIE* **811**, 22–30 (1987).
7. C. H. F. Velzel and J. L. F. de Meijere, "Sensitivity Analysis of Optical Systems Using the Angle Characteristic, *SPIE* **1168**, 164–175 (1989).
8. P. R. Yoder, Jr., "Lens-Mounting Techniques," *SPIE* **770**, 155–164 (1988).
9. G. Jaunet, Y. P. Mariage, F. Farfal, M. Mullot, and B. Bonino, "Procédés de Centrage des Surfaces Optiques," *Nouv. Rev. Optique* **9**, 31–44 (1977).
10. R. W. Clary and P. J. Thompkins, "Performance of a KrF Excimer Laser Stepper," *SPIE* **922** (1988).
11. D. M. Williamson, Optical Reduction System. European Patent Application 0 350 955.

12. T. E. Jewell, M. M. Becker, J. E. Bjorkholm, J. Bokor, L. Eichner, R. R. Freeman, W. M. Mansfield, A. A. MacDowell, M. L. O'Malley, E. L. Raab, W. T. Silfvast, L. H. Szeto, D. M. Tennant, W. K. Waskiewicz, D. L. White, D. L. Windt, O. R. Wood II, and J. H. Bruning, "20:1 Projection Soft X-Ray Lithography Using Tri-Level Resist," *SPIE* **1263**, to be published.
13. J. L. F. de Meijere and C. H. F. Velzel, "Dependence of Third- and Fifth-Order Aberration Coefficients on the Definition of Pupil Coordinates," *J. Opt. Soc. Amer.* **A6**, 1609–1617 (1989).
14. J. Sánchez Mondragon and K. B. Wolf (eds.), "Lie Methods in Optics," in "Lecture Notes in Physics 250," Springer, Berlin, 1986.
15. D. Shafer, "Optical Design Methods: Your Head as a Personal Computer," *SPIE* **351**, 49–58 (1985).
16. O. Stavroudis, "Modular Optical Design (Series in Optics 28)," Springer, Berlin, 1982.
17. B. J. Hoenders and C. H. Slump, "On the Calculation of the Exact Number of Zeroes of a Set of Equations," *Computing* **30**, 137–147 (1983).
18. B. Kruizinga, Delft Instruments, Delft, the Netherlands, private communication.
19. N. Maan and R. M. Zwiers, "Accurate High-Speed Assembly of Optical Components in Optical Scanning Systems Using UV-Curing. Adhesives: A Positional and Stability Study," *J. Adhesion* **26**, 85–100 (1988).
20. C. J. Nonhof, "Material Processing with Nd-Lasers," Electrochemical Publications Ltd., Ayr, Scothland, 1988.
21. K. Creath, "Phase-Measurement Interferometry Techniques," in "Progress in Optics XXVI" (E. Wolf, ed.), North-Holland, Amsterdam, 1987.
22. P. Clark (ed.), "Meeting the Challenges of Optical Design," *Optics News* **14**, no. 6 (1988).

CHAPTER 24

Interferometry: What's New Since Michelson?

P. Hariharan
CSIRO Division of Applied Physics, Lindfield, Australia

1. Introduction

A. A. Michelson was one of the greatest experimental physicists of his time. The applications he found for interferometric measurements, apart from the celebrated experiment that "disproved" the existence of a luminiferous ether, included determinations of the thickness of soap films, indices of refraction, coefficients of expansion, the gravitational constant, the wavelengths of spectral lines, the fine structure of spectral lines, the diameters of stars, and the measurement of the meter bar in wavelengths. It was quite proper that the citation to his Nobel Prize in 1907 referred to "his precision optical instruments and the spectroscopic and metrological investigations conducted therewith."

Michelson's contributions to interferometry during the half century from 1880 to 1930, which were summarized in his two books [1, 2], dominated the field to such an extent that they created the impression that there was little of any importance left for his successors. This chapter will review a few selected developments in interferometry to show that this pessimistic view was unjustified, in fact, and will speculate on some future possibilities.

2. Interferometric Metrology

Michelson's imaginative proposal to use the wavelength of a spectral line as a standard of length finally was adopted in 1960, when the meter was

ISBN 0-12-289690-4

defined in terms of the wavelength of the orange line from a ^{86}Kr discharge lamp. The meter has been redefined since in terms of the second and the best available value for the speed of light [3], but practical methods for realizing the meter are based on the vacuum wavelengths of lasers.

Michelson had to go through a laborious series of comparisons to measure the number of wavelengths of a spectral line in the standard meter. The great coherence length of laser radiation has made possible measurements of very large distances in a single step. Lasers also have made electronic fringe-counting a practical technique for length interferometry. Typically, an interferometer is used that provides two interference fields, in one of which an additional phase difference of $\pi/2$ has been introduced. Two detectors viewing these fields yield signals in quadrature to drive a bidirectional counter.

The very narrow spectral linewidths of lasers also make it possible to observe beats between two laser frequencies. One system for length measurements based on heterodyne techniques, shown in Fig. 1, uses a He-Ne laser that is forced to oscillate simultaneously at two frequencies separated by about 2 MHz [4]. Normally, the frequencies of the outputs from the two detectors are the same and no net count accumulates. However, if one of the reflectors is moved, the net count gives the change in the optical path in wavelengths.

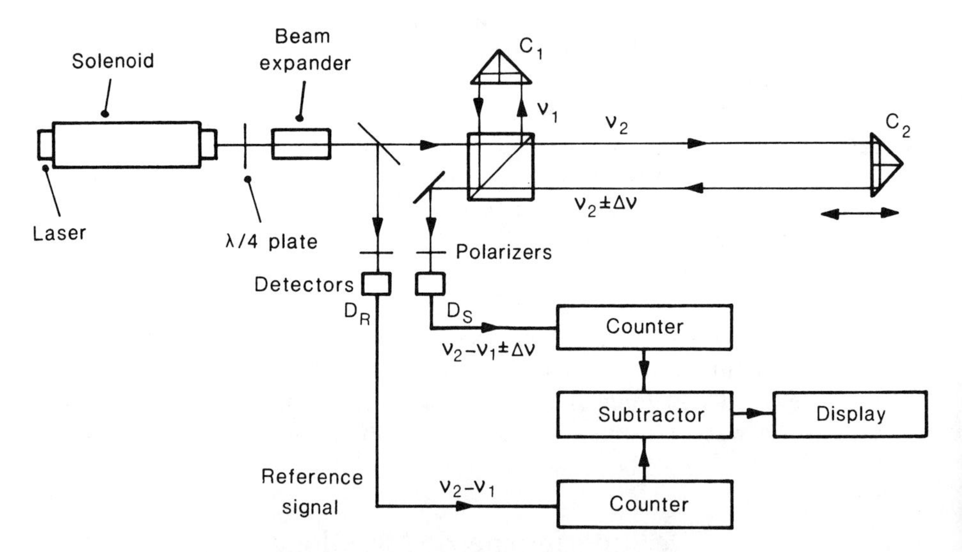

FIG. 1. Heterodyne fringe-counting interferometer (Dukes and Gordon, 1970 © Hewlett-Packard Company. Reproduced with permission).

Small changes in length also can be measured very accurately by heterodyne techniques [5]. The two mirrors of a Fabry-Perot interferometer are attached to the two points between which measurements are to be made, and the wavelength of a laser is locked to a transmission peak of the interferometer. A displacement of one of the mirrors results in a change in the wavelength of the laser and hence, in its frequency. These changes are measured by mixing the beam from the laser with the beam from a reference laser and measuring the beat frequency.

New techniques for distance measurements are possible with semiconductor lasers that can be tuned electrically over a range of wavelengths. One way is to sweep the frequency of the laser linearly with time [6]. If the optical path difference between the two beams in an interferometer is L, they reach the detector with a time delay L/c, where c is the speed of light, and interfere to yield a beat signal with a frequency $f = (L/c)(df/dt)$, where df/dt is the rate at which the laser frequency varies with time.

3. Optical Testing

A closely related application of interferometry has been in testing optical components and optical systems. The only interferometers used for this purpose for many years were the Twyman-Green and the Fizeau, and the first significant advance was the development of shearing interferometers that eliminated, in many cases, the need for a reference surface of similar dimensions.

Lasers have led to the development of several new types of interferometers for optical testing, as well as new techniques for measuring wavefront errors directly with very high precision. One method, based on optical heterodyne techniques, involves introducing a frequency difference between the two beams by diffraction at a Bragg cell. The intensity in the interference pattern then varies at the difference frequency, and the phase difference at any point can be determined by comparing the phase of the output from a movable detector with that from a fixed reference detector [7].

Alternatively, the phase difference between the interfering beams is varied with time, by changing one optical path or the wavelength of the laser, and the values of the intensity at each data point in the interference pattern for a set of equally spaced values of the additional phase difference are measured and stored. The values of the intensity at each point then can be represented by a Fourier series whose coefficients can be evaluated to obtain the original phase difference between the interfering wavefronts [8, 9].

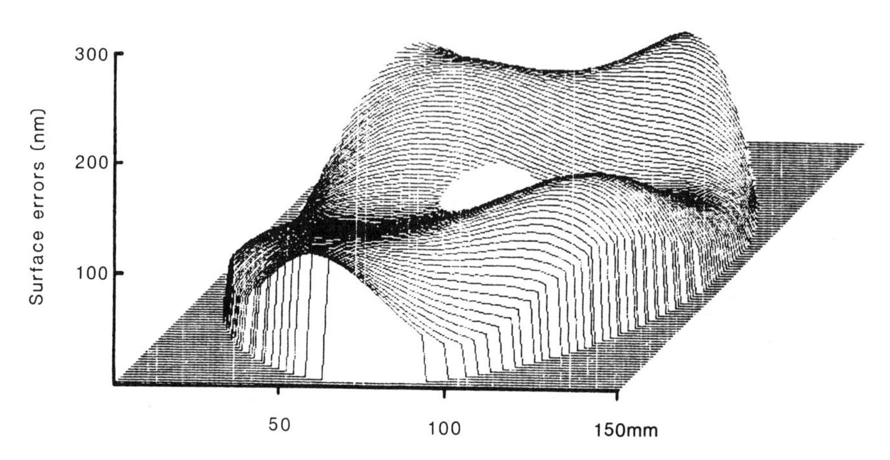

FIG. 2. Isometric plot of the errors of a concave mirror obtained with a digital phase-shifting interferometer.

Since measurements can be made rapidly and the data can be stored, the effects of vibrations and air currents can be minimized by averaging a number of readings. Errors due to the interferometer optics can be subtracted from the readings and the results can be presented as an isometric plot, as shown in Fig. 2. Laser interferometers with digital phase measurement systems are used extensively now in the production of high-precision optical components. Another application has been in studies of surface structure and surface roughness, where rms deviations as small as 0.01nm have been measured.

4. Fiber Interferometers

With the development of lasers, it became possible to build analogs of conventional interferometers using single-mode fibers. *Fiber interferometers* permit very long paths in a small space. In addition, because of the very low noise level, sophisticated detection techniques can be used.

One of the first applications of fiber interferometers was in rotation sensing. The possibility of using an interferometer as a rotation sensor dates back to experiments by Sagnac and by Michelson to detect rotation in an inertial frame with an interferometer in which the two beams traveled around the same closed circuit in opposite directions. Much higher sensitivity can be obtained if a closed multi-turn loop made of a single fiber is used instead of a conventional mirror interferometer.

Fiber-optic rotation sensors have the advantages of small overall size and low cost. Performance close to the limit set by shot noise is possible if

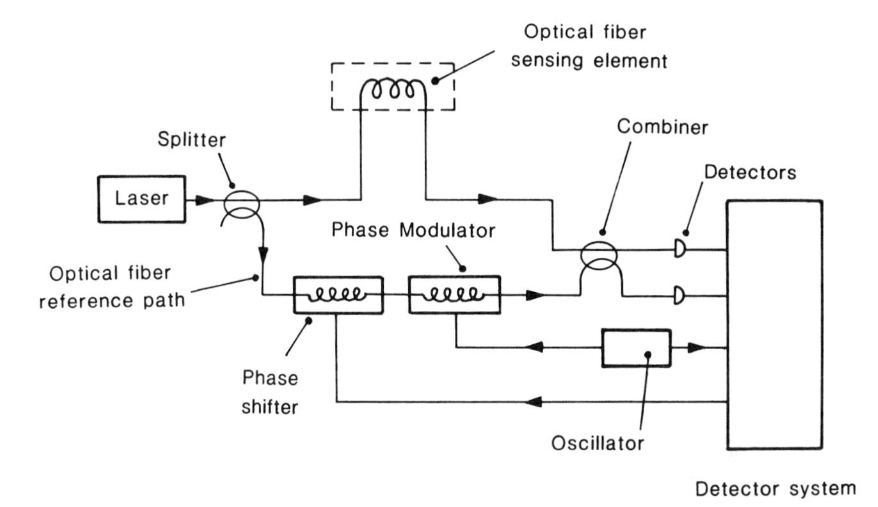

FIG. 3. Fiber-optic interferometric sensor [11] (© 1982 IEEE).

precautions are taken to minimize the effects of local temperature variations, vibration, and external magnetic fields [10].

Since the optical path in a fiber changes when the fiber is stretched, and also is affected by the ambient temperature and pressure, fiber interferometers can be used as sensors for a number of physical quantities [11, 12]. Figure 3 is a schematic of a typical fiber interferometer sensor using a layout analogous to a Mach-Zehnder interferometer with optical fiber couplers replacing beamsplitters. Optical phase shifts as small as 10^{-6} radian can be detected. Such an interferometer also can be used to measure magnetic or electric fields by bonding the fiber sensor to a magnetostrictive or piezoelectric element.

5. Laser-Doppler Interferometry

Laser-Doppler interferometry makes use of the fact that light scattered from a moving particle undergoes a frequency shift. This frequency shift can be measured using the beats produced by interference between the scattered light and a reference beam, or between the scattered light from two illuminating beams incident at different angles. With the optical system shown in Fig. 4, the frequency of the beat signal is given by the relation

$$f = \frac{2v \sin \theta}{\lambda}, \tag{1}$$

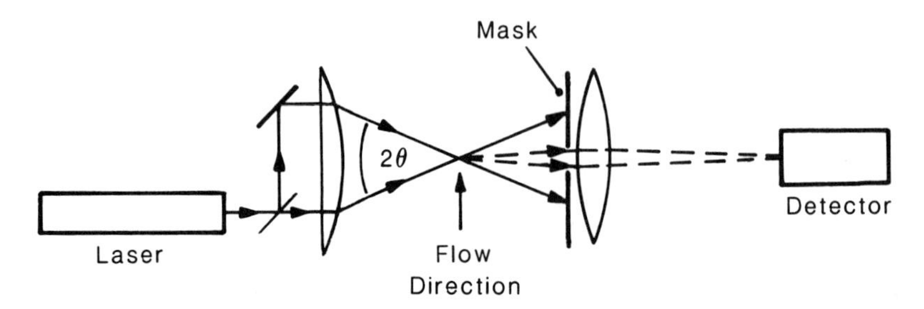

FIG. 4. Laser-Doppler interferometer for the measurement of flow velocities.

where v is the component of the velocity of the particle in the plane of the beams at right angles to the direction of observation. Laser-Doppler interferometry is widely used now in industry as well as in research to measure flow velocities [13].

Laser-Doppler techniques also can be used to study vibrating objects. Typically, one of the beams is reflected from a point on the vibrating surface, while the other is reflected from a fixed mirror. If the frequency of one beam is offset a known amount by diffraction at a Bragg cell, the output from a detector consists of a component at the offset frequency (the carrier) and two sidebands. The amplitude of the vibration can be determined by a comparison of the amplitudes of the carrier and the sidebands, while the phase of the vibration can be obtained by comparison of the carrier with a reference signal [14]. Vibration amplitudes down to a few thousandths of a nanometer can be measured by such techniques.

6. High-Resolution Spectroscopy

Michelson was the first person to derive the structure of spectral lines from observations of the visibility of the fringes in his interferometer as a function of the optical path difference. His work laid the foundation for the modern technique of Fourier transform spectroscopy, which is used widely for studies on faint sources as well as to obtain very high resolving power. Other new instruments that are used for high-resolution spectroscopy include the spherical Fabry-Perot interferometer and the multiple-pass Fabry-Perot interferometer.

A completely different method, which has been used effectively to obtain a resolution in excess of 10^6, is based on laser heterodyne techniques. An interesting application has been to measure the profiles of absorption lines due to molecular carbon dioxide in planetary atmospheres. Light from a telescope is mixed with the output from a carbon-dioxide laser at a high-speed photodetector, and the difference frequency spectrum, which lies in the radio-frequency range, is analyzed [15].

Another application of such techniques has been to measure the extremely narrow width of laser lines. In this case, light from the laser is mixed at a detector with a frequency-shifted reference beam derived from the same laser. A delay τ is introduced between the two beams that is much larger than $1/\Delta\nu$, where $\Delta\nu$ is the width of the laser line. The spectral width of the beat signal then is twice the laser line width [16].

7. Stellar Interferometry

Since a star can be regarded as a small incoherent light source, its angular diameter can be calculated from observations of the visibility of the fringes in an interferometer that uses light from the star reaching the surface of the earth at points separated by different distances.

Michelson's first stellar interferometer used a pair of mirrors mounted at the ends of a 6m-long support on the 2.5m telescope at Mt. Wilson. A later instrument with a baseline of 15m had to be abandoned because of problems with mechanical stability and atmospheric turbulence. Different approaches have been followed since to overcome these problems.

One approach that was remarkably successful involved measurements of the degree of correlation between the outputs of two photodetectors separated by a suitable distance [17]. The *intensity interferometer* has the advantage that atmospheric turbulence, which only affects the phase of the incident wave, has no effect on the measured correlation. In addition, because of the narrow bandwidth of the electronics, it is necessary to equalize the paths only to within a few centimeters. A resolution of 0.42×10^{-3} second of arc was obtained with a baseline of 188m.

The low sensitivity of the intensity interferometer limits measurements to stars brighter than +2.5 magnitude. As a result, construction of a new version of Michelson's stellar interferometer using modern detection, control, and data-handling techniques has been taken up [18]. This instrument is expected to make measurements over baselines up to 1,000m.

A completely different approach, which is discussed elsewhere in this volume, is speckle interferometry. Yet another, made possible by the development of lasers, is based on heterodyne techniques [19]. In this case, the light from the star is mixed with light from a laser at two photodetectors, and the resulting heterodyne signals are multiplied in a correlator. The output signal from the correlator is a measure of the degree of coherence of the wave fields at the two photodetectors.

As with the intensity interferometer, the two optical paths need be equalized only to within a few centimeters. However, the sensitivity is much higher than with direct detection because the output is proportional to the product of the intensities of the laser and the star.

A two-element heterodyne stellar interferometer made up of two telescope units with an aperture of 1.65m has been constructed. This interferometer is expected to map a few thousand stellar objects with an angular resolution of 0.001 second of arc [20].

8. Relativity and Gravitational Waves

Laser heterodyne techniques also have been used to verify the null result of the Michelson-Morley experiment and to confirm Einstein's special theory of relativity to a very high degree of accuracy. In one experiment, the output frequency of a He–Ne laser was locked to a resonance of a very stable Fabry-Perot cavity mounted along with it on a rotating horizontal granite slab. The frequency of this laser was monitored by comparing it with a stabilized reference laser. However, no frequency shifts greater than 5×10^{-7} of that predicted by classical theory were observed [21].

A phenomenon predicted by the general theory of relativity is gravitational radiation from binary systems of neutron stars or black holes and collapsing supernovas. Early attempts to detect gravitational waves used short resonant detectors, but it soon became evident that a long wide-band antenna offered advantages. Such an antenna, as shown in Fig. 5, could consist of a Michelson interferometer in which the beamsplitter and the end reflectors are attached to separate, freely suspended masses. The arm lying along the direction of propagation does not experience a strain, whereas the strain produced by the gravitational wave acts on the other arm. With a gravitational wave propagating at right angles to the plane of the interferometer, the changes in length of the two arms have opposite signs.

Theoretical estimates of the intensity of bursts of gravitational radiation suggest that a sensitivity to strains of the order of a few parts in 10^{21} is needed. One way to obtain the required sensitivity is to use two identical Fabry-Perot interferometers at right angles whose mirrors are mounted on freely suspended masses. The frequency of a laser is locked to a transmission peak of one interferometer, while the separation of the mirrors in the other is adjusted continually so that its peak transmittance is at this frequency. The corrections applied to the second interferometer are used to monitor changes in the relative lengths of the two interferometers [22].

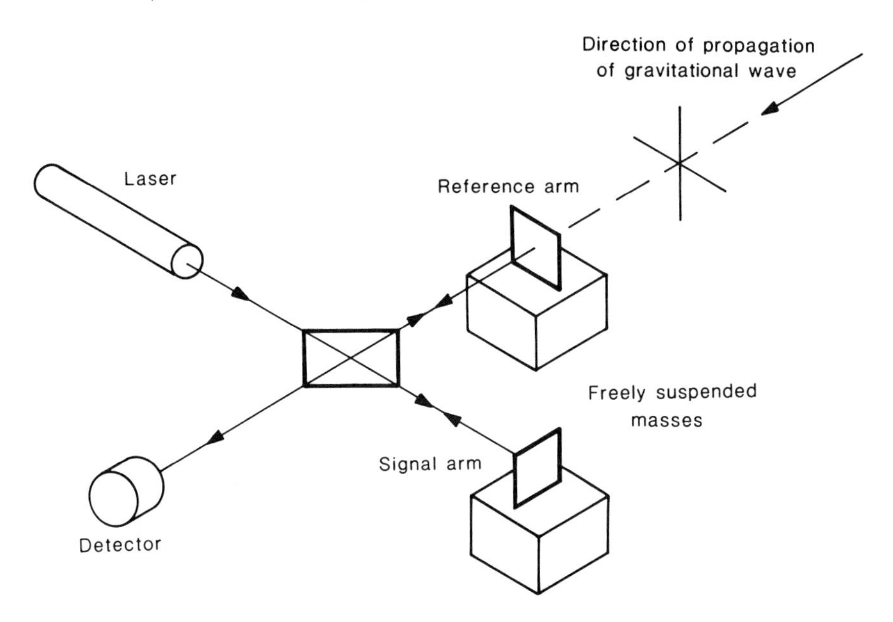

FIG. 5. Schematic of an interferometer for detecting gravitational waves.

Prototype interferometers with arm lengths of 10–40 m have achieved strain sensitivities of 3×10^{-18}, and several gravitational wave detectors with arm lengths of 1–3 km are under construction.

9. Nonlinear Interferometers

The availability of lasers capable of producing light of extremely high intensity has opened up completely new areas of interferometry based on the use of nonlinear optical elements. One such area is *phase-conjugate interferometry*.

In phase-conjugate interferometry, the wavefront under study is made to interfere with its complex conjugate. As a result, no reference wavefront is necessary, and the sensitivity is doubled. Figure 6 is a schematic of an interferometer in which the conjugate wave is generated by four-wave mixing [23]. In this arrangement, the crystal located in the upper arm of the interferometer also is irradiated by a pump beam derived from the same laser. The conjugate wave generated by the crystal travels down the lower arm of the interferometer to the second beamsplitter, where it interferes with the signal beam.

It also is possible to replace the mirrors in a conventional interferometer with phase-conjugating mirrors and obtain systems with unique properties.

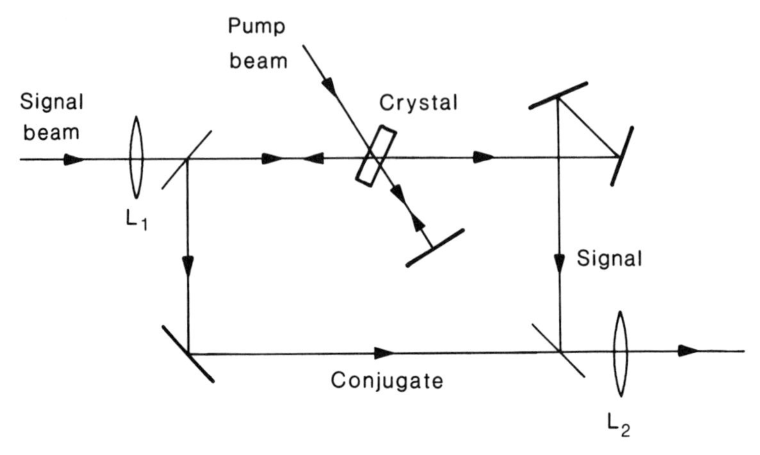

FIG. 6. Phase-conjugate interferometer using four-wave mixing [23].

The interference pattern obtained in such an interferometer is unaffected by misalignment of the mirrors, and disturbances due to air currents are cancelled out [24].

10. Future Directions

This review shows that the post-Michelson era has seen major advances in interferometry. It is an interesting, though risky, exercise to speculate on the directions of future growth in this field.

Some trends, of course, are obvious. With the rapid progress being made in the development of solid-state lasers, we can expect them to be used to an ever-increasing extent. The wider use of digital signal processing should result in further increases in the scope, speed, and accuracy of measurements. We also can expect an increasing trend towards the use of infrared wavelengths, following the replacement of the human eye by photodetectors. New devices will be built incorporating fiber interferometers. All these developments should lead to the use of interferometric techniques for a very wide range of measurements in science and industry.

Other likely advances can be linked to specific new techniques. One is the use of phase-conjugating mirrors in interferometry. The delay in the response of a phase conjugator such as barium titanate forms the basis of an interferometer that displays only dynamic changes in the optical path difference [25]. In the steady state, the interferometer is insensitive to spatial variations in the optical path difference, and the field of view is completely dark. Any sudden local change in the optical path difference results in a bright spot that then fades away. We can look forward to more new techniques stemming from the availability of improved nonlinear materials.

A breakthrough that opens up new possibilities in interferometry has been the successful generation and detection of squeezed states of light with lower quantum noise than normal laser light [26]. We can represent the electric field of a light wave in the form

$$E = E_0[X_1 \cos \omega t + X_2 \sin \omega t] \tag{2}$$

where X_1 and X_2 are field quadrature operators whose variances obey the uncertainty relationship. For normal coherent light, the variances are equal, but for a squeezed state the variances can be different, though their product remains the same. Typically, phase fluctuations can be reduced at the expense of a corresponding increase in the amplitude fluctuations.

Squeezed states can be generated by four-wave mixing in a medium with a nonlinear susceptibility. Energy transfer from two strong pump beams establishes correlations between the photons in two weak signal beams, which then are combined to produce light that exhibits the characteristics of squeezed states.

The use of squeezed light could lead to major improvements in the performance of high-precision interferometers, such as those under construction for the detection of gravitational waves. The fractional error of such measurements due to shot noise is $1/\sqrt{N}$, where N is the number of photons detected. With squeezed light, the measurement uncertainty could be reduced to $1/N$. Successful detection of gravitational waves, besides helping to decide between competing theories of gravity, would open up a completely new field in observational astronomy.

References

1. A. A. Michelson "Light Waves and Their Uses," The University of Chicago Press, Chicago, 1907.
2. A. A. Michelson "Studies in Optics," The University of Chicago Press, Chicago, 1927.
3. B. W. Petley *Nature* **303**, 373–6 (1983).
4. J. N. Dukes and G. B. Gordon *Hewlett-Packard Journal* **21**, no. 12, 2–8 (1970).
5. S. F. Jacobs and D. Shough *Appl Opt* **20** 3461–3 (1981).
6. T. Kubota, M. Nara, and T. Yoshino, *Opt Lett* **12**, 310–2 (1987).
7. R. Crane, *Appl Opt* **8**, 538–42 (1969).
8. J. H. Bruning, D. R. Herriott, J. E. Gallagher, D. P. Rosenfeld, A. D. White, and D. J. Brangaccio, *Appl Opt* **13**, 2693–703 (1974).
9. K. Creath "Progress in Optics," Vol. XXVII (E Wolf, ed.), pp. 349–93, Elsevier, Amsterdam, 1988.
10. S. Ezekiel, *Proc SPIE* **487**, 13–20 (1984).

11. T. G. Giallorenzi, J. A. Bucaro, A. Dandridge, G. H. Sigel Jr., J. H. Cole, S. C. Rashleight, and R. G. Priest, *IEEE J Quant Electron* **QE – 18**, 626–65 (1982).
12. Culshaw B. "Optical Fibre Sensing and Signal Processing," Peregrinus, London, 1984.
13. F. Durst, A. Melling, and J. H. Whitelaw, "Principles and Practice of Laser-Doppler Interferometry," Academic Press, London, 1976.
14. W. Puschert *Opt Commun* **10** 357–61 (1974).
15. A. L. Betz, E. C. Sutton, and R. A. McLaren, "Laser Spectroscopy III" (J. L. Hall and J. L. Carsten eds.) pp. 31–8, Springer, Berlin, 1977.
16. C. Abitbol, P. Gallion, H. Nakajima, and C. Cabran, *J Opt* (Paris) **15**, 411–8 (1984).
17. R. Hanbury Brown, "The Intensity Interferometer: Its Application to Astronomy," Taylor and Francis, London, 1974.
18. J. Davis, "Proc. Internatl. Symp. on Measurements and Processing for Indirect Imaging" (J Roberts, ed.) pp. 125–41, Cambridge University Press, Cambridge, 1984.
19. M. A. Johnson, A. L. Betz, and C. H. Townes, *Phys Rev Lett* **33**, 1617–20 (1974).
20. C. H. Townes *J. Astrophys Astron* **5**, 111–30 (1984).
21. A. Brillet and J. L. Hall *Phys Rev Lett* **42,** 549–52 (1979).
22. R. W. P. Drever, "Interferometric Detection of Gravitational Radiation" (N. Deruelle and T. Pitan, eds.), pp. 321–38, North-Holland, Amsterdam, 1982.
23. F. A. Hopf, *J Opt Soc Am* **70**, 1320–2 (1980).
24. J. Feinberg, *Opt Lett* **8**, 569–71 (1983).
25. D. Z. Anderson, D. M. Lininger, and J. Feinberg, *Opt Lett* **12**, 123–5 (1987).
26. R. E. Slusher, L. W. Hollberg, B. Yurke, J. C. Mertz, and J. F. Valley, *Phys Rev Lett* **55** 2409–12 (1985).

CHAPTER 25

Current Trends in Optical Testing

Daniel Malacara*
The Institute of Optics
University of Rochester
Rochester, New York

1. Introduction

Optical surfaces most of the time have a flat or spherical shape, but they also may be toroidal or rotationally symmetrical aspheric. Although optical surfaces generally are spherical, aspherical surfaces are being used more and more frequently because of their great design advantages. The reason is that one aspherical surface can eliminate or reduce aberrations as well as three or four spherical surfaces can. In most cases, an aspherical surface is a conic of revolution, like a paraboloid, hyperboloid, or ellipsoid.

On the other hand, as can be understood easily, an aspherical surface is more difficult to manufacture. In this process, the most difficult step is its testing. It has been said, with good reason, that an aspherical surface can be made only as well as it can be tested. Thus, good testing methods always have been important to develop. In this chapter, we will describe briefly the history and present status of the field of optical testing. For general coverage of this subject, the reader may consult the book edited by Malacara [1]. In this chapter, we also will speculate about the future of this field.

* On leave from the Centro de Investigaciones en Optica, León, Gto. México.

ISBN 0-12-289690-4

2. Qualitative Tests

Some traditional tests never will be obsolete because they provide qualitative results almost instantly about the shape of the optical surface or wavefront. These tests, in general, also are very cheap and easy to perform because no special and sophisticated equipment is required. Examples of these tests are the Foucault or knife-edge test, the Ronchi test, and the star test. Now we will describe these tests briefly.

2.1. Foucault Test

The *Foucault test*, also known as the *knife-edge test*, was invented by Leon Foucault in France [2]. The test may be considered a method for detecting the presence of transverse aberrations. This is accomplished by intercepting the rays deviated from their ideal trajectory, as shown in Fig. 1. Then, the observer is behind the knife, looking at the illuminated surface. The regions corresponding to the intercepted rays will appear dark, as in Fig. 2.

This test is extremely sensitive. If the wavefront is nearly spherical, irregularities as small as a fraction of the wavelength of the light may be detected easily. This is probably the most powerful qualitative test for observing small irregularities and evaluating the general smoothness of the surface being tested.

2.2. Ronchi Test

The *Ronchi test* was invented by Vasco Ronchi in Italy in 1923 [3]. A coarse ruling is placed in the convergent light beam coming out from the surface being tested, near its focus. The observer places an eye behind the ruling, as shown in Fig. 3. The dark bands in the ruling then are projected

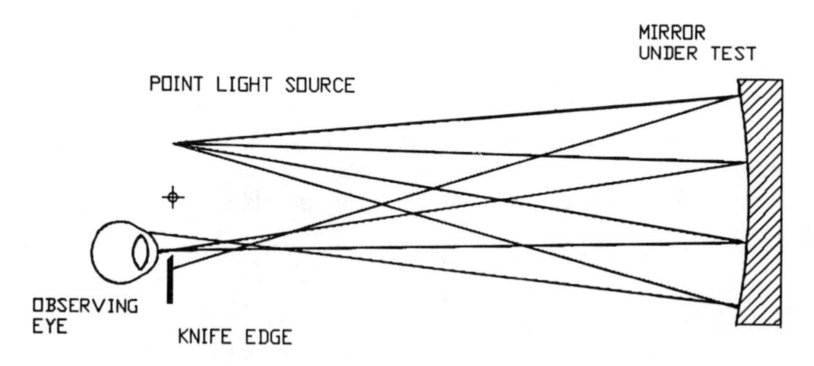

FIG. 1. Optical schematics for the Foucault test of a spherical mirror.

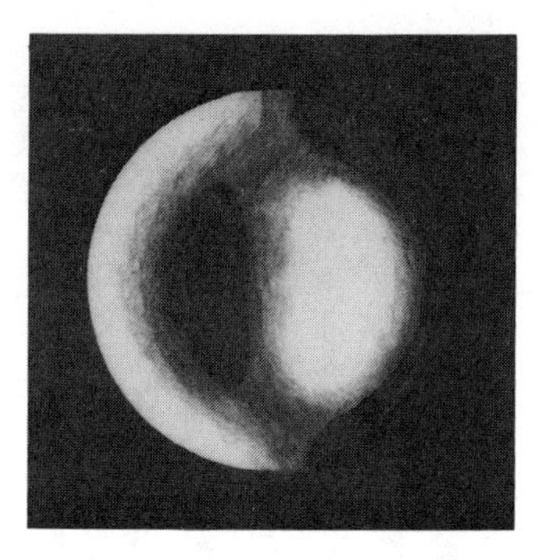

FIG. 2. An optical surface being examined by the Foucault test.

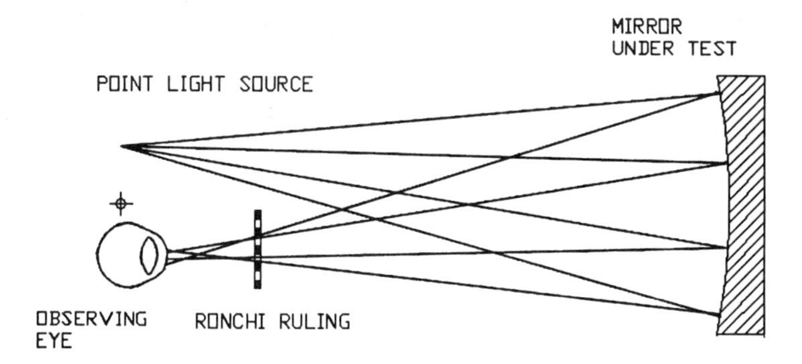

FIG. 3. Testing a wavefront by means of the Ronchi test.

back to the surface being tested as shadows, as in Fig. 4. These shadows will be straight and parallel only if the wavefront is perfectly spherical. Otherwise, the fringes will be curves whose shape and separation will depend on the wavefront deformations.

This test, like the Foucault test, allows us to determine the degree of smoothness of the wavefront, but with less sensitivity. The main virtue of this test is that each type of aberration wavefront has a characteristic Ronchi pattern. Thus, the aberrations in the optical system can be identified easily and their magnitude estimated. It is important to point out that the Ronchi fringes may be interpreted not only as geometrical shadows, but also as interferometric fringes.

2.3. Star Test

The *star test* is performed by direct observation of the image of an artificial star, by means of an eyepiece. The size and shape of this image

FIG. 4. A typical Ronchi pattern.

gives an immediate indication of the magnitude and type of aberration. However, this test gives no direct information about the local wavefront deformations, especially if they are not due to any of the primary aberrations.

3. Null Tests

The most common type of interferometer, with the exception of a lateral or rotational shearing interferometer, produces interference patterns in which the fringes are straight, equidistant, and parallel, when the wavefront being tested is perfect and spherical with the same radius of curvature as the reference wavefront.

If the surface being tested does not have a perfect shape, the fringes will not be straight and their separations will be variable. The deformations of the wavefront can be determined by a mathematical examination of the shape of the fringes. By introducing a small spherical curvature on the reference wavefront or by changing its angle with respect to the wavefront being tested, the number of fringes in the interferogram can be changed to reduce its number as much as possible. The greater the number of fringes, the smaller the sensitivity of the test.

If the wavefront being tested is not spherical, the fringes in the interferogram will not be straight, nor will they have constant separations between them. In an interferogram of an aspherical surface, the number of fringes may be adjusted by changing the angle between the reference wavefront and the wavefront being tested (tilt), and the curvature of the spherical reference wavefront (focusing), but this number of fringes can not be smaller than a certain minimum, which, in general, is a very large number. The larger the asphericity, the greater this minimum number of fringes. Since the fringe separations are not constant, the fringes will be quite spaced apart in some places, but too close together in others.

The sensitivity of the test depends on the separation between the fringes because an error of one wavelength in the wavefront distorts the fringe shape by an amount equal to the separation between the fringes. Thus, the sensitivity is directly proportional to the fringe separation. Where the fringes are widely separated, the sensitivity of the test is about one-tenth of a wavelength, but the sampled points will be quite separate from each other, leaving many zones without any information. On the other hand, where the fringes are very close to each other, there is a high density of sampled data points, but the sensitivity is lower.

It is desirable then that the spherical aberration of the wavefront under testing be compensated in some way, so that the fringes appear straight, parallel, and equidistant for a perfect wavefront. This may be accomplished by means of some special configurations.

These special configurations can be used to perform a null test of a conic surface. These are described in Appendix 4 of the book by Malacara [1]. Almost all of these surfaces have rotational symmetry. If the testing configuration has spherical aberration, then refractive, reflective, or diffractive compensators can be used, as will be described now.

3.1. Refractive Compensators

Frequently, a null test can not be obtained without some additional optical elements, which compensate the spherical aberration of the optical element being tested. Many different types of compensators have been invented. Some are reflective elements, or lenses. Others are diffractive elements, such as real or computer-generated holograms.

For example, the simplest way to compensate the spherical aberration of a paraboloid or hyperboloid, tested at the center of curvature, is a single convergent lens placed near the point of convergence of the rays, as shown in Fig. 5. This lens is called a *Dall compensator*. Unfortunately, the correction due to a single lens is not complete, so a system of two lenses must be used to obtain better compensation. This system is called an *Offner compensator* and is shown in Fig. 6. The field lens L_2 is used to image the surface being tested on the plane of the compensating lens L_1.

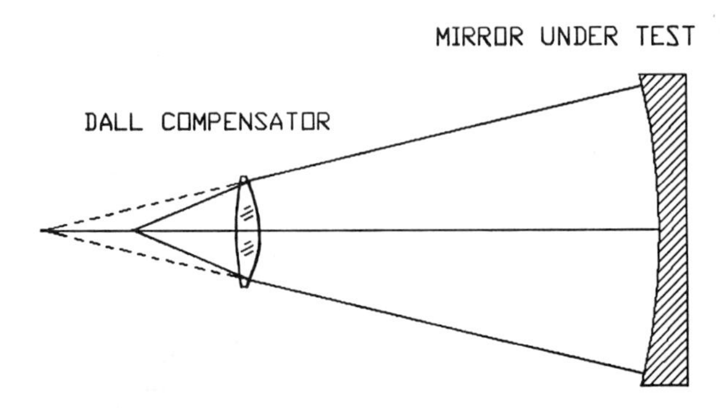

FIG. 5. The Dall compensator. Only the reflected beam is shown.

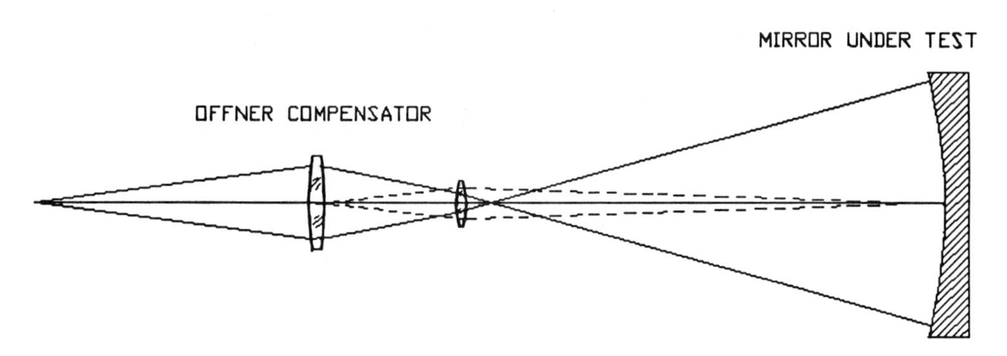

FIG. 6. The Offner compensator. Only the reflected beam is shown.

3.2. Holographic Compensators

Diffractive holographic elements also may be used to compensate for the sperical aberration of the system and to obtain a null test. The hologram may be real; that is, produced by photographing an interferometric pattern. This pattern has to be formed by superimposing on the screen a wavefront like the one we plan to test, and a perfectly flat or spherical wavefront. The only problem with this procedure is that a perfect wavefront with the same shape as the wavefront to be tested has to be produced first, and that is not always easy.

A better approach is to simulate the holographic interference pattern in a computer, as in Fig. 7 [4]. Then this image is transferred to a small photographic plate with the desired dimensions. There are many experimental arrangements to compensate the aspherical wavefront aberration by means of a hologram. One of these is illustrated in Fig. 8.

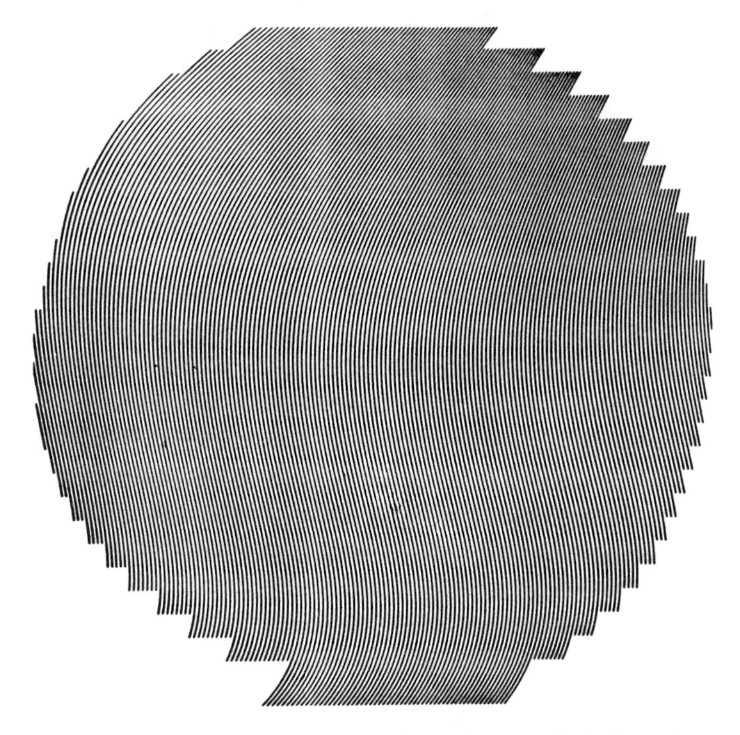

FIG. 7. Computer-generated hologram for testing an aspherical wavefront. (From Ref. 4.)

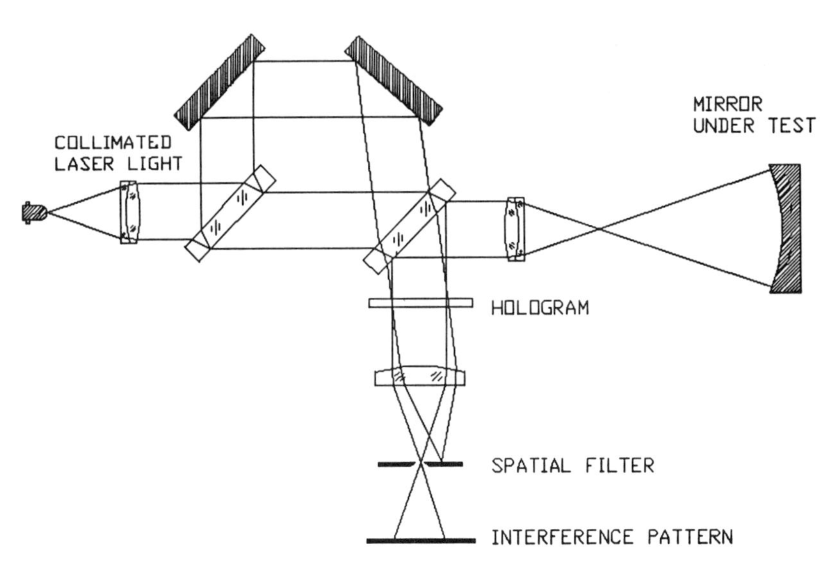

FIG. 8. An optical arrangement for testing an aspherical wavefront with a computer-generated hologram.

4. Quantitative Tests

In high-quality optical systems, it is necessary to make a quantitative evaluation of the wavefront deformations. This can be done by means of a geometrical test, like the Hartmann test, or by interferometry, as we will see in the next sections.

4.1. Hartmann Test

The *Hartmann test* was invented in Germany by J. Hartmann [5]. It is one of the most powerful methods used to determine the figure of a concave aspherical mirror. Figure 9 shows the optical configuration of this test, where a point light source illuminates the optical surface, with its Hartmann screen in front of it. Then the positions of the beams reflected through each hole on the screen are measured to determine the value of the transverse aberration at each point.

If the screen has a rectangular array of holes, as shown in Fig. 10, the typical Hartmann plate image for a parabolic mirror is similar to Fig. 11.

After numerical integration of the values of the transverse aberrations, this test gives the mirror shape with a very high accuracy. Extended, localized errors, as well as symmetric errors like astigmatism, are detected with this test. However, an important problem is that small, localized defects are not detected if they are not covered by the holes on the screen. Not only is this information lost, but the integration results will be false if the localized errors are large. The second important problem of the Hartmann test is that it is very time-consuming, due to the time used in measuring all the data points on the Hartman plate.

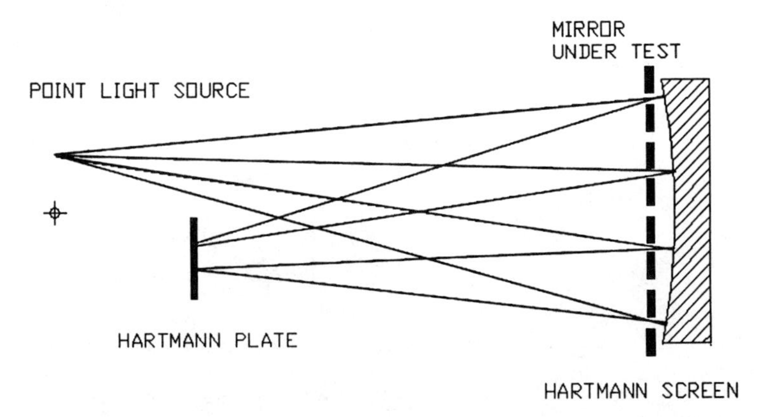

FIG. 9. Optical arrangement for performing the Hartmann test.

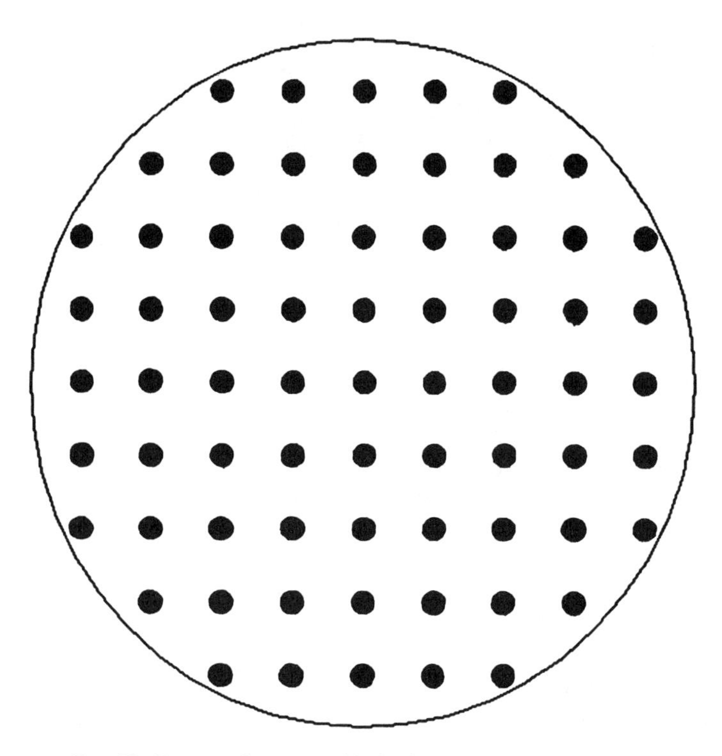

Fig. 10. Rectangular array of holes in a Hartmann test screen.

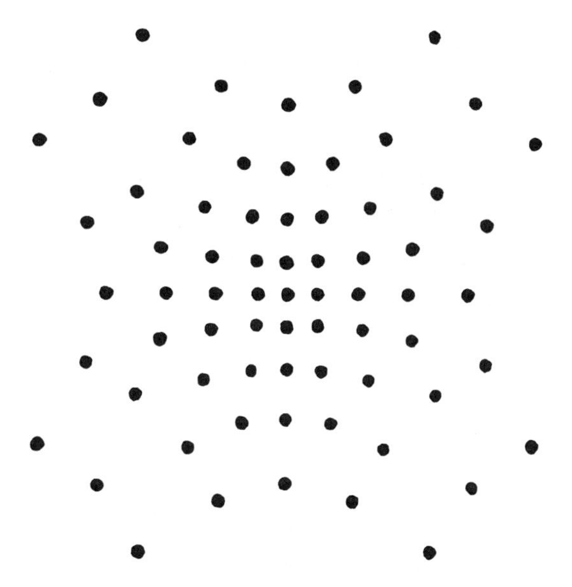

Fig. 11. Array of spots in a Hartmann plate of a parabolic mirror.

These problems are avoided by complementing this test with the Foucault test, using an Offner compensator, to be sure about the smoothness of the surface.

4.2. *Fixed Interferogram Evaluation*

The traditional interferogram analysis method requires the measurement of the position of several data points located on top of the fringes, as shown in Fig. 12. These measurements are made in many ways; for example, by means of a measuring microscope, a digitizing tablet, or a video camera connected to a computer.

After the measurements are made, the wavefront is computed. If only the measured points are used, the density of the data will not be uniform unless we perform a null test. Thus, as in the Hartmann test, information about many large zones is lost. A way to overcome this problem is to interpolate intermediate values by any of several existing methods. One method is to fit the wavefront data to a two-dimensional polynomial by means of least-squares fitting, or by the use of splines. Unfortunately, this procedure also has many problems if the wavefront is very irregular. The value obtained with the polynomial may be completely wrong, especially near the edge if the wavefront is too irregular.

4.3. *Phase-Shifting Interferometry*

All these problems of the traditional interferometric methods have been overcome by phase-shifting interferometric techniques, where the density

FIG. 12. Interferogram to be digitized.

of sampled data points—as well as the sensitivity and accuracy of the test—is constant over the wavefront.

In addition, *phase-shifting interferometry* is simple and fast, thanks to modern tools like array detectors (CAD) and microprocessors. Most of the conventional interferometers—like the Fizeau, Twyman-Green, etc.—have been adapted to perform the phase-shifting techniques to be described here. Since the current trend in interferometry is to use phase-shifting techniques, we now will describe them. Additional details can be found in the review article by Creath [6].

Phase-shifting interferometric techniques have their first indirect antecedent in the works by Carré [7], but they really started in the past 20 years or so with Crane [8], Moore [9], Bruning *et al.* [10, 11], and many others. In phase-shifting interferometers, the reference wavefront is moved along the direction of propagation with respect to the wavefront being tested, changing their phase difference in this manner. The interference fringes then change their positions, hence the initial term *fringe scanning* for these techniques.

By measuring the irradiance changes for different phase shifts, it is possible to determine the difference in phase between the wavefront being tested and the reference wavefront for that measured point over the wavefront. By obtaining this phase difference for many points over the wavefront, the complete wavefront shape thus is determined.

If we consider any fixed point in the interferogram, the phase difference between the two wavefronts has to be changed to make several measurements. This change in the phase may be accomplished only if the frequency of one of the beams is modified to form beats. Of course, as we will see shortly, this may be done in a continuous fashion using certain devices, but only for a relatively short period of time with some other devices. Since the frequency can be modified in a permanent way, this technique frequently is called *AC interferometry, heterodyne interferometry*, or using a *frequency-shift interferometer*. However, phase-shifting interferometry is the more common term.

The procedure just described can be implemented in almost any kind of two-beam interferometer; for example, as in the Twyman-Green shown in Fig. 13. The phase may be shifted or, equivalently, the frequency of one of the beams may be changed in many ways. As pointed out before, the phase may be shifted in a continuous fashion by introducing a permanent frequency shift in the reference beam, or in a discontinuous manner by periodically increasing and decreasing the optical path difference with an oscillation of the phase.

The first method that can be used to shift this phase is by moving the mirror for the reference beam along the light trajectory in a continuous or discrete manner, as shown in Fig. 14A. This can be done in many ways,

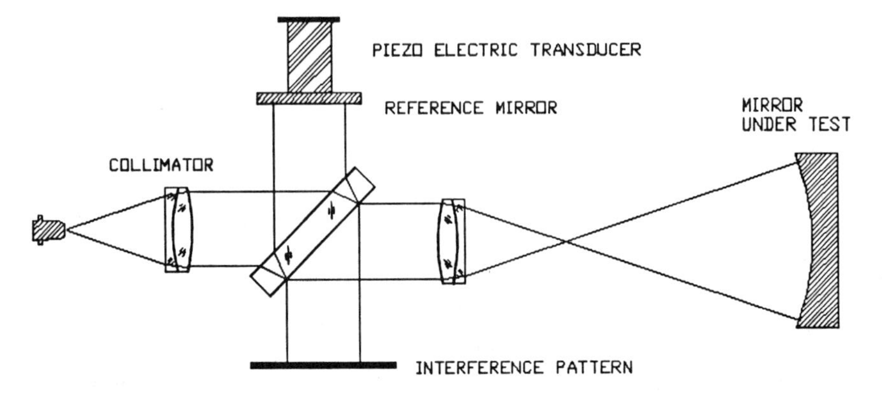

FIG. 13. Twyman-Green interferometer.

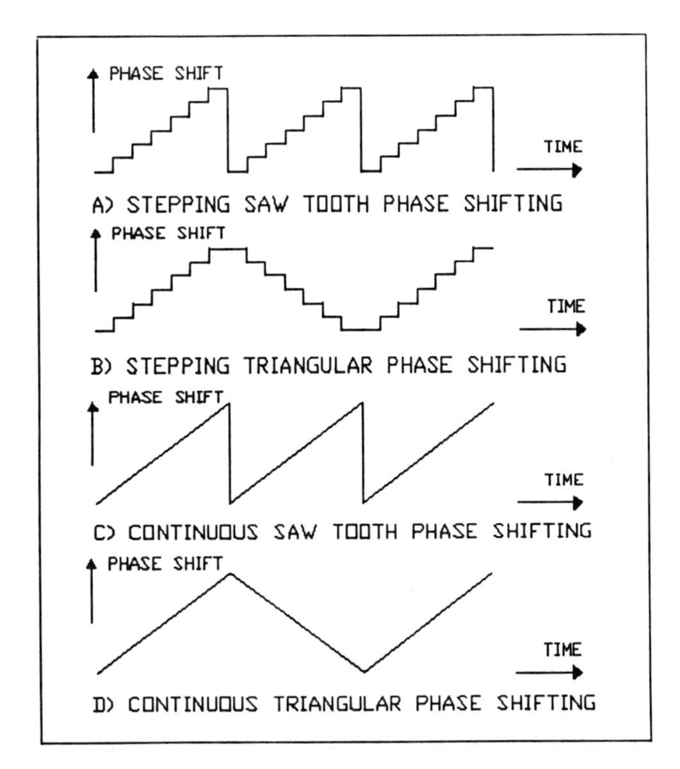

FIG. 14. Four different methods for doing the phase shifting, when the frequency of the reference beam is not permanently modified.

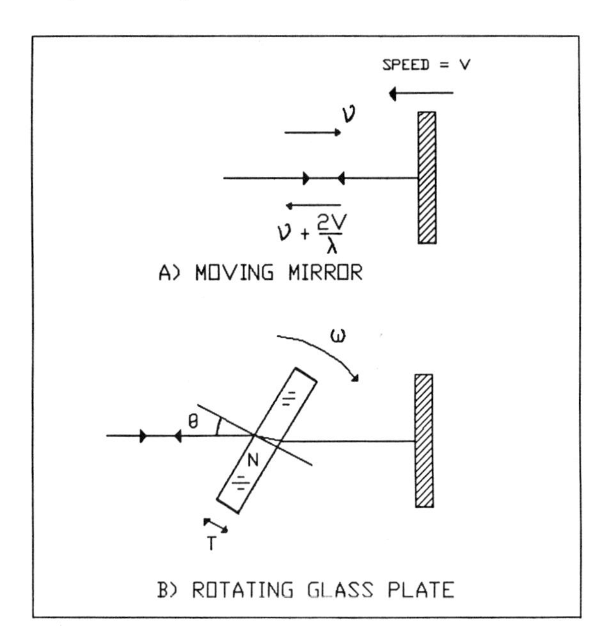

FIG. 15. Obtaining the phase shift by means of a moving mirror or a rotating glass plate.

such as with a piezoelectric crystal or with a coil in a magnetic field. If the mirror moves with a speed V, the frequency of the reflected light is shifted by an amount equal to $\Delta\nu = 2V/\lambda$.

Another method to shift the phase is by inserting a plane parallel glass plate in the light beam, as shown in Fig. 15B. The phase is shifted when the optical path changes. The optical path difference (OPD) introduced by this glass plate, when tilted an angle θ with respect to the optical axis, is given by

$$\mathrm{OPD} = T(N\cos\theta' - \cos\theta), \tag{1}$$

where T is the plate thickness and N is its refractive index. The angles θ and θ' are the angles between the normal to the glass plate and the light rays outside and inside the plate, respectively. A rotation of the plate increasing the angle θ increases the optical path difference. Thus, if the plate is rotated with an angular frequency ω, the frequency ν of the light passing through it is shifted by an amount $\Delta\nu$ given by

$$\Delta v = \frac{Tv\omega}{c}\left[1 - \frac{\cos\theta}{N\cos\theta'}\right]\sin\theta, \tag{2}$$

where c is the speed of the light. The only requirement in this method is

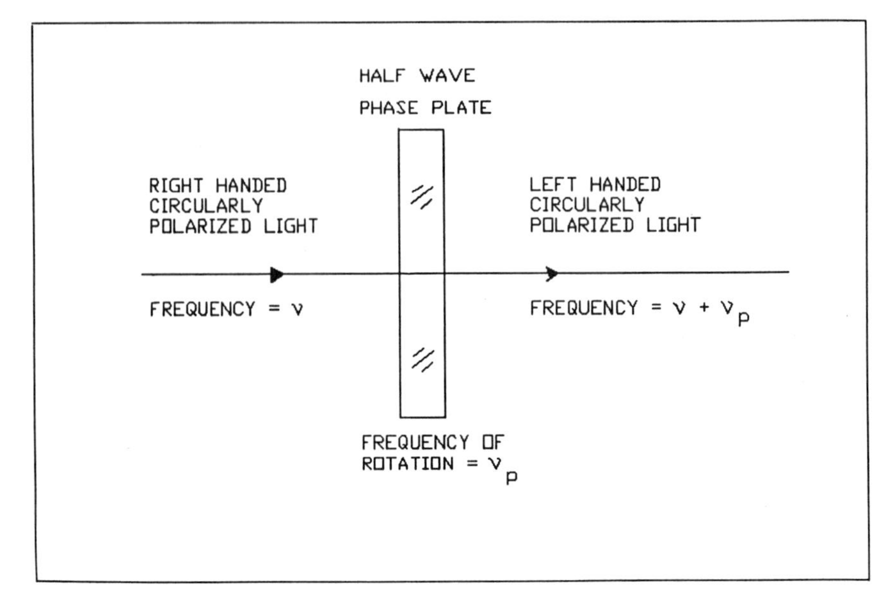

FIG. 16. Obtaining the phase shift by means of phase plates and polarized light, with a single pass of the light beam.

that the plate be inserted in a collimated light beam to avoid introducing aberrations.

The phase also may be shifted by means of the device shown in Fig. 16. If a beam of circularly polarized light goes through a half-wave-retarding phase plate, the handiness of the circular polarization is reversed. If the half-wave phase plate rotates, the frequency of the light changes. This frequency change $\Delta\nu$ is equal to twice the frequency of rotation of the phase plate ν_p; that is, $\Delta\nu = 2\nu_p$. This arrangement works if the light goes through the phase plate only once. However, in a Twyman-Green interferometer, the light passes through the system twice. Thus, it is necessary to use the configuration in Fig. 17. The first quarter-wave-retarding plate is stationary, with its slow axis at 45 degrees with respect to the plane of polarization of the incident linearly polarized light. This plate also transforms the returning circularly polarized light back to linearly polarized light. The second-phase retarder also is a quarter-wave plate; it is rotating and the light goes through it twice. Hence, it really acts as a half-wave plate. The shift in frequency is limited by the maximum mechanical speed of rotation of the phase plate, which is only 1 or 2 kHz.

Another way to obtain the shift of the phase is by means of a diffraction grating moving perpendicularly to the light beam, as shown in Fig. 18A. It is easy to notice that the phase of the diffracted light beam is shifted *m*

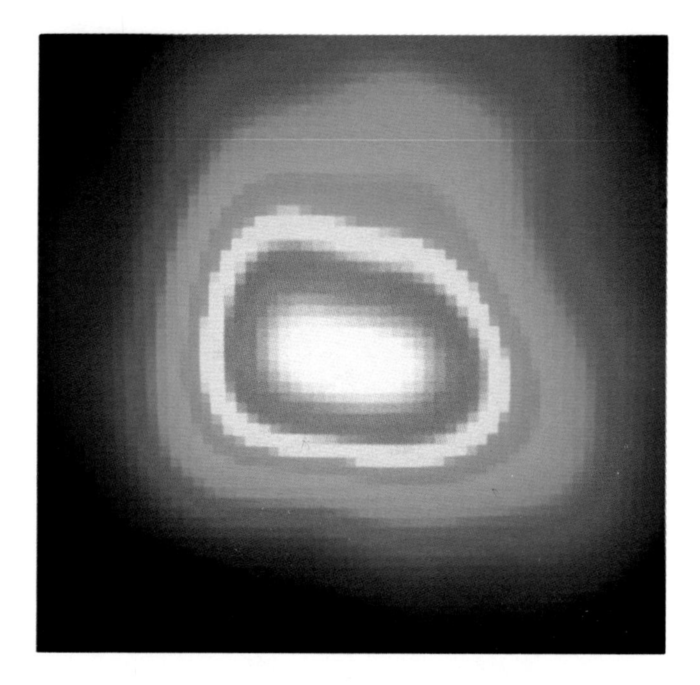

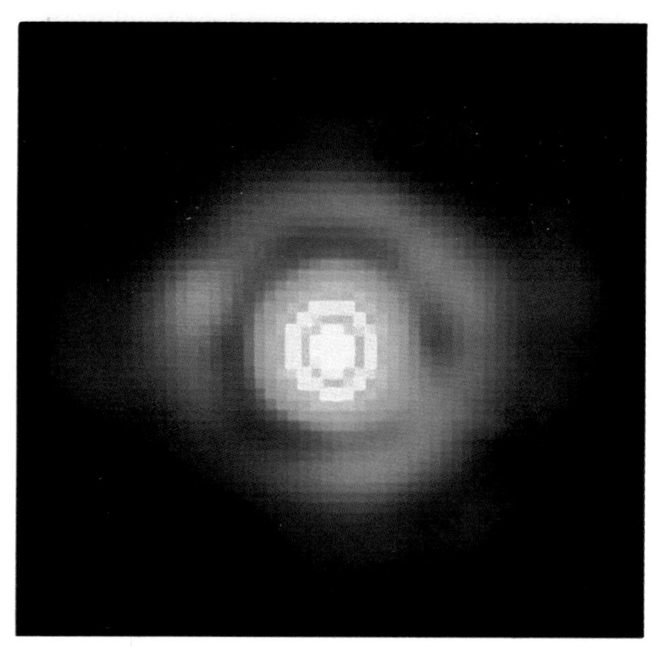

FIG. 7. Image of the unresolved star HR6519 uncorrected (top) and corrected (bottom) in real time by adaptive optics (D = 3.6m, r_0 = 13 cm. 19-actuator deformable mirror) at 3.8μm wavelength. The asymmetry in the uncorrected image is due to a guiding problem of the telescope that also is corrected by the system. The image FWHM is reduced from 0.7 arcseconds to 0.23 arcseconds; the diffraction ring clearly is visible.

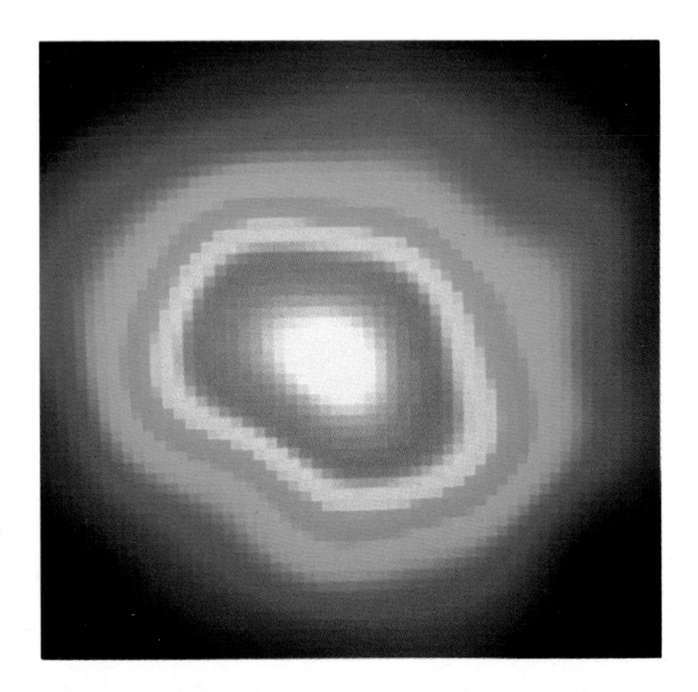

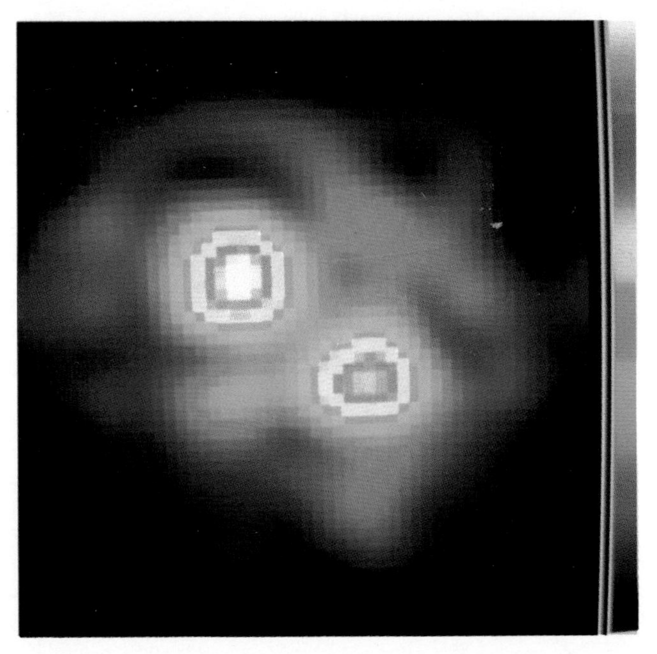

FIG. 7. *(Continued)* Image of the binary star HR6658 uncorrected (top) and corrected (bottom) in real time by the same adaptive optics system. The separation of the two components is 0.38 arcseconds. The conditions are as in the preceding case.

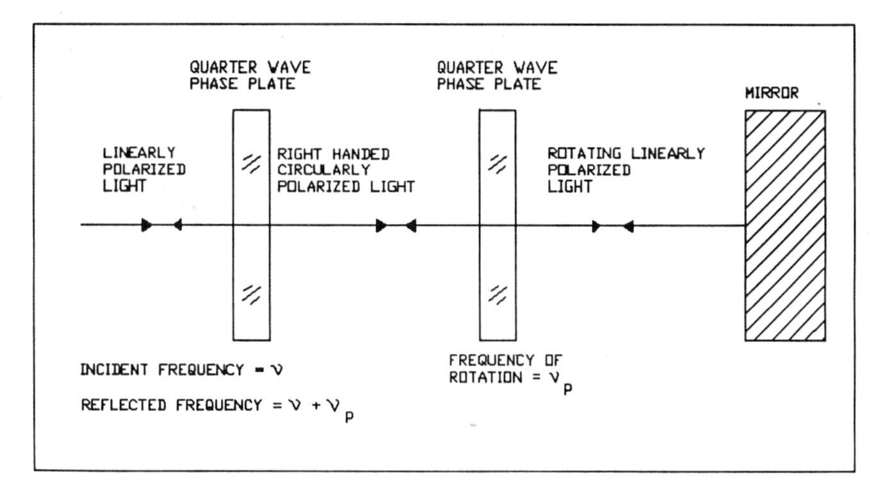

FIG. 17. Obtaining the phase shift by means of phase plates and polarized light, with a double pass of the light beam.

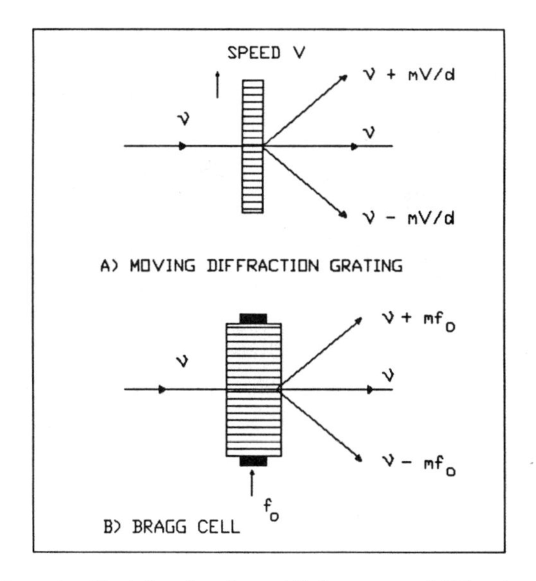

FIG. 18. Obtaining the phase shift by means of diffraction.

times the number of slits that pass through a fixed point. The letter m represents the order of diffraction. Thus, the shift in the frequency is equal to m times the number of slits in the grating that pass through a fixed point per unit of time. To say it in a different manner, the shift in the frequency is equal to the speed of the grating, divided by period d of the grating. Thus,

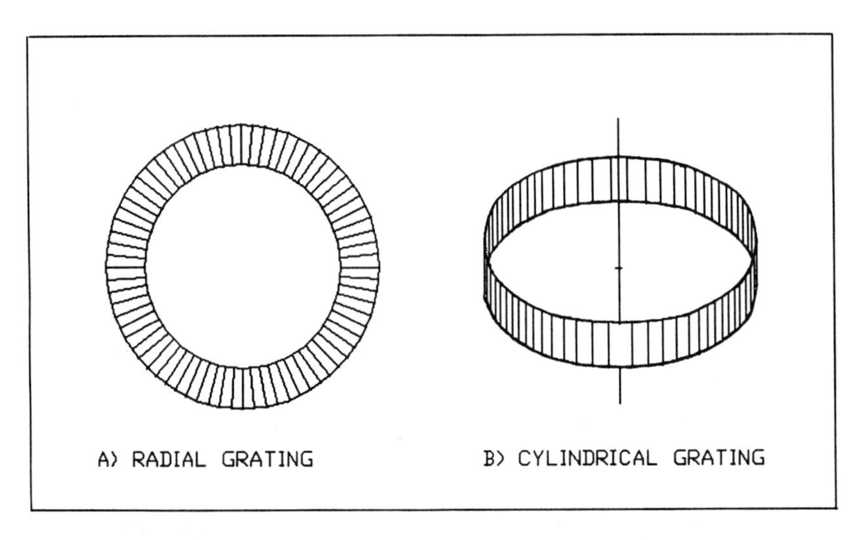

FIG. 19. Two continuously moving diffraction gratings.

we finally may write: $\Delta\nu = mV/d$. It is interesting to notice that the frequency is increased for the light beams diffracted in the same direction as the movement of the grating. The light beams diffracted in the direction opposite to the movement of the grating decrease their frequency. This method for shifting the phase may be implemented by means of oscillating gratings, using a rotating grating, with the slits in the radial direction, or a cylindrical grating, as illustrated in Fig. 19. As is to be expected, the direction of the beam is changed because the first-order beam has to be used, and the zero-order beam must be blocked by means of a diaphragm.

Diffraction of light also may be used to shift the frequency of the light by means of an acousto-optic Bragg cell, as shown in Fig. 18B. In this cell, an acoustic transducer produces ultrasonic vibrations in the liquid of the cell. These vibrations produce periodic changes in the refractive index, moving in the liquid. These periodic changes in the refractive index act as a diffraction grating, diffracting the light. The change in the frequency is equal to the frequency f of the ultrasonic wave, multiplied by the order of diffraction m. Thus, we may write: $\Delta\nu = mf$.

The last method to be mentioned here for shifting the frequency of the light beam is a laser emitting two frequencies, ν and $\nu + \Delta\nu$. This is a *Zeeman split line*, in which the frequency separation $\Delta\nu$ may be controlled with a magnetic field. The two frequencies have orthogonal polarizations, so to be able to interfere, their plane of polarization is made to coincide. The non-shifted frequency is used for the testing wavefront and the shifted frequency is used for the reference wavefront.

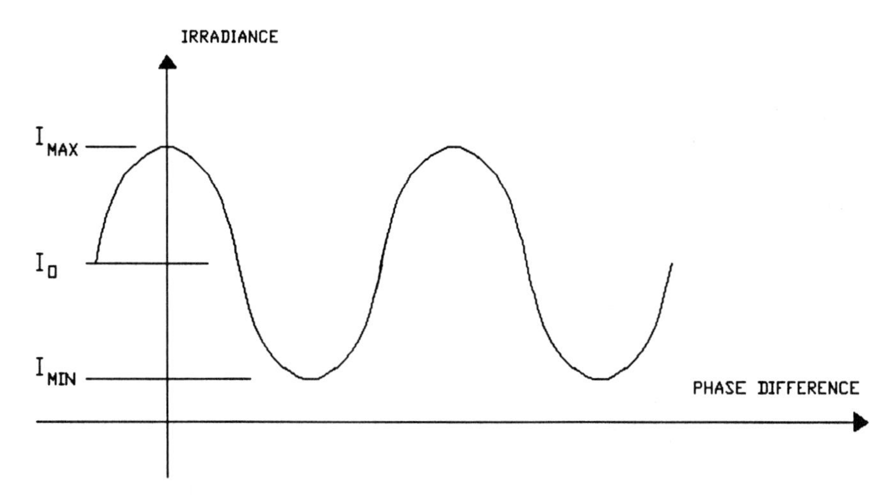

FIG. 20. Irradiance signals in a given point in the interference pattern, as a function of the phase difference between the two interfering waves.

Once the phase-shift method is implemented, it is necessary then to determine the procedure by which the non-shifted relative phase of the two interfering wavefronts is going to be measured. This is done by measuring the irradiance with several pre-defined and known phase shifts. Let us assume that the irradiance of each of the two interfering light beams at the point x, y in the interference pattern is $I_1(x, y)$ and $I_2(x, y)$, and that their phase difference is $\phi(x, y)$; it is easy then to show that

$$I(x,y) = I_1(x,y) + I_2(x,y) + 2\sqrt{I_1(x,y)I_2(x,y)} \cos \phi(x,y) . \qquad (3)$$

This is a sinusoidal function describing the phase difference between the two waves, as shown in Fig. 20. It has maximum and minimum values given by

$$I_{\max}(x,y) = I_1(x,y) + I_2(x,y) + 2\sqrt{I_1(x,y)I_2(x,y)} \qquad (4)$$

and

$$I_{\min}(x,y) = I_1(x,y) + I_2(x,y) - 2\sqrt{I_1(x,y)I_2(x,y)} , \qquad (5)$$

respectively, and an average value given by $I_0 = I_1 + I_2$.

As explained before, the basic problem is determining the non-shifted phase difference between the two waves with the highest possible precision. This may be done by any of several different procedures to be described next. The best method to determine the phase depends on how the phase or frequency shift has been made.

4.3.1. Zero Crossing

This method detects when the irradiance plotted in Fig. 8 passes through zero when changing the phase difference between the two interfering waves. This zero does not really mean zero irradiance, but the axis of symmetry of the function—which has a value equal to I_0—or almost any other intermediate level between the maximum and minimum irradiance. The points crossing the axis of symmetry can be found very easily by amplification of the irradiance function to saturation levels. In this manner, the sinusoidal shape of the function becomes a square function. The *zero-crossing* technique has been used both by Crane and Moore [8, 9].

The wavefront shape is obtained by measuring the phase Φ at some reference point on the wavefront (reference signal) and then at several other points on the wavefront (test signal). A practical implementation of this method uses a clock that starts when the reference signal passes through zero and stops when the test signal passes through zero. The ratio of the time the clock runs to the known period of the irradiance signal gives the wavefront deviation with respect to the reference point.

4.3.2. Phase Lock

The *phase-lock method* can be explained with the help of Fig. 21 [12, 13, 14]. Let us assume that an additional phase difference is added to the initial phase $\Phi(x, y)$. The phase being added has two components, one with a fixed value and the other with a sinusoidal time shape. Both components can have any predetermined desired value. Thus,

$$\phi = \Phi(x, y) + \delta(x, y) + A \sin \omega t. \tag{6}$$

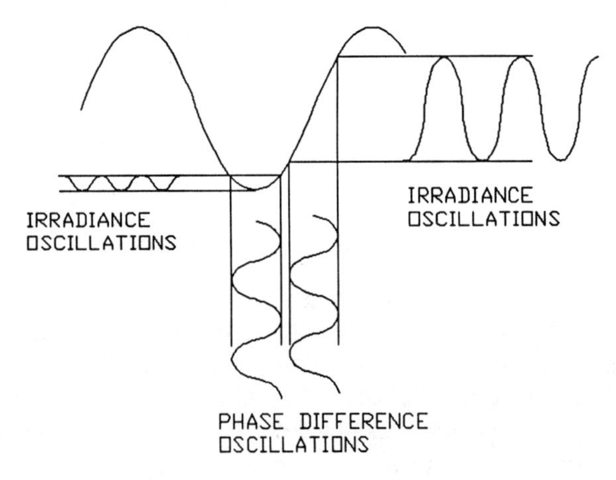

FIG. 21. Methods to find the phase with a small sinusoidal modulation of the phase.

Then, the irradiance $I(x, y)$ would be given by

$$I = I_1 + I_2 + 2\sqrt{I_1 I_2}\cos[\Phi + \delta + A \sin \omega t], \tag{7}$$

and this function can be expanded in a series as follows:

$$\begin{aligned} I = I_1 + I_2 + 2\sqrt{I_1 I_2}\, \{&\cos(\Phi + \delta)[J_0(A) + 2J_2(A)\cos 2\omega t + \cdots] \\ &- \sin(\Phi + \delta)[2J_1(A)\sin \omega t + 2J_3(A)\sin 3\omega t + \cdots]\}. \end{aligned} \tag{8}$$

Here, J_n is the Bessel function of order n. The first part of this expression represents harmonic components of even order, while the second part represents harmonic components of odd order.

Let us assume now that the amplitude of the phase oscillations $A \sin \omega t$ are much smaller than π. If we now adjust the fixed phase δ to a value such that $\Phi + \delta = n\pi$, then $\sin(\Phi + \delta)$ is zero. Hence, only even harmonics remain. The effect is shown in Fig. 22, near one of the minima of this function. This is done in practice by slowly changing the value of the phase δ, while maintaining the oscillation $A \sin \omega t$, until the minimum amplitude of the first harmonic, or fundamental frequency, is obtained. Then, we have $\delta + \Phi = n\pi$. Since the value of δ is known, the value of Φ has been determined.

An equivalent method would be to find the inflection point for the sinusoidal function, as shown in Fig. 11, by changing the fixed component of the phase until the first harmonic has its maximum amplitude. Then, $\sin \delta = 1$.

4.3.3. Phase Stepping

This method consists of measuring the irradiance values for several known increments of the phase. There are several versions of this method, which will be described later. The measurement of the irradiance for any given phase takes some time, since there is a time response for the detector. Hence, the phase has to be stationary during a short time to take the

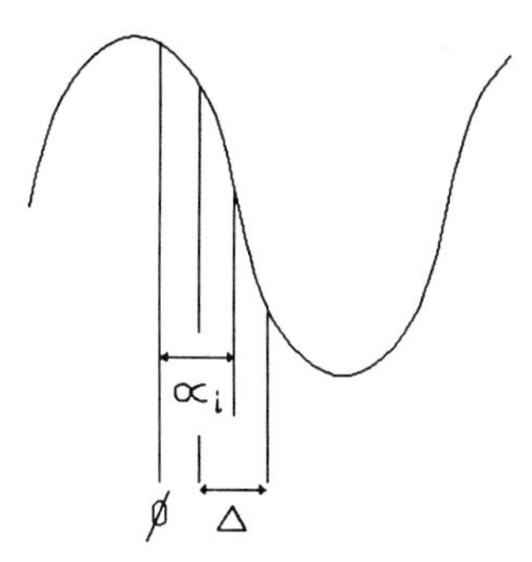

Fig. 22. Averaged signal measurements with the integrating phase-shifting method.

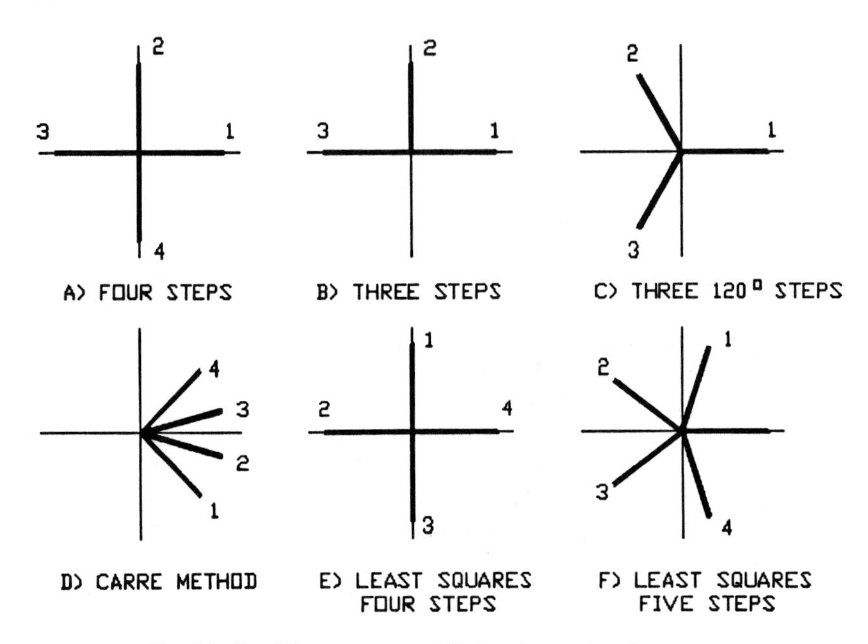

FIG. 23. Six different ways to shift the phase using phase steps.

measurement. Between two consecutive measurements, the phase may change as rapidly as desired to get to the next phase with the smallest delay. Let us assume that an irradiance I_i is measured when the phase has been incremented from its initial value Φ by an amount α_i, as shown in Fig. 23. Thus,

$$I_{\mathrm{i}} = I_1 + I_2 + 2\sqrt{I_1 I_2}\cos(\Phi + \alpha_i)\,. \tag{9}$$

The mathematical treatment will be completed in the next section, when describing the integrating phase-shifting method, because the *phase-stepping method* may be considered as a particular example of that method.

4.3.4. Integrating Phase-Shifting

This method, also called *integrating bucket*, is very similar to the phase-stepping method, with the only difference that the phase changes continuously and not by discrete steps. The problem with the phase-stepping method is that the sudden changes in the mirror position may introduce some vibrations to the system.

In the *integrating phase-shifting method*, the detector measures the irradiance continuously during a fixed time interval without stopping the phase, but since the phase changes continuously, the average value during the measuring time interval is measured. Thus, the integrating phase-stepping method may be mathematically considered a particular case of the phase-stepping method, if the detector has an infinitely short time response, such that the measurement time interval is reduced to zero. As in the phase-stepping method, the phase may be shifted using a sawtooth profile as in Fig. 2C, or a triangular profile, as in Fig. 2D, to avoid sudden changes.

Let us assume, as in Fig. 22, that the average measurement is taken from $\alpha = \alpha_i - \Delta/2$ to $\alpha = \alpha_i + \Delta/2$, with center value α_i. Then, the average value of the irradiance would be given by

$$\begin{aligned} I_i &= \frac{1}{\Delta} \int_{\alpha-\Delta/2}^{\alpha+\Delta/2} [I_1 + I_2 + 2\sqrt{I_1 I_2} \cos(\Phi + \alpha)]\, d\alpha \\ &= I_1 + I_2 + 2\sqrt{I_1 I_2}\, \mathrm{sinc}(\Delta/2) \cos(\Phi + \alpha_i)\,. \end{aligned} \tag{10}$$

As expected, this expression is the same as Eq. 9 for the measurement with phase steps when $\Delta = 0$, since the sinc function has unit value. If the integration interval Δ is different from zero, the only difference is that the apparent contrast of the fringes is reduced. We will consider this expression to be the most general.

In general, in the phase-stepping as well as the integrating phase-shifting methods, the irradiance is measured at several different values of the phase α_i. We are not going to describe these algorithms in detail, but basically, there are six different methods, as shown in Fig. 23, with different numbers of measurements of the phase.

4.4. Two-Wavelength Interferometry

In phase-shifting interferometry, each detector must have a phase difference smaller than π with the closest neighboring detector to avoid 2π phase ambiguities and ensure phase continuity. In other words, there should be at least two detector elements for each fringe. If the slope of the wavefront is very large, the fringes will be too close together and the number of detector elements would be extremely large.

A solution to this problem is to use two different wavelengths, λ_1 and λ_2, simultaneously. The group wavelength or equivalent wavelength λ_{eq} is much longer than either of the two components and is given by

$$\lambda_{eq} = \frac{\lambda_1 \lambda_2}{|\lambda_1 - \lambda_2|} \tag{11}$$

Under these conditions, the requirement to avoid phase uncertainties is that there should be at least two detectors for each fringe produced if the wavelength were λ_{eq}.

The great advantage of this method is that we may test wavefronts with large asphericities, limited by the group wavelength and accuracy limited the shortest wavelength of the two components.

5. The Future of Optical Testing

As with anything else in this world, it is very difficult to predict the future of the optical testing field. However, the present trend towards automatization of the whole process is likely to continue. The new electronic phase-shifting digital techniques have several distinct advantages, among which we may mention: a) There is improved accuracy of the interferometric measurements by at least one order of magnitude. b) With the help of new imaging devices and microcomputers, not only is the precision better, but also, the quantitative results can be obtained in real time. c) There is a smaller possibility that the interference pattern might be misinterpreted. Unfortunately, there also are some important disadvantages: a) The equipment requirements are considerably greater than for the traditional interferometric methods, making them at least one or two orders of magnitude more expensive. b) To get the experimental setup working takes longer than traditional methods. c) The testing instruments are less portable.

Summarizing, it is expected that automatic methods will be improved and that new research will make phase-shifting interferometry cheaper and easier to perform. However, in spite of these improvements, traditional methods also will continue to be used. Nothing is cheaper and easier to use than a knife edge! If great accuracy is desired and/or many identical optical systems are to be tested, then the newer automatic methods are ideally suited.

References

1. Daniel. Malacara (ed.), "Optical Shop Testing," John Wiley and Sons, New York, 1978.
2. L. M. Foucault, "Description des Procedes Employes pour Reconnaitre la Configuration des Surfaces Optiques," *C. R. Acad. Sci. Paris* **47**, 958 (1852). Reprinted in "Classiques de la Science," Vol. II, by Armand Colin.
3. V. Ronchi, "Le Franque di Combinazione Nello Studio Delle Superficie e Dei Sistemi Ottici," *Riv. Ottica mecc. Precis.*, 2, 9 (1923).
4. J. C. Wyant, "Holographic and Moiré Techniques," in "Optical Shop Testing" (D. Malacara, ed.), John Wiley and Sons, New York (1978).
5. J. Hartmann, "Bemerkungen uber den Ban und die Justirung von Spektrographen,[N]ZT," *Instrumentenkd.* **20**, 47 (1900).
6. Katherine Creath, "Phase-Measurement Interferometry Techniques," in "Progress in Optics XXVI" (E. Wolf, ed.), Elsevier Science Publishers, 1988.
7. P. Carré, "Installation et Utilisation du Comparateur Photoelectrique at Interferentiel du Bureau International des Poids et Measures," *Metrologia*, **2**, 13 (1966).

8. R. Crane, "Interference Phase Measurement," *Appl. Opt.*, **8**, 538 (1969).
9. Duncan T. Moore, "Gradient Index Optics Design and Tolerancing," Ph.D. Thesis, University of Rochester, 1973.
10. J. H. Bruning, D. J. Herriott, J. E. Gallagher, D. P. Rosenfeld, A. D. White, and D. J. Brangaccio, "Digital Wavefront Measurement Interferometer," *Appl. Opt.* **13**, 2693 (1974).
11. J. H. Bruning, "Fringe Scanning Interferometers," in "Optical Shop Testing" (D. Malacara, ed.), John Wiley and Sons, New York, 1978.
12. Glen W. Johnson, Dennis C. Leiner, and Duncan T. Moore, "Phase-Locked Interferometry," *Proc. SPIE.*, **126**, 152 (1977).
13. Glen W. Johnson, Dennis C. Leiner, and Duncan T. Moore, "Phase-Locked Interferometry," *Opt. Eng.*, **18**, 46 (1979).
14. Duncan T. Moore, "Phase-Locked Moiré Fringe Analysis for Automated Contouring of Diffuse Surfaces," *Appl. Opt.* **18**, 91 (1979).

CHAPTER 26

Adaptive Optics

Fritz Merkle
European Southern Observatory
Garching, Germany

1. Introduction

Image quality degradation due to atmospheric turbulence is one of the main limitations for imaging, laser propagation, and communication through the atmosphere. Adaptive optics is a technology to overcome this problem by real-time phase compensation. The main components for such systems—like wavefront correctors, wavefront sensors, and dedicated computers—exist and complete systems have been tested successfully. This chapter describes the principles of adaptive optics, concentrates on its applications for astronomical imaging, and presents some early observing results. The performance in image quality that can be expected from adaptive optics and the requirements for its application are discussed. Examples are given and perspectives for the future indicated.

2. The Principle of Adaptive Optics

A perfect imaging system has a theoretical angular resolution of

$$\Delta\alpha = 1.22\frac{\lambda}{D}$$

(λ: wavelength, D: aperture diameter), where $\Delta\alpha$ is the radius of the first dark ring of the diffraction pattern. However, in practice, image quality is

ISBN 0-12-289690-4

degraded by constant and random, spatially and temporally varying wavefront perturbations, originating from the aberrations of the optical system, and from atmospheric turbulence and other optical inhomogeneities like thermal blooming in the air path in front of and inside the optical system. These types of aberrations are well-known for optical systems like astronomical telescopes, cameras for reconnaissance, laser beam directors, and similar applications.

All wavefront perturbations of an optical system taken together result in a complex phase aberration,

$$\phi(\mathbf{r}, t) - iA(\mathbf{r}, t).$$

The real part $\phi(\mathbf{r}, t)$ represents the phase distortion of the wavefront, usually called *seeing*, while the imaginary part $A(\mathbf{r}, t)$ is a measure for the intensity fluctuations across the aperture plane, called *scintillation*.

It is possible to correct the phase distortion term with *adaptive optics* [1, 2]. The basic principle of this technique is to measure the aberrated wavefront and to apply this data to a phase-shifting optical element, which can be spatially and temporally controlled to compensate the aberrations, including the rapidly varying atmospheric phase distortions in a closed-loop feedback system (see Fig. 1). A correction of the scintillation would require a spatially and temporally controlled apodization element.

An adaptive system then will have to perform the following correction:

$$-\phi(\mathbf{r}, t) + \mu iA(\mathbf{r}, t)$$

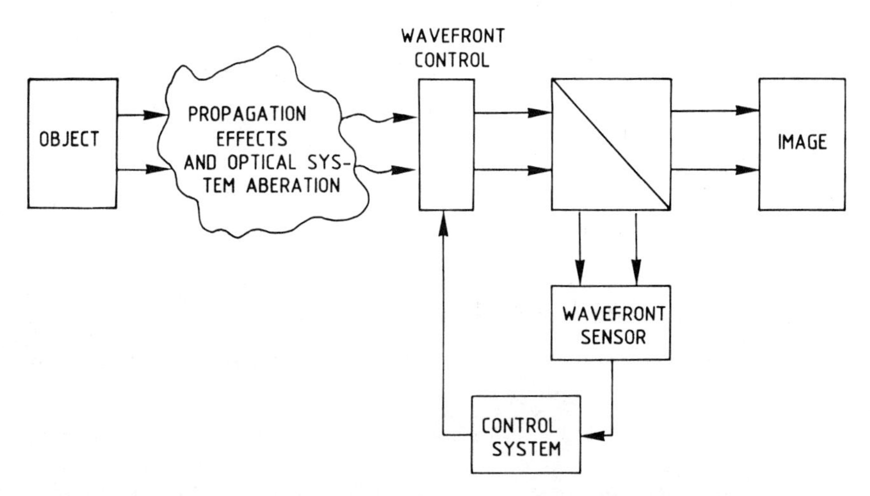

FIG. 1. Principle of an adaptive optical system. It is a closed-loop wavefront correction system.

(where the factor μ takes care of the physical units). For systems with large apertures, the phase-correction part usually is sufficient, due to the averaging of the intensity fluctuations over the pupil.

This type of optical phase compensation often is referred as *conventional adaptive optics* to discriminate it from adaptive correction by phase conjugation, which will not be addressed here.

In addition, adaptive optics often is confused with *active optics*. Active optics can be considered as a special case of adaptive optical correction. It refers to the correction of aberrations occurring only inside the optical system (like the real-time correction of telescope aberrations), while adaptive optics covers all aberrations along the optical path. In another definition, active optics is used for open-loop systems, while adaptive optics characterizes closed-loop systems. A standardization in terminology would be helpful.

The first suggestions for the construction of an adaptive optical correction device came from astronomy. In 1953, H. W. Babcock published a paper on "The Possibility of Compensating Astronomical Seeing"—a possibility that would involve using a surface-deformable oil film [3]. This proposal seemed to be forgotten until the early 1970s when the application of laser propagation for defense purposes became of interest. During the last 20 years, vast development took place in this area, but in a nearly fully confidential environment. Only recently, the civil branch of adaptive optics has reached a recognizable level.

3. The Needs for Adaptive Optics in Astronomy

The needs for adaptive optics in astronomy can be grouped in the following application categories (and they are not very different from other applications):

- Concentration of light energy.
- High-resolution imaging.

The first application is of major importance in the area of high-resolution spectroscopy. It allows reduction of the entrance slit width of a spectrograph and thus an increase in its resolution.

The second application—high resolution imaging—can be divided into two parts. For single-telescope imaging, adaptive optics will allow diffraction-limited—or at least close to diffraction-limited—imaging in the infrared; and in the future, it is very likely also for the visible wavelength range. Since adaptive optics suffers under sensitivity problems, as shown later, it may be a very high potential in combination with speckle-imaging methods.

For multiple-telescope applications in long baseline interferometry, adaptive optics can be considered as mandatory to obtain the adequate signal-to-noise ratio in reconstructed high-resolution images [4].

4. The Design Parameters for an Adaptive Optical System

An adaptive optical system (see Fig. 2) contains three major components: a phase-shifting optical element, a wavefront sensor, and a servo-control system. The distortion of the received wavefront usually is compensated by reflecting the light wave from deformable mirror. The surface of this mirror is adjusted in real time to compensate the path length aberrations. The information required to deform the mirror is obtained by analyzing the light wave with a wavefront sensor. A map of wave-

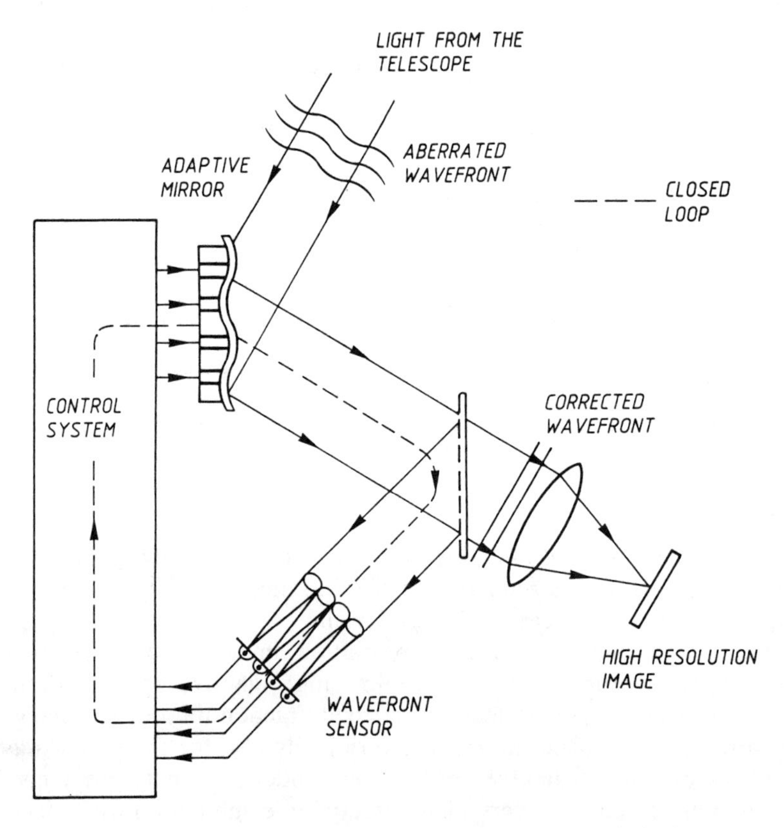

FIG. 2. Generic adaptive optics system for use in an astronomical telescope.

front errors then is derived at each instant of time. Using this error map, the control system determines the signals required to drive the phase-shifting optical element and to null the phase aberrations by closing the adaptive loop. The phase correction values can be obtained by expanding the phase-correction function

$$\phi(\mathbf{r}, t) = \sum_{n=1}^{N} a_n(t) f_n(\mathbf{r})$$

in spatially ($f_n(\mathbf{r})$) and temporally ($a_n(t)$) dependent functions. The complexity and design of an adaptive system depends on the atmospheric conditions, the aperture size of the telescope, and the direction of the optical path through the atmosphere. The atmospheric conditions usually are characterized by the so-called seeing or Fried's parameter r_0, which is the correlation length of the wavefront projected on the ground, and the correlation time τ_0 [5, 6]. r_0 scales with the wavelength as

$$r_0 \approx \lambda^{1.2}.$$

The properties of the atmosphere can be approximated by Kolmogorov statistics [7].

The sampling of the distorted wavefront or the number of subapertures (N) (degrees of freedom), the correction rate ($\approx 1/\tau_0$), and the wavefront amplitude range (stroke Δz) are the basic design parameters for an adaptive system.

The number of subapertures can be approximated by

$$N \approx \left(\frac{D}{r_0}\right)^2$$

and the coherence time by

$$\tau_0 \approx \frac{r_0}{\nu},$$

where D is the aperture diameter and v the mean windspeed in the turbulent air layers.

Typical values of these parameters for different wavelengths (assuming one of the future large astronomical telescopes of the 8-meter class, of which several currently are under design or construction) are given in Table I with respect of an average seeing quality described by a correlation length r_0 of 10cm at 0.5μm wavelength. It is obvious that for short wavelengths and especially in the visible range, an adaptive optics system becomes extremely complex.

A perfect adaptive optical system will perform diffraction-limited imaging only on the optical axis. Off-axis parts in the image have different

Table I

λ	0.5 μm	2.2 μm	5.0 μm	10 μm
r_0	10 cm	60 cm	160 cm	360 cm
N	6400	180	12	4
τ	6 ms	35 ms	95 ms	220 ms
Θ	1.8″	10″	30″	70″

viewing angles through the atmosphere. The angular range wherein the light suffers from nearly the same atmospheric disturbance is called *isoplanatic angle* Θ (see Table I). It is only a few arcseconds in the visible but grows to the range of arcminutes in the medium infrared. This parameter together with r_0 and τ_0 completely determines the optical propagation effects of the turbulent layers projected on the ground.

5. Strategy for Seeing Optimization

There are two ways to develop the phase correction function, using either a *modal* or a *zonal* (*nodal*) set of polynomials.

The shape of an optical wavefront often is represented by a set of orthogonal whole-aperture modal functions, like the Zernike polynomials, which correspond to systematic optical aberrations, such as defocus, astigmatism, etc. [8].

In the modal concept, the single steps of a typical feedback procedure are:

- Measurement of the wavefront by determining the local slope (zonal procedure).
- Computation of a wavefront map.
- Computation of the modal coefficients (e.g., Zernike coefficients).
- Computation of the control signals for the wavefront correction device.
- Conversion of control signals to the required drive signal for each subaperture of the correction device (zonal procedure).

If the modal approach is used rather than a zonal decomposition for $f_n(\mathbf{r})$, image improvement is possible even with a limited number of modes, because the correction of each mode individually contributes already to the image quality improvement. This is one of the essential advantages of the modal correction strategy.

6. Elements of an Adaptive Optical System

The main elements of an adaptive optical system are the wavefront correction device, the wavefront sensor, and the control computer.

5.1. Wavefront Correctors

The distorted wavefront can be controlled by changing either the velocity of propagation or the optical path length. The former is achieved by varying the refractive index of an optical medium, while the latter can be implemented by moving a reflective surface, such as a mirror.

At the present time, reflective devices are used most commonly as wavefront correctors. The problems with the other devices mainly are the limited range of refractive index change, spectral absorption, and nonuniform transmission. Because the optical path is confined to one side of a mirror surface, a great variety of substrates and methods for deforming a mirror are available. Another advantage is that the wavefront deformation is a true optical path length change, independent of wavelength. Figure 3 shows the basic types of deformable mirrors [9]. The actuators for these mirrors can be based on piezoelectric, electrostrictive, electrostatic, magnetostrictive, electromagnetic, or other effects. A large variety of mirrors has been developed, and constant progress is taking place towards the integration of hundreds and even thousands of actuators. The thin faceplates of these mirrors vary from glass, metal, and ceramics to silicon and other materials.

In practice, a deformable mirror often is combined with a tip-tilt mirror. This latter mirror corrects all wavefront tilts that usually dominate in strength [8], and relaxes the requirements for the stroke of the deformable mirror.

5.2. Wavefront Sensors

It is not possible to measure directly the phase of an optical wavefront, as no existing detector will respond to the temporal frequencies involved. Four techniques commonly are used to overcome this problem:

(1). Measurements can be made on the intensity distribution of the image produced by the entire wavefront.

(2). A reference wavefront of the same or slightly different wavelength can be combined with the wavefront to be measured to produce interference fringes.

(3). The wavefront slope of small zones of the wavefront can be measured. This can be achieved by using a shearing interferometer or the Hartmann test.

(4). The local wavefront curvature can be measured with a so-called curvature sensor.

Each of these four approaches has its own advantages and disadvantages. A realization of the first technique is the so-called *multi-dither technique* [2], which requires bright sources, and it is only applicable,

- Segmented mirrors

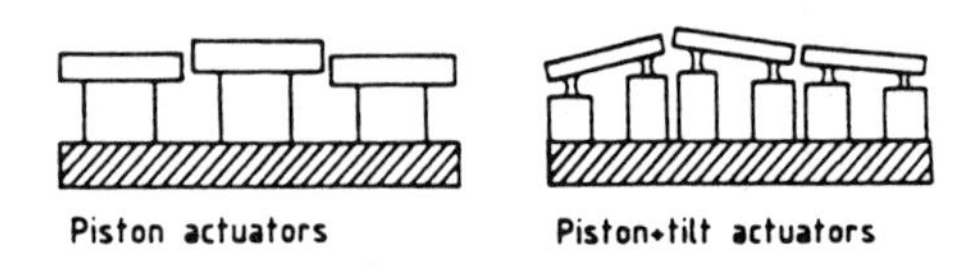

- Continuous thin-plate mirrors

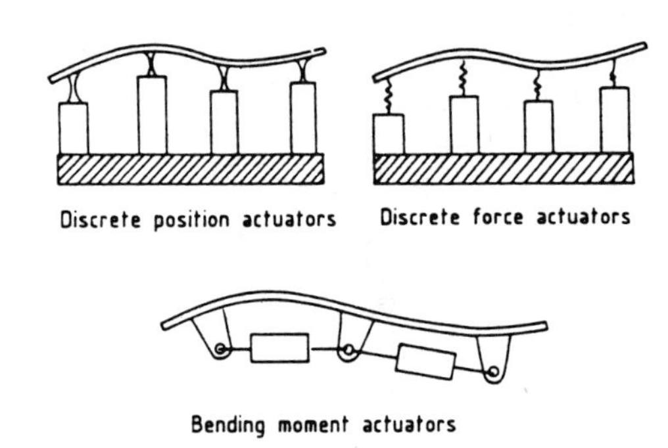

- Monolithic mirrors

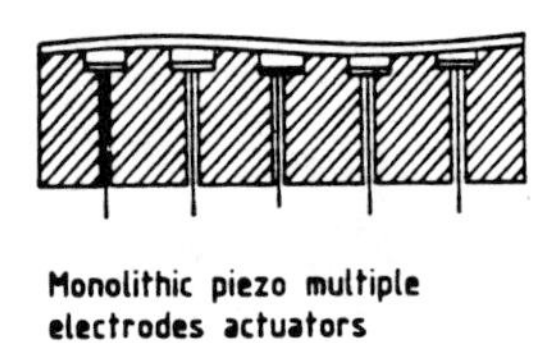

- Membrane or pellicle mirrors.

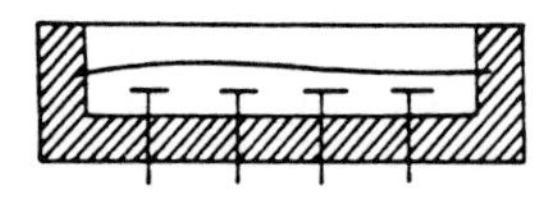

FIG. 3. Typical examples of deformable mirrors.

therefore, for compensation in laser systems. The second technique needs monochromatic light sources. Therefore, it is excluded from astronomical applications because of the nature of the astronomical light sources. Its main application again is in laser propagation and communication.

For application in astronomy with very weak sources, only the third and fourth techniques are efficient. The following approaches have been tested successfully: the shearing interferometer, the Shack-Hartmann sensor, and recently, the curvature sensor.

Shearing Interferometer. In a shearing interferometer [10, 11], the wavefront to be measured is amplitude-divided into two components, which are mutually displaced and recombined with each other to generate an interference pattern. If the path lengths of the two beams are equal, then fringes are generated even with incoherent light sources. Several methods of producing sheared wavefronts exist. One of the most useful is a moving grating located at the focus of the beam. After recollimation, the intensity in the detector plane is modulated with twice the temporal frequency of the moving grating (see Fig. 4). The phase of this light modulation is proportional to the slope of the optical wavefront in the corresponding zone of the aperture. In practice, rotating gratings with radial

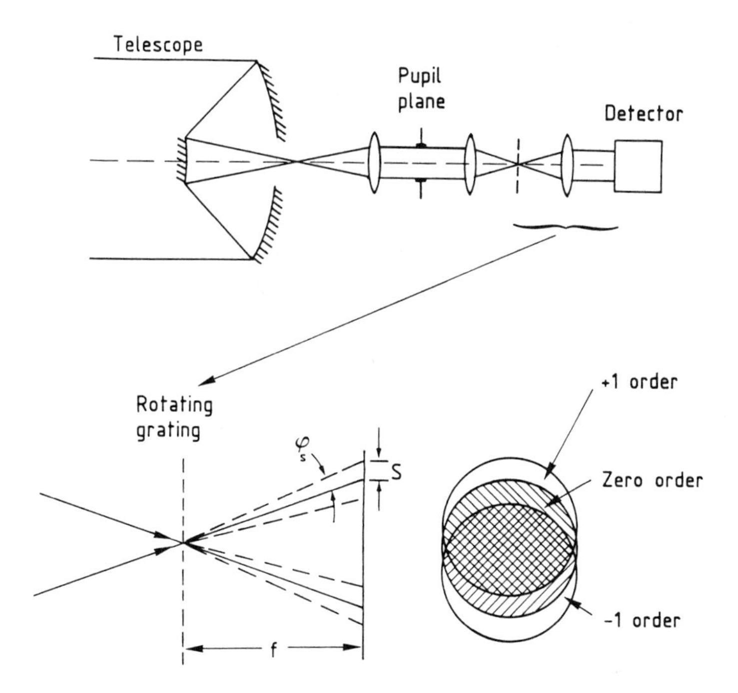

FIG. 4. Principle of the rotating grating lateral shear interferometer.

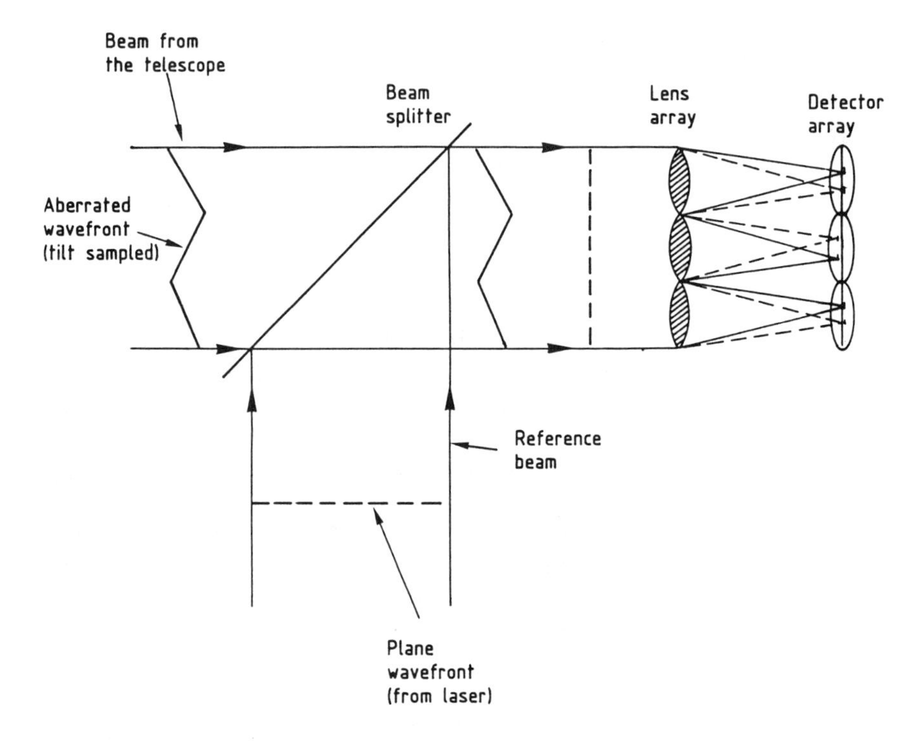

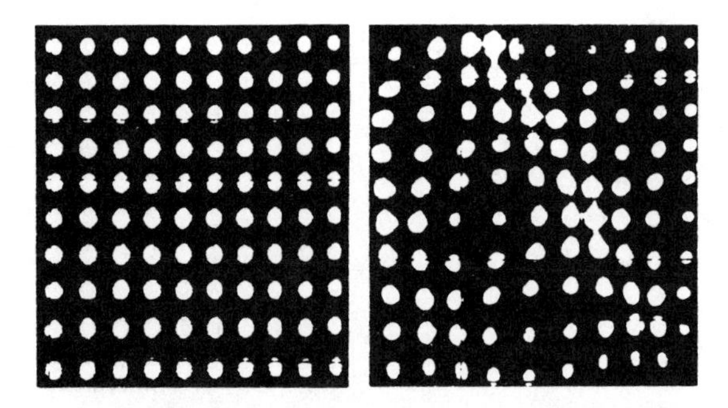

FIG. 5. Principle of the Shack-Hartmann wavefront sensor with a reference beam for self-calibration and a typical Shack-Hartmann focal pattern with and without wavefront aberrations.

patterns are used to shear the wavefront in two orthogonal directions to provide two orthogonal sets of slope measurements.

Shack-Hartmann Wavefront Sensor. The Shack-Hartmann wavefront sensor is based on the classical Hartmann test [12, 13]. Figure 5 shows the schematic of a Shack-Hartmann sensor. The wavefront is divided into a number of zones, which are contiguous and of equal size. The light from

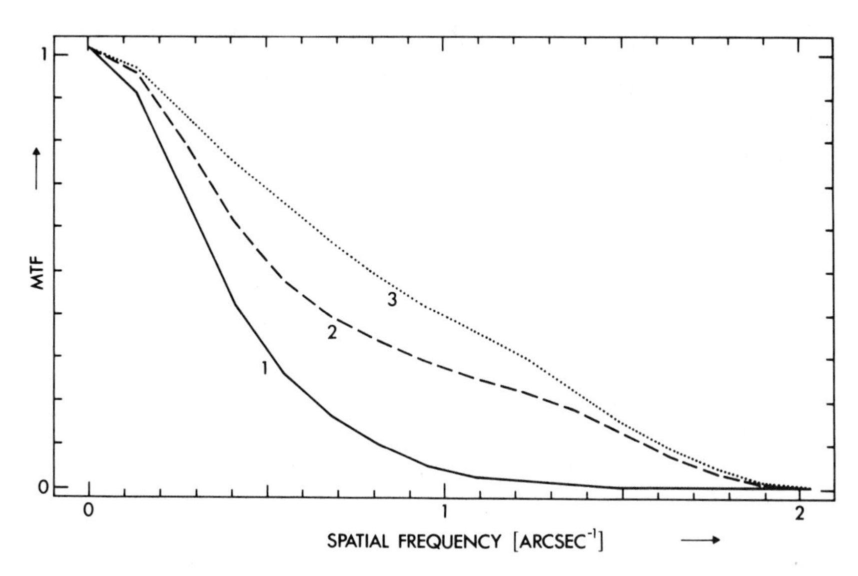

FIG. 6. Measured modulation transfer function (2) of an adaptive optics system with a 19-actuator deformable mirror applied to a 1.52-meter telescope for 3.5 μm wavelength (L-Band) in comparison with the uncorrected (1) and theoretical (3) behavior [17, 18].

each zone is brought to a separate focus with a lenticular array, and the position of the centroid of each focus is measured in two dimensions. Figure 6 also displays a typical focal pattern of a Shack-Hartmann lenticular array for a wavefront with and without aberrations.

Curvature Sensor. The curvature sensor is based on mapping the local wavefront curvature. One possibility is to sense the difference in irradiance distribution over two cross sections of the beam at equal distance on each side of the image plane. This difference has been shown to be a measure of the local wavefront Laplacian [14], the curvature.

5.3. Control System

Slope-measuring wavefront sensors usually require a reconstruction of the wavefront itself. In general, two orthogonal wavefront slope measurements are made for each subaperture. In other words, there are twice as many measurements as unknowns, so that a least-squares can be performed [9].

All algorithms that perform the control loop require very high computation powers to meet the temporal and spatial sampling requirements. With special dedicated hardware or hybrid systems, this problem has been approached successfully.

In case a curvature sensor is used together with a mirror of the bending actuator type, a direct analog feedback loop may be possible.

7. Accuracy of the Adaptive Optics Correction

The main sources for errors in an adaptive optical system are wavefront fitting errors (σ_F), which depend on how closely the wavefront corrector can match the actual wavefront error; the detection error (σ_D), which essentially is reciprocal to the signal-to-noise ratio of the wavefront sensor output; and the prediction error (σ_P), which is due to the time delay between the measurement of the wavefront disturbances and their correction. The overall residual error then is given by

$$\sigma_R^2 = \sigma_F^2 + \sigma_D^2 + \sigma_P^2.$$

However, even if the residual errors do not lead to the diffraction limit, it is possible to reach a partial correction with adaptive optics [15, 16, 17]. Especially in astronomy, this effect will be of high importance for improving the imaging in the visible and near ultraviolet. Further investigations are needed to show what can be achieved.

8. First Results with Adaptive Optics

Figure 6 shows the measured modulation transfer functions (MTF) [17] of a 1.52-meter telescope equipped with an adaptive optics system [18] with a thin deformable mirror with 19 actuators at 2.2 μm wavelength. The uncorrected (curve 1) and corrected (curve 2) cases are compared with the theoretical (curve 3) behavior. It is obvious that this system allows reconstruction of the MTF to a large extent.

Figure 7 shows the resulting image of real-time correction with adaptive optics for a single and a binary star observed with a 3.6-meter telescope with the same adaptive optics system as mentioned earlier, at 3.8 μm wavelength. The corrected images show clearly their diffraction-limited nature [19] with a diffraction pattern of 0.22 arcseconds FWHM (full width at half maximum).

9. The Reference Source Problem

Any correction requires a measurement of the effect to be corrected. This is one of the major problems in applying adaptive optics to astronomical observations. The observed sources are too faint in most cases, so that their light is not sufficient for the correction. A brighter nearby reference source within the same isoplanatic patch rarely is available. Only for infrared wavelengths does the situation become more favorable because of the increase of the isoplanatic angle, as shown in Table II.

The infrared/visible correlation of the atmospheric modulation transfer funtions [20] opens the possibility of measuring the wavefront in the visible

Table II

λ	0.5 μm	2.2 μm	5.0 μm	10 μm
r_0	10 cm	60 cm	160 cm	360 cm
m_{lim}	7	13	15.5	17
C_P	$\simeq$0%	0.1%	30%	100%
C_E	$\simeq$0%	0.3%	100%	100%

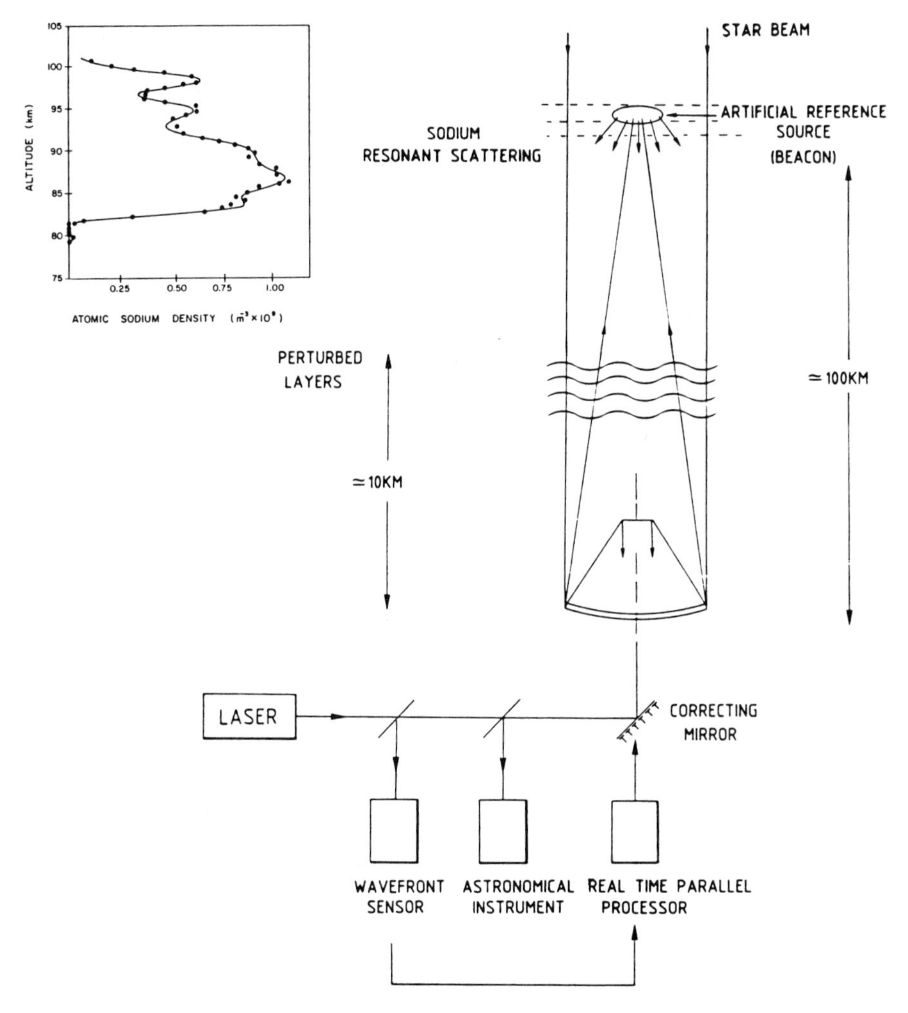

FIG. 8. Generation of an artificial reference source by resonance scattering of a laser beam in the mesospheric sodium layer. The insert shows the sodium concentration vs. the altitude.

range and compensating for infrared wavelengths (polychromatic adaptive correction). This method also has been applied in the aforementioned system. Table II gives the limitation for adaptive optics in astronomy for the visible and for selected infrared wavelengths, and the resulting sky coverage at the galactic pole (C_P) and the galactic equator (C_E).

Recently, a new technique to overcome the reference source problem even in the visible range has been proposed [21] and a first test indicates its feasibility [22]. An artificial reference source is generated by using, e.g., resonance scattering of laser light in a high-altitude atmospheric layer (see Fig. 8). These artificial reference sources have to be generated within the isoplanatic angle of the astronomical source at a repetition rate synchronous to the adaptive correction rate. It would require gating the astronomical detector during the wavefront sensing time not to pollute the detector with scattered light from lower altitude, which means a small loss compared to the high gain of the adaptive correction. It is limited by the need for multiple reference sources due to the fact that the light passes on a non-parallel path back through the atmosphere because of the limited altitude of the artificial source. Multiple spots on the Shack-Hartmann sensor will make the detection problem more complex. Higher-altitude layers have the advantage of a reduced number of spots, but then the light travel becomes the limitation (e.g., for the sodium layer of 80 to 100 km height it is in the range of 2 times 300 μsec), which reduces the duty cycle for the real astronomical observation.

10. What Comes Next?

Artificial reference star generation would mean a great step forward towards adaptive optics even at visible wavelengths. It would make adaptive optics applicable under nearly any condition.

The technology applied for constructing the deformable mirrors sets limitations to the maximum number of controlled subapertures. With the currently used discrete piezoelectric actuators, mirrors with several hundreds or thousands of subapertures may create severe technical difficulties. An alternative could be the use of a three-stage system instead of the two-stage correction with a tip-tilt and a normal deformable mirror. The first stage will correct tilt errors that have the largest amplitudes, the second will correct low and medium spatial frequency aberrations with medium amplitudes and frequency response, and the third will handle high spatial frequency aberrations (see Fig. 9). The correcting element for this third stage could be based on electro-optical or similar effects. These devices would have to have very high spatial resolution, but could have limited wavefront modulation capabilities in the range of only a fraction of a wavelength for visible wavelengths.

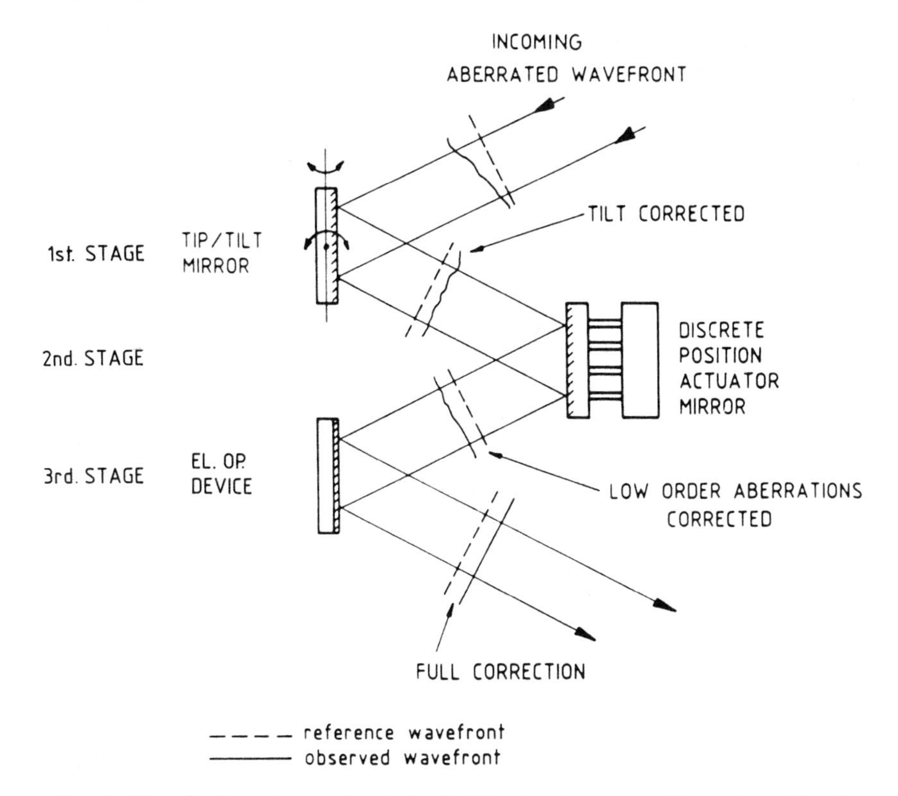

FIG. 9. Use of a three-stage system instead of the two-stage correction system with only a tip-tilt and a deformable mirror. Here, the first stage will correct tile errors that have the largest amplitudes; the second stage, low and medium spatial frequency aberrations with medium amplitudes and frequency response; and the third stage, high spatial frequency aberrations.

If adaptive optics is to be applied in the visible range for large telescopes, the control loop will have to cope with several hundreds or thousands of loop cycles per second for a very large number of parallel channels. The necessary electronics hardware and software can be identified as major critical items for large-scale adaptive systems, and will require further development. Only dedicated processors seem to satisfy the needs for the future. Neural networks [23] or optical computers may become part of future adaptive optics systems.

To overcome the problems of isoplanicity, it has been proposed to use cascaded adaptive optics systems, which individually correct the aberrations of the various height layers of the atmosphere, similar to reconstructing structures by tomography [24].

It is obvious that defense-related developments in the area of adaptive optics are ahead of the technology status described here. In these programs, huge effort has been invested in systems of several hundreds or even thousands of subaperatures and correction rates of several tens of thousands, but no details or results are available in public literature.

It can be concluded that adaptive optics is a very powerful technique for real-time imaging through the turbulent atmosphere or other aberrating media. Successfully operating adaptive optical systems may become standard astronomical telescope equipment in the near future, and will revolutionize ground-based astronomical observations. Major developments still are necessary to solve the reference source and data processing problems and to extend the application range of adaptive optical systems for large apertures into the visible wavelength range.

The astronomical application is only one application in which adaptive optics proves its potential. In other areas, applications of adaptive optics are under development, e.g., in ophthalmology and laser machining. More applications may come in the future, when adaptive optics will bring performance to the diffraction limits.

References

1. J. W. Hardy, *Proc. IEEE* **66**, 651 (1978).
2. J. E. Pearson, R. H. Freeman, and H. C. Reynolds, in "Applied Optics and Optical Engineering," **VII**, 245, Academic Press, New York, 1979.
3. H. W. Babcock, *Publ. Astron. Soc. Pac.* **65**, 229 (1953).
4. F. Merkle, *J. Opt. Soc. Am. A* **5**, 904 (1988).
5. D. L. Fried, *J. Opt. Soc. Am.* **55** 1427.
6. F. Roddier, "The Effect of Atmospheric Turbulence in Optical Astronomy," in "Progress in Optics," (E. Wolf, ed.), North-Holland (1981).
7. V. I. Tatarskii, "Wave Propagation in a Turbulent Medium," McGraw-Hill, New York (1961).
8. R. J. Noll, *J. Opt. Soc. Am.* **66**, 207 (1976).
9. J. Hardy, *AGARD (Advisory Group for Aerospace Research & Development) Conf. Proc.* No. 300, 49–1 (1980).
10. C. L. Koliopoulos *Applied Optics* **19**, 1523 (1980).
11. J. C. Wyant *Applied Optics* **14** 2622 (1975).
12. J. M. Feinleib, US Patent No. **4,141,652**, 1975.
13. L. Noethe, F. Franza, P. Giordano, and R. N. Wilson, *IAU (International Astronomical Union) Colloquium*, no. 79, 67 (1984).
14. F. Roddier, *Applied Optics* **27**, 1223 (1988).
15. R. W. Smithson, and M. Peri, *Applied Optics* **27** (1988).
16. Barakat Nisenson, *J. Opt. Soc. Am. A* **4**, 12 (1987).
17. G. Rousset, J. C. Fontanella, P. Kern, P. Gigan, F. Rigaut, P. Léna, C. Boyer, P. Jagourel, J. P. Gaffard, and F. Merkle, *Astronomy and Astrophysics* **230**, L-29 (1990).
18. F. Merkle, G. Rousset, P. Kern, and J. P. Gaffard, *Proc. SPIE* **1236**, 193 (1990).
19. F. Merkle, G. Gehring, F. Rigaut, P. Kern, P. Gigau, G. Rousset, and C. Boyer, *The Messenger* (ESO) **60**, 9 (1990).
20. F. Roddier, C. Roddier, *Proc SPIE* **322**, 252 (1986).
21. R. Foy and A. Labeyrie, *Astronomy and Astrophysics* **152**, L-29 (1985).
22. L. A. Thompson and C. S. Gardner, *Nature* **328**, 229 (1987).
23. R. C. Smithson, *Lockheed Horizon* **28**, 24 (1990).
24. J. M. Beckers, *Proc SPIE* **1114**, 215.

CHAPTER 27

Triple Correlations and Bispectra in High-Resolution Astronomical Imaging

Gerd Weigelt
Max-Planck-Institut für Radioastronomie
Bonn, Federal Republic of Germany

1. Introduction

Refractive index fluctuations in the atmosphere of the earth restrict the resolution of large, ground-based telescopes to about 0.5 arcsec. This limitation is very frustrating, since the theoretical diffraction limit λ/D (λ = wavelength, D = telescope diameter) of a 5 m telescope is about 0.02 arcsec at $\lambda \sim 500$ nm. Fortunately, it is possible to overcome atmospheric image degradation by various interferometric speckle techniques. The speckle interferometry method can reconstruct the diffraction-limited autocorrelation of astronomical objects [1]. True images with diffraction-limited resolution can be reconstructed by the Knox-Thompson method [2] and by the speckle masking method [3–6], also called bispectrum or triple correlation processing. An important advantage of speckle masking is that it can measure all *closure phases*. Therefore, it is possible to apply speckle masking to both single-dish telescopes and to optical long-baseline interferometers.

In Section 2, we describe the theory and astronomical applications of the speckle-masking method. Two different speckle spectroscopy methods are discussed in Sections 3 and 4. Finally, optical long-baseline interferometry is discussed in Section 5.

ISBN 0-12-289690-4

2. Speckle Masking: Bispectrum or Triple-Correlation Processing

The raw data for all speckle techniques are *speckle interferograms*. These are short-exposure photographs recorded with an exposure time of about 0.05 sec or shorter. This short exposure time is necessary to "freeze" the atmosphere during the exposure time. Figure 1a shows that speckle interferograms consist of many small bright dots, called *speckles*. Speckles are interference maxima with an average diameter of about $\lambda/D \sim 0.02$ arcsec for $\lambda \sim 500$ nm and $D \sim 5$ m. In other words, speckles are as small as the theoretical Airy pattern. Therefore, speckle interferograms contain high-resolution object information. For recording high-contrast speckle interferograms, one has to use interference filters with a bandwidth of about 30 nm (quasi-monochromatic light). Image intensifiers have to be used because of the required short exposure time. For very faint objects, high-gain image intensifiers are required that can record individual photon events.

The intensity distribution $i_n(\mathbf{x})$ of the nth recorded speckle interferogram can be described by the incoherent, space-invariant imaging equation

$$i_n(\mathbf{x}) = o(\mathbf{x}) * p_n(\mathbf{x}) \qquad n = 1, 2, 3, \ldots N(N \sim 10^3 \text{ to } 10^5), \tag{1}$$

where $o(\mathbf{x})$ is the object intensity distribution, $*$ denotes the convolution operator, $p_n(\mathbf{x})$ is the point spread function of the atmosphere/telescope, $\mathbf{x}$ is a one- or two-dimensional space vector, and N is the total number of recorded speckle interferograms. In the text that follows, the index n of the random functions $i_n(\mathbf{x})$ and $p_n(\mathbf{x})$ and of their Fourier transforms will be omitted. The goal of the speckle methods is to reconstruct a diffraction-limited image of the object $o(\mathbf{x})$ from a sequence of recorded speckle interferograms $i(\mathbf{x})$.

Speckle masking consists of the following processing steps:

Image processing step 1: There are two possibilities here: Calculation of the *ensemble average triple correlation*

$$\langle T(\mathbf{x}, \mathbf{y}) \rangle = \left\langle \int\!\!\int i(\mathbf{x}')i(\mathbf{x}' + \mathbf{x})i(\mathbf{x}' + \mathbf{y})\, d\mathbf{x}' \right\rangle \tag{2}$$

or calculation of the *ensemble average bispectrum*

$$\langle B(\mathbf{u}, \mathbf{v}) \rangle = \langle I(\mathbf{u})I(\mathbf{v})I^*(\mathbf{u} + \mathbf{v}) \rangle, \tag{3}$$

where * denotes conjugate complex, $\langle \ldots \rangle$ denotes ensemble average over all N speckle interferograms, and $\mathbf{u}$ and $\mathbf{v}$ are one- or two-dimensional coordinate vectors in Fourier space. $I(\mathbf{u})$, $I(\mathbf{v})$, and $I^*(\mathbf{u} + \mathbf{v})$ are Fourier

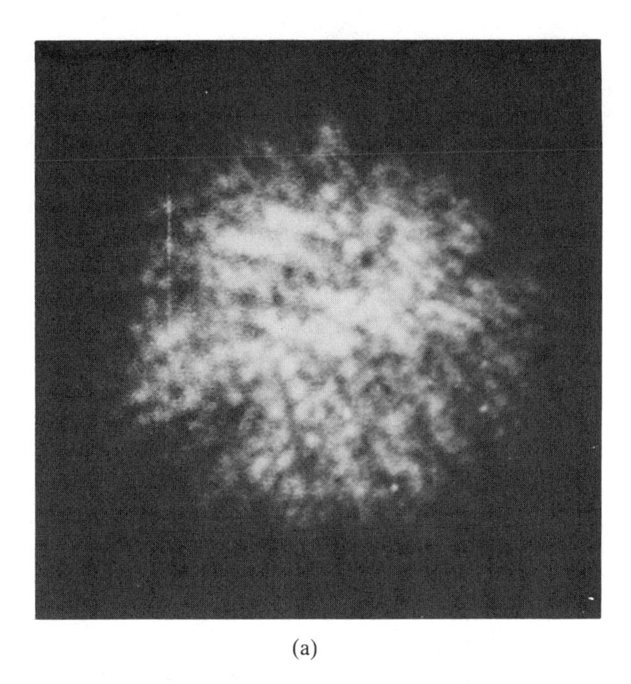

(a)

(b)

FIG. 1. Speckle-masking observation of the close spectroscopic double star ψ Sagittarii: (a) One of 150 speckle interferograms recorded with the 3.6m ESO telescope. (b) High-resolution image reconstructed from 150 speckle interferograms by speckle masking. The separation of the two stars is about 0.18 arcsec (epoch 1982.378) (From [4]).

transforms of $i(\mathbf{x})$, i.e.,

$$I(\mathbf{u}) = \int i(\mathbf{x}) \exp(-2\pi i \mathbf{x} \cdot \mathbf{u})\, d\mathbf{x},$$

$$I(\mathbf{v}) = \int i(\mathbf{x}) \exp(-2\pi i \mathbf{x} \cdot \mathbf{v})\, d\mathbf{x}, \text{ and} \tag{4}$$

$$I^*(\mathbf{u}+\mathbf{v}) = \int i(\mathbf{x}) \exp[2\pi i \mathbf{x} \cdot (\mathbf{u}+\mathbf{v})]\, d\mathbf{x}.$$

The bispectrum $B(\mathbf{u}, \mathbf{v})$ is the Fourier transform of the triple correlation $T(\mathbf{x}, \mathbf{y})$ [5]. If $\mathbf{x}$, $\mathbf{y}$, $\mathbf{u}$, and $\mathbf{v}$ are one-dimensional vectors (e.g., for one-dimensional infrared data), the triple correlation $T(\mathbf{x}, \mathbf{y})$ and the bispectrum $B(\mathbf{u}, \mathbf{v})$ are two-dimensional functions. If $\mathbf{x}$, $\mathbf{y}$, $\mathbf{u}$, and $\mathbf{v}$ are two-dimensional vectors, the triple correlation $T(\mathbf{x}, \mathbf{y})$ and the bispectrum $B(\mathbf{u}, \mathbf{v})$ are *four-dimensional functions*. In other words, the dimension of the triple correlation and the bispectrum always is *twice* the dimension of the speckle interferograms.

In most applications, it is sufficient to work with small subsets of the average bispectrum [6]. Then the data processing can be performed easily with a small computer. An alternative is to evaluate many different one-dimensional projections of each two-dimensional speckle interferogram. In this case, the bispectra obtained are two-dimensional. A one-dimensional image of the object can be reconstructed for each projection direction. From many different one-dimensional projections of the two-dimensional object, a two-dimensional image can be reconstructed by tomography [7]. This technique is very instructive, since the modulus of the two-dimensional bispectra can be displayed easily on the computer monitor [5, 7]. In most applications, it is advantageous to use bispectrum processing. The advantages of the triple correlation are that it can be visualized easily and that it can be used for photon-counting triple correlation techniques [7]. In the text that follows, we will discuss the theory of bispectrum processing.

Image processing step 2: Compensation of the photon bias in the ensemble average bispectrum $\langle B(\mathbf{u}, \mathbf{v}) \rangle$ [8].

Image processing step 3: Compensation of the speckle-masking transfer function. From $i(\mathbf{x}) = o(\mathbf{x}) * p(\mathbf{x})$ follows for the Fourier transform $I(\mathbf{u})$ of $i(\mathbf{x})$:

$$I(\mathbf{u}) = O(\mathbf{u}) P(\mathbf{u}), \tag{5}$$

where $O(\mathbf{u})$ and $P(\mathbf{u})$ are the Fourier transforms of $o(\mathbf{x})$ and $p(\mathbf{x})$, respectively. If we insert Eq. (5) into Eq. (3), we obtain for the ensemble average bispectrum $\langle B(\mathbf{u},\mathbf{v})\rangle$,

$$\begin{aligned}\langle B(\mathbf{u},\mathbf{v})\rangle &= \langle O(\mathbf{u})P(\mathbf{u})O(\mathbf{v})P(\mathbf{v})O^*(\mathbf{u}+\mathbf{v})P^*(\mathbf{u}+\mathbf{v})\rangle \\ &= O(\mathbf{u})O(\mathbf{v})O^*(\mathbf{u}+\mathbf{v})\langle P(\mathbf{u})P(\mathbf{v})P^*(\mathbf{u}+\mathbf{v})\rangle.\end{aligned} \tag{6}$$

$\langle P(\mathbf{u})P(\mathbf{v})P^*(\mathbf{u}+\mathbf{v})\rangle$ is called the *speckle-masking transfer function.* It can be derived from the speckle interferograms of a point source or it can be calculated theoretically. Since it is greater than zero up to the diffraction cutoff frequency [5], Eq. (6) can be divided by the speckle-masking transfer function, and we obtain for the bispectrum $B_o(\mathbf{u},\mathbf{v})$ of the object $o(\mathbf{x})$,

$$B_o(\mathbf{u},\mathbf{v}) \equiv O(\mathbf{u})O(\mathbf{v})O^*(\mathbf{u}+\mathbf{v}) \tag{7}$$

$$= \langle I(\mathbf{u})I(\mathbf{v})I^*(\mathbf{u}+\mathbf{v})\rangle \,/\, \langle P(\mathbf{u})P(\mathbf{v})P^*(\mathbf{u}+\mathbf{v})\rangle. \tag{8}$$

Image processing step 4: Derivation of modulus and phase of the object Fourier transform $O(\mathbf{u})$ from the object bispectrum $B_o(\mathbf{u},\mathbf{v})$. We denote the phase of the Fourier transform of the object by $\varphi(\mathbf{u})$ and the phase of the bispectrum of the object by $\beta(\mathbf{u},\mathbf{v})$, i.e.,

$$O(\mathbf{u}) = |O(\mathbf{u})| \exp[i\varphi(\mathbf{u})] \text{ and} \tag{9}$$

$$B_o(\mathbf{u},\mathbf{v}) = |B_o(\mathbf{u},\mathbf{v})| \exp[i\beta(\mathbf{u},\mathbf{v})]. \tag{10}$$

Inserting Eqs. (9) and (10) into Eq. (7) yields

$$B_o(\mathbf{u},\mathbf{v}) = |B_o(\mathbf{u},\mathbf{v})| \exp[i\beta(\mathbf{u},\mathbf{v})]$$

$$= |O(\mathbf{u})| \exp[i\varphi(\mathbf{u})] |O(\mathbf{v})| \exp[i\varphi(\mathbf{v})] |O(\mathbf{u}+\mathbf{v})| \exp[-i\varphi(\mathbf{u}+\mathbf{v}) \tag{11}$$

$$\rightarrow \exp[i\beta(\mathbf{u},\mathbf{v})] = \exp[i\varphi(\mathbf{u})] \exp[i\varphi(\mathbf{v})] \exp[-i\varphi(\mathbf{u}+\mathbf{v})], \tag{12}$$

$$\rightarrow \beta(\mathbf{u},\mathbf{v}) = \varphi(\mathbf{u}) + \varphi(\mathbf{v}) - \varphi(\mathbf{u}+\mathbf{v}), \tag{13}$$

$$\rightarrow \varphi(\mathbf{u}+\mathbf{v}) = \varphi(\mathbf{u}) + \varphi(\mathbf{v}) - \beta(\mathbf{u},\mathbf{v}). \tag{14}$$

Equation (14) is a recursive equation for calculating the phase of the object Fourier transform at coordinate $\mathbf{w} = \mathbf{u} + \mathbf{v}$ if the phase of the object Fourier transform is known at coordinates $\mathbf{u}$ and $\mathbf{v}$. The phase $\beta(\mathbf{u},\mathbf{v})$ is known from Eq. (8). Equation (13) is called a *closure phase relation.* In other words, we have shown that speckle masking can reconstruct closure phases. Closure phases play a very important role in radio interferometry [9, 10] and optical long-baseline interferometry [11]. They can overcome aberrations of interferometers and telescopes [5]. Therefore, speckle masking can be applied both to large single-dish telescopes and optical long-baseline interferometers. In other words, image reconstruction is possible even if the optical transfer function (uv coverage) consists of isolated dots with large gaps between the dots (as in the case of the ESO

Very Large Telescope). In this case, the Knox-Thompson method cannot be applied.

For the recursive calculation of the phase $\varphi(\mathbf{w}) \equiv \varphi(\mathbf{u}+\mathbf{v})$ of the Fourier transform of the object, we need, in addition to the bispectrum phase $\beta(\mathbf{u},\mathbf{v})$, the starting values $\varphi(0,0)$, $\varphi(0,1)$, and $\varphi(1,0)$. (We assume now that $O(\mathbf{u})$ is two-dimensional.) Since $o(\mathbf{x})$ is real, $O(\mathbf{u})$ is hermitian. Therefore, $O(\mathbf{u}) = O^*(-\mathbf{u})$, $O(0,0) = O^*(0,0)$, and, therefore, $\varphi(0,0) = 0$. $\varphi(0,1)$ and $\varphi(1,0)$ can be set to zero, since we are not interested in the absolute position of the reconstructed image. With these starting values, we obtain, for example:

$$\begin{aligned}
\varphi(0,2) &= \varphi(0,1) + \varphi(0,1) - \beta[(0,1),(0,1)],\\
\varphi(0,3) &= \varphi(0,2) + \varphi(0,1) - \beta[(0,2),(0,1)],\\
\varphi(0,4) &= \varphi(0,3) + \varphi(0,1) - \beta[(0,3),(0,1)],\\
&\dots\\
\varphi(2,0) &= \varphi(1,0) + \varphi(1,0) - \beta[(1,0),(1,0)],\\
\varphi(3,0) &= \varphi(2,0) + \varphi(1,0) - \beta[(2,0),(1,0)],\\
\varphi(4,0) &= \varphi(3,0) + \varphi(1,0) - \beta[(3,0),(1,0)],\\
&\dots\\
\varphi(1,1) &= \varphi(1,0) + \varphi(0,1) - \beta[(1,0),(0,1)],\\
\varphi(2,1) &= \varphi(2,0) + \varphi(0,1) - \beta[(2,0),(0,1)],\\
\varphi(3,1) &= \varphi(3,0) + \varphi(0,1) - \beta[(3,0),(0,1)],
\end{aligned} \tag{15}$$

The advantage of this recursive phase calculation is the fact that for each element of the object Fourier phase $\varphi(\mathbf{w}) = \varphi(\mathbf{u}+\mathbf{v})$, there are many different recursion paths, and that it is possible to average over all $\varphi(\mathbf{w})$-values to improve the signal-to-noise ratio. For example, for the element $\varphi(3,2)$ there are eight recursion paths; for $\varphi(6,4)$, there are 64 paths. Averaging over all paths yields

$$\varphi(\mathbf{w}) \equiv \varphi(\mathbf{u}+\mathbf{v}) = (1/N) \sum_{0\langle \mathbf{u}\cdot\mathbf{w}/|\mathbf{w}| \leq |\mathbf{w}|/2} \varphi(\mathbf{u}) + \varphi(\mathbf{w}-\mathbf{u}) - \beta(\mathbf{u},\mathbf{w}-\mathbf{u}), \tag{16}$$

where N is the number of recursion paths. In actual applications, the phase calculation is performed with complex exponential functions:

$$\exp[i\varphi(\mathbf{w})] = (1/N) \sum_{0\langle \mathbf{u}\cdot\mathbf{w}/|\mathbf{w}| \leq |\mathbf{w}|/2} \times \exp[i\varphi(\mathbf{u})] \exp[i\varphi(\mathbf{w}-\mathbf{u})] \exp[-i\beta(\mathbf{u},\mathbf{w}-\mathbf{u})]. \tag{17}$$

Not all recursion paths for the same $\varphi(\mathbf{w})$-value yield the same signal-to-noise ratio. Therefore, different weight functions have to be chosen for different paths. Another possibility is to use only a few short **v**-vectors instead of all [6].

The modulus of the object Fourier transform can be derived from the object bispectrum in two different ways. From Eq. (7) follows for $\mathbf{u} = 0$:

$$B_o(0, \mathbf{v}) = O(0)O(\mathbf{v})O^*(0 + \mathbf{v}) = \text{const. } |O(\mathbf{v})|^2. \tag{18}$$

The second way is the *recursive* calculation. From Eq. (11) follows:

$$\begin{aligned} &|B_o(\mathbf{u}, \mathbf{v})| = |O(\mathbf{u})|\,|O(\mathbf{v})|\,|O(\mathbf{u} + \mathbf{v})| \text{ or} \\ &|O(\mathbf{u} + \mathbf{v})| \equiv |O(\mathbf{w})| = |B_o(\mathbf{u}, \mathbf{v})| \,/\, [|O(\mathbf{u})|\,|O(\mathbf{v})|] \text{ for } \mathbf{u}, \mathbf{v} \text{ with} \\ &|O(\mathbf{u})|\,|O(\mathbf{v})| \neq 0. \end{aligned} \tag{19}$$

Modifications of speckle masking are photon-counting triple correlation processing [7] and tomographic speckle masking [7]. A pupil plane bispectrum method is triple-shearing interferometry [12]. A simple special case, where speckle masking has been applied, is the case where there is no atmospheric image degradation but only *random image motion*. The general case is that all the following types of image degradation exist simultaneously: atmospheric turbulence (speckle noise), image motion (e.g., tracking errors of the telescope in addition to atmospheric image motion of the speckle interferograms), stationary or slowly changing telescope aberrations, photon noise, and other types of noise (e.g., dark current).

Figures 1 to 4 show speckle-masking observations of the spectroscopic double star Psi SGR [4], Eta Carinae [13], the central object in NGC 3603 [14], and the Seyfert galaxy NGC 1068 [15].

3. Objective-Prism Speckle Spectroscopy

Imaging spectroscopy plays an important role in astronomy. We have developed two speckle methods that can reconstruct both a high-resolution image and the spectrum of each resolution element of the object. The two methods are called *objective-prism speckle spectroscopy* [16, 17] and *projection speckle spectroscopy* [18, 19]. The raw data for objective prism spectroscopy are objective prism speckle spectrograms $i_s(\mathbf{x})$, which are obtained by inserting a prism or grating into a pupil plane in the speckle camera. Then each speckle is dispersed in a linear spectrum. The instantaneous intensity distribution $i_s(\mathbf{x})$ of an objective prism speckle spectrogram can be described by

$$i_s(\mathbf{x}) = \sum_m o_m(\mathbf{x} - \mathbf{x}_m) * s_m(\mathbf{x}) * p(\mathbf{x}), \tag{20}$$

where $o_m(\mathbf{x} - \mathbf{x}_m)$ denotes the mth resolution element of the object, $\mathbf{x}$ is a two-dimensional vector in object space, $s_m(\mathbf{x})$ is the spectrum of the

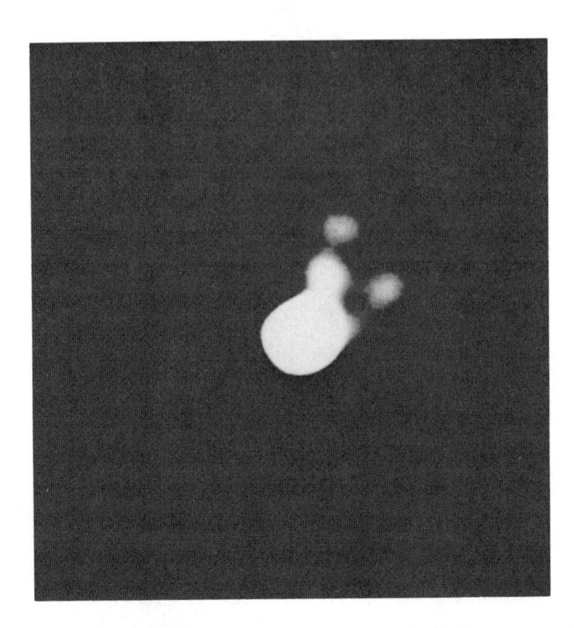

FIG. 2. High-resolution image of the variable star η Carinae reconstructed from 300 speckle interferograms by speckle masking. The diffraction-limited image shows that η Carinae consists of a dominant star and three close objects at separations of 0.11 arcsec, 0.18 arcsec and 0.21 arcsec (From [13]).

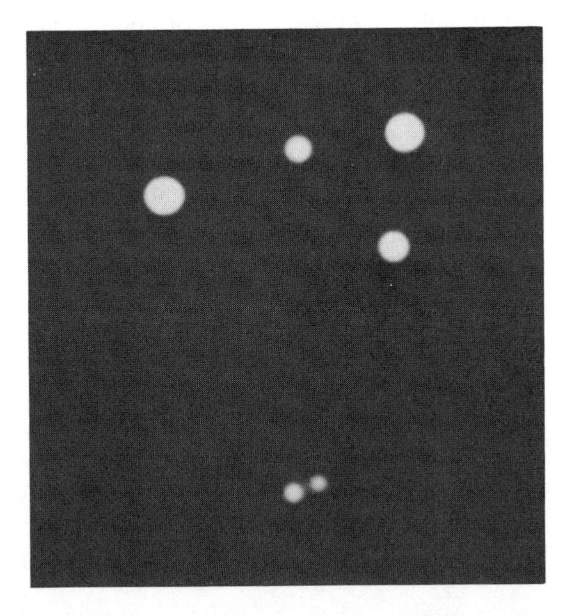

FIG. 3. Diffraction-limited image of the central object in the HII region NGC 3603 reconstructed from 300 speckle interferograms by speckle masking. The image shows that the object is a star cluster consisting of six stars. The stars have astronomical magnitudes (brightness) in the range of 12 to 15. The separation of the closest pair is $\sim$ 0.09 arcsec (From [14]).

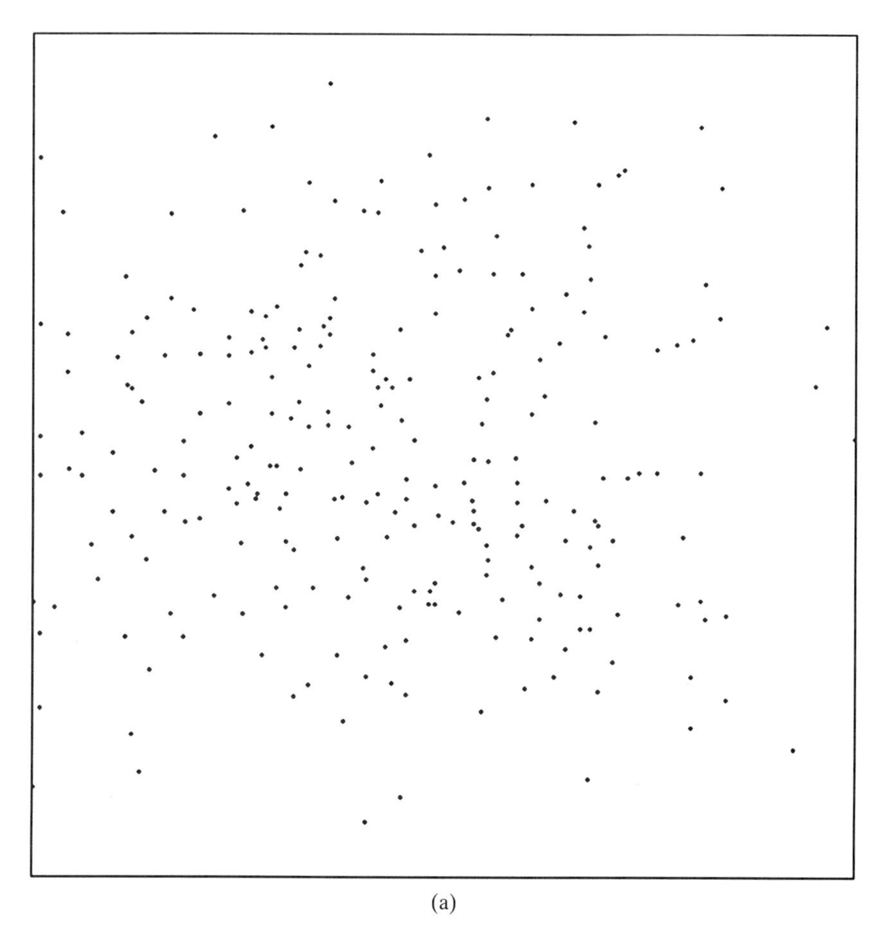

(a)

FIG. 4. Speckle-masking observation of the Seyfert galaxy NGC 1068: (a) One of 10,000 photon-counting speckle interferograms recorded with the 1.5 m Danish/ESO telescope. Each black dot is a photon event. (b) Long-exposure image of NGC 1068 calculated by averaging all 10,000 re-centered speckle interferograms. (c) High-resolution image of NGC 1068 reconstructed from the same 10,000 speckle interferograms by speckle masking. The image shows that NGC 1068 consists of one dominant cloud and two clouds at a separation of about 0.5 arcsec (From [15]).

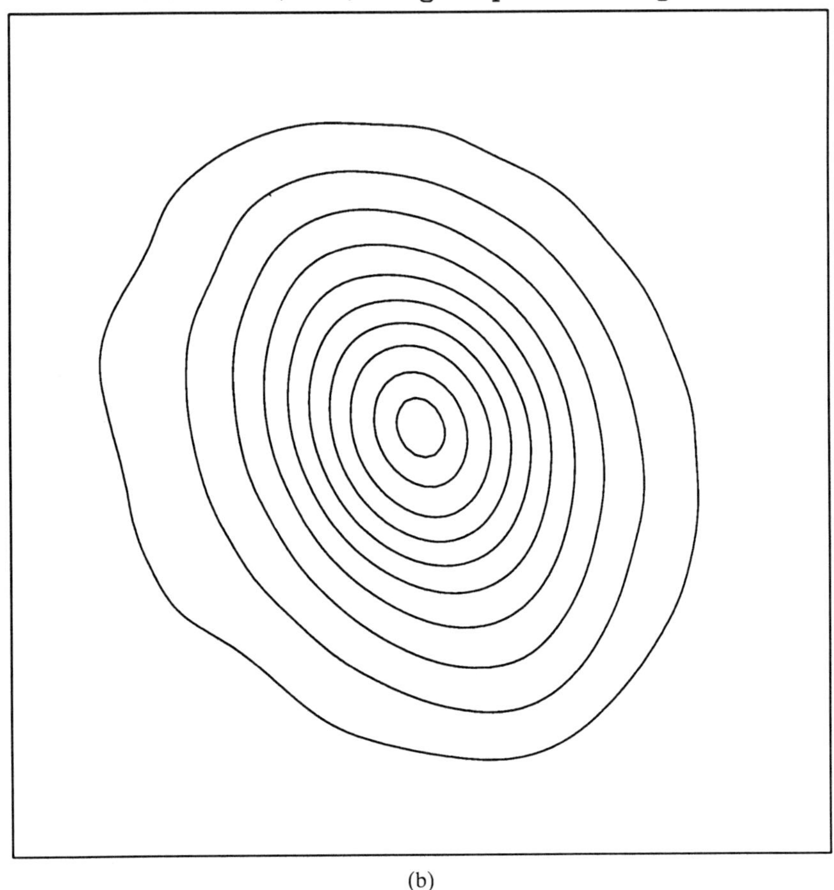

(b)

FIG. 4 (*Continued*)

mth resolution element, $p(\mathbf{x})$ is the instantaneous point-spread function of the atmosphere/telescope, and $*$ is the convolution operator. $p(\mathbf{x})$ is wavelength-independent in small (or large) wavelength bands, as discussed in [18] in more detail. From a sequence of speckle spectrograms $i_s(\mathbf{x})$, the desired high-resolution objective prism spectrum.

$$\sum_m o_m(\mathbf{x} - \mathbf{x}_m) * s_m(\mathbf{x}) \tag{21}$$

can be reconstructed by speckle masking. A laboratory simulation of objective-prism speckle spectroscopy is reported in [17].

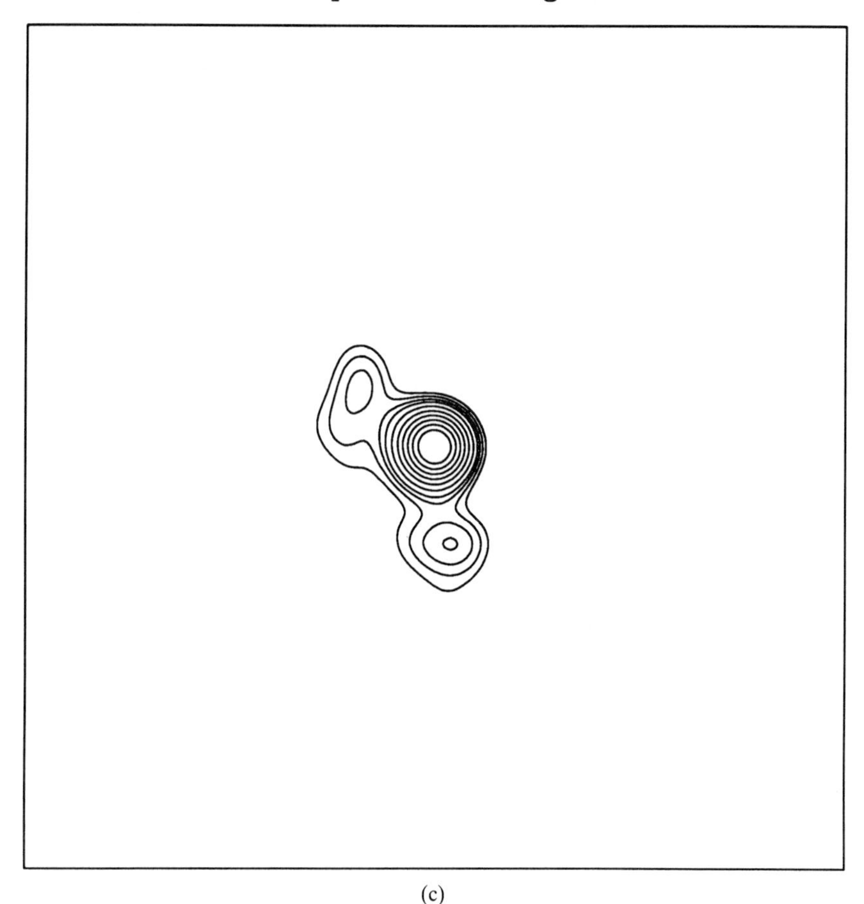

(c)

FIG. 4 (*Continued*)

4. Wideband Projection Speckle Spectroscopy

Projection speckle spectroscopy has the two additional advantages that (1) it can be applied to general objects and (2) the whole spectrum from 350nm to 850nm can be obtained simultaneously [18, 19]. The principle of projection speckle spectroscopy is summarized in Fig. 5. The figure shows from top to bottom:

Image 1: Two-dimensional object, a triple star.
Image 2: Two-dimensional speckle interferogram of the object.

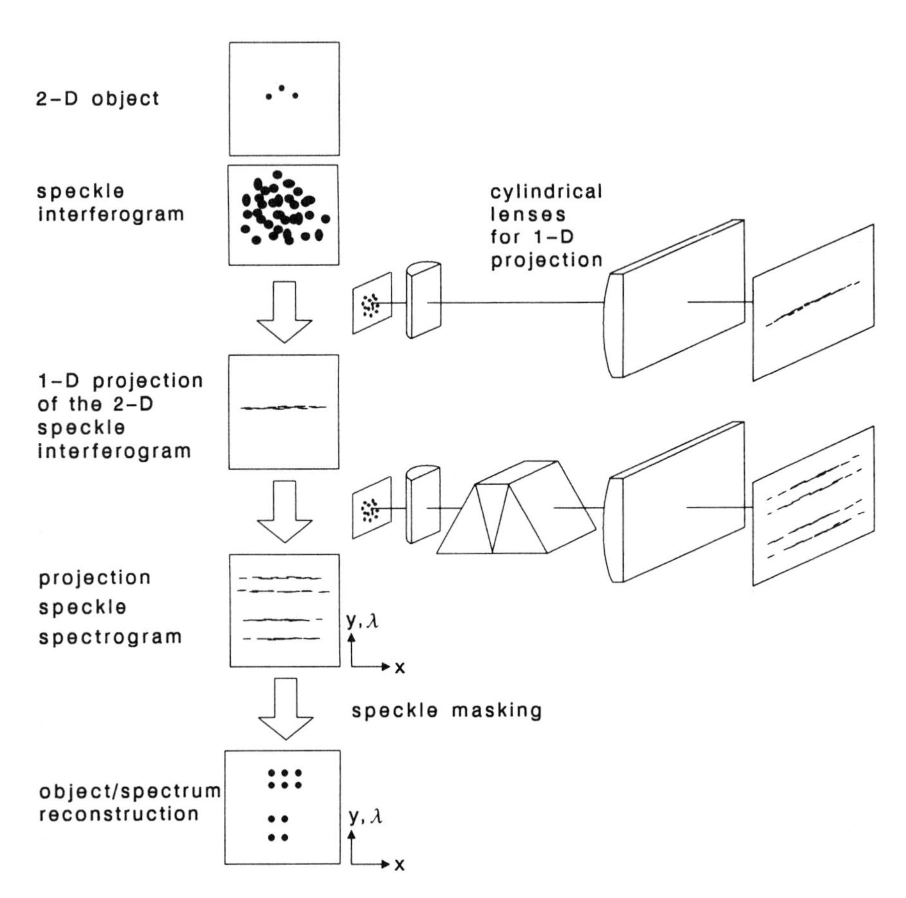

FIG. 5. Principle of projection speckle spectroscopy.

Image 3: One-dimensional projection of the two-dimensional speckle interferogram. The projection can be performed by an anamorphic imaging system of two crossed cylindrical lenses. The 1-D projection of the 2-D speckle interferogram is equal to the convolution of the 1-D projection of the object intensity distribution and the 1-D projection of the atmospheric point-spread function [18].

Image 4: Spectrally dispersed image $d(x, \lambda)$ of the one-dimensional speckle interferogram. The spectral dispersion can be performed, for example, by a non-deviating prism. The spectrograms $d(x, \lambda)$ are the raw data for projection speckle spectroscopy.

Image 5: From the spectrograms $d(x, \lambda)$, the object/spectrum reconstruction $o'(x, \lambda)$ can be reconstructed by one-dimensional speckle-masking processing of all individual lines of $d(x, \lambda)$. The object/spectrum reconstruction $o'(x, \lambda)$ is a dispersed version of a one-dimensional projection of the two-dimensional object. High-resolution spatial object information is contained in the horizontal x-direction. The spectral information is found in the y-direction. Three-dimensional data cubes $o''(x, y, \lambda)$ can be obtained if many two-dimensional object/spectrum reconstructions are made using different projection and dispersion directions, and if tomographic techniques are applied. A laboratory simulation of the projection speckle spectroscopy method is described in [18]. A first astronomical application is reported in [19].

5. Optical Long-Baseline Interferometry and Aperture Synthesis

The great advantage of *optical long-baseline interferometry* is the fact that it can yield images and spectra with fantastic angular resolution. For example, at $\lambda \sim 600$nm and with a baseline of 150m, a resolution of 0.001 arcsec can be obtained. The largest planned optical interferometer is the ESO Very Large Telescope Interferometer (VLTI). Construction started in 1990. It will consist of four 8m telescopes and at least two 2m telescopes. The longest baseline will have a length of about 75 to 150m. The VLTI interferograms will consist of many *fringed speckles* [20]. The size of each speckle will be about 0.01 arcsec, the width of each fringe about 0.001 arcsec. Image reconstruction from VLTI interferograms will be a great challenge, since the instantaneous optical transfer function (uv coverage) will consist of a few isolated dots only.

A complete computer simulation of optical long-baseline interferometry with the VLTI is described in [20]. In this computer experiment, an interferometer consisting of four 8m telescopes and a distance of 25m between the telescopes was simulated. The total number of calculated long-baseline speckle interferograms was 34,000. The simulation included a model with many turbulence cells in front of each telescope. Therefore, the interferograms consist of fringed speckles. Photon noise corresponding to a mean count number of 6,000 photon events per interferogram was simulated. Finally, the simulation involved interferograms recorded during 12 hours observing time (aperture synthesis by earth rotation). The experimental results in [20] show that it was possible to reconstruct a diffraction-limited image of the object (triple star with ~3 milli-arcsec separation) from the interferograms by speckle masking and the image-processing method CLEAN.

This computer simulation of the ESO VLTI and other computer simulations (see references in [20]) show that speckle masking can be applied to optical long-baseline interferometers in spite of the fact that the instantaneous optical transfer function consists of a few isolated dots with large gaps between the dots. The Knox-Thompson method [2] cannot be applied in this case. The application of the radio phase closure method [9, 10] is difficult because of the large number of turbulence cells in front of each telescope of the VLTI (for Fried parameter $r_o = 16$ cm: $\sim 50 \times 50 = 2{,}500$ turbulence cells; *multi-speckle case*). For the application of the radio phase closure method, so-called non-redundant masks in front of each 8m telescope (see references in [10]) or similar techniques would be required, which would absorb a large fraction of the light, as discussed in [20] in more detail.

Acknowledgment

We thank ESO for observing time. The results shown in Figs. 1 to 4 are based on data collected at the European Southern Observatory, La Silla, Chile.

References

1. A. Labeyrie, *Astron. Astrophys.* **6**, 85 (1970).
2. K. T. Knox and B. J. Thompson, *Astrophys. J. Lett.* **193**, L 45 (1974).
3. G. Weigelt, *Optics Commun.* **21**, 55 (1977).
4. G. Weigelt and B. Wirnitzer, *Optics Lett.* **8**, 389 (1983).
5. A. W. Lohmann, G. Weigelt, and B. Wirnitzer, *Appl. Opt.* **22**, 4028 (1983).
6. K. H. Hofmann and G. Weigelt, *Astron. Astrophys.* **167**, L 15 (1986).
7. D. Schertl, F. Fleischmann, K. H. Hofmann, and G. Weigelt, *Soc. Photo-Opt. Instr. Eng.* **808**, 38 (1987).
8. B. Wirnitzer, *J. Opt. Soc. Am. A* **2**, 14 (1985).
9. R. C. Jennison, *Mon. Not. Roy. Astr. Soc.* **118**, 276 (1958).
10. T. J. Cornwell, *Science* **245**, 263 (1989).
11. T. Reinheimer and G. Weigelt, *Astron. Astrophys.* **176**, L 17 (1987).
12. K. H. Hofmann and G. Weigelt, *Appl. Opt.* **25**, 4280 (1986).
13. K. H. Hofmann and G. Weigelt, *Astron. Astrophys.* **203**, L 21 (1988).
14. G. Baier, J. Eckert, K. H. Hofmann, W. Mauder, D. Schertl, H. Weghorn, and G. Weigelt, *The Messenger* (European Southern Observatory) **52**, 11 (1988).
15. K. H. Hofmann, W. Mauder, and G. Weigelt, "Photon-Counting Speckle Masking," in *Proc. of the 15th Congress of the International Commission for Optics,* SPIE Proc. Vol. 1319 (1990), p. 442.
16. G. Weigelt, "Speckle Interferometry, Speckle Holography, Speckle Spectroscopy, and Reconstruction of High-Resolution Images from HST Data," in "Proc. Scientific Importance of High-Angular Resolution at Infrared and Optical Wavelengths" (M. H. Ulrich and K. Kjär, eds.), p. 95 ESO, Garching, Germany, 1981.
17. G. Weigelt, G. Baier, J. Eberberger, F. Fleischmann, K. H. Hofmann, and R. Ladebeck, *Opt. Engineering* **25**, 706 (1986).

18. F. Grieger, F. Fleischmann, and G. Weigelt, "Objective Prism Speckle Spectroscopy and Wideband Projection Speckle Spectroscopy," in "Proc. High-Resolution Imaging by Interferometry," (F. Merkle, ed.), p. 225 ESO, Garching, 1988.
19. F. Grieger and G. Weigelt, "Projection Speckle Spectroscopy and Objective Prism Speckle Spectroscopy," in *Proc. of the 15th Congress of the International Commission for Optics*, SPIE Proc. Vol. 1319 (1990), p. 440.
20. T. Reinheimer and G. Weigelt, "Optical Long-Baseline Interferometry in Astronomy," in *Proc. of the 15th Congress of the International Commission for Optics*, SPIE Proc. Vol. 1319 (1990), p. 678.

CHAPTER 28

Phase-Retrieval Imaging Problems

J. R. Fienup
Optical Science Laboratory, Advanced Concepts Division
Environmental Research Institute of Michigan
Ann Arbor, Michigan

1. Introduction

In many areas of physics and engineering, one is given the modulus $|F(u,v)|$ of the Fourier transform of an object, $f(x,y)$, where

$$F(u,v) = |F(u,v)|e^{i\Psi(u,v)} = \int_{-\infty}^{\infty} f(x,y)e^{-i2\pi(ux+vy)}dx\,dy, \tag{1}$$

and one wishes to retrieve the Fourier phase $\Psi(u,v)$ or, equivalently, reconstruct $f(x,y)$. This also is equivalent to reconstructing the object from its autocorrelation function, which can be computed easily by inverse Fourier transforming the square of the Fourier modulus. As will be explained shortly, phase retrieval is particularly valuable because it enables one to obtain fine-resolution images with imaging systems of poor quality or that do not make the measurements one ordinarily must make to form an image.

There is one phase retrieval problem—that of finding $\Psi(u,v)$ given the moduli, $|F(u,v)|$ and $|f(x,y)|$, in both domains, as is found in electron microscopy and some wavefront sensing problems—that is considered to be fairly well solved by the Gerchberg-Saxton algorithm [1], and will not be covered here. Rather, we will explore the more difficult problem of reconstruction when the only available data is $|F(u,v)|$ and one has, in addition, some constraints on $f(x,y)$.

ISBN 0-12-289690-4

The phase can be retrieved from the modulus only if one has sufficiently powerful constraints on the image. Two constraints are commonly available. The first is a nonnegativity constraint, which is true for the case of incoherent imaging by interferometry. The second is a support constraint (i.e., the object is known to be zero outside some area), which is true for both incoherent and coherent imaging problems for the case of bright objects on dark backgrounds. Such cases include, for example, imaging of space objects (earth-orbiting satellites or astronomical objects) or imaging when one has the ability to illuminate only a desired region (e.g., laser illumination of the ground from an aircraft). The support constraint can be obtained from prior knowledge or estimated from the autocorrelation function [2].

In this chapter we review several applications of phase retrieval to imaging, discuss the most promising phase retrieval algorithms, and comment on the state of knowledge of the uniqueness of the solution to the phase retrieval problem. We will point out what the most difficult outstanding phase retrieval problem is, and what difficult associated problems must also be solved. The purpose here is not to present a comprehensive review, but to bring forth some of the most interesting recent developments and to suggest areas of research which deserve further effort.

2. Imaging Applications of Phase Retrieval

To obtain fine-resolution imagery, the limit of diffraction forces one to employ short wavelengths and large apertures in an imaging system. For earthbound observations, the turbulence of the atmosphere typically limits resolution to about one second of arc, the resolving power of a telescope of diameter 0.1m at optical wavelengths. That is, existing 4-meter ground-based telescopes potentially could achieve 40 times better resolution than the atmosphere presently permits. Large, lightweight telescopes that could be deployed economically in space will suffer from severe aberrations due to a warping of the primary mirror of the telescope or from a misalignment of the segments of a segmented telescope. Phase retrieval can allow one to obtain fine-resolution images despite these aberrations. Here, we consider three major types of imaging: *passive incoherent imaging*, *active coherent imaging*, and *imaging correlography*, a hybrid of the two. As will be seen, active coherent imaging poses the most difficult phase-retrieval problem.

2.1. Passive Incoherent Imaging

First, consider the case of passive imaging of an incoherently illuminated or self-emissive object. An alternative to a conventional telescope that gathers images directly in the focal plane is *aperture-plane* (*Michelson stellar) interferometry* [3], in which the coherence function of the optical field is measured in the aperture plane of the telescope. For an incoherent object, the coherence function is the Fourier transform $F_I(u, v)$ of the incoherent image $f_I(x, y)$ of the object, by virtue of the van Cittert-Zernike theorem. The aberrations cause the phase $\Psi_I(u, v)$ of the complex-valued $F_I(u, v) = |F_I(u, v)| \exp[i\Psi_I(u, v)]$ to be lost. Then one needs to retrieve $\Psi_I(u, v)$ from the modulus $|F_I(u, v)|$ to reconstruct the image by inverse Fourier transformation. Phase-retrieval algorithms have been shown to be fairly robust and reliable for this case, in which one has both a support constraint and a nonnegativity constraint.

If one has the opportunity to view the object through many different realizations of atmospheric turbulence, then another option is to gather many different short-exposure blurred images of the object in the conventional focal plane of the telescope. From these many blurred images, one can arrive at $|F_I(u, v)|$ using Labeyrie's technique [4] and then use phase retrieval to reconstruct an image. Alternatively, there are many additional reconstruction approaches that use the collection of blurred images to compute an image more directly. The reader is referred to Section 7.5A of [5] for a discussion of them. In particular, the preferred approach seems to be to use extended Knox-Thompson or triple-correlation (bispectrum) [6]. The phase retrieval algorithms that will be described later still can play a valuable role in this case as well. The Knox-Thompson and bispectrum reconstruction algorithms do not make use of the available constraints of support and nonnegativity when computing an image. The phase-retrieval algorithms, which could start close to the correct solution by using this image as an initial estimate, should produce an improved image by forcing the solution to be consistent with the constraints.

2.2. Active Coherent Imaging

Next, consider the case of active, coherent imaging. This is mostly of interest for imaging satellites. A lightweight, inexpensive type of imaging sensor is possible for the case of the object being illuminated by a coherent laser. For coherently illuminated objects, the laws of optical propagation tell us that the complex-valued optical field $F(u, v)$ in the plane of the aperture is related to the complex-valued optical field $f(x, y)$, reflected from the object, by a Fourier transform (or by a Fresnel transform; but the approach is essentially the same in both cases, so for simplicity, we will consider only the Fourier transform). Unfortunately, optical fields are very difficult to measure by heterodyning techniques. Furthermore, detection of the optical field would suffer from severe phase errors unless no atmosphere is present and the positions of the detectors are known to within a small fraction of a wavelength. A much easier measurement to make is of the intensity of the optical field in the aperture plane. The square root of this intensity measurement is $|F(u, v)|$, the Fourier modulus, which is not affected by phase errors. Then, similar to the case of interferometry, one needs to retrieve the phase $\Psi(u, v)$ from the modulus $|F(u, v)|$ of $F(u, v)$ to reconstruct a fine-resolution image.

In this case, however, the reconstruction is much more difficult. For a rough, coherently illuminated object, $f(x, y)$ is complex-valued and can have a random phase; consequently, the nonnegativity constraint, which has proven to be so powerful for the reconstruction of incoherent images, is not available. Nevertheless, if the support of the object is well known *a priori* and is of a favorable type (polygonal with no parallel sides or having separated parts), then robust, reliable phase retrieval is possible [7, 8]. However, for many applications, the support of the object is not known *a priori*, and one has available only a coarse upper bound on the support, derived from the Fourier modulus data [2]. This phase-retrieval problem is so difficult that current algorithms are adequate only in special cases.

There are two special cases for which phase retrieval for complex-valued objects without *a priori* known support is relatively easy. The first is when the object has well-separated parts. Then the support-reconstruction methods work particularly well [2], and phase retrieval with such support constraints usually is successfully [7]. The second is when the object has a bright glint (or a small number of bright glints). These glints do not have to satisfy the holographic separation condition. Then a combination of three phase-retrieval algorithms that make use of the glints has been found to be successful. The first algorithm is the triple-autocorrelation product method [9], which finds the positions and complex values of the glints. Then the Baldwin-Warner algorithm is used to reconstruct much of the rest of the object [10]. Finally, the iterative Fourier transform algorithm is used to complete the reconstruction [7].

Reconstruction of other types of complex-valued objects from their Fourier modulus (without the support known *a priori*) is much more difficult. It is particularly difficult if the object has a convex support (e.g., a circle) because in that case, the upper bound on the support (i.e., the support constraint) computed from the support of the autocorrelation function can be considerably larger than the object [9]. An object with a centro-symmetric support also is more difficult to reconstruct because it encourages the problem of stagnation of the iterative transform algorithm at an image having components of both the upright image and its twin (the image rotated by 180°) [11]. Phase retrieval is particularly difficult when the object has tapered edges. By having tapered edges, we mean that the intensity of the illuminated object falls slowly to zero at its edges; that is, the edges are not sharp. Phase retrieval for such objects is difficult even if the support is known *a priori* [8].

One of the most difficult phase-retrieval problems also is one that is of practical interest. If an object is illuminated by a laser that is near the receiving aperture (i.e., monostatic imaging), then the $\cos^2\theta$ fall-off of a Lambertian scattering surface tends to give the object tapered edges. Also, for many objects of interest, the support is convex. There has been no reliable phase-retrieval algorithm published for this situation—a complex-valued object with convex support, not known *a priori*, with tapered edges. This is the case even for noise-free data, and even more so for the case of low light levels for which the Fourier modulus data is noisy.

There is another problem, associated with this difficult phase-retrieval problem, that also is extremely important and has received insufficient attention. Recall from the discussion earlier that it usually is necessary to infer the support of the object from the support of its autocorrelation function, and some ways of doing this have been devised. It is essential that this be done well, since for complex-valued objects, the support constraint is the only constraint we have in the object domain. Then the support constraint is all the leverage we have to retrieve the Fourier phases. However, it is impossible to estimate the object support well unless the support of the autocorrelation is estimated well; and, as will be seen, it can be difficult to estimate accurately the support of the autocorrelation from the given autocorrelation function.

The difficulty in estimating the autocorrelation support is illustrated in Fig. 1, which shows slices through the magnitude of the autocorrelation of a computer-simulated complex-valued image. The object was a square of side 16 pixels (embedded in a field of zeros in a square array of side 64) filled with identically distributed complex Gaussian random numbers. The diffraction-limited image of the object was formed by Fourier-transforming the object, multiplying the 64×64 Fourier transform by a square binary aperture of side 16, and inverse transforming. Then the autocorrelation of the image was computed by inverse transforming the squared modulus of the apertured Fourier transform. The autocorrelation, then, is of width 31 pixels, in the interval 17 to 47 in each of the slices shown in the figure. If the object were incoherent and unspeckled, and if there were no diffraction effects, the slices through the autocorrelation would be simple triangle functions of width 31 pixels. Because of the variability associated with the coherent speckles in the complex-valued autocorrelation in combination with the diffraction sidelobes caused by the aperture, the width of the autocorrelation is difficult to determine. It would be difficult to specify a simple rule, such as thresholding at a certain percentage of the peak, that would accurately and consistently predict the width from the slices through the autocorrelation function. Note that, at the expense of resolution, the sidelobes can be reduced by aperture weighting, making the task easier. However, when photon noise is added to the Fourier data, or if the object

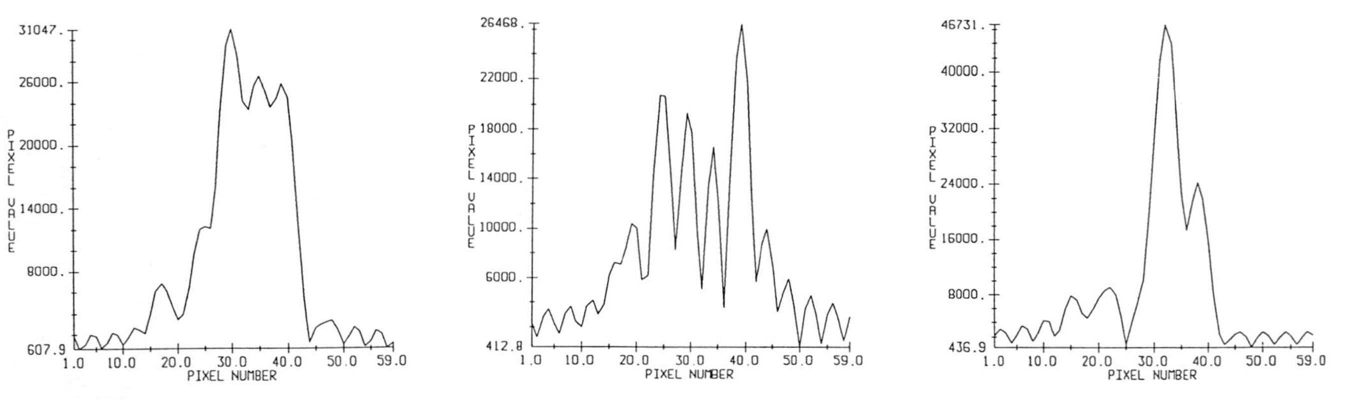

FIG. 1. Horizontal slices through the autocorrelation of a complex-valued square object of width 16 pixels, illustrating the difficulty of estimating the width of the autocorrelation (which is of width 31 pixels).

has tapered edges, then the task becomes more difficult than for the example shown here.

2.3. Imaging Correlography

Yet another imaging possibility for a laser-illuminated object is available if one can collect many realizations of $|F(u, v)|$, as would be available if there is a relative motion between the object and the imaging system (as the speckles of the optical field move across the aperture). Then from the average autocovariance of the speckle intensities $|F(u, v)|^2$, one can estimate $|F_I(u, v)|$ [12]. That is, from many realizations of the intensity of the aperture-plane coherent optical field, one can determine the modulus of the Fourier transform of the incoherent object (the object as though it were illuminated incoherently). This also can be shown to be completely analogous to intensity interferometry [13]. Then one can perform phase retrieval using a nonnegativity constraint as well as a support constraint to reconstruct an incoherent image of the object. This mode of imaging is called imaging correlography. Unfortunately, the signal-to-noise ratio of the estimate of $|F_I(u, v)|$ is low unless a large number (several hundred to several thousand) of independent realizations of the aperture-plane intensity are collected [12].

3. Phase-Retrieval Algorithms

Many algorithms have been proposed for retrieving the Fourier phase, and thereby reconstructing an image, from the Fourier modulus. Sections 7.3 and 7.4 of [5] summarize them. Despite the fact that many other algorithms have been developed after it, the iterative Fourier transform algorithm, first demonstrated in 1974, is still the most widely used approach, since it is relatively fast, robust, and general. In this section, we will describe briefly the iterative transform algorithm and some recent variations that allow it to take advantage of additional information and cause it to converge more reliably. See [14] and Section 7.4 of [5] for a more complete description of the basic algorithm. In addition, we will discuss the use of simulated annealing, maximum entropy, and blind deconvolution algorithms as alternatives.

3.1. The Iterative Transform Algorithm

A block diagram of the *iterative Fourier transform algorithm* is shown in Fig. 2. It is a generalization of—and an improvement on—the Gerchberg-Saxton algorithm [1]. It involves the transformation back and forth between the Fourier domain, where the measured Fourier modulus data is

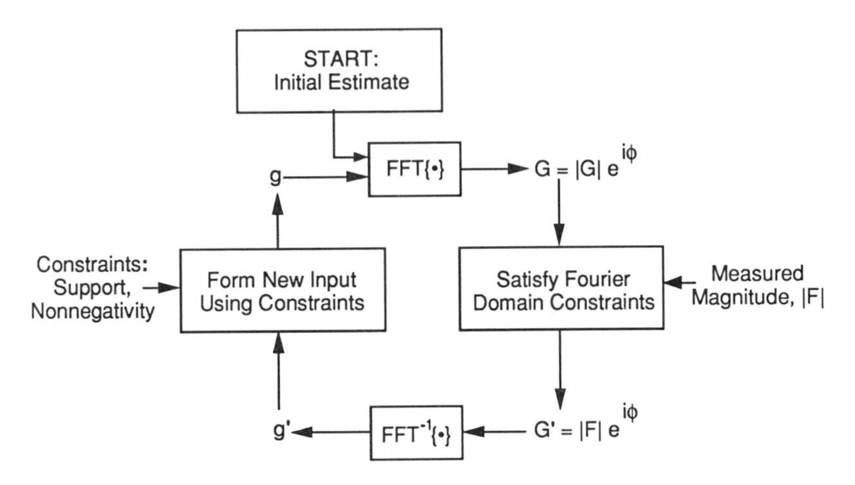

FIG. 2. Block diagram of the iterative Fourier transform algorithm.

enforced, and the image domain, where the nonnegativity and/or support constraints are used. Iterations are continued until a solution is found that agrees (within the limits allowed by noise and diffraction) simultaneously with both the Fourier modulus data and the constraints. Several different versions of the algorithm have been developed for different types of data that might be available [14, 15, 16] and in order to speed convergence [11, 14]. For most versions of the algorithm, the Fourier-domain operation is to replace the modulus of the Fourier transform of the input image by the measured Fourier modulus. If additional information in the Fourier domain, such as partial phase information, is known, then that information is enforced as well. The object-domain operation typically is what is varied to arrive at different versions of the algorithm [14]. The hybrid input-output version of the algorithm has proven to be the most useful. It employs the image-domain operation

$$g_{k+1}(x,y) = \begin{cases} g'_k(x,y), & (x,y) \notin \gamma \\ g_k(x,y) - \beta g'_k(x,y), & (x,y) \in \gamma \end{cases}, \tag{2}$$

where $g_k(x,y)$ is the input image at the kth iteration, $g'_k(x,y)$ is the output image from the inverse Fourier transform, and γ is the set of points for which $g'_k(x,y)$ violates the object-domain constraints. The hybrid input-output algorithm converges much more rapidly and avoids getting trapped in local minima as compared with the simplest form of the algorithm, the error-reduction algorithm, for which the object-domain operation is to satisfy the object-domain constraint (which is a projection-onto-sets algorithm).

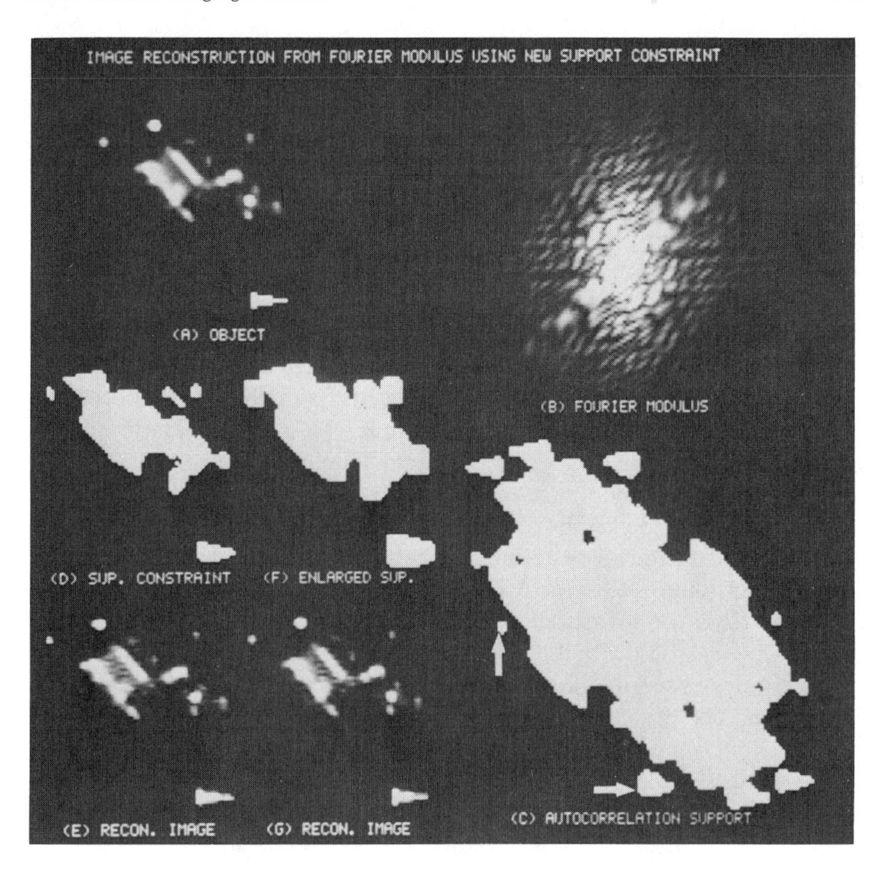

FIG. 3. Phase-retrieval example: (A) Object, consisting of two separated parts; (B) Fourier modulus; (C) thresholded autocorrelation (estimated autocorrelation support) computed from the Fourier modulus; (D) support constraint computed by intersecting three translates of the autocorrelation support (centered at the origin and at the two points indicated by the arrows in (C) [2]); (E) image reconstructed from the Fourier modulus with 10 iterations of the iterative transform algorithm using the support constraint shown in (D), and starting with the support constraint filled with random numbers; (F) support constraint enlarged to ensure that the image fits within it; (G) image reconstructed from 10 more iterations using the enlarged support constraint.

Figure 3 shows an example of support estimation by triple intersection of the autocorrelation support [2] and image reconstruction from the Fourier modulus by the iterative transform algorithm. This is a particularly easy case, since the object is real-valued and nonnegative, and since it consists of two separated parts. This reconstruction produced a good image after fewer than a dozen iterations. For most nonnegative objects, a few tens to a few hundreds of iterations are required more typically. For complex-valued objects, a few hundred to thousands of iterations may be required.

Next, we mention two modifications to the iterative transform algorithm that take advantage of additional phase information if it is available. The first uses a noisy phase estimate over the entire Fourier aperture. The second uses a good phase estimate that is known over a small part of the Fourier aperture.

3.1.1. The Phase-Variance Algorithm

One recent modification to the iterative transform algorithm is to allow a noisy estimate of the Fourier phase to be used. Examples of the source of a noisy phase estimate include, for example, (1) a Knox-Thompson or bispectrum reconstruction of an incoherent image when one has multiple blurred focal-plane images or (2) coherent aperture-plane data taken by an array of imperfectly phased heterodyne detectors. If the Fourier modulus is measured more accurately than the phase, then one would want to force the solution to agree with the modulus data but to be able to wander away from the measured phase values, but not wander too far. This can be accomplished with the phase variance algorithm [17]. The algorithm can take any of several different forms. One that has been shown to be useful is to replace the computed Fourier phase $\phi(u, v)$ by

$$\phi'(u,v) = \begin{cases} \phi_{\text{est}}(u,v) - c\sigma, & [\phi(u,v) - \phi_{\text{est}}(u,v)]_{\text{mod}\,2\pi} < -c\sigma \\ \phi(u,v), & -c\sigma \leq [\phi(u,v) - \phi_{\text{est}}(u,v)]_{\text{mod}\,2\pi} \leq c\sigma, \\ \phi_{\text{est}}(u,v) + c\sigma, & [\phi(u,v) - \phi_{\text{est}}(u,v)]_{\text{mod}\,2\pi} > c\sigma \end{cases} \tag{3}$$

where $\phi_{\text{est}}(u, v)$ is the noisy estimate of the Fourier phase, σ is the estimated standard deviation of $\phi_{\text{est}}(u, v)$, c is a constant (on the order of unity), and the subscript mod2π denotes modulo 2π in the range $(-\pi, \pi)$. In Eq. (3), the new phase is given by ϕ, the computed phase, if it is within $c\sigma$ of ϕ_{est}, but the new phase is not allowed to deviate from ϕ_{est} by more than $c\sigma$. It has been found to be advantageous, after performing a couple of cycles of iterations with the phase-variance algorithm (which approximately reinforces the noisy phase estimate), to eliminate the phase constraint and continue on with the usual phase-retrieval algorithm (reinforcing only the Fourier modulus).

If there is a situation in which the Fourier phase is measured more accurately than the Fourier modulus, then it would be advantageous to enforce the Fourier phase exactly and allow the Fourier modulus to wander. Alternatively, both the Fourier phase and Fourier modulus can be allowed to wander according to the standard deviation of the error of each.

3.1.2. The Expanding Weighted Modulus Algorithm

A second possible situation is to know the Fourier phase over a small part of a larger aperture, while the Fourier modulus is known over the entire aperture [16]. This could be the case, for example, in coherent imaging if one gathers the Fourier modulus data with an array of light-bucket detectors and has a small diffraction-limited imaging system embedded in the array. Intensity measurements would be made of the low-resolution, diffraction-limited image in the back focal plane of the lens and of an image of the aperture plane. Then the phase over the small diffraction-limited aperture can be retrieved using the Gerchberg-Saxton algorithm. A support constraint can be estimated either from the low-resolution, diffraction-limited image or from the autocorrelation function computed from the intensity measurements over the entire aperture. All this information can be combined by the iterative transform algorithm. The phase over the small aperture is utilized most effectively if the Fourier modulus is multiplied initially by a weighting function that goes to zero over an area only slightly larger than the area over which the phase is known. At this stage, only a thin annulus of additional phase information is retrieved, which is accomplished easily. Then the weighting function is expanded in area, and further iterations are performed. This is continued until the entire Fourier phase is retrieved. This approach has been shown to be very effective for reconstructing complex-valued images that otherwise are very difficult to reconstruct [16].

3.2. Simulated Annealing

A phase-retrieval algorithm developed recently that deserves attention is *simulated annealing* [18]. It involves computing an error metric (how well the autocorrelation or Fourier modulus of a guess agrees with the measured data), then changing the value of one pixel in the image and recomputing the error metric. If the error metric decreases, then the change is kept. If the error metric increases, then the change is kept or rejected with a probability that depends on the current value of the *temperature*. A change that increases the error metric is more likely to be accepted for higher values of the temperature. As the iterations progress, the temperature lowered slowly. This method, like the hybrid input-output algorithm, is able to climb out of local minima in which the error-reduction algorithm stagnates. It is reported to perform better in the presence of noise than does the iterative tranform algorithm. Unfortunately, it is many times slower to converge than the hybrid input-output version of the iterative transform algorithm. Consequently, a logical strategy is to reconstruct an image first with the iterative transform algorithm, then perform further iterations using the simulated annealing algorithm.

3.3. Tracking Zero Sheets

Another interesting approach to phase retrieval is by tracking the zero sheets of the Fourier intensity $|F(u, v)|^2 = F(u, v)F^*(u, v)$ analytically extended into two-dimensional complex space (four real dimensions). In theory, the zero sheet of F can be separated from the zero sheet of F^*, allowing F to be retrieved. This is discussed by R. H. T. Bates and H. Jiang in the next chapter of this book. While mathematically elegant, this approach, like other methods that employ the complex zeros, tends to be computationally intensive and sensitive to noise. It is also notable because it can be used to solve the blind deconvolution problem, of which phase retrieval can be viewed as a special case.

3.4. Iterative Blind Deconvolution Algorithm

The Ayers-Dainty iterative algorithm [19, 20] is another blind deconvolution algorithm that can be applied to phase retrieval [21]. Unfortunately, it has proven to have convergence properties that are very similar to the error-reduction algorithm, which is to say that it converges very slowly. However, owing to a built-in Wiener filter, it appears to handle noise better than the iterative transform algorithm [21].

3.5. Maximum Entropy

Maximum entropy is another algorithm that has been applied to phase retrieval [22]. One way to think about its relationship to other phase-retrieval algorithms is as follows. In the absence of noise, for a real, nonnegative 2-D object, there usually is only one unique image consistent with a given Fourier modulus (aside from the twin image and translations). An algorithm such as the iterative transform algorithm will arrive at this image unless it stagnates at a local minimum. Maximum entropy, on the other hand, tries to maximize a metric proportional to the log of the image values. Since for negative values of the image, the entropy is undefined, maximum entropy in practice enforces a nonnegativity constraint. Therefore, if there is a unique nonnegative solution, then a maximum entropy solution would have to coincide with the solution arrived at by the iterative transform algorithm. This point of view is supported by digital experiments on a simple object in which the entropy was computed as a function of two Fourier phase values [23]. Various biases were added to the image by adding constants to the origin of the Fourier transform, and plots were made of the entropy as a function of the two phase values for images with the various biases. As the added bias approached zero, the space over which the entropy was defined (i.e., over which the image was nonnegative) decreased to a single pair of points, the solutions corresponding to the object and its twin image. Then the problem of finding the maximum entropy image was reduced to that of finding the only image for which the image was nonnegative, and the entropy could be defined.

A disturbing aspect of maximum entropy reconstruction is what it will do in the event that there are two or more solutions consistent with the data and constraints. It has been shown that if there are multiple solutions, then the iterative transform algorithm, after multiple trials with different random starting guesses, will tend to reconstruct each of the multiple solutions. However, maximum entropy will presumably reconstruct only the solution with the maximum entropy, and one would not even be aware that the other solutions exist.

4. Uniqueness of Phase Retrieval

It is well-known now that for 2-D objects, the phase-retrieval problem, in the absence of noise, almost always is unique. That is, aside from translations $f(x - x_o, y - y_o)$, twin images $f^*(-x - x_o, -y - y_o)$, and multiplicative constants, usually only one image $f(x, y)$ is consistent with a given Fourier modulus $|F(u, v)|$. This is understood most easily from the point of view that the Fourier transform of a discrete object (defined on a grid of points) is a polynomial, and the solution of the phase-retrieval problem is ambiguous if the polynomial is factorable [24]. A second, ambiguous solution is computed by complex-conjugating (flipping) one of the Fourier factors. Since it is unusual for a polynomial of two complex variables to be factorable, the 2-D phase-retrieval problem almost always is unique.

It also is known that the "uniqueness condition is stable in the sense that it is not sensitive to noise" [25]. However, the practical question can be asked: "Given a certain noise level in a noisy estimate of $|F(u, v)|$, how likely is it that an image, very different from the object $f(x, y)$, will be consistent with the estimate of $|F(u, v)|$ to within the noise level?" The answer to this practical uniqueness question is largely unknown.

An initial attempt to answer the practical uniqueness question was made by numerically exploring the space of all objects of a certain size [26]. Consider, for example, the 3×2 object $f(x, y)$ represented by the array of real numbers

$$\begin{bmatrix} a & b & c \\ d & e & f \end{bmatrix}.$$

For this object, it can be shown that the solution is ambiguous if

$$(af - cd)^2 - (ae - bd)(bf - ce) = 0. \tag{4}$$

This equation defines a five-dimensional surface embedded in the six-dimensional space of real-valued 3×2 objects. The ambiguous object closest (measured in Fourier modulus space) to a given object can be found by a reduced-gradient search along this 5-D surface. The ambiguous counterpart to this ambiguous object can be computed by flipping one of its factors, and it can be computed how different it is from the given object. If the Fourier modulus of the ambiguous object is within the noise level of the Fourier modulus of the given object, and if its ambiguous counterpart (which has the same Fourier modulus as the ambiguous object) is very different from the given object, then the object is ambiguous in the practical sense. A Monte Carlo experiment involving a large number of objects can be performed to determine the probability of ambiguity in the practical sense for a certain class of objects [26].

Unfortunately this approach to determining the probability of ambiguity seems practical only for small sampled objects due to the large amount of computation required. A comprehensive approach to this question, practical for large objects, is not known presently.

5. Conclusion

While phase retrieval for real-valued, nonnegative objects is fairly reliable with available algorithms, the reconstruction of complex-valued objects with poorly known convex support and noisy Fourier modulus data requires much better algorithms than are available currently. Only for some special cases, such as objects with separated parts or with glints or when the phase information is partially known, is phase retrieval for complex-valued objects reliable. Phase-retrieval algorithms must not only be able to find solutions, but must do so robustly in the presence of noise for general objects (in addition to the special cases), and not require an unbearable computational burden. An area in which more effort should be expended is the development of algorithms that explicitly take into account the amount of and the statistics of the noise present in the Fourier modulus estimate. The phase-variance algorithm does this in a crude way for the noisy phase data, but it does not arrive at an ideal result such as a least-squares estimate or a maximum likelihood reconstruction. An auxiliary problem also deserves further research: How does one estimate the support of the autocorrelation function from noisy, speckled, diffraction-limited data. Finally, the probability of uniqueness in a practical sense is understood only for very small objects, and an approach applicable to larger objects needs development.

As this book goes to press, several groups of researchers are applying a number of different phase-retrieval algorithms to the problem of determining precisely the aberrations of the Hubble Space Telescope from measured images of stars. Here, $f(x, y)$ is the aberrated pupil function with an approximately known support (shape) constraint, and $|F(u\, v)|^2$ is the star image. Variations on old algorithms and some new algorithms are being developed for this purpose.

References

1. R. W. Gerchberg and W. O. Saxton, "A Practical Algorithm for the Determination of Phase from Image and Diffraction Plane Pictures," *Optik* **35**, 237–246 (1972).
2. T. R. Crimmins, J. R. Fienup, and B. J. Thelen, "Improved Bounds on Object Support from Autocorrelation Support and Application to Phase Retrieval," *J. Opt. Soc. Am. A* **7**, 3–13 (1990).
3. D. G. Currie, S. L. Knapp, and K. M. Liewer, "Four Stellar-Diameter Measurements by a New Technique: Amplitude Interferometry," *Astrophys. J.* **187**, 131–144 (1974).
4. A. Labeyrie, "Attainment of Diffraction Limited Resolution in Large Telescopes by Fourier Analysing Speckle Patterns in Star Images," *Astron. and Astrophys.* **6**, 85–87 (1970).
5. J. C. Dainty and J. R. Fienup, "Phase Retrieval and Image Reconstruction for Astronomy," (H. Stark, ed.), Chapter 7 in "Image Recovery: Theory and Application," pp. 231–275, Academic Press, 1987.
6. A. W. Lohmann and B. Wirnitzer, "Triple Correlation, " *Proc. IEEE* **72**, 889–901 (1984).
7. J. R. Fienup, "Reconstruction of a Complex-Valued Object from the Modulus of Its Fourier Transform Using a Support Constraint," *J. Opt. Soc. Am. A* **4**, 118–123 (1987).
8. R. G. Paxman, J. R. Fienup, and J. T. Clinthorne, "Effect of Tapered Illumination and Fourier Intensity Errors on Phase Retrieval," in *Digital Image Recovery and Synthesis, Proc. SPIE* **828**, 184–189 (1987).
9. J. R. Fienup, T. R. Crimmins, and W. Holsztynski, "Reconstruction of the Support of an Object from the Support of Its Autocorrelation," *J. Opt. Soc. Am.* **72**, 610–624 (1982).
10. J. E. Baldwin and P. J. Warner, "Phaseless Áperture Synthesis," *Mon. Not. R. Astr. Soc.* **182**, 411–422 (1978).
11. J. R. Fienup and C. C. Wackerman, "Phase Retrieval Stagnation Problems and Solutions," *J. Opt. Soc. Am. A* **3**, 1897–1907 (1986).
12. P. S. Idell, J. R. Fienup, and R. S. Goodman, "Image Synthesis from Nonimaged Laser Speckle Patterns," *Opt. Lett.* **12**, 858–860 (1987).
13. R. Hanbury Brown and R. Q. Twiss, "Correlation Between Photons in Two Coherent Beams of Light," *Nature* **177**, 27–29 (1956).
14. J. R. Fienup, "Phase Retrieval Algorithms: A Comparison," *Appl. Opt.* **21**, 2758–2769 (1982).
15. J. R. Fienup, "Reconstruction and Synthesis Applications of an Iterative Algorithm," in "Transformations in Optical Signal Processing," (W. T. Rhodes, J. R. Fienup, and B. E. A. Saleh, eds.), *Proc. SPIE* **373**, 147–160 (1981).
16. J. R. Fienup and A. M. Kowalczyk, "Phase Retrieval for a Complex-Valued Object by Using a Low-Resolution Image," *J. Opt. Soc. Am. A* **7**, 450–458 (1990).

17. J. R. Fienup, "Image Reconstruction Using the Phase Variance Algorithm," (A. F. Gmitro, P. S. Idell and I. J. LaHaie, eds.), in *Digital Image Synthesis and Inverse Optics*, *Proc. SPIE* **1351**, 652–660 (July., 1990).
18. M. Nieto-Vesperinas and J. A. Mendez, "Phase Retrieval by Monte Carlo Methods," *Opt. Commun.* **59**, 249–254 (1986).
19. G. R. Ayers and J. C. Dainty, "An Iterative Blind Deconvolution Method and its Applications," *Opt. Lett.* **13**, 547–549 (1988).
20. B. L. K. Davey, R. G. Lane and R. H. T. Bates, "Blind Deconvolution of Noisy Complex-Valued Image," *Opt. Commun.* **69**, 353–356 (1989).
21. J. H. Seldin and J. R. Fienup, "Iterative Blind Deconvolution Algorithm Applied to Phase Retrieval," *J. Opt. Soc. Am. A* **7**, 428–433 (1990).
22. R. K. Bryan and J. Skilling, "Maximum Entropy Image Reconstruction from Phaseless Fourier Data," *Opt. Acta* **33**, 287–299 (1986).
23. R. Narayan, "Phase Retrieval with the Maximum Entropy Method," (J. W. Goad, ed.), *Interferometric Imaging in Astronomy,* pp. 183–186, ESO/NOAO, Oracle, AZ (Jan. 12–15, 1987).
24. Yu. M. Bruck and L. G. Sodin, "On the Ambiguity of the Image Reconstruction Problem," *Opt. Commun.* **30,** 304–308 (1979).
25. J. L. C. Sanz, T. S. Huang, and F. Cukierman, "Stability of Unique Fourier-Transform Phase Reconstruction," *J. Opt. Soc. Am.* **73**, 1442–1445 (1983).
26. J. H. Seldin and J. R. Fienup, "Numerical Investigation of the Uniqueness of Phase Retrieval," *J. Opt. Soc. Am. A* **7**, 412–427 (1990).

CHAPTER 29

Blind Deconvolution—Recovering the Seemingly Irrecoverable!

R. H. T. Bates and Hong Jiang
Electrical and Electronic Engineering Department, University of Canterbury, Christchurch, New Zealand

1. Introduction

About a century ago, one of the Duke of Hamilton's retainers on the Isle of Skye in the Inner Hebrides was asked how he knew that the Inaccessible Pinnacle in the Black Cuillins was aptly named. "Because I climbed it," he replied, but in the singular vernacular of his misty northern homeland. Blind deconvolution has a similar flavour. While obviously impossible, it can nevertheless be done!

An ideally blurred version of some original image $f(\mathbf{x})$ can be expressed as

$$b(\mathbf{x}) = f(\mathbf{x}) \odot h(\mathbf{x}), \tag{1}$$

where $h(\mathbf{x})$ is an isoplanatic (or space invariant) point spead function (psf), $\mathbf{x}$ is the position vector of an arbitrary point in K-dimensional image space, and $\odot$ is the K-dimensional convolution operator. The goal of *deconvolution* is to recover $f(\mathbf{x})$ from $b(\mathbf{x})$. This is the *ideal deconvolution problem.* However, there is (thankfully!) nothing ideal about this imperfect world of ours, so that any blurred image that actually is given to us must be of the form

$$g(\mathbf{x}) = b(\mathbf{x}) + c(\mathbf{x}), \tag{2}$$

where the contamination $c(\mathbf{x})$ encompasses every departure of the given blurred image from the ideal convolution model (1). The imperfections can

ISBN 0-12-289690-4

include assorted nonlinearities, nonisoplanatic blurring effects, and involuntary movements of image capture apparatus, besides the ubiquitous recording noise.

It can happen, of course, that $c(\mathbf{x})$ swamps $b(\mathbf{x})$ in Eq. (2), implying there is no hope of recovering $f(\mathbf{x})$ from $g(\mathbf{x})$. If so, it just has to be accepted, like the other minor disappointments of life. It is true nevertheless that $b(\mathbf{x})$ holds its own sufficiently in a great many situations of practical importance for it to be possible to recover $f(\mathbf{x})$ reasonably faithfully. Of course, this accounts for the huge body of literature on all the variegated aspects of deconvolution.

The *realistic deconvolution problem* is as follows: Given $g(\mathbf{x})$, recover an estimate $\hat{f}(\mathbf{x})$ of $f(\mathbf{x})$. There are two different manifestations of this general problem. For the older and more conventional of these, here called the *clear-sighted deconvolution problem*, we are given the psf $h(\mathbf{x})$ in addition to $g(\mathbf{x})$. The other problem is much more interesting. It is the *blind deconvolution problem*, for which we are expected to recover $\hat{f}(\mathbf{x})$ without prior knowledge of the details of $h(\mathbf{x})$.

2. Clear-Sighted Deconvolution

It is by crossing over into Fourier space that we are convinced most easily of the possibility of performing deconvolution successfully. Applying the convolution theorem to (1) gives

$$B(\mathbf{u}) = F(\mathbf{u})H(\mathbf{u}), \tag{3}$$

where $\mathbf{u}$ is the position vector of an arbitrary point in K-dimensional Fourier space, and upper-case letters, expressed as functions of $\mathbf{u}$, represent K-dimensional Fourier transforms, or spectra, of quantities written as corresponding lowercase letters, expressed as functions of $\mathbf{x}$ [1, 2].

Given $h(\mathbf{x})$, we can compute $H(\mathbf{u})$ and divide it into $B(\mathbf{u})$ to obtain $F(\mathbf{u})$, as indicated by Eq. (3). The ideal clear-sighted deconvolution problem thus is solved immediately and trivially.

In the real-world, division by $H(\mathbf{u})$ merely amplifies the presence of the inescapable contamination. Any spatial frequencies for which $|C(\mathbf{u}) / H(\mathbf{u})|$ is large are distorted out of recognition, implying that the estimate of $f(\mathbf{x})$ defined by

$$\hat{f}(\mathbf{x}) \leftrightarrow G(\mathbf{u})/H(\mathbf{u}), \tag{4}$$

where $\leftrightarrow$ interconnects members of a K-dimensional Fourier transform pair, can be expected to be disfigured by unacceptably pronounced artifacts [3, 4].

There are several ways of avoiding undue amplification of contamination. In many applications, the most appropriate way is to multiply $G(\mathbf{u})$ by the Wiener filter,

$$W(\mathbf{u}) = H^*(\mathbf{u})/[|H(\mathbf{u})|^2 + \Phi], \tag{5}$$

where the asterisk denotes complex conjugation and the filter constant Φ characterizes the average spatial frequency content of the contamination. The filter constant can be generalised to the filter function $\Phi(\mathbf{u})$. Note that $W(\mathbf{u})$ reduces to the inverse filter $1/H(\mathbf{u})$ wherever $|H(\mathbf{u})|^2 \gg \Phi(\mathbf{u})$. On the other hand, $W(\mathbf{u})$ becomes increasingly negligible the more $\Phi(\mathbf{u})$ dominates $|H(\mathbf{u})|^2$. Consequently, a stable estimate of $f(\mathbf{x})$ is obtained by replacing $1/H(\mathbf{u})$ in Eq. (4) by $W(\mathbf{u})$. In Section 5, we present several examples of Wiener filtering, for $K = 2$. It is worth recognising that Wiener filtering is one of the simplest examples of what has become known as *regularisation* [5].

3. Structure of Convolution

In the real world, all images are compact, in the sense that both they and their supports are finite (an image's support is the region of K-dimensional image space that it occupies). Since sophisticated image processing is feasible only when implemented with digital computers, images necessarily must be sampled at discrete points constituting, most conveniently, a rectangular grid in image space. Because each such sample now is generally called a *pixel* (i.e., picture element), we say that a sampled image is *pixellated*. We find that no confusion need result from using the term pixel to denote, interchangeably, the location in image space and the amplitude of the sample.

It transpires that the spectrum $F(\mathbf{u})$ of the compact pixellated K-dimensional image $f(\mathbf{x})$ is defined, up to an arbitrary complex multiplicative constant, by its zero-sheet $Z(F, K)$ [6]. To specify the latter, we generalise each of the K real Cartesian coordinates comprising $\mathbf{u}$ to a complex variable, so that the real position vector $\mathbf{u}$ is replaced by

$$\mathbf{w} = \mathbf{u} + \mathbf{i}\mathbf{v}, \tag{6}$$

where the kth of the real coordinates comprising $\mathbf{u}$ and $\mathbf{v}$ combine to form the kth complex variable

$$w_k = u_k + iv_k, \tag{7}$$

implying that complex Fourier space, in which $\mathbf{w}$ is the position vector of an arbitrary point, is $2K$-dimensional. $Z(F, K)$ is the $(2K-2)$-dimensional surface on which $F(\mathbf{w}) = 0$, where $F(\mathbf{w})$ is the analytical continuation of $F(\mathbf{u})$ into complex Fourier space.

The aforementioned continuation is straightforward conceptually for compact images, as is seen from replacing $\mathbf{u}$ by $\mathbf{w}$ in the Fourier integral

$$F(\mathbf{w}) = \int_{S(f,K)} f(\mathbf{x}) \exp(i2\pi\mathbf{w}\cdot\mathbf{x})\, dV(\mathbf{x}), \tag{8}$$

where $dV(\mathbf{x})$ is the K-dimensional volume element in image space, wherein $S(f, K)$ represents the support of $f(\mathbf{x})$. Provided $|\mathbf{w}| < \infty$, there is no way the integral in Eq. (8) can "blow up." When $f(\mathbf{x})$ is pixellated, it happens to be much more convenient computationally to evaluate $Z(F, K)$ via the Z-transform than the Fourier transform, but nothing of theoretical or practical consequence is altered thereby [7].

Because $2K - 2 = 0$ when $K = 1$, we see that $Z(F, 1)$ is a set of discrete points, called the zeros of $F(w)$ or of $Z(F, 1)$, in the complex w-plane (subscripts can be discarded without fear of ambiguity when $k = K = 1$) [6].

On account of the convolution theorem, as expressed by Eq. (3) for instance, the zero-sheet of the spectrum of the ideal blurred image $b(\mathbf{x})$ can be written as

$$Z(B, K) = Z(F, K) \cup Z(H, K), \tag{9}$$

where $\cup$ is the set-theoretic union operator [7]. We say that an image is irreducible if it can not be expressed as the convolution of two other images, neither of which occupies only a single pixel. If $f(\mathbf{x})$ and $h(\mathbf{x})$ both are irreducible, then Eqs. (1) and (3) indicate that $g(\mathbf{x})$ is reducible. It is worth emphasising that, when $K = 1$, all pixellated images (occupying more than one pixel) effectively are reducible.

When $f(\mathbf{x})$ and $h(\mathbf{x})$ are both compact and pixellated, then $g(\mathbf{x})$, despite being a convolution, also is compact and pixellated. When $K = 1$, therefore, $Z(B, K)$ is a set of discrete zeros. Consequently, in the absence of extra *a priori* information, there is no preferred way of partitioning the zeros of $Z(B, 1)$ amongst $Z(F, 1)$ and $Z(H, 1)$.

When $K > 1$, the situation is transformed completely because, provided $f(\mathbf{x})$ is irreducible, $Z(F, K > 1)$ almost always—meaning always except in contrived circumstances—is continuous and analytic, or smooth as we say, throughout complex Fourier space. Because it transpires that $Z(F, K)$ and $Z(H, K)$, as related in Eq. (9), can be expected to intersect at many points in complex Fourier space, $Z(B, K)$ almost always is rough (i.e., it ceases to be analytic) at such points of intersection. Wherever the two zero-sheets cross over, the slope of $Z(B, K)$ changes abruptly if one hops from the generator of $Z(F, K)$ across to that of $Z(H, K)$, or vice versa. Not only are the points of intersection recognisable, therefore, but we also can identify (almost) always separately the smooth parts of $Z(B, K > 1)$ at each roughness. The general inference is that, while blind deconvolution almost is never possible when $K = 1$, it in principle is feasible almost always when $K > 1$ [6, 7].

When contamination is present as it always is in the real-world, the zero-sheet $Z(G, K)$ of the spectrum of the given blurred image $g(\mathbf{x})$ can be expected to be smooth, so that it is impossible to separate $Z(F, K)$ from $Z(H, K)$ in the manner intimated earlier. This does not imply that blind deconvolution is impracticable, but it does mean we have to fall back on iterative techniques (such as are described in Sections 4.2 and 4.3). Although the zero-sheet concept cannot be taken practical advantage of as yet (but keep in mind our several speculations offered in Section 7), it is so graphic that it bolsters our confidence that we are not pursuing a mere chimera by seeking to devise blind deconvolution algorithms that eventually might be incorporated into routinely invokable image-processing software.

4. Blind Deconvolution and Phase Retrieval

The term *blind deconvolution* was introduced originally to denote signal or image recovery from a large collection of differently blurred versions of a particular signal or image [8]. We now refer to this type of deblurring as ensemble blind deconvolution (see Section 4.1) to distinguish it from pure blind deconvolution (Section 4.3). We find it appropriate to relate the latter to the former through phase retrieval (Section 4.2).

There is a delightful historical coincidence associated with ensemble blind deconvolution. The most striking of Stockham's applications of this concept is his recovery of (hopefully!) pristine vocalisations of erstwhile famous opera singers from ancient gramophone records [9]. Across the Atlantic in France, virtually simultaneously, Labeyrie was devising his speckle interferometry [10] (outlined in Section 4.2) for overcoming the optical astronomical seeing problem [11], which ever since Sir Isaac Newton's invention of the reflecting telescope had seemed to set an inescapable limit on our ability to resolve details of heavenly bodies, at least for as long as we remain earthbound. Both Stockhalm and Labeyrie required references to complete their deconvolutions. The former invoked a modern recording of the same aria by a singer of similar quality, while the latter observed a bright isolated star under similar seeing conditions to those pertaining when the celestial object of interest was viewed. So, a current "star" like Dame Joan Sutherland can rescue Dame Nellie Melba from the blaring and squeaks disfiguring her recorded voice. Many other speckle imaging techniques have been introduced during the past 20 years [12], one of them being described in Section 4.1.

All of the methods described in detail in this section are illustrated by example, for $K = 2$, in Section 5.

4.1. Ensemble Blind Deconvolution

The *ensemble blind deconvolution* problem is as follows: Recover $\hat{f}(\mathbf{x})$ from a given set (or emsemble) $\{g_m(\mathbf{x});\ m = 1, 2, \ldots, M\}$ of independently blurred versions of $f(\mathbf{x})$, where

$$g_m(\mathbf{x}) = f(\mathbf{x}) \odot h_m(\mathbf{x}) + c_m(\mathbf{x}), \tag{10}$$

with $h_m(\mathbf{x})$ and $c_m(\mathbf{x})$ being statistically independent members of the ensembles of, respectively, psf's and contaminations. An ensemble is understood here to be a collection of entities, all of which possess the same statistics. We call the $g_m(\mathbf{x})$ speckle images [12].

Real-world ensembles of blurred images, while never satisfying the previous conditions perfectly, of course, nevertheless often satisfy them closely enough for the various speckle-imaging techniques to be useful. We now describe *shift-and-add*, which is the easiest of these techniques to implement in practice [13].

The brightest point $\mathbf{x}_m$ in each $g_m(\mathbf{x})$ is defined by $|g_m(\mathbf{x}_\mathrm{m})| > |g_m(\mathbf{x} \neq \mathbf{x}_m)|$. The simplest form of shift-and-add image is

$$f_{\mathrm{sa}}(\mathbf{x}) = \langle g_m(\mathbf{x} + \mathbf{x}_m) \exp(-i\phi_m) \rangle \tag{11}$$

where the angled brackets denote the average over all given values of m, and $\phi_m = \text{phase}\ \{g_m(\mathbf{x}_m)\}$.

A *shift-and-reference* can be obtained from an ensemble $\{g_{r,m}(\mathbf{x});\ m = 1, 2, \ldots, M_0 \approx M\}$ of blurred versions of an image comprising a single pixel (note that this corresponds to an unresolvable object), captured under conditions statistically similar to those pertaining when the $g_m(\mathbf{x})$ are recorded. It follows that each $g_{r,m}(\mathbf{x})$ can be expressed as

$$g_{r,m}(\mathbf{x}) = h_{r,m}(\mathbf{x}) + c_{r,m}(\mathbf{x}), \tag{12}$$

where the $h_{r,m}(\mathbf{x})$ and $c_{r,m}(\mathbf{x})$ are statistically similar to, but independent of, the $h_m(\mathbf{x})$ and the $c_m(\mathbf{x})$, respectively. The *reference shift-and-add image is*

$$f_{r,\mathrm{sa}}(\mathbf{x}) = \langle g_{r,m}(\mathbf{x} + \mathbf{x}_{r,m}) \exp(-i\phi_{r,m}) \rangle \tag{13}$$

where corresponding quantities in Eqs. (11) and (13) are similarly defined.

Because a shifted convolution remains a convolution, we see from Eqs. (10) and (11) that

$$f_{\mathrm{sa}}(\mathbf{x}) = f(\mathbf{x}) \odot \tilde{h}(\mathbf{x}) + \tilde{c}(\mathbf{x}), \tag{14}$$

where we call $\tilde{h}(\mathbf{x})$ and $\tilde{c}(\mathbf{x})$ the composite psf and contamination, respectively. It follows similarly that

$$f_{r.\mathrm{sa}}(\mathbf{x}) = \tilde{h}_r(\mathbf{x}) + \tilde{c}_r(\mathbf{x}), \tag{15}$$

where the form of $c_r(\mathbf{x})$ usually can be expected to be similar to that of $c(\mathbf{x})$. On the other hand, $\tilde{h}(\mathbf{x})$ is likely to be significantly different from $\tilde{h}_r(\mathbf{x})$ unless $f(\mathbf{x})$ possesses a dominatingly bright pixel, by which is meant that $|f(0)| \gg |f(\mathbf{x} \neq 0)|$, with the origin of image space chosen where $f(\mathbf{x})$ is brightest [12]. We have demonstrated how the blurring due to $\tilde{h}(\mathbf{x})$ can be removed from $f_{\mathrm{sa}}(\mathbf{x})$ by pure blind deconvolution [14].

4.2. Phase Retrieval

If $h(\mathbf{x})$ is specialised to $f^*(-\mathbf{x})$, it follows from Eqs. (1) and (2) that $g(\mathbf{x})$ is a contaminated version of $ff(\mathbf{x})$, which is the notation we adopt here for the autocorrelation of $f(\mathbf{x})$. Since $|F(\mathbf{u})|^2$ is the spectrum of $ff(\mathbf{x})$, blind deconvolution reduces to retrieving *phase* $\{F(\mathbf{u})\}$ from $|F(\mathbf{u})|$ [7].

As Fienup explains elsewhere in this volume, such *phase retrieval* can be accomplished readily when *phase* $\{f(\mathbf{x})\}$ is constant. When *phase* $\{f(\mathbf{x})\}$ varies arbitrarily with $\mathbf{x}$, Fienup's algorithms tend to be much more expensive computationally, although they can still be successful, albeit under stricter constraints [15].

The essence of practical phase retrieval [16] is as follows: (a) The given samples of $|F(\mathbf{u})|$ are spaced close enough to permit $ff(\mathbf{x})$ to be reconstructed without aliasing; (b) the support $S_f(\mathbf{x})$ is deduced from inspection of $ff(\mathbf{x})$, (c) a pseudo-random initial estimate of $f(\mathbf{x})$ is constructed within $S_f(\mathbf{x})$, and (d) by appropriate iteration between image space and Fourier space, $\hat{f}(\mathbf{x})$ is "urged" (in the subtle manner devised by Fienup for overcoming the stagnation difficulties associated with "forcing" $\hat{f}(\mathbf{x})$ to satisfy the support constraint) to be confined within $S_f(\mathbf{x})$ simultaneously with $|\hat{F}(\mathbf{u})|$ being constrained to coincide with the given $|F(\mathbf{u})|$. After each iteration, the success of the algorithm is gauged by the smallness of the ratio ε_f of the fractions of $|\hat{f}(\mathbf{x})|^2$ lying outside and within $S_f(\mathbf{x})$.

It is appropriate here to present the essentials of Labeyrie's speckle interferometry [10, 12]. We define the *Labeyrie transform* $T_L(\mathbf{u})$ as the average of the squared magnitudes of the spectra of the speckle images $g_m(\mathbf{x})$:

$$T_L(\mathbf{u}) = \langle |G_m(\mathbf{u})|^2 \rangle. \tag{16}$$

The *Labeyrie reference transform* $T_{r,L}(\mathbf{u})$ is defined similarly in terms of the spectra of the $g_{r,m}(\mathbf{x})$. We then see, from the squared magnitudes of the Fourier transforms of Eqs. (10) and (12), that an estimate of the squared magnitude of $F(\mathbf{u})$ is provided by

$$|\hat{F}(\mathbf{u})|^2 = T_L(\mathbf{u}) W_L(\mathbf{u}), \tag{17}$$

where the *Labeyrie Wiener filter* $W_L(\mathbf{u})$ is defined by Eq. (5) with $H(\mathbf{u})$ replaced by $T_{r,L}(\mathbf{u})$.

An estimate of $S_f(\mathbf{x})$ is obtained from inspection of the estimate of $ff(\mathbf{x})$ generated by Fourier transforming $|\hat{F}(\mathbf{u})|^2$. The latter also can be subjected to phase retrieval, thereby generating an estimate $\hat{f}(\mathbf{x})$ of $f(\mathbf{x})$.

4.3. Pure Blind Deconvolution

The *pure blind deconvolution* problem is as follows: Recover $\hat{f}(\mathbf{x})$ from $g(\mathbf{x})$ as defined by Eqs. (1) and (2). Iterative algorithms [17, 18], inspired by Fienup's approach to phase retrieval (see Section 4.2), have been developed to accomplish blind deconvolution of contaminated blurred images. The essential strategy is outlined here, on the assumption that $S_f(\mathbf{x})$ is given in addition to $g(\mathbf{x})$. We have shown how the algorithms can be extended to deduce $S_f(\mathbf{x})$, or equivalently, $S_h(\mathbf{x})$, when only $g(\mathbf{x})$ is given [18].

The first step is to deduce $S_h(\mathbf{x})$ from $S_f(\mathbf{x})$ and $S_g(\mathbf{x})$, the latter, of course, being immediately inferrable from the given $g(\mathbf{x})$. A Wiener filter $W_f(\mathbf{u})$ is formed from a pseudo-random image confined to $S_f(\mathbf{x})$. An estimate $\hat{H}(\mathbf{u})$ of $H(\mathbf{u})$ is defined by $(G(\mathbf{u})\ W_f(\mathbf{u}))$. The Fourier transform of $\hat{H}(\mathbf{u})$ is urged (in the sense this term is used in Section 4.2) to be confined within $S_h(\mathbf{x})$, thereby generating an estimate $\hat{h}(\mathbf{x})$ of $h(\mathbf{x})$. A Wiener filter $W_h(\mathbf{u})$ is formed from $\hat{h}(\mathbf{x})$. An estimate $\hat{F}(\mathbf{u})$ of $F(\mathbf{u})$ is defined by $(G(\mathbf{u})\ W_h(\mathbf{u}))$. The Fourier Transform of $\hat{F}(\mathbf{u})$ is urged to be confined within $S_f(\mathbf{x})$, thereby generating a new estimate $\hat{f}(\mathbf{x})$ of $f(\mathbf{x})$. A new version of $W_f(\mathbf{u})$ then is formed from the spectrum of this $\hat{f}(\mathbf{x})$. The preceding steps are retraced, repeatedly generating new versions of $\hat{H}(\mathbf{u})$, $\hat{h}(\mathbf{x})$, $W_h(\mathbf{u})$, $\hat{F}(\mathbf{u})$, $\hat{f}(\mathbf{x})$, and $W_f(\mathbf{u})$.

5. Illustrative Example

In this section, we illustrate the different approaches to deconvolution described in Sections 2 and 4, emphasising interrelationships among the various techniques. To add a palpable touch of realism to each restored image, contamination is added to the data before restoration is attempted. We say that the data are contaminated to a level of $\alpha\%$ when pseudo-random noise, having a root mean square (rms) value of $-20(2 - \log_{10}\alpha)$ dB with respect to the average pixel value of the data, is added to each pixel of the data. The average pixel value of the data is the rms of all non-zero pixels of the data. We refer to each of the blurred images shown in Fig. 1 as a *contaminated convolution*.

We base our illustrative example on the original image shown in Fig. 1a, which is a Chinese ideogram representing luck. It has been demonstrated that blind deconvolution, of both the ensemble and pure varieties, can be

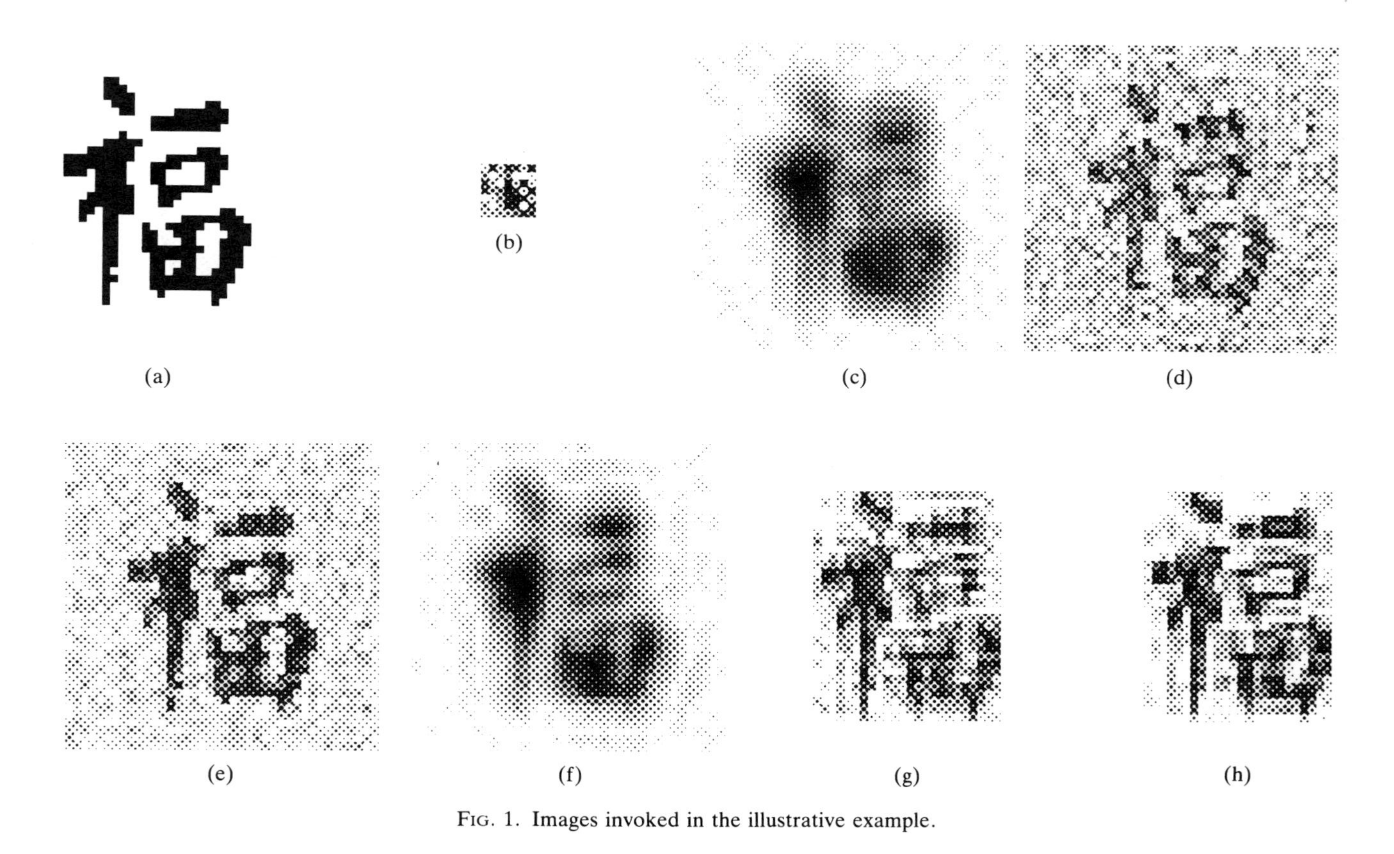

FIG. 1. Images invoked in the illustrative example.

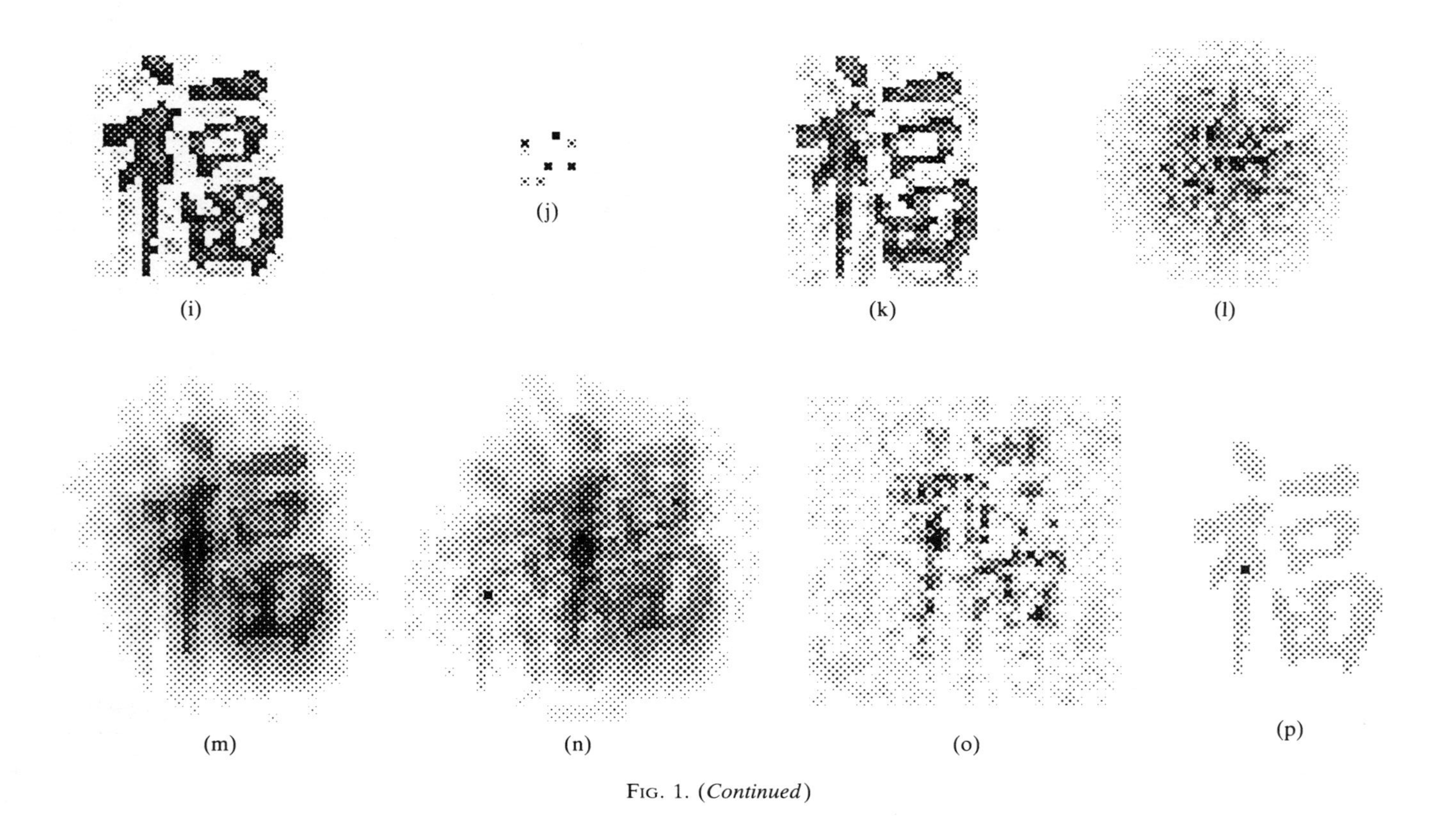

FIG. 1. (*Continued*)

accomplished successfully for complex-valued images. For convenience of display, however, we choose our original images to be positive (i.e. real and non-negative). We employ both complex-valued and positive psf's and contaminations.

The positive image depicted in Fig. 1b is an idealised version of a short (by comparison with the fluctuation time of the medium) exposure of an unresolvable object viewed through a randomly fluctuating medium. Figure 1c shows the contaminated ($\alpha = 10$) convolution of the images depicted at Fig. 1a and 1b, which respectively represent $f(\mathbf{x})$ and $h(\mathbf{x})$, in the notations adopted in Sections 2 and 4.3. Figures 1d, 1e, and 1f illustrate how the faithfulness of the versions of $\hat{f}(\mathbf{x})$ obtained by Wiener filtering $g(\mathbf{x})$, represented by Fig. 1c, depends upon the filter constant Φ. When Φ is too small (Fig. 1d), $\hat{f}(\mathbf{x})$ reveals detail resolved comparatively to that apparent in $f(\mathbf{x})$ (Fig. 1a), but the artifact level is high. When Φ is too large (Fig. 1f), $\hat{f}(\mathbf{x})$ is virtually free of artifacts, but the resolution is comparatively poor. When Φ is optimised (Fig. 1e), there is an acceptable compromise between resolution and artifact level. Keep in mind that a contamination level of 10% is appreciable in practical terms. We think it worthwhile including these few illustrations of clear-sighted deconvolution because Wiener filtering features twice in each iteration of the pure blind deconvolution algorithm outlined in Section 4.3.

In many practical applications, a complete specification for $h(\mathbf{x})$ is unavailable, implying that pure blind deconvolution should be attempted. Figure 1g shows the version of $\hat{f}(\mathbf{x})$ obtained by applying the algorithm outlined in Section 4.3 to the contaminated convolution shown in Fig. 1c. While Fig. 1g suffers by comparison with Fig. 1e, it actually does reveal most of the detail exhibited by Fig. 1a. The contamination level of 10% represents a severe challenge to pure blind deconvolution, the performance of which improves noticeably as the contamination level is reduced, as demonstrated by Figs. 1h and 1i ($\alpha = 3$ and 1, respectively). Pure blind deconvolution is less sensitive to the contamination level when the psf has a higher spatial frequency content, as is true of the psf shown in Fig. 1j. Even when the convolution of Figs. 1a and 1j is contaminated to 10%, the image restored by pure blind deconvolution (Fig. 1k) loses little by comparison with Figs. 1e and 1i.

When many $g_m(\mathbf{x})$ are recorded (see Section 4.1), $\hat{f}(\mathbf{x})$ can be reconstructed by ensemble blind deconvolution, the most versatile form of which is that often called bispectral processing. (See discussions by Dainty and Weigelt elsewhere in this volume.) It is highly demanding computationally, however, especially by comparison with shift-and-add processing (see Section 4.1). Besides being very simple, the latter itself is quite versatile when combined with pure blind deconvolution, in which case a faithful $\hat{f}(\mathbf{x})$ can be reconstructed even when $f(\mathbf{x})$ does not possess a single dominatingly bright pixel [14]. However, for $f_{sa}(\mathbf{x})$ to be a reasonably faithful version of $f(\mathbf{x})$, the latter must possess such a bright pixel, which we ensure by highlighting (as explained shortly) the original image $f(\mathbf{x})$ shown in Fig. 1a. In the remainder of this section, both the psf's and the contaminations are chosen to be complex-valued, as applies when shift-and-add images are formed from coherent wave fields [12]. The phases are distributed uniformly from 0 to 2π. The psf's, the magnitude of a typical one of which is depicted in Fig. 1l, are generated by Fourier transformation of randomly-phased, constant-magnitude spectra.

When $f(\mathbf{x})$ is highlighted with a single pixel, 10 times brighter than the other pixels of $f(\mathbf{x})$, shift-and-add processing of 200 statistically independent contaminated convolutions $g_m(\mathbf{x})$ generates the $f_{sa}(\mathbf{x})$ shown in Fig. 1m. Note that Fig. 1l typifies the $h_m(\mathbf{x})$. Because of the limited dynamic range with which images can be reproduced in a book such as this, we have thresholded the highlighted pixel (but not so severely as to obscure its location) in Fig. 1m, revealing most of the detail exhibited by Fig. 1a, but clouded by the fog that is well-known to practitioners of shift-and-add [12]. To illustrate how $f_{sa}(\mathbf{x})$ is affected by $f(\mathbf{x})$ possessing more than one bright pixel, we highlighted $f(\mathbf{x})$ with two pixels, each 10 times brighter than the other pixels. The resulting thresholded version of $f_{sa}(\mathbf{x})$, formed from 200 contaminated convolutions, is shown in Fig. 1n, which reveals superimposed, shifted versions of $f(\mathbf{x})$. These are the shift-and-add ghosts [12], which can be removed by pure blind deconvolution [14]. We remark that the highlighted pixels in $f(\mathbf{x})$ are accorded the same brightness as the other pixels when computing the contamination level, which is 10% for the speckle images giving rise to Figs. 1m and 1n.

It is instructive to compare shift-and-add (Section 4.1) with Labeyrie's speckle interferometry combined with Fienup's phase retrieval (Section 4.2). Fig. 1o shows the $\hat{f}(\mathbf{x})$ so obtained by invoking, for the contaminated ($\alpha = 10$) speckle images giving rise to the singly highlighted version of $f(\mathbf{x})$ whose shift-and-add image is shown in Fig. 1m, what we currently hold to be the optimum form of Fienup's hybrid-input-output algorithm. Much less detail is apparent in Fig. 1o than in Fig. 1m, probably because of speckle interferometry's greater sensitivity to noise. It is worth emphasising that speckle interferometry is second order interferometry, whereas shift-and-add is first order interferometry. When α is reduced to zero, the Labeyrie and Fienup algorithms together generate Fig. 1p, which reveals essentially all the detail apparent in Fig. 1a.

6. Can the Blindfold Really Be Removed?

There is no way of restoring spatial frequency information that has been obliterated irretrievably. Suppose, for instance, that the psf is Gaussian.

Then so is its spectrum. Because a Gaussian function falls smoothly to zero away from its peak, the spatial frequency content of the blurred image must be below the contamination level for all spatial frequencies greater than some limiting value. Even clear-sighted deconvolution is incapable of recovering spatial frequency information above this limit.

Despite that, the effectiveness of pure blind deconvolution presently is only comparable to that of clear-sighted deconvolution when the psf either has a "spiky" structure or is sharply truncated. The "smoother" the psf, the poorer is the performance of pure blind deconvolution by comparison with clear-sighted deconvolution. Another somewhat puzzling, current deficiency of pure blind deconvolution is that its performance falls off, the more nearly equal in size are the supports of $f(\mathbf{x})$ and $h(\mathbf{x})$. While these inadequacies of blind deconvolution algorithms doubtless will be ameliorated by the kind of steady sequence of minor improvements to be expected from ongoing research, we do not think the blindfold will be lifted fully until the major advance presaged in the next section has been realised.

7. Towards Clear-Sighted Blindness

To make blind deconvolution as effective as its clear-sighted counterpart, we need to learn how to unravel contaminated zero-sheets. An essential preliminary is to gain a much clearer understanding of how the intersections of the zero-sheets of the spectra of the original image and the psf are deformed by the addition of contamination to the ideally blurred image. The question is: How do we estimate, by inspection of $Z(G,2)$, where the intersections of $Z(F,2)$ and $Z(H,2)$ would lie in four-dimensional complex Fourier space if there were no contamination? If this question could be answered successfully (i.e., if the intersection could be located accurately enough to be useful), $Z(F,2)$ and $Z(H,2)$ could be estimated separately, presumably allowing the best possible version of $\hat{f}(\mathbf{x})$ and $\hat{h}(\mathbf{x})$ to be reconstructed.

It seems to us that this is mainly a computer graphics problem. Sufficiently powerful and versatile hardware and software need to be combined to allow the deformation of zero-sheets, caused by varying both the level and the type of contamination, to be displayed virtually instantaneously. Only the leading information-processing laboratories in the world's richest countries probably are capable of mounting research efforts having any realistic hope of solving this problem.

We are presently addressing the simpler, but still decidedly challenging, related problem: Given a positive quantity $I(\mathbf{u})$, which is a contaminated version of a particular spectral intensity $|F(\mathbf{u})|^2$, construct an alternative spectral intensity that differs as little from $I(\mathbf{u})$ as permitted by the contamination level.

The above problem is solved readily for $K = 1$. The reason is that $Z(I, 1)$ necessarily comprises pairs of discrete zeros. Those pairs, whose members are complex conjugates of each other, satisfy the requirements of the problem. While any pairs, whose members are real but non-coincident, do not satisfy their requirements, they easily can be made to do so by the following simple adjustment procedure. The real zeros are ordered along the u-axis, in the direction of increasing u, say, such that the first pair are those zeros having the first and second most negative values of u, and the second pair are those zeros having the third and fourth most negative values of u, etc. If each of these pairs of zeros is replaced by a double zero at the point on the u-axis mid-way between the members of the pair, then all zeros (both complex and real) satisfy the requirements of the problem.

Things are different altogether when $K = 2$. We can choose any particular complex value for w_2 and carry out the procedure just described for the discrete zeros of $I(w_1, w_2)$, with w_1 being treated as a complex variable. The procedure can be rerepeated, of course, for any desired number of different particular values of w_2. The trouble is, there is no guarantee that, when w_2 is treated as complex variable with w_1 fixed, the discrete zeros in the w_2-plane will satisfy the requirements of the problem.

Nevertheless, the restricted problem posed here seems much more tractable than the more general problem of estimating, from inspection of $Z(G, 2)$, the locations of points in complex Fourier space where $Z(F, 2)$ and $Z(H, 2)$ intersect. We intend to pursue this restricted problem and the preliminary problem of how best to implement *zero-and-add* in two dimensions. The principle of zero-and-add is that, in the absence of contamination,

$$Z(F, K) = Z(G_m, K) \cap Z(G_n, K) \tag{18}$$

for any pair of integers m and n, both in the range of 1 to M [19]. So, in the inevitable presence of contamination, superposition of several of the $Z(G_m, K)$ should help to provide an estimate of $Z(F, K)$. Our recently introduced zero-tracks may permit a useful version of $\hat{f}(\mathbf{x})$ to be obtained, with the aid of two-dimensional zero-and-add, from far fewer speckle images than are needed to implement any of the established speckle imaging techniques [20].

References

1. R. N. Bracewell, "The Fourier Transform and Its Applications," McGraw-Hill, New York, 1978.
2. R. H. T. Bates and M. J. McDonnell, "Image Restoration and Reconstruction," Clarendon Press, Oxford, 1986.
3. A. Rosenfeld and A. C. Kak, "Digital Picture Processing," Vol. 1 of "Computer Science and Applied Mathematics," 2 ed., Academic Press, New York, 1982.

4. B. R. Hunt, "Digital Image Processing," *Proceedings IEEE* **63**, 693–708 (1975).
5. M. Bertero, C. de Mol, and G. A. Viano "The Stability of Inverse Problems," in "Inverse Scattering Problems in Optics" (H. P. Baltes, ed.), Springer-Verlag, Berlin, pp. 161–214, 1980.
6. R. G. Lane and R. H. T. Bates, "Automatic Multi-Dimensional Deconvolution," *Journal of the Optical Society of America A*, **4**, 180–188 (1987).
7. R. G. Lane, W. R. Fright, and R. H. T. Bates, "Direct Phase Retrieval," *IEEE Transactions on Acoustics Speech and Signal Processing*, **ASSP-35**, 520–526 (1987).
8. T. G. Stockham, Jr., T. M. Cannon, and R. B. Ingebretson, "Blind Deconvolution through Digital Signal Processing," *Proceedings IEEE* **63**, 678–692 (1975).
9. T. G. Stockham, Jr., *Audio Engineering Society Convention Abstracts* 10 (1971), ibid, 12 (1972).
10. A. Labeyrie, "Attainment of Diffraction Limited Resolution in Large Telescopes by Fourier Analyzing Speckle Patterns in Star Images," *Astronomy & Astrophysics* **6**, 85–87 (1970).
11. F. Roddier, "Interferometric Imaging in Optical Astronomy," *Physics Reports* **170**, 97–166 (1988).
12. R. H. T. Bates, "Astronomical Speckle Imaging," *Physics Reports* **90**, 203–297 (1982).
13. R. H. T. Bates, W. R. Fright, F. M. Cady, and G. J. Berzins, "Speckle Processing, Shift-and-Add, and Compensating for Instrument Aberrations," in "Transformations in Optical Signal Processing", (W. T. Rhodes, J. R. Fienup, and B. E. A. Salah, eds.), *Proceedings SPIE* **373**, 197–202 (1981).
14. R. H. T. Bates and B. L. K. Davey, "Deconvolution Ancient and (Very) Modern," in "Diffraction-Limited Imaging with Very Large Telescopes," (D. Alloin and J. M. Mariotti, eds.), pp. 293–303, Kluwer Academic Publishers, Drodrecht, 1989.
15. J. R. Fienup, "Reconstruction of a Complex-Valued Object from the Modulus of Its Fourier Transform Using a Support Constraint," *Journal of the Optical Society of America A*, **4**, 118–123 (1987).
16. R. H. T. Bates and D. Mnyama, "The Status of Practical Fourier Phase Retrieval," in "Advances in Electronics and Electron Physics" (P. W. Hawkes, ed.), Vol. 67, pp. 1–64, Academic Press, New York, 1986.
17. G. R. Ayers and J. C. Dainty, "An Iterative Blind Deconvolution Algorithm and Its Applications," *Optics Letters* **13**, 547–549 (1988).
18. B. L. K. Davey, R. G. Lane, and R. H. T. Bates, "Blind Deconvolution of Noisy Complex-Valued Images," *Optics Communications* **69**, 353–356 (1989).
19. B. L. K. Davey, A. M. Sinton, and R. H. T. Bates, "Zero-and-Add," *Optical Engineering* **26**, 765–771 (1986).
20. R. H. T. Bates, B. K. Quek, and C. R. Parker, "Some Implications of Zero-Sheets for Blind Deconvolution and Phase Retrieval," *Journal of the Optical Society of America A* **7**, 468–478 (1990).

CHAPTER 30

Pattern Recognition, Similarity, Neural Nets, and Optics

Henri H. Arsenault and Yunlong Sheng
Centre d'Optique, Photonique et Laser, Université Laval
Québec, Canada

1. Introduction

As computers have increased in power and failed to live up to some of their promises, the capability of computers to perform intelligent operations increasingly has been put into question. For example, Hilary Putnam—the founder of functionalism, the mainstream theory of the mind, and the principal proponent of the idea of the mind as a Turing machine, in which mental states are considered as the functional states of an abstract digital computer—recently has refuted his own theory [1]. The well-known mathematician, Roger Penrose, has written a book with the title, *The Emperor's New Mind*; the title says it all [2]. The January, 1990 issue of *Scentific American* contains two articles arguing both sides of the question of whether or not intelligent machines are possible [3, 4]. Some researchers in neural networks are describing some operations of their models with a vocabulary that suggests intelligent operations, although, at least in some cases, that is unjustified and misleading. What are the implications of this on pattern recognition and on optics?

We will not attempt here to resolve the interesting and important question of whether machines are capable of intelligent operations or not. What we shall do is examine the nature of the concepts of similarity and invariance, and how their arbitary nature can be exploited to improve the performance of systems.

ISBN 0-12-289690-4

The concept of *invariance* in pattern recognition is related closely to that of *similarity*. If an object is considered to be invariant under some transformation, it means that after the transformation, it is considered similar to itself before the transformation. Implemention of a pattern recognition method that is invariant under some transformation of the object must be based upon some measure that does not change under this transformation. For example, a rotated geometrical figure would be considered similar to an unrotated one if certain metrical properties of interest are preserved under the rotation.

The mathematical classification of objects usually is based on a distance between points in some feature space. The feature space is divided into sub-spaces by means of hypersurfaces, and objects are classified according to their positions within the sub-spaces. In simple cases like matched filtering, the features are the pixel values of the objects, the hyperspace is simply a line, and a threshold divides the line into two parts: Objects that are on one side of the threshold are put into one group (recognized), and those below threshold are put into another group (non-recognized).

In cases where the hyperspace has two or more dimensions, objects that cluster together are classified usually as belonging to the same class of objects. Sometimes two different clusters are considered equivalent. Much of the problem of pattern recognition consists of finding appropriate surfaces to segment the feature space into sub-spaces so that objects will be classified together in a useful way.

Most pattern recognition techniques are relatively *ad hoc*, so little attention has been paid to some of the unstated assumptions that underlie many methods.

A problem related to classification is the extraction of semantic information from data such as images. A simple example of extracting meaning from an image could be phrased as follows:

> This is an image with a view looking down from about 1,000 meters, and the scene contains background consisting of about half forest, one-quarter town, and one-quarter crop fields, in the late summer. There are two main roads that intersect in the middle of the town, which has closely packed rectangular, modest houses and a plaza near the intersection, and there is a narrow bridge east of the town. There are 10 cars, two large trucks, and one bus on the roads; two pickup trucks, four cars, and five bicycles are parked on a parking lot near a baseball field at the southern edge of town.

Although all of the objects in this scene probably can be recognized by automated pattern-recognition techniques, computers are incapable of recognizing them in context; that is, recognizing the relationships between the objects.

A few years ago, S. Watanabe discovered an important theorem that has important consequences in pattern recognition, a statement that he dubbed "the theorem of the ugly duckling." This theorem states: *From a logical point of view, similarity is a totally arbitrary concept* [5]. Although on the face of it, this is preposterous, it becomes obvious when one realizes that similarity is measured by comparing weighted features, and that objects can be made more or less similar by changing the weights attributed to the features. If all the weight is put on one or more features that two objects have in common, the two objects are considered identical, and if all the weight is put on one or more features that the two objects do not have in common, then the two objects are considered totally different. The degree of similarity between objects thus depends on the weights attributed to the features of the objects.

2. Invariance Is Arbitrary

What are the consequences of this theorem on invariance? We have stated that invariance means that one or more parameters of objects do not change under some transformation. However, if no weight is attributed to the unchanging parameters, the objects before and after the transformation will not be considered similar. For example, take the classical matched filter. If an aircraft is rotated, it is still an aircraft, but the correlation between two aircraft rotated with respect to each other will be low (the corresponding pixels do not match), and the decision operation taken by the threshold operation will not consider them to be identical.

Such a filter will be unable to recognize correctly a rotated aircraft without setting the threshold so low that many other objects such as cars and trucks that are obviously not aircraft will be incorrectly classified as aircraft. Only if some invariant parameters are considered will the recognition be invariant. But it is important to notice that from a purely logical point of view, the weights given to those invariant parameters are a totally arbitrary matter. So we see: *The concept of invariance also is arbitrary.*

What are the consequences of the fact that invariance is a logically arbitrary concept? It means that in principle, it should be possible to classify objects as belonging to any arbitrary classes. It is an unstated assumption in most pattern recognition work that the classification of objects corresponds to some underlying reality, but the preceding discussion shows that this idea is not a logical necessity, but a projection of our own human programmed weights onto reality. There is nothing wrong with that, if we are aware of it and if that is what we want to do. But given the limitations of present pattern-recognition techniques, it may be worthwhile to look at other possibilities.

How can we classify objects that differ by at least one feature into completely arbitrary classes? We shall discuss at the end of this chapter how that can be achieved both for matched filters and neural nets.

3. How Is Similarity Measured?

3.1. Inner Product

Many similarity measures are carried out by weighted products. If the jth features of all members of the training set or prototypes X_k consists of a weight vector with elements x_{jk}, an unknown object with weights y_i is classified by comparing its weights with those of the training set. This can be done by means of the inner product

$$\mathbf{w} = \mathbf{X}^t\mathbf{y}, \tag{1}$$

where $\mathbf{X}$ is a matrix whose rows contain the weight vectors x_k. The vector $\mathbf{w}$ contains the inner product of the unknown vector with all of the weight vectors, and the largest element of the vector $\mathbf{w}$ corresponds to the prototype whose position in feature space is closest to that of the unknown object, since the product of two vectors is maximum when they are parallel. This measure is important because it corresponds to operations carried out both by neural nets and by matched filters. For objects stored in computers, the basic features $\mathbf{x}_k$ are the pixel values.

Why is the product of Eq. (1) a good measure of similarity? Because for binary objects with pixel values $(0, 1)$, it is equal to the number of pixels that are equal to one in both objects involved in the product. We shall see later in the context of neural nets that this measure is not always good, and can lead to pathological behavior of a network.

3.2. Distance Functions

A related approach and one that is frequently used is pattern classification by distance functions [6]. There are a number of distance functions, the simplest of which is the Euclidean distance between the prototype and the unknown vector,

$$\mathbf{D}_i = \|\mathbf{x}_i - \mathbf{y}\|. \tag{2}$$

With the so-called minimum-distance method, the unknown object is attributed to the class containing the prototype having the minimum distance to the unknown object. It is interesting to note that the least-squares difference between two objects is equal to the square of the distance between the objects in a Euclidean feature space with dimensions equal to the number of pixels of the objects, so methods that minimize the least-squares difference are minimizing the distances between points in this space. There are other useful measures of similarity based on distance. One is the Mahalanobis distance,

$$\mathbf{D} = (\mathbf{x} - \mathbf{y})\mathbf{C}^{-1} (\mathbf{x} - \mathbf{y}), \tag{3}$$

where $\mathbf{C}$ is the covariance matrix of a pattern population; this measure, based on a Bayes classifier, minimizes the classification error under certain conditions that assume the population is randomly distributed with a normal distribution, which is not a very realistic assumption for most practical cases of interest, but which has the advantage of fitting into a nice theoretical mold.

The important point for the purposes of this discussion is that all these measures of similarity modify the weights of the object features in some way to obtain a measure of similarity. The features used are pixel values usually, but other features can be used as well.

3.3. Non-Numeric Measures

Similarity concepts may be applied to non-numeric features. Such features usually are compared by attributing numeric labels to them. The labels then can be manipulated by various techniques such as the *ID3 Approach* [7], which is a procedure for pattern recognition and classification for synthesizing an efficient discrimination tree for classifying patterns with non-numeric features. The features are examined in sequence by means of an information-theoretical approach to find a combination of feature values that suffice to determine class membership and that has minimum entropy. This assumes that the features are distributed randomly with a known or assumed probability distribution, so the presence of two objects in the same class means only that this grouping minimizes the number of steps required for classification, and not necessarily that the two objects have many properties in common.

There are other non-numeric measures of similarity [8], but the comments just made apply to them as well; that is to say, objects are put in the same class to optimize some classification scheme. Such methods are useful when objects have many non-numeric features, but it is debatable whether objects that are put in the same class would be considered similar by one who was not concerned with the time required to classify large numbers of such objects.

3.4. Matched Filter

The *matched filter* or *correlation filter* was developed originally for radar applications, and applied later to images. It has the property that under certain statistical conditions, it is the correlation filter that maximizes the output signal-to-noise ratio. It has played an important role in optics because, in addition to being shift invariant, it can be implemented easily in an optical correlator [9]. The matched filter carries out the correlation operation between an unknown object $f(x, y)$ and a prototype $h(x, y)$,

$$c(x, y) = \int_{-\infty}^{\infty} \int_{-\infty}^{\infty} f(\zeta,\eta)h(\zeta + x, \eta + y)\, d\zeta\, d\eta. \tag{4}$$

The correlation of interest is the value at the origin $c(0,0)$, which is equal to

$$c(0,0) = \int_{-\infty}^{\infty} \int_{-\infty}^{\infty} f(\zeta, \eta)h(\zeta, \eta)\, d\zeta\, d\eta. \tag{5}$$

If the pixel values of the object $f(x,y)$ and of the prototype $h(x,y)$ are arranged lexicographically into vectors $\mathbf{y}$ and $\mathbf{x}$, respectively, we see that the correlation at the origin is equal to the scalar product $c = \mathbf{x} \cdot \mathbf{y}$ between the object and the prototype, so the measurement of similarity for the correlation and for the inner product is the same.

4. Invariance and Normalization

There are many kinds of invariance that are useful for pattern recognition: Invariance under translation, rotation, scale, noise, and contrast are of the greatest interest in pattern recognition. We have discussed elsewhere how such invariance can be achieved with matched filters and other techniques, such as invariant moments [10].

It should be noted that as a matched filter is made more invariant, it becomes less sensitive to changes in the objects it is trying to recognize; In the extreme, a filter invariant to everything could recognize nothing. So, there is a trade-off between invariance and discrimination ability. The loss of discrimination ability can be offset by using multiple filters, so time can be traded off to gain discrimination ability.

Some measures of similarity are not normalized. Sometimes additional invariance can be obtained by normalization. For example, normalizing all vectors in a feature space can make the recognition invariant to the contrast of the objects. Normalization in a vector space throws away some information to gain invariance, as the vectors become distinct only by virtue of their orientations.

In neural nets, many measures of similarity are not normalized. Despite this fact, good results are obtained in the early stages of model development and simulation.

We have studied in some detail a neural net base on the *Hamming model* [11], which uses a winner-take-all slab call MAXNET, where each neuron is connected to its neighbors in an on-center off-surround manner; this uses lateral inhibition to drive all the neurons in the slab to zero except the one that has the greatest value, corresponding to the inner product of the input vector with the stored vector that is most similar according to some criterion.

The Hamming distance between two vectors is equal to the number of bits that are different. Hamming distances can be calculated by the inner product only when the two vectors are bipolar $(-1, 1)$. In this case, the Hamming distance is equal to the number of elements $\mathbf{N}$ in the vectors minus the inner product between the two vectors. Larger values of inner products correspond to shorter Hamming distances.

In Table I, we assume binary inputs and stored values. The base input and store vectors $\mathbf{I}$ and $\mathbf{S}$ are assumed to have pixel values of $(1, 0)$, and the overbars indicate the complements of the vectors. The table shows the results for input values of $(1, 0)$ and $(1, -1)$; in the latter case, the input vector is $\mathbf{I} + \overline{\mathbf{I}}$. In addition to those values, the elements of the stored vectors can have the normalized values $(\mathbf{S}/\mathbf{S}^t\mathbf{S}, 0)$ and $(\mathbf{S}/\mathbf{S}^t\mathbf{S}, \overline{\mathbf{S}}/\overline{\mathbf{S}}^t\overline{\mathbf{S}})$. The second line corresponds to the Hamming model, the third line to the model earlier proposed by us [12], and the last line to what we call the *Balanced Weights Model* (BWM) [13], where the positive pixels and the negative pixels are normalized separately in such a way that the sum of the pixel

Table I. Measures of similarity for various values of stored and input vectors

	input	input
stored	1, 0	1, −1
1, 0	I^tS	$I^tS - \overline{I}^tS$
1, −1	$I^tS - I^t + \overline{S}$	$I^tS + \overline{I}^t\overline{S} - (I^t\overline{S} + \overline{I}^tS)$
$\frac{S}{S^tS}$	$\frac{I^tS}{S^tS}$	$\frac{I^tS - \overline{I}^tS}{SS}$
$\frac{S}{S^tS}, \frac{\overline{S}}{\overline{S}^t\overline{S}}$	$\frac{I^tS}{S^tS} - \frac{I^t\overline{S}}{\overline{S}^t\overline{S}}$	$\frac{I^tS}{S^tS} + \frac{\overline{I}^t\overline{S}}{\overline{S}^t\overline{S}} - \left(\frac{I^t\overline{S}}{\overline{S}^t\overline{S}} + \frac{\overline{I}^tS}{S^tS}\right)$

values is equal to 0, and the inner product of both the positive part and the negative part of the object with itself is equal to 1. The idea of this model is that each pixel should have a weight inversely proportional to the number of pixels having the same value. For example, if the input vector is equal to $(1, 0, 0, 1)$, then $\mathbf{I} = (1, 0, 0, 1)$ and $\overline{\mathbf{I}} = (0, 1, 1, 0)$. The products in the table are inner products written in matrix notation, so, for example, the value of $\mathbf{I}^t\mathbf{S}$ is equal to the number of ones in common to the input vector and the stored vector. It is easy to see that the Hamming distance using this notation is equal to

$$\mathbf{H} = \mathbf{I}^t\overline{\mathbf{S}} + \overline{\mathbf{I}}^t\mathbf{S}. \tag{6}$$

We see from the table that no combination of input values or stored values allows this system to measure the Hamming distance—not even the Hamming model, whose measure is equal to the length of the vectors minus twice the Hamming distance between the vectors. We have found that all similarity measures except the BWM lead to networks that have some pathological behavior (from the point of view of a human observer only; from a purely logical point of view, there can be no such thing as pathological behavior because similarity between objects is arbitrary). An example of pathological behavior is shown in Fig. 1.

In the case at the top, the network had stored interconnects of $(1, -1)$ and input values of $(1, 0)$, and was trained as an autoassociative memory [11]. The input to the network having 16 stored letters is a slightly distorted binary letter 3, but after a few iterations, the output converges to the letter 8. This phenomenon of the network being "impressed" by the larger number of non-zero pixels in the letter 8 is caused by this measure of similarity's overemphasizing certain kinds of errors in the input. The Hamming model with inputs and stored values $(1, -1)$ corrects this

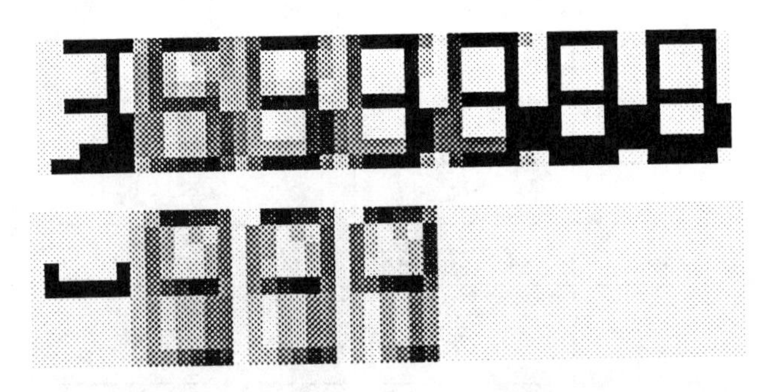

FIG. 1. Pathological behavior of a neural net. Top: A deformed 3 at the input is classified as an 8. Bottom: A piece of a 9 cannot be classified. The BWM model classifies both objects correctly.

kind of error, but it has difficulties in recognizing objects from small parts of them, as shown in the bottom of Fig. 1. The input on the left is a part of the number 9, which was stored in the network along with other letters, but the Hamming network could not decide whether the object was a 9 or a 2, and finally did not classify the object, although it was a part of no other stored object.

The different kinds of errors are discussed in our recent paper [13], which showed that the Balanced Weights Model, using weights normalized separately for positive and negative values, eliminates all the kinds of pathological behavior associated with the other models.

This has important practical consequences because a similar model is used for the feedforward part of networks that are capable of unsupervised learning, such as the Adaptive Resonance Theory (ART) model [14], which will tend to manifest the same kinds of pathological behavior unless these kinds of normalised measures of similarity are used.

5. Achieving Arbitrary Classification

5.1. Matched Filters

It is possible to design matched filters that yield arbitrary outputs for any given input. This is achieved by means of *composite filters* [15], also known as *synthetic discriminant filters* (SDF) [16].

A composite filter is a linear combination of matched filters, whose coefficients are chosen to obtain a given correlation value. If the correlation values $C_{ij}(0,0)$ between lexicographically arranged objects X_i are put into the columns of a two-dimensional matrix $\mathbf{C}$, and if the required outputs for a linear filter with impulse response $h(x, y)$ are arranged into the vector $\mathbf{y}$, the required coefficients $\mathbf{z}$ for the filter $h(x,y)$ expressed as a linear combination of the training set $X_i(x,y)$ are found by solving the matrix equation

$$\mathbf{y} = C^t\mathbf{z}. \tag{7}$$

The solution is

$$\mathbf{z} = [\mathbf{C_{ji}}^{\mathbf{t}}\mathbf{C_{ji}}]^{-1}\mathbf{C_{ji}}^{\mathbf{t}}{}_{\mathbf{z}} = \mathbf{C}^{\dagger}\mathbf{y}, \tag{8}$$

where $\mathbf{C}^{\dagger}$ is the pseudoinverse of matrix $\mathbf{C}$.

The composite filter determines the correlation values only for the principal correlation peaks. Unfortunately, composite filters usually are associated with high sidelobes, and much effort has been expended to reduce the effects of sidelobes. The more components a composite filter has, the more sensitive to noise it is. In practice, at least for optical implementations, no more that five or six components should be used.

5.2. Neural Nets

How can arbitrary classification be achieved by means of a neural net? In view of our earlier discussion on the equivalence of correlation and inner products, the solution is rather straightforward. We can use the Balanced Weights Model, and set the output weights (between each neuron of the MAXNET slab and the output slab) equal to the elements of the required output, so that when only one neuron remains on after the MAXNET has driven the outputs of all the others to zero, this neuron will set each neuron of the output slab equal to the required value. This can be either an associated prototype or a class index. Indeed, in our experiments with this model, the output weights were chosen so that the system would function as an autoassociative memory, but it is no more difficult to set the weights so that the system will function as a hetero-associative memory with any *a priori* association between the stored objects and the outputs. For example, we can train the system in such a way that when a letter of the alphabet is input, the output will be the letter L (for letter), and when the input is a number, the output will be the letter N (for number). For the network to do this, it probably is necessary to train it with all 26 letters and 10 digits, because the difference between a letter and a number usually is purely conventional, and a network can be trained only by learning all the examples.

The arbitrary classification also could be achieved by a linear feedforward net similar to the Balanced Weights Model, but with the MAXNET slab replaced by a slab of neurons connected only to the input slab, and to the output slab by weights equal to those of the BWM model; but the neurons would have no thresholds, and the relation between the input and the output of the network would be

$$\mathbf{y} = [\mathbf{C}_{ji}{}^{t}\mathbf{C}_{ji}]^{-1}\mathbf{C}_{ji}{}^{t}\mathbf{T}\mathbf{x}, \tag{9}$$

where $\mathbf{C}$ is the correlation matrix between the training set, $\mathbf{x}$ is an input vector, and $\mathbf{T}$ is the interconnection matrix between the intermediate slab of neurons and the output slab, whose columns contain the required output vectors for each stored object. However, such a network could not correct errors in the inputs.

6. Conclusion

We have shown how the fact that similarity is an arbitrary concept can be exploited to achieve invariance in pattern recognition. Objects can be classified into arbitrary classes by means of composite matched filters, or by means of neural nets using the BWM approach. We have seen that normalization of similarity measures can reduce pathological behavior of pattern recognition systems, in particular of neural nets. These ideas can be exploited to improve the performance of neural nets—like the ART model—that are capable of unsupervised learning.

For optical and computer images obtained from cameras or scanners, the basic features are the pixel values, so the theorem of the ugly duckling and its consequences obviously hold for all features that can be derived from those elementary features. For object features that cannot be derived from pixel values (and there are many who would argue that such features exist—for instance, beauty), it is an open question as to whether the theorem applies; the resolution of the question is related to whether or not information and meaning can be reduced without loss to elements directly related to the experiment. Thirty years ago, most scientists would have agreed without hesitation that this is possible, but in recent years, the positivist philosophy of which this hypothesis is the cornerstone has lost almost all its followers, so in harmony with the times, we leave this question open. After all, many advocates of intelligent machines claim that a neural network is more than the sum of its parts (a belief that is not yet supported by evidence), perhaps without realizing that this implies that the network could not be understood by decomposing it into its components.

The discussions of this chapter raise some interesting questions with respect to the behavior of neural networks that use unsupervised learning. It usually is assumed that such systems classify object according to some intrinsic underlying properties belonging to the objects. Because similarity is a logically arbitrary concept, and because of the debatable capability of computers to extract such information, it is necessary to investigate whether such systems are not using hidden underlying assumptions about the object features used for the classification, rather that extracting hidden underlying properties. This may be exploited to improve our capability to get machines to do what we want them to do. In any case, questioning the underlying assumptions of this complex field is a healthy activity; and whatever is found to be possible, we like to think that optics can make it better.

References

1. H. Putnam, "Representation and Reality," MIT Press, Cambridge, MA, 1988.
2. R. Penrose, "The Emperor's New Mind," Oxford University Press, Oxford, 1989.
3. J. R. Searle, "Is the Brain's Mind a Computer Program?" *Scientific American* **262**, 26–31 (1990).
4. P. M. Churchland and P. Smith Churchland, "Could a Machine Think?" *Scientific American* **262**, 32–39 (1990).
5. S. Watanabe, "Pattern Recognition, Human and Mechanical," John Wiley and Sons, New York, 1985.

6. J. T. Tou and R. C. Gonzalez, "Pattern Recognition Principles," Addison-Wesley, Reading, MA, 1974.
7. J. R. Quinlan, "Learning Efficient Classification Procedures and Their Application to Chess End-Games," in "Machine Learning" (R. S. Michalski, J. G. Carbonell, and T. M. Mitchell, eds.), Tioga, Palo Alto, CA, 1983.
8. Y.-H. Pao, "Adaptive Pattern Recognition and Neural Networks," Addison-Wesley, Reading, MA, 1989.
9. A. VanderLugt, "Signal Detection by Complex Filtering," *IEEE Trans. Inform. Theory* **IT–10**, 139 (1964).
10. H. H. Arsenault, "Pattern Recognition, Similarity, Neural Nets and Optics," in Optics in Complex Systems, Proc. SPIE **1319**, paper 426, Garmisch-Partenkirchen, Federal Republic of Germany, (Aug. 5–10, 1990).
11. H. H. Arsenault and Bohdan Macukow, "Neural Network Model for Fast Learning and Retrieval," *Opt. Eng.* 28, 506–512 (1989).
12. H. H. Arsenault, Y. Sheng, A. Jouan, and C. Lejeune, "Improving the Performance of Neural Networks," *Proc. SPIE* **882**, 75–82 (1988).
13. H. H. Arsenault, "Pathological Behavior in a Neural Net and Its Correction," *Proc. 1989 IEEE Conf. on Society, Man and Cybernetics*, Cambridge, MA, 401–404 (Nov. 14–17, 1989).
14. G. A. Carpenter and S. Grossberg, "Category Learning and Adaptive Pattern Recognition, a Neural Network Model," *Proc. Third Army Conference on Applied Mathematics and Computing, ARO Report* **86–1**, 37–56 (1985).
15. H. J. Caulfield and W. T. Maloney, "Improved Discrimination in Optical Character Recognition," *Appl. Opt.* **8**, 2354 (1969).
16. C. F. Hester and D. Casasent, "Multivariant Technique for Multiclass Pattern Recognition," *Appl. Opt.* **19**, 1758 (1980).

CHAPTER 31

Towards Nonlinear Optical Processing

Tomasz Szoplik and Katarzyna Chalasinska-Macukow
Warsaw University, Institute of Geophysics
Warsaw, Poland

1. Introduction

Linearity of the *Fourier transform* (FT) implemented with a lens constitutes a basis of optical linear shift-invariant imaging systems. For the last few decades, optical information processing techniques were based on two linear operations: convolution and correlation, which are available in a 4-f-type correlator. Optical linear shift-invariant systems are useful in a lot of applications, where through modification of a modulation transfer function of a system, one arrives at a desired result. However, in Fourier filtering and pattern recognition, it was hard to get the performance level high enough for practical use. To widen the spectrum of possible optical operations, two important nonlinear image-processing techniques were introduced: halftone screen processing and intensity-to-spatial frequency conversion [1, 2]. Thus, nonlinear operations such as logarithms, power laws, and thresholding became available. Another and simpler possibility for introducing nonlinearities into linear processors is through the use of *spatial light modulators* (SLMs). Computer-controlled SLMs arbitrarily modulate the amplitude or phase of an optical signal. The accomplishment of this promising method has depended on the development of fast, multi-element SLMs. SLMs can be used to insert input plane, output plane, and Fourier plane nonlinearities. The excellent and light-efficient idea of Fourier plane phase-only filters has led to important improvements in

ISBN 0-12-289690-4

correlator performance. Recent progress in Fourier plane nonlinear operations is described in Section 3 [3–5].

In the last 20 years, a lot of digital nonlinear filters have been designed and used successfully in 1-D and 2-D signal processing [6]. These include order statistics, median, and morphological filters. Recently, a considerable effort has been initiated to develop some of these filters for optical processing as part of input and output plane nonlinear operations. Section 4 presents our conviction that the combination of these filters and improved SLMs will establish new frontiers in nonlinear optical processing.

In the section that follows, we start with the description of useful nonlinear dependence, which can be introduced into the classical FT.

2. Nonlinear Angular Magnification of Anamorphic Fourier Spectra

Coherent illumination of a 2-D object function $f(x,y)$, placed in the x, y front focal plane of a spherical lens, results in an object spectrum $F(u,v)$ in the u, v back focal plane of a lens. The 2-D FT represented in Cartesian coordinates,

$$\mathcal{F}\{f(x,y)\} = F(u,v) = \iint_{-\infty}^{\infty} f(x,y)\exp[-2\pi i(xu+yv)]\,dx\,dy \tag{1}$$

is space-variant; that is, the integral kernel

$$h(u,v;x,y) = \exp[-2\pi i(xu+yv)]$$

depends both on where the input signal is applied and where the output arrives. In polar coordinates, the 2-D FT kernel $h(\rho,\theta-\phi;r) = \exp[-2\pi i r\rho\,\cos(\theta-\phi)]$ is rotation-invariant and radially shift-variant, where r, θ are input polar coordinates and ρ, ϕ are output spatial frequencies.

The 2-D FT also can be implemented optically by means of perpendicularly oriented cylindrical lenses of different focal lengths [7]. The 2-D anamorphic FT is defined by

$$\begin{aligned}\mathcal{F}_{\mathrm{A}}\{f(x,y)\} = F_{\mathrm{A}}(u,v) = F(u,Mv) = \\ \iint_{-\infty}^{\infty} f(x,y)\exp[-2\pi i(xu+Myv)\,dx\,dy\end{aligned} \tag{2}$$

in Cartesian coordinates, and

$$\begin{aligned}\mathcal{F}_{\mathrm{A}}\{f(r,\theta)\} = F_{\mathrm{A}}(\rho,\phi) = F(\rho,\zeta) = \\ \int_0^{\infty}\int_0^{2\pi} f(r,\theta)\exp[-2\pi i r\rho(\cos\theta\cos\phi + \mathrm{M}\sin\theta\sin\phi)]\,r\,dr\,d\theta\end{aligned} \tag{3}$$

in polar coordinates. M is the coefficient of anamorphism of the transformer, which is given by the ratio of the focal lengths of the cylindrical lenses active in the x and y directions, and ζ is the azimuthal frequency in the anamorphic spectrum. To get an anamorphic Fourier spectrum, the back focal planes of cylindrical lenses must coincide. The anamorphic FT is linear.

Azimuthal frequency in the anamorphic spectrum ζ is related to that in the classical spectrum ϕ as follows:

$$\zeta = \arc\tan\,(M\tan\phi), \tag{4}$$

and the corresponding plots are shown in Fig. 1. The nonlinear dependence between azimuthal spatial frequencies ϕ and ζ in the FT and anamorphic FT, respectively, comes directly from full space variance of the anamorphic FT in polar coordinates. From Eq. (4), we arrive at the coefficient of angular magnification $K(\alpha, M)$ of anamorphic azimuthal frequencies ζ with respect to those in a classical spectrum ϕ, which is defined as

$$K(\alpha, M) = \frac{d\zeta}{d\phi} = \frac{M}{1 + (M^2 - 1)\sin^2\alpha}. \tag{5}$$

The angular magnification is very helpful when fine operations are to be performed on the angular spectrum. In the magnified part of a spectrum, a

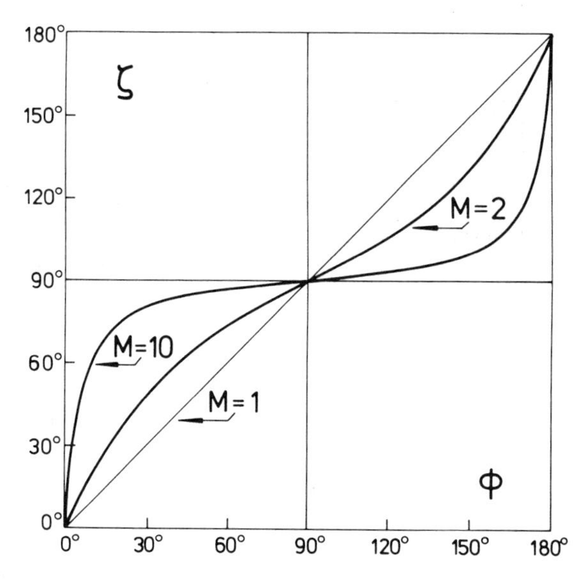

FIG. 1. Plots of azimuthal frequencies ζ in the anamorphic Fourier spectrum versus azimuthal frequencies ϕ in the classical Fourier spectrum calculated for three values of the coefficient of anamorphism $M = 1$, 2, and 10.

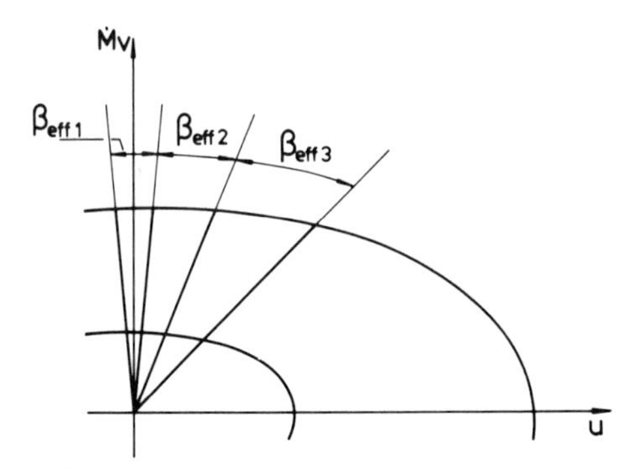

FIG. 2. Scheme of different effective angular extents of the same wedge filter placed at different orientations in the anamorphic spectrum.

wedge filter of angular extent β is equivalent to a wedge with a smaller effective extent,

$$\beta_{\text{eff}} = \frac{\beta}{K(\alpha, M)} \tag{6}$$

This idea was used to improve the accuracy of angular analysis of Fourier spectra [8].

Another application of nonlinear magnification of anamorphic spectra was found in directional pseudocolor encoding of spatial frequencies [9]. The progress with respect to previous work was due to the small effective extents of chromatic sector filters, which are shown schematically in Fig. 2.

Recently, a holographic optical element in the form of an elliptical zone plate was proposed to simplify anamorphic systems [10]. Future anamorphic correlators should be composed of computer-generated optical elements with compensated aberrations.

3. Nonlinear Correlators

Matched filtering is a parallel technique of searching for a desired pattern in an input scene by optical comparison of the Fourier spectra of both the input scene and the target. Recent development of nonlinear procedures makes this technique very powerful.

Nonlinear procedures can be introduced into optical correlators in different ways. A nonlinear operation can be applied in the filter plane as a nonlinear preprocessing of the target spectrum (see Fig. 3). When N_2 is a nonlinearity and N_1 is identity function, we have the case of the *phase-only filter* (POF) or *binary phase-only filter* (BiPOF) [3–5]. When nonlinear

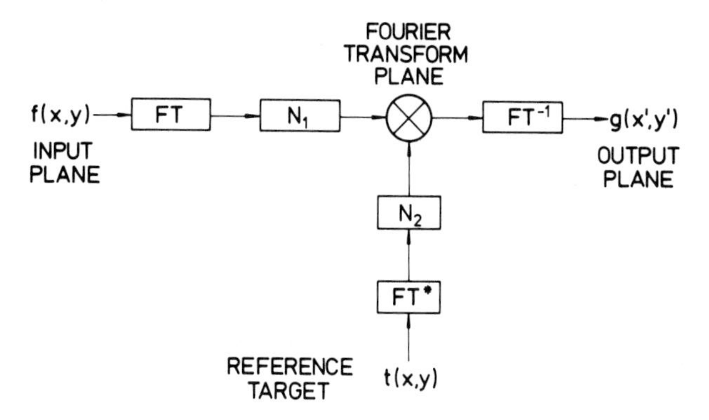

FIG. 3. Nonlinear optical correlator.

functions N_1 and N_2 are identical, we have, for example, phase-only information processing.

The applications that have been found for such nonlinear systems indicate a great potential for future research and development of optical and hybrid information processing methods.

3.1. Phase-Only Filter and Its Binary Version

For a long time, optical correlation methods suffered from very inefficient use of light. Recent work has shown that the problem of low light efficiency of correlators can be solved with the help of various POFs [3–5].

In general, the FT of an object $t(x, y)$ is complex

$$\mathscr{F}\{t(x,y)\} = A(u,v)\exp[i\varphi(u,v)], \tag{7}$$

where $A(u, v)$ and $\varphi(u, v)$ are spectral amplitude and phase, respectively. The nonlinear operation realizing the POF with transmittance T_φ is defined as

$$T_\varphi(u,v) = \begin{cases} \exp[-i\varphi(u,v)], & \text{for } A(u,v) \neq 0 \\ 0, & \text{for } A(u,v) = 0. \end{cases} \tag{8}$$

The light efficiency of this filter is high and tends to the 100% theoretical limit independent of the object function. The POF properties—better discrimination ability between similar objects than that of a classical matched spatial filter (MSF) and very narrow output signal in the correlation plane (see Figs. 4a,b)—result from the domination in the impulse response of high spatial frequencies enhanced by the nonlinear operation of Eq. (8). Simultaneously, the large spectral filter bandwidth reduces the

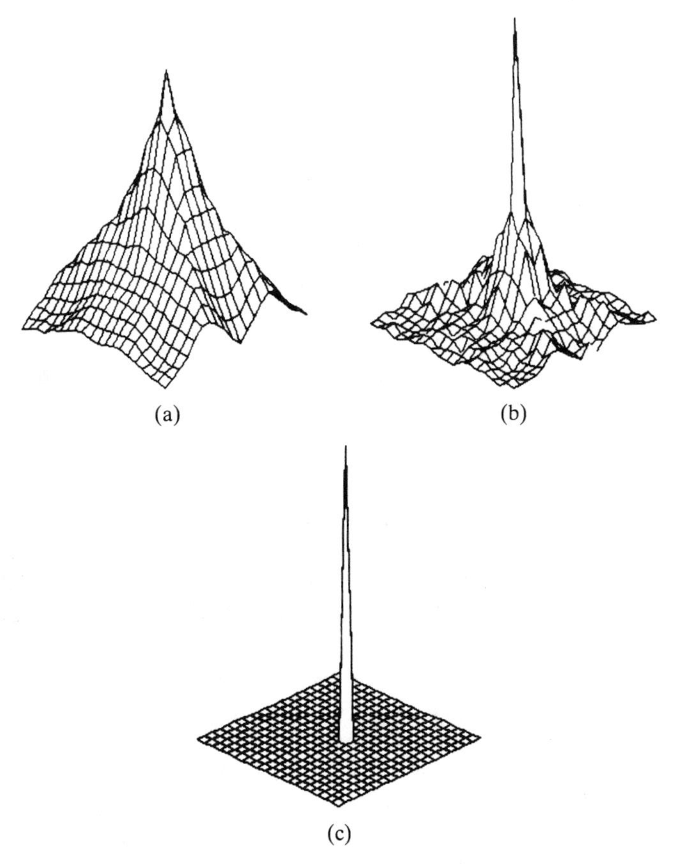

FIG. 4. Output signals from : (a) matched spatial filter (MSF); (b) phase-only filter (POF); and (c) phase-only correlaion (POC).

signal/noise ratio (SNR) when stochastic noise is present. Thus, in the presence of noise, the POF bandwidth has to be optimized, and for each recognition problem, a compromise between the sharpness of the correlation peak and the SNR value must be made.

The continuous POF defined by Eq. (8) reproduces accurately the reference phase-only function. In some applications, however, restricting the filter function only to binary values is advantageous. Binary nonlinear operations realizing the BiPOFs are defined as

$$T'_{\varphi}(u,v) = \begin{cases} 1, & \text{for } \mathrm{Im}[T_{\varphi}(u,v)] > 0 \\ -1, & \text{otherwise} \end{cases} \tag{9}$$

or

$$T'_{\varphi}(u,v) = \begin{cases} 1, & \text{for } \mathrm{Re}[T_{\varphi}(u,v)] > 0 \\ -1, & \text{otherwise}. \end{cases} \tag{10}$$

The method of constructing the BiPOF using Eq. (9) is equivalent to a binary sine transform, while the method of Eq. (10) is equivalent to a binary cosine transform [11].

From the viewpoint of applications in optical correlators, BiPOFs are the most attractive filters because of easy implementation with SLMs. All recently proposed phase-only methods work well when binary phase quantization is used.

Despite their great potential, BiPOFs suffer from certain drawbacks, such as inherent noise and the possibility of large false alarms. These difficulties depend on the techniques employed in BiPOF synthesis. Both versions of BiPOFs essentially contain only half of the information required for unambiguous correlation of an object. As a result, a BiPOF made for an arbitrary object recognizes both the object and its inverse. Optimization of BiPOFs in the SNR sense—and to avoid false alarms—was studied recently [4].

The general case of the nonlinear matched filter can be described by the formula proposed by Javidi,

$$g(E) = \begin{cases} E^k, & \text{for} \quad E \geq 0 \\ -|E|^k, & \text{for} \quad E < 0, \end{cases} \tag{11}$$

where $g(E)$ represents the transfer characteristics of the nonlinearity, E is the linear matched filter input transmittance of the nonlinearity, and k is a nonnegative real number [12]. According to this formula, various types of nonlinear filters can be produced by varying the degree of nonlinearity. The hard-clipping nonlinear filter, with $k = 0$, produces the BiPOF, and its first-order harmonic term corresponds to the continuous POF [12].

3.2. Pattern Recognition Using Phase Information Only

Pattern recognition using phase information only is a generalization of nonlinear matched filtering in which both the filter function and the input scene spectrum pass through the same nonlinearity defined by Eq. (8) before they are multiplied in the spectral domain (the case of $N_1 = N_2$ in Fig. 3) [13–15]. As a result, only phase information is kept about both the target and the input scene. This nonlinear procedure assures larger spectral bandwidth of the system than in the POF case. As a consequence, the output peak is a diffraction-limited delta function (see Fig. 4c). It is worth mentioning that this nonlinear system is equivalent to a three-layer neural network [15].

The pure *phase-only correlation* (POC) method has high capability to discriminate between different objects [14, 15]. An additional advantage of the method is its invariance with respect to the contrast of one-element

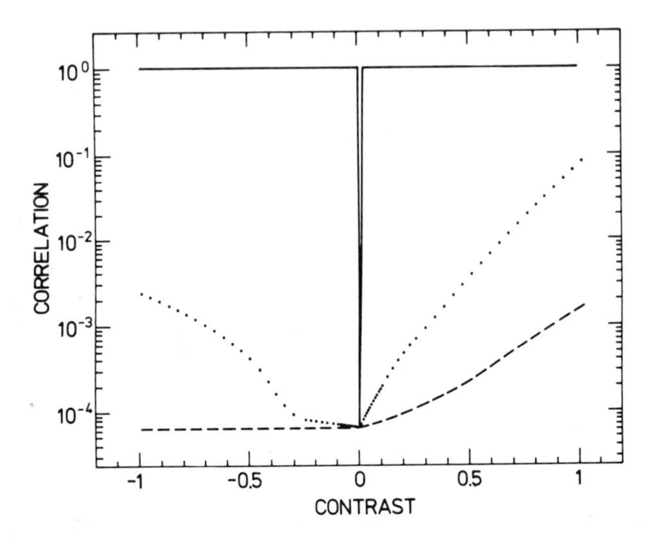

FIG. 5. Contrast sensitivity in case of a binary input scene: MSF— dashed line; POF—dotted line; and POC—continuous line [16].

input scenes. Figure 5 presents a comparison of contrast sensitivities in the case of a one-element binary input scene calculated for the MSF, POF, and POC [16]. In the case of a multi-object input scene, the POC method works well but is contrast-variant. Its sensitivity to intermodulation noise, which can produce false alarms and misses, depends on the input scene geometry [16].

Phase-only correlators were not optically implemented yet and still are under study.

3.3. Nonlinear Joint Transform Correlator

The *nonlinear joint transform correlator* is an optical-digital system that, without any additional filters, realizes various functions depending on the nonlinearity applied in the FT plane. Different implementations of such a system using both high-contrast optically and electrically addressed SLMs are possible. The input plane contains the reference and the input signals displayed on the SLM (see Fig. 6). The images are Fourier-transformed by lens FTL_1 and the interference between the FTs of the input and reference functions appears in the Fourier plane. The interference intensity is detected nonlinearly either by a high-contrast optically addressed SLM or by a CCD array connected to a computer. The interference intensity, in the first case, is optically thresholded by a high-contrast SLM; and in the second case, is digitally thresholded by a computer, and an SLM is only

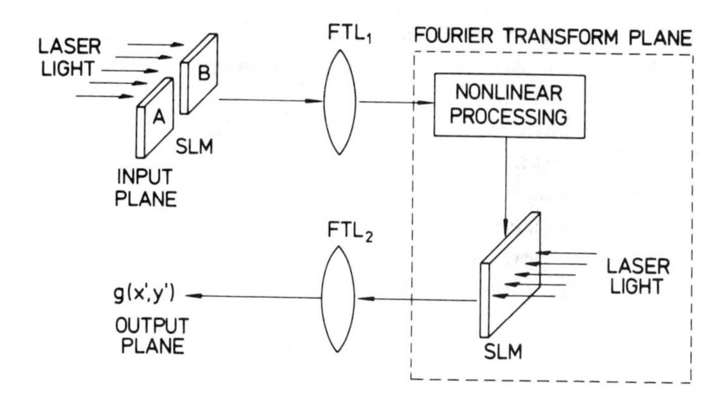

FIG. 6. Nonlinear joint transform correlator.

used for readout. The correlation function is produced in the output plane by the inverse FT of the thresholded interference intensity distribution.

To find the effects of nonlinearity on the correlation signal, a general type of kth-law device in the FT plane was considered [17]. The transfer characteristic $N(I)$ of the device has the form

$$N(I) = I^k, \qquad \text{for } k < 1. \tag{12}$$

The case of $k = 1$ corresponds to a linear device (in the intensity sense) and describes the conventional joint transform correlator. The case of $k = 0$ corresponds to a hard-clipping nonlinearity and describes a binary joint transform correlator.

The high value and good definition of the correlation peak are due to the fact that the increase of nonlinearity causes an increase of the system spectral bandwidth. However, in multiple object scenes, high nonlinearity can produce false alarms and misses [18].

4. Morphological Image Processing

Morphological image processing is an image-plane nonlinear local filtering technique [6]. It was developed during the last 20 years as a digital technique. At present, a considerable effort is being made to use it in optical parallel processors. This means that instead of digital processing of discrete images, we are shifting to morphological processing of continuous images in holographic systems and discrete images in systems with SLMs.

Morphological filters (MFs) locally modify an input image form and structure by interaction with a finite structuring element [6, 19, 20]. The filtration result depends on the structuring element size and shape, which are chosen arbitrarily to get a desired effect. There are no analytic criteria to design an MF. The MFs are useful in edge detection, noise supression, and shape modification and smoothing.

There are four basic MF transformations: dilation, erosion, opening, and closing. The morphological transformations are invariant under translation; that is, the translation of the image transform is equivalent to the transform of the translated image. They are compatible under change of scale. The third property results from the assumed equivalence of local structures of an object and is called the *local knowledge principle*. It says that transforms made piecewise in a bounded image subset are equivalent. Finally, morphological transformations satisfy the *semi-continuity principle*, which says that the boundary of the transform is equal to the transform of the boundaries [6].

In digital applications, MFs are defined for three cases: as set-processing filters for binary images and binary structuring elements; as function-processing filters for gray-scale images and gray-scale structuring elements; and as function- and set-processing filters for gray-scale and binary images and binary structuring elements [19, 20]. From the point of view of optical nonlinear processing, function-processing and the function- and set-processing filters are the most interesting. They have different properties, however.

The function-processing filters are based on morphological convolutions. Dilation is defined as a maximum of the sum of function f and a structuring function g of finite region of support,

$$(f \oplus g^s)(x) = \max\{f(y) + g(y - x) : y \in Z^2\}, \tag{13}$$

where $g^s(x) = g(-x)$ is a function symmetric with $g(x)$ and Z^2 is a 2-D set of integers. Erosion is defined as a minimum of the difference of functions f and g,

$$(f \ominus g^s)(x) = \min\{f(y) - g(y - x) : y \in Z^2\}. \tag{14}$$

Opening is defined as erosion followed by dilation, and closing is defined as dilation followed by erosion.

On the other hand, due to the binary structuring element, the function- and set-processing fillters commute with thresholding. Thus the gray-scale image problem can be reduced easily to a binary one via a so-called threshold decomposition [20].

Recent analysis has shown that morphological operations can be used for the description of independently developed rank *order statistics filters* (OSFs) [19, 20]. Use of OSFs in digital nonlinear processing is based on the discrete finite area convolution operator approach. These filters are very efficient in spiky-noise removal and the smoothing of images. OSFs locally interact with an image in such a way that each pixel is replaced by a filtered value taken from the close neighborhood contained within the sliding filter window [21]. At any 2-D image matrix element x_{kl}, a filter window w_{rs} with odd pixel number $r = 1, \ldots, 2p + 1$ and $s = 1, \ldots, 2q + 1$ can be centered. We have $x_{kl} = w_{p+1,q+1}$ and $(2p + 1)(2q + 1) = 2N + 1 = T$. The filter window elements denoted by Z_{kl} can be ranked, for example, in an ascending series as

$$Z_{kl}(1) \leq Z_{kl}(2) \leq \cdots \leq Z_{kl}(T). \tag{15}$$

The output filtering operation $y_{kl} = f(x_{kl})$ is given as a function of the whole filter window content,

$$y_{kl} = f\{Z_{kl}(1), Z_{kl}(2), \cdots, Z_{kl}(T)\}, \tag{16}$$

which combines linear operations, like averaging; and nonlinear ones, like thresholding or taking ith-order rank values.

OSFs are divided into three classes. L-type OSFs give an output value as a linear combination of the order statistics

$$y_{kl} = \sum_{i=1}^{T} a(i) Z_{kl}(i), \tag{17}$$

where $a(i)$ is a data-dependent coefficient. This linear expression becomes nonlinear when in the summation, we reject α extreme values on both sides of the ranked series. Thus, we arrive at the α-trimmed mean filter. Another possible way to make the L-type filter nonlinear is through proper thresholding.

R-type filters give an output as the ith-ranked order values of input data

$$y_{kl} = \text{order statistics of} \{g(w_{rs})\}, \tag{18}$$

where $g(\cdot)$ denotes any linear function of the window content.

In the M-type filters, maximum-likelihood estimates are taken into account before calculating the output value. Extreme input values are replaced by low and high limit values of the estimated data range. Then the output is calculated as an average or ith order rank value of modified input data. The median filter is a special case of L−, R−, and M-type filters.

OSFs commute with thresholding and, therefore, are closely related to function- and set-processing MFs [20].

So far, a small number of optical experiments with MFs and OSFs have been made (e.g., [22–24]). Figure 7 shows a scheme of the recently proposed programmable real-time optical morphological image processor [24]. The experiments prove the potential usefulness of optically implemented filters for spiky-noise removal and image smoothing with simultaneous edge preserving. MFs and OSFs will be used in optical parallel

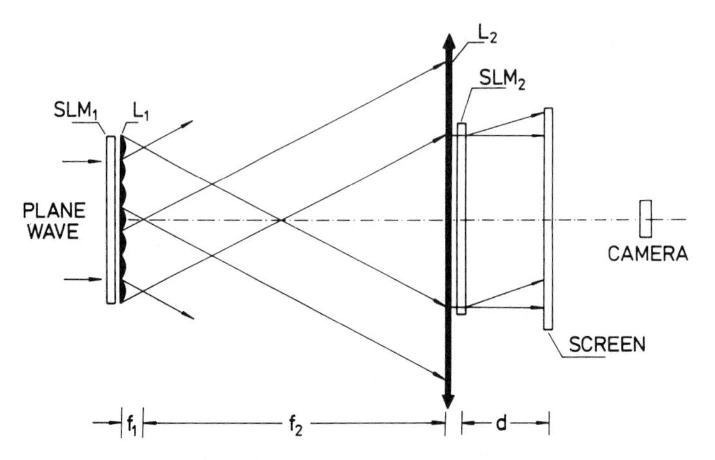

FIG. 7. Scheme of the real-time programmable morphological image processor [24]. SLM_1 illuminates chosen L_1 lenslet elements to form a structuring element. Lenses L_1 and L_2 direct a structuring element beam onto an image displayed on SLM_2. Screen and camera collect all beams arriving at different angles and result is thresholded.

processors for similar purposes as their digital forerunners. This includes the processing of images with distinctive geometrical structure, such as microscope images of mineral and biomedical samples, meteorological radar images, and other images.

5. Conclusions

The future usefulness of optical nonlinear processing with phase-only, binary phase-only, morphological, and order statistics filters will depend on the successful solution of a few technical problems. The most important issue is the existence of multi-element, fast, and cheap SLMs. This need has been present in the development of optical image processing for many years. Nowadays, however, the feasibility of good SLMs is becoming more probable. The next issues are connected with thresholding and uniform illumination of processed images. Thresholding is the most important operation involved in nonlinear processing. The threshold level depends on the amount of light in a system and requires uniform illumination conditions. The latter demand is difficult to fulfill in holographic systems for morphological filtering. The aforementioned two questions have not been studied yet in detail.

In the previous sections, we presented three separate problems in different stages of development and of diverse importance. The idea of the anamorphic FT has improved the accuracy of angular operations on

Fourier spectra. Phase-only filtering is a well-developed technique that has enhanced the value of the optical correlation method. Subsequent optical implementations of nonlinear joint transform correlators and nonlinear matched filtering depend on the development of SLMs. Morphological filtering is a technique in an early stage of development. It is difficult to estimate its future impact on nonlinear image processing. However, we should take care to check its possibilities.

References

1. S. H. Lee, "Nonlinear Optical Processing," in "Optical Information Processing—Fundamentals" (S. H. Lee, ed.), Springer, Berlin, 1981.
2. A. A. Sawchuk and T. C. Strand, "Fourier Optics in Nonlinear Signal Processing," in "Applications of Optical Fourier Transform" (H. Stark, ed.), Academic Press, New York, 1982.
3. H. Bartelt, "Unconventional Correlators," in "Optical Signal Processing" (J. L. Horner, ed.), Academic Press, San Diego, 1987.
4. D. L. Flannery and J. L. Horner, "Fourier Optical Signal Processors," *Proc. IEEE* **77**, 1511–1527 (1989).
5. K. Chalasinska-Macukow, "Generalized Matched Spatial Filters with Optimum Light Efficiency," in "Optical Processing and Computing" (H. H. Arsenault, T. Szoplik, and B. Macukow, eds.), Academic Press, Boston, 1989.
6. J. Serra, "Image Analysis and Mathematical Morphology," Academic Press, London, 1982.
7. T. Szoplik, "Line Detection and Directional Analysis of Images," in "Optical Processing and Computing" (H. H. Arsenault, T. Szoplik, and B. Macukow, eds.), Academic Press, Boston, 1989.
8. T. Szoplik, K. Chalasinska-Macukow, and J. Kosek, "Accuracy of Angular Spectral Analysis with an Anamorphic Fourier Transformer," *Appl. Opt.* 25, 188–192 (1986).
9. M. S. Millan, C. Ferreira, A. Pons, and P. Andrés, "Application of Anamorphic Systems to Directional Pseudocolor Encoding," *Opt. Eng.* **27**, 129–134 (1988).
10. S. Bará and C. Gomez-Reino, "A Holographic Optical Element for Nonsymmetric Fourier Transform Systems," *J. Mod. Opt.* **36**, 21–30 (1989).
11. J. A. Davis, D. M. Cottrell, G. W. Bach, and R. A. Lilly, "Phase-Encoded Binary Filters for Optical Pattern Recognition," *Appl. Opt.* **28**, 258–261 (1989).
12. B. Javidi, "Generalization of the Linear Matched Filter Concept to Nonlinear Matched Filters," *Appl. Opt.* **29**, 1215–1224 (1990).
13. J. J. Pearson, D. C. Hines, Jr., S. Golosman, and C. D. Kuglin, "Video-Rate Image Correlation Processor," *Proc. SPIE* **119**, 197–205 (1977).
14. K. Chalasinska-Macukow and E. Baranska, "Phase-Only Discrimination with Preprocessing," *Proc. SPIE* **1134**, 93–95 (1989).
15. O. K. Ersoy and M. Zeng, "Nonlinear Matched Filtering," *JOSA* **A6**, 636–648 (1989).
16. K. Chalasinska-Macukow, F. Turon, M. J. Yzuel, and J. Campos, "Performance of the Pure Phase-Only Correlation Method for Pattern Recognition," *Proc. SPIE* **1347**, 262–273 (1990).
17. B. Javidi, "Nonlinear Joint Power Spectrum Based Optical Correlation," *Appl. Opt.* **28**, 2358–2367 (1989).
18. F. T. S. Yu, F. Cheng, T. Nagata, and D. A. Gregory, "Effects of Fringe Binarization of Multiobject Joint Transform Correlation," *Appl. Opt.* **28**, 2988–2990 (1989).

19. P. Maragos, "Tutorial on Advances in Morphological Image Processing and Anaylsis," *Opt. Eng.* **26**, 623–632 (1987).
20. P. Maragos and R. W. Schafer, "Morphological Filters—Part I: Their Set-Theoretic Analysis and Relations to Linear Shift-Invariant Filters; Part II: Their Relations to Median, Order-Statistics, and Stack Filters," *IEEE Trans. Acoust., Speech, Signal Processing* **ASSP-35**, 1153–1184 (1987).
21. Y. S. Fong, C. A. Pomalaza-Ráez, and X. H. Wang, "Comparison Study of Nonlinear Filters in Image Processing Applications," *Opt. Eng.* **28**, 749–760 (1989).
22. J. M. Hereford and W. T. Rhodes, "Nonlinear Optical Image Filtering by Time-Sequential Threshold Decomposition," *Opt. Eng.* **27**, 274–279 (1988).
23. E. Botha, J. Richards, and D. P. Casasent, "Optical Laboratory Morphological Inspection Processor," *Appl. Opt.* **28**, 5342–5350 (1989).
24. Y. Li, A. Kostrzewski, D. H. Kim, and G. Eichmann, "Compact Parallel Real-Time Programmable Optical Morphological Image Processor," *Opt. Lett.* **14**, 981–983 (1989).

CHAPTER 32

New Aspects of Optics for Optical Computing

Valentin Morozov*
*P.N. Lebedev Physical Institute of the USSR Academy of Sciences
Moscow, USSR*

1. Introduction

The problem of designing an optical computer is akin to the question of what is possible and what is impossible in optics, and has been the focus of attention of investigators for a long time. The advent of lasers stimulated optical research significantly, and many things are realized today that recently looked impossible. Implementation of a laser, for example, is impossible from the standpoint of classical physics. Over many centuries, inventors in various countries were attracted by the problem of designing an optical system capable of collecting a divergent optical beam into a narrow, directed light beam. The classical theory of optical devices, however, asserts that no optical system can increase the light beam's brightness. The proof of this assertion, as based on the Lagrange-Helmholtz law, may be found in any manual on optics. This law holds for any optical system composed of any number of reflecting and refracting elements arranged in an arbitrary manner; but as soon as we turn to laser systems, we question the validity of this assertion. Indeed, what is the difference between a ruby rod with mirrors and xenon lamp and conventional optical devices?

* Current address: Opticomp Corporation, P. O. Box 10779, Zephyr Cove, Lake Tahoe, NV 89448

ISBN 0-12-289690-4

Nevertheless, is ruby, neodimium, and other lasers with optical pumping, brightness increases tremendously. This example demonstrates that light brightness can be increased by transformation of light beams; i.e., the laws governing transformation of light beams may be of a more fundamental nature.

In optical computers, information is carried by optical beams and logical operations are carried out by optical logic gates. Information about two-dimensional images (pictures), matrices, etc. is stored in holographic form. As applied to optical computers, the question can be formulated as follows: Is our knowledge of optics sufficient for the determination of optical computer parameters such as power consumption of logical gates or optical memory size? Design of optical computers should take account of the experience gained from electronic computer engineering. Today, tens and hundreds of thousands of different logic gates are integrated on a single chip. Is it possible to integrate holograms with other elements of optical beam control, i.e., to make in planar form the existing three-dimensional optical schemes?

This, of course, is not an exhaustive list of questions arising from optical computer design. From the standpoint of modern knowledge, we try to answer the following questions:

- What is the maximal efficiency of transforming one light beam into another, i.e., eventually, the efficiency of an optical logic gate?
- What maximal amount of information can be accommodated in a hologram?
- Are planar counterparts of three-dimensional optical schemes plausible?

Not all the considerations that are presented in this chapter are implemented in practice, but they will illustrate the great potentialities of optics.

2. Thermodynamics of Light-Beam Transformation

In digital computers, each logic element can be in one of two states, only one of which is regarded as 1 and another as 0. The optical logic gate performs Boolean logic operations over two or more light beams. In current electronic circuits, the power for switching one gate within a wide frequency range 10^6 through 10^9 cps is approximately the same and is 10^{-11} to 10^{-13} J. It is very important to know for optical computers the minimal power consumption of one switching event. Derivation of these estimates seemingly should abstract the performances of the existing devices that are dictated by their design and technology levels. The realizability of low power consumption for switching is defined by numerous factors, such as the physical phenomenon used, the characteristics of material, the technological level, etc. It seems that forecasting the characteristics of optical logic gates, transistors, etc. on the basis of existing designs would resemble an attempt to forecast the development of electronic computer hardware on the basis of an improved tube. It is more correct to consider only the physical limitations following from thermodynamics because their validity is of an absolute nature.

Let a light beam fall onto an optical system, by which any material medium is meant, including an optical logic gate. Upon transformation of the light beam by the system, a portion of the beam energy δE is liberated within the system, i.e.

$$E = \delta E + E',$$

where E and E' are the light beam densities, respectively, before and after the optical system, with the system input and output areas assumed to be equal. The optical system transformation factor

$$\eta = \frac{E'}{E} = \frac{E - \delta E}{E}$$

indicates which portion of the initial beam energy is used for its transformation, i.e., in the case of an optical logic gate, for execution of a logic operation. For a linearly polarized light beam propagating within the solid angle $\Delta\Omega$ and concentrated in the spectral interval ω, $\omega + \Delta\omega$, the entropy density is as follows:

$$\Delta S = \Delta\rho\,\Delta\Omega\,[(n+1)\ln(n+1) - n\ln n],$$

where $\Delta\rho = \omega^2\,\Delta\omega/(2\pi c)^3$ is the number of field oscillators in the frequency interval $\omega, \omega + \Delta\omega$, and n is the mean number of quanta per field oscillator. In this case, the light beam energy density is

$$E = \Delta\rho\,\Delta\Omega h\omega \cdot n.$$

If the light beam temperature is defined as usual as

$$kT = \frac{h\omega}{\ln(1 + n^{-1})},$$

and if one takes into account that in the optical portion of the spectrum, radiation always features high temperature, i.e., $h\omega/kT \ll 1$, the light beam entropy

$$\Delta S = \frac{E}{kT}\left(1 + \ln\frac{kT}{h\omega}\right) \qquad \text{if } \frac{h\omega}{kT} \ll 1.$$

Assuming that δE is liberated in an optical system at temperature T_0 and that

$$\delta S = \frac{\delta E}{kT_0},$$

we obtain

$$\eta \leq 1 - \frac{T_0}{T}\left(1 + \ln\frac{kT}{h\omega}\right) \quad \text{for} \quad \frac{h\omega}{kT} \ll 1.$$

The expression for the light beam transformation efficiency resembles the well-known expression for the Carnot cycle efficiency, the light beam temperature T corresponding to that of the working medium and the optical system temperature T_0 to that of the coolant, which usually is the room temperature. Depending on the parameters of radiation power, divergence, and spectrum width, the radiation temperature estimate for a typical single-mode semiconductor laser is $T = 10^{12}$ to 10^{14} K. For visible light, $h\omega \sim 10^{-19}$ J, for a semiconductor laser, $h\omega/kT \sim 10^{-9}$, and the minimal energy loss that is fundamentally necessary for light transformation is practically negligible, i.e.,

$$\frac{\delta E \min}{E} \geq \frac{T_0}{\mathrm{T}} \ln\frac{kT}{h\omega} \sim 10^{-10}.$$

About 10^8 optical logic elements of 1 μm^2 size may be located within 1 cm^2, and, for beam power 1 W, the minimum energy absorbed for switching one element as defined by thermodynamic considerations is 10^{-18} J.

The discrepancy between the real values and those obtained through thermodynamic considerations for electronic gates is considerable, which is due to a number of technological and system engineering reasons. The same situation, obviously, will take place for optical logic elements. For electronic circuits, the increase of fluctuation levels with the reduction of electronic gate volumes and improvement of their packaging density, however, is unavoidable. Currents in gates lead to crosstalk noise, which amounts to an increase of effective noise temperature. The number of neighbors effectively influencing a given gate grows with packaging density. Let the effective noise temperature be

$$T_{\text{eff}} = T + \alpha \Delta E_S,$$

where ΔE_S is signal energy, and α the coefficient that allows for the portion of energy of neighboring elements generating noise for the element of interest. Taking into account the fact that the signal energy should be greater than that of the effective noise temperature T_{eff}, assume that $\Delta E_S = \beta T_{\text{eff}}$, where $\beta > 1$. Hence,

$$T_{\text{eff}} = \frac{T}{1 - \alpha\beta}.$$

As the coefficient of mutual crosstalk noise α approaches β, higher effective noise temperature results, as compared with T, and higher switching energy will be observed. This fact is unavoidable and is related to the charge of an electron carrying information in electronic circuits. Since the photons are neutral and light interaction with an optical system can go on without current, the influence of this constraint in optical logic gates may be minimized.

An optical logic gate can perform coherent light beam transformations of amplitude, phase, polarization, and frequency. A wider variety of physical phenomena can be used for optical signal transformation as compared with electronic signal processing. General thermodynamic considerations, of course, cannot suggest specific ways to implement effective gates, but the difference in the physical nature of information carriers (photons and electrons) can be decisive in the determination of power loss per switching event, and the preceding thermodynamic relations stimulate design of optical logic elements.

3. What Maximal Amount of Binary Information Can Be Stored in a Hologram?

Optical computer engineering makes wide use of holographic storage of binary information. Let us estimate the amount of information that can be stored in a thin hologram. Of course, any optical system has limited resolution owing to the wave nature of light. This limitation is due to diffraction and is defined by the wavelength of radiation used and the numerical aperture of the optical system. The *Rayleigh resolution* is defined by the minimal resolved distance between two point light sources of the same intensity. For holograms of diameter a, the resolvable angular distance between two points is

$$\zeta \sim \frac{\lambda}{a}$$

where λ is the light wavelength. A "point" image reconstructed from a hologram will have at distance F from it the following size

$$\Delta\chi = F\zeta \sim \frac{\lambda F}{a}.$$

If D is the size of the image field, the number of points that can be placed in it is

$$n = \frac{Da}{\lambda F},$$

and the total number of points in the area D^2 is

$$n^2 = \frac{D^2 a^2}{\lambda^2 F^2}.$$

If the complete number of resolved points is related to the hologram area, we obtain the density of information written in the hologram

$$C = \frac{n^2}{a^2} = \frac{\Omega}{\lambda^2},$$

where Ω is the angular aperture of the optical system. For $\Omega = 0.2$ steradians and $\lambda = 10^{-4}$ cm we obtain $C \sim 2.10^5$ bit/mm^2. The actual experimental information density is somewhat less and is about $\sim 10^4$ bit/mm^2 because rather high signal/noise ratio is required in neighboring image elements. The relations for information density in holograms have been derived for classical relations and indicate that the amount of information in a thin hologram can be increased only by increasing the optical system aperture or reducing the radiation wavelength.

The founder of holography, D. Gabor, established, however, that the total number of degrees of freedom of a wave field is

$$N = S\Omega/\lambda^2,$$

with S being the object's area and Ω the optical system angular aperture, is the fundamental invariant of a stationary wave field . Within the total number of informational degrees of freedom of the wave field, one can vary relations between spatial, temporal, and polarizational degrees of freedom, maintaining their total number N. It is very important that the total number of degrees of freedom N is the optical system invariant and not the bandwith of the transmitted spatial frequencies Ω/λ^2. This fact suggests that effective redistribution of informational degrees of freedom could improve the maximal transmitted spatial frequency of a given optical system. In particular, reduction of the useful field of object S brings about the same increase of bandwith of the transmitted spatial frequencies Ω/λ^2 and, thus, allows one to surpass the classical limit of information density on the hologram. The gain in resolution is inversely proportional to the reduction of useful object area, since

$$N = \left(\frac{S}{K}\right)\frac{K\Omega}{\lambda^2},$$

with K for the super-resolution factor.

The aforementioned is illustrated by the optical scheme of Fig. 1. In the conventional optical scheme of Fig. 1a, the point source located in the object plane generates a spherical wave whose spatial spectrum cannot be transmitted completely into the image space because of limited hologram aperture. As a result, only a part of the spatial spectrum is transmitted into the image space, thus confining the size of point source image to the classical relation. The wave nature of light, however, prohibits only one thing—generating a point source image whose size is smaller than the light wavelength. If the size of the point source image appreciably exceeds the

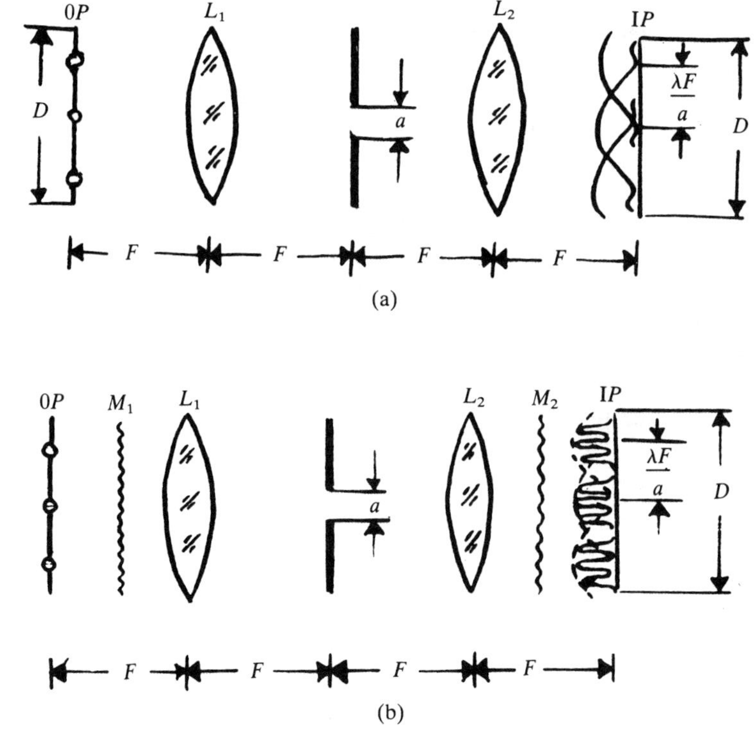

FIG. 1. Optical scheme for point-source imaging: a) OP—object plane; IP—image plane, L_1, L_2—lenses, a-hologram aperture. b) with super-resolution; M_1, M_2—diffraction gratings

wavelength—i.e., $F/a \gg 1$—the resolution of the given optical system can be improved in several ways. Figure 1b illustrates expansion of the bandwidth of the transmitted spatial frequencies at the expense of narrowing the useful object field. Identical diffraction gratings M_1 and M_2 are placed in the object and image spaces. Owing to the multiplicative properties of M_1, small but different areas of the original spatial spectrum are transmitted into the image space. Upon diffraction on M_2, the original direction of the beams is restored in the image space; and in the image plane, an image of the point source occurs featuring better resolution than the scheme of Fig. 1a. Diffraction at M_2, however, also gives rise to waves of undesirable directions, which reduce the useful object field. Actually, optical synthesis of the aperture is performed by the first diffraction grating M_1. Each diffraction order generated by this grating basically is a separate optical system with its own aperture. The optical system of Fig. 1b, therefore, corresponds in terms of capacity to a scheme where optical information is transmitted through multiple hologram apertures rather than a single one, their total number corresponding to the number of orders generated by M_1 at light diffraction.

Figure 2 depicts the results of an experimental study of one-dimensional optical aperture synthesis. The scheme of hologram writing and imaging is identical to that of Fig. 1b. The holograms were recorded by a helium-neon laser, with a hologram diameter of 12 μm, and an objective focus distance of 250 mm. Point-source holograms were recorded and reconstructed. One can see in Fig. 2 that within the Airy disk point source image, super-resolution is observed together with additional structural maxima. By selecting the diffraction grating parameters, these structural maxima may be removed from the Airy disk, but, nevertheless, they will reduce the useful field of view in the object plane.

In a conventional scheme with the same components but without diffraction gratings, the hologram writing density estimated by the classical formulas is 10^4 bit/mm^2. The writing density obtained with aperture synthesis in terms of two-dimensional synthesis is 2.5×10^6 bit/mm^2, which is much higher than the classical diffraction limit.

When applying the method of optical aperture synthesis to storage of binary information in holograms, one must bear in mind two important points enabling one to surpass the classical limit of information writing density. First, the information is stored as a two-dimensional set of signal points, and it is only the point's presence or absence that counts, and not its brightness or form. Second, the positions of signal points in the two-dimensional set are known in advance, and the only problem is to detect information in one or another position. These factors allow one to place in the observation plane a non-transparent mask (with openings) that blocks the additional structural maxima and passes light only where information

FIG. 2. Experimental results for imaging of a point source in an optical scheme with super-resolution: The central peak is the image of the point source; sidelobes are additional structural maxima. The hologram diameter is 12 μm, and the objective focal length is 250 mm. The gain in one-dimensional super-resolution is 16.

signals are located. In a synthetic aperture system, the hologram information writing density may be upper-bounded as follows.

If l is the distance between signal points and k is reduction of useful image area, the following relation,

$$\frac{\lambda^2 F^2}{a^2} = kl^2,$$

should be satisfied because k points are observed instead of a single one. The number of points seen in the image plane after the mask is

$$n^2 = \frac{D^2}{l^2},$$

where D is the linear dimension of the observed picture, and the hologram information writing density is

$$C = \frac{n^2}{a^2} = \frac{D^2}{a^2 l^2}.$$

Since for aperture synthesis, the maximal gain is equal to the change of the object's useful area,

$$k_{\max} = \frac{l^2}{\lambda^2},$$

and the limiting hologram writing density is

$$C = \frac{\Omega}{\lambda^2} \frac{F}{a},$$

where Ω is the optical system angular aperture. For $F/a \gg 1$, an essential gain in the hologram information writing density is possible. With an optimum $F/a \sim a/\lambda$, the hologram capacity is $n^2 = \Omega a^3/\lambda^3$ instead of $n^2 = \Omega a^2/\lambda^2$. Application of the optical aperture synthesis method to a traditional area such as holography thus demonstrates that the possibilities of optics in dense information storage are far from being exhausted. Practical implementation of super-dense thin holograms requires further theoretical and experimental research, but it is important to note that well-known optical principles may lead to absolutely unexpected results in practical applications.

4. Is It Possible to Integrate Holograms with Integrated Optical Circuits?

The contribution of optics to information processing is appreciable. Two approaches to optical computers currently exist: *free-space optics* and *integrated optics.* Today, however, free-space optics features wider functional possibilities due to the ability of lenses to perform Fourier transforms and to the ability of holograms to store information on the amplitude and phase of a light wave. Methods have been developed in integrated optics that enable one to create, in waveguides, lenses performing a one-dimensional Fourier transform. Therefore, it is very important to get insight into the possibility of writing and reconstructing holograms in optical waveguides. The answer to this question has become clear only recently.

For hologram writing, light-sensitive areas are required in the waveguide. Figure 3 shows three versions of waveguiding hologram writing. In the first case, the reference and object waves propagate along the waveguide. In the second case, the reference wave travels along the waveguide, and the object wave is a volume one. In the third case, both reference and object waves are volume ones. In each cases, a hologram is reconstructed by the wave propagating along the waveguide. It is preferable to write

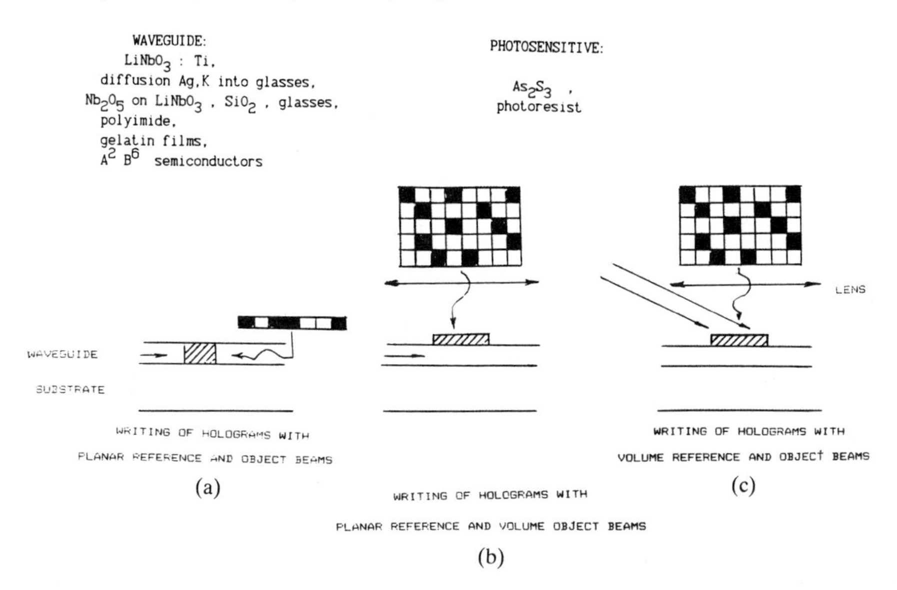

FIG. 3. Basic schemes for the writing of waveguide holograms: a) writing with waveguiding modes. b) writing with a reference waveguiding mode and an object volume wave. c) writing with volume reference and object beams.

waveguide holograms by volume waves because this allows the use of a two-wave write/read technique, i.e., to write holograms by means of gas laser radiation in the visible spectrum where a wide range of photosensitive media exist, and to reconstruct them by radiation of semiconductor lasers that are the most miniature and effective sources of coherent radiation. Figure 4 shows a scheme for writing waveguide holograms by volume waves. An image reconstructed by semiconductor laser radiation propagating in a waveguide is shown in Fig 5. The hologram was written by radiation of 0.44 μm wavelength and reconstructed by 0.85 μm radiation. The waveguide was implemented on a glass substrate by means of Ag or K ion diffusion. As_2S_3 0.5-1 μm-thick films were used as the light sensitive medium evaporated on the waveguide through a mask corresponding to the hologram dimensions. The diffraction efficiency of the holograms was over 50%, and hologram information density was about 5.10^4 bit/mm^2. Two images, real (or virtual) and mirror images, are observed upon hologram reconstruction, with the zero order missing. Thanks to the waveguide propagation of the reconstructing beam, all the beam energy is transformed into images. Importantly, waveguiding holograms can be written on photoresist or As_2 S_3 films by an electron beam that allows integration of high-quality hologram production directly into the integrated circuit

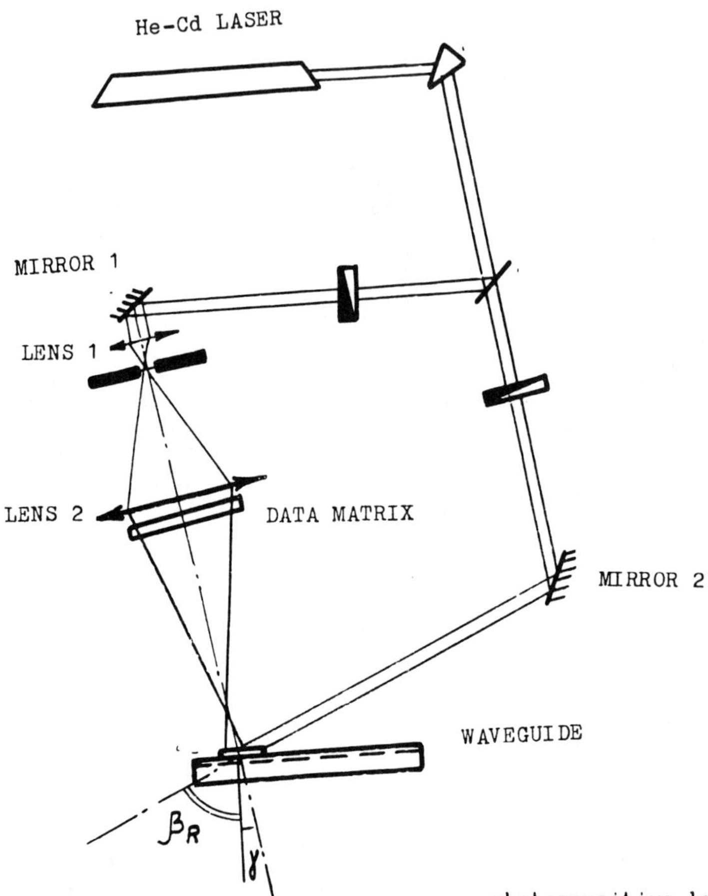

He-Cd laser, λ=0.44μm
P=10mW/mm^2, exposure time-
a few seconds,
waveguide:
- ion exchange of Ag in
glasses,

photosensitive layer:
-As_2S_3, h=0.5-1μm,
after etching:
NaOH : C_2H_5OH
$(C_2H_5)_2NH$, $(CH_3)_2NH$
efficiency of hologram:
up to 80%

FIG. 4. Writing of waveguide holograms with volume beams; the photosensitive layer is on top of the waveguide.

FIG. 5. Example of an image reconstructed from waveguide hologram with a diode laser.

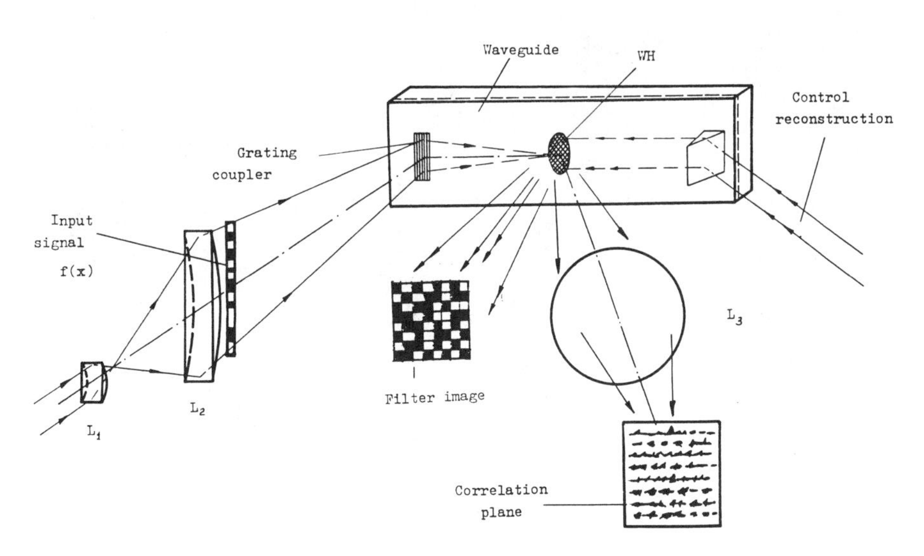

FIG. 6. Correlator scheme for one-dimensional words based on a waveguide hologram.

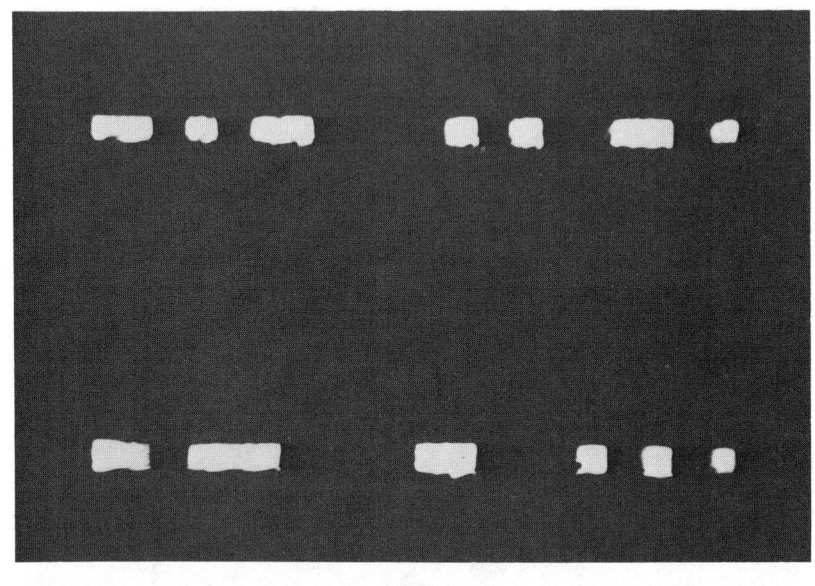

(a)

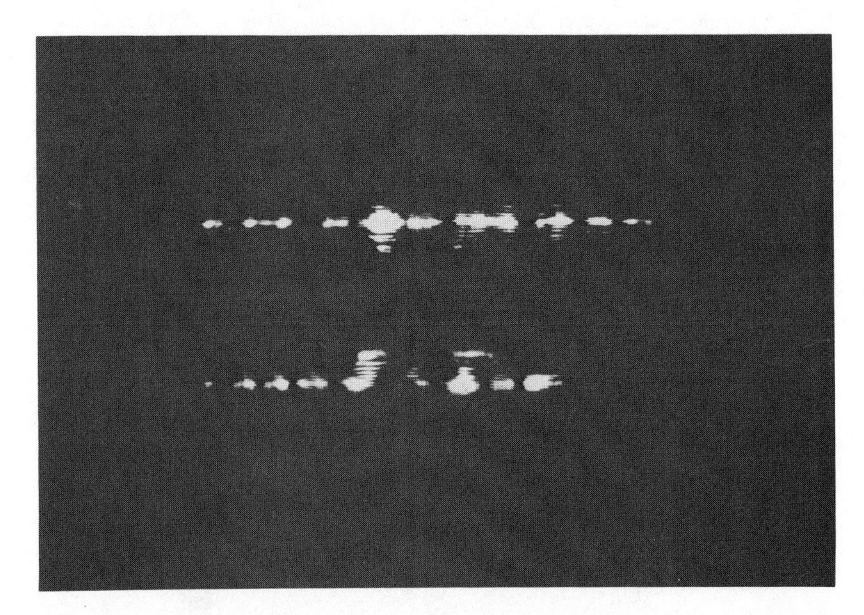

(b)

FIG. 7. a) Content of the filter in the correlation scheme: 20-bit words. b) Correlation signals for upper input words: autocorrelation and crosscorrelation.

production process. The next step in this direction, creation of waveguide holograms in semiconductor waveguides, would improve the functional capabilities of integrated optoelectronic circuits.

Holographic techniques are widely used in optical computing, e.g., for design of associative memory, optical interconnections, correlation analysis, data storage, etc. Waveguide holograms can be used to advantage in practically any application where the usual holograms are used. Figure 6 presents an example of the design of a one-dimensional signal correlator built around waveguide holograms. In the waveguide hologram, a Fourier transform of several one-dimensional signals is written. The input-word Fourier transform is input into the waveguide so that it is situated in the domain of the waveguide hologram. The lens L_3 forms the correlation signal of the input and reference signals. Figure 7 shows two 20-bit words written in the hologram and the correlation signals obtained in a search for the upper word. Waveguide holograms also may be used for optical interconnection. Optical interconnections between electronic integrated circuits is exemplified by the scheme shown in Fig. 8. The ICs with optical I/O are situated on the lower side of the substrate, and the waveguide with holograms on the upper one. The first hologram transforms the volume wave into a waveguide mode, which reconstructs from another hologram a one- or two-dimensional image projected on a line or array of photodiodes. Interconnections can be controlled both by varying the radiation wavelength and by switching on the appropriate laser. Waveguide holograms can be used for radiation I/O and, therefore, enable optical interconnections in free space.

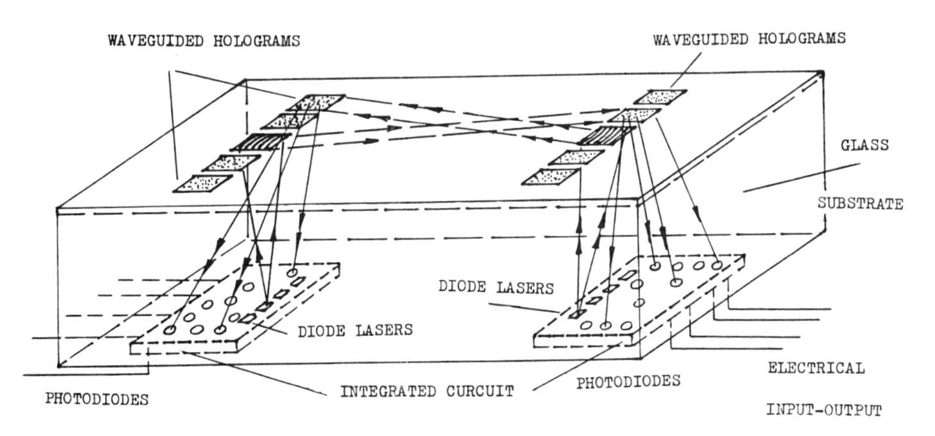

FIG. 8. Optical interconnection scheme for integrated circuits with optical input/output based on waveguide holograms and tunable diode lasers.

5. Conclusion

What answer can be given to the questions put at the beginning of chapter? The estimates of energy losses for optical logic gate switching and information density on plane holograms indicate that the state-of-the-art in this domain is far from fundamental optical limits. The methods of waveguide holography reveal new ways for implementation of optical computers. It thus follows that the decade to come will witness significant advances in gates and systems for both optical and optoelectronic computers.

References

1. L. Landau and E. Lifshitz, "Statistical Physics," Pergamon Press 1980.
2. M. Born and E. Wolf, "Principles of Optics," Pergamon Press, 1968.
3. P. Hariharan, "Optical Holography: Principles, Techniques and Applications," Cambridge University Press, Cambridge, 1984.
4. R. G. Hunsperger, "Integrated Optics: Theory and Technology," Springer-Verlag, 1984.

CHAPTER 33

Digital Optical Computing

S. D. Smith
Department of Physics, Heriot-Watt University
Riccarton, Edinburgh, Scotland

E. W. Martin
High Technology Center, Boeing Aerospace and Electronics Inc.
Seattle, WA

1. Introduction

Propositions about the possibilities of *optical computers* require qualification in that the term encompasses analogue, threshold, and digital schemes, and is sometimes interpreted as implying the emergence of general-purpose computers using optical logic elements. If research is to be believed as leading to a possible technology, it will be necessary to justify a potential contribution to a more restricted area of information processing. Generalised claims have left the subject open to criticism [1, 2] when compared to the proven principles and the future possibilities of silicon technology [3]. Keyes' objections to optical logic form a useful starting point. For example, it is common ground that long sequences of digital operations are obligatory for most computation. Standardised output and restoring logic are a consequence. Fan-out and fan-in are a necessity for any plausible architecture. Table I shows how the subject has advanced with respect to some of these features, which previously were thought to be impossible or difficult to achieve when using optical elements for nonlinear processing.

The conclusion is that , in terms of laboratory proof-of-principle demonstrations, nearly all these points now have been met.

ISBN 0-12-289690-4

Table I. Keyes' Objections to Optical Logic 1984

	Comments
1. Large component no. mandatory thus:	Thin film technology.
—cost/cpt low	
—power/cpt low	1 mW–1 μW.
2. Standardised output	Achieved 1987.
Large noise margin	Achieved 1982.
Gain	Achieved 1982.
3. Long sequence of digital operations thus: separate power and signal, cascadable	Achieved 1987.
4. Input/output isolation: unidirectional processing	Achieved 1988.
Further Requirements 1989	
5. Relatable to a plausible (practical) architecture	Lock-and-clock circulation; optical cellular logic image processor architecture formulated 1987.
6. Capable of fan-out and fan-in	Demonstrated.
three-port capability	Available via BEAT and with two-element standardised input 1989.
7. Uniform so that arrays may speak to arrays	225 channels demonstrated by BEATS cascading with > 90% accuracy demonstrated: 1989.
8. Data rate ~ 10^9 bits/sec.	Circa 6×10^7 currently by SEEDS—32 channels 1990.
9. Realisable—so that digital optical circuit development can proceed, compatible with laser sources	Loop, full adder, CLIP circuit Diode laser and Nd: YAG

The preceding discussion is a necessary but not sufficient condition for belief that digital optical methods can yield a useful information technology. The further condition is that the method can be combined with a plausible architecture leading to the solution of significant problems found difficult by digital electronics. There is the view now emerging that the natural parallelism of free-space optics can be combined with a facility of connectivity between processing elements to provide a new freedom of architectural design. This naturally leads to consideration of 2-D computing problems such as image recognition, edge extraction, noise suppression, and parallel (and hence, faster) acquisition of data from memory for use in computing generally. Other requirements for massively parallel computing exist in fluid dynamics, structural stress analysis, and elementary particle physics [4]. In all these problems, the 2-D nature of light seems to be capable of producing advantage.

Arguments sometimes are advanced that optical schemes will have very high data rates. Indeed, if parallelism ~ 10^6 can be used and combined with cycle times ~ 1 ns, data rates ~ 10^{15} bits/s are implied. However, the example of the human eye/brain interface that accomplishes so much with time constants ~ 10 ms and parallelism ~ 10^7 demonstrates that , given the correct architecture, much can be achieved with input data rates ~ 1 GHz. The example, of course, involves neural networks and, it is thought, threshold operations. These arguments set the stage for designing research demonstration of digital optical logic schemes working towards an initial parallelism of about 10^4 and a modest (e.g., microsecond) cycle time. Performance of optical logic elements and, indeed, the micro-optics itself also suggests at this stage that fan-out and fan-in should be limited to one-to-few, thus setting some limitations on possible architectures that could be demonstrated in the next few years. There are a number of possibilities to take advantage of the combination of 2-D parallelism and interconnecting freedom through 3-D. They are all based upon a similar topology: (a) a 2-D array of logic units, each mapped to a pixelled position of the 2-D data image to be processed; (b) one-to-one or one-to-few fan-out and fan-in, and (c) iteration around a looped optical circuit with memory capability; and (d) operation in a single-instruction multiple-data-stream architecture with electronic control. Such considerations give rise to a wish list for optical logic plane arrays, as shown in Table II. This table also indicates the need to interface with serial electronics and shows that these target specifications steadily are approaching feasibility.

Practical realisation of such optical circuits and computational sub-systems require the development of several sub-technologies; e.g., the

Table II. Wish List for Optical Logic Plane

	Comments
1. 2-D Array 10^4–10^6 elements	2×10^2 achieved; 4×10^4 pump beams demonstrated.
2. Hold power/element 1mW–10 μW	Achieved.
3. Cycle time 1 μs–10 ns	1 μs achieved.
4. Switch energy 1 nJ–1 pJ	Achieved.
5. Stability $< 1\%$ over hours	Achieved.
6. Uniformity 1–2%	Achieved.
7. Contrast $> 10:1$	Achieved.
8. Good throughput: $T > 50\%$	Close.
9. Can be made for various wavelengths	Achieved.
10. Insensitive to small wavelength change	Satisfactory.
11. Data processing rate $> 1000 \times$ LCLV SLM	Reasonable prospects of 10^9 bit/second write rate with e-beam tuned device.

logic planes themselves, the optical power sources, the interconnects, the control and clocking systems. To progress on this, the focus of attempting to construct and test simple digital optical circuits and looped interative computer sub-systems has the virtue of defining the requirements on the components. In particular, the laboratories at Heriot-Watt, Boeing High Technology Center, and AT & T Bell have pursued such projects in the last couple of years. The successful demonstration of electronically controlled looped circuits, and the digital data transfer between moderately parallel (225 at Heriot-Watt, 64 at Bell) optical logic planes provide useful milestones of achievement. In all these programmes, the discipline of attempting the demonstration is proving useful in many ways. As we shall relate, it exposes the limitation of many of the devices and points the way to future research. The impression already grows that the software aspects will be at least as important as device developments. This, in turn, points up the coming urgency of identifying likely application. Computers already are used more for non-arithmetical tasks than for computing; 2-D-image-type processing computers, judging from human behaviour, are likely to find a ready market once they are achievable in real time.

2. Architecture

2.1. *The Optical Cellular Image Logic Processor (O-CLIP)*

There have been many suggestions of architectures for digital optical logic computational devices [5]; many of these are impracticable in terms of existing devices. The *electronic cellular logic image processing* (CLIP) *computers* developed by Duff and co-workers [6] , however, do provide a relatively simple basis that can be adapted towards an optical demonstration scheme as suggested by Wherrett *et al* [7].

Fig. 1 primitively describes an optical circuit currently being constructed based on a simplified CLIP processing cell element. It contains six functional modules:

(a) A spatial light modulator (SLM) as an image input device.
(b) A programmable logic unit that receives simultaneously a pair of images and can output a chosen logical combination of the pair; determined at each cycle by a single-instruction programming of the unit.
(c) An optical interconnect; in the first instance, a nearest-neighbour fan-out.
(d) A threshold element that converts the fanned-in combination of inputs at each pixel into binary output levels.

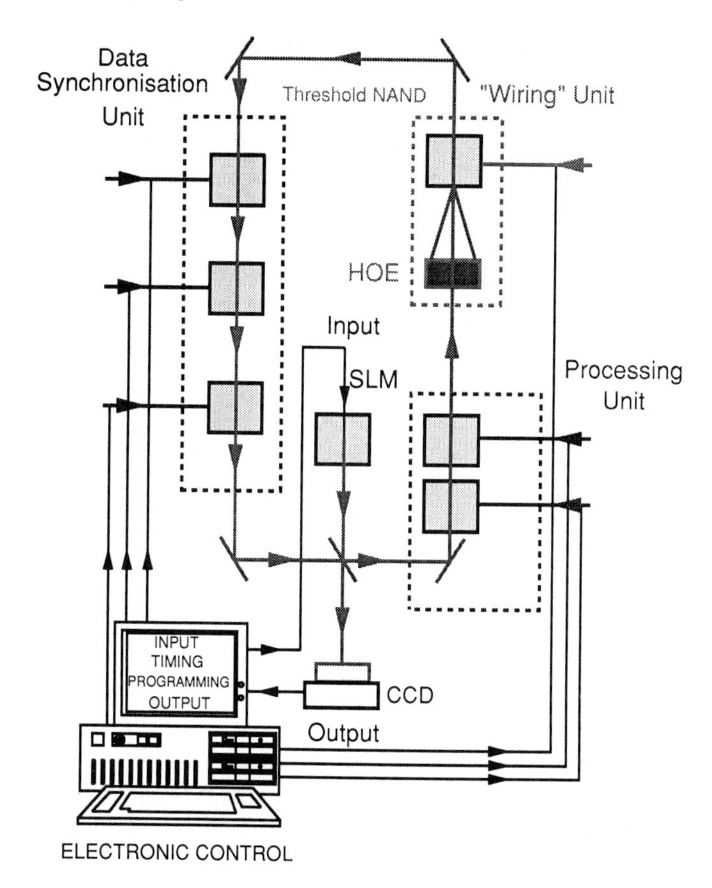

FIG. 1. Schematic diagram of CLIP architecture.

(e) A clocking unit that acts as a temporary memory to synchronise the flow of iterated images.
(f) A second programmable logic unit leading directly to an array output or to further iterative circuits.

These functional units then are able to use a 1–4 nearest-neighbour interconnection to implement all those binary image processing functions for which the four directions (N, S, E, W) are not distinguished. Examples are: noise removal, dilation, contraction, and labelling. There are a number of advantages to this approach. Firstly, the scheme can be modelled electronically on a parallel machine such as a distributed array processor (DAP). Secondly, by the use of optical logic elements, which are naturally bistable and controllable by the setting of a hold or bias beam synchronisation is achievable using lock-and-clock architecture previously developed and demonstrated [8]. Additionally, the hold-and switch procedure that allows gain, cascadability and restoring logic to be achieved also, by a programmed control of hold-beam power, enables function to be changed cycle by cycle. The control of the necessary iterations required for the various processing functions then are included in a programme for the hold beams and stored in the electronic control computer. Thus, a complete programmable iterative computing sub-system is constituted. We review progress towards such a demonstration in Sections 3 and 4.

2.2. Symbolic Substitution

An architecture based upon *symbolic substitution* contains many of the functional modules described in Section 2.1 but also includes a shift function [9]. This approach can be shown to make possible any generalised computing process. It will have to be proved that for such generalised computing, the parallelism gives sufficient increase in data rate to be competitive. Tests at wish-list specification would be informative.

2.3. Switching Networks

One of the more promising interconnection schemes proposed for *photonic switching* is the 3-D crossover network, which is equivalent topologically to a shuffle network [10]. To utilise the parallelism offered by free-space optics, the crossover interconnects must provide 3-D connectivity between planar (2-D) arrays of optical logic devices. As in 2.1 and 2.2, such schemes currently are being demonstrated.

The examples of architecture referred to throughout Section 2 all provide practical routes to near-future demonstrations. They call for a common set of necessary sub-technologies that we will review. Although limited in their scope, they nevertheless address potentially interesting computational problems and certainly can be expanded greatly via, for example, more complicated optically achieved interconnection. A further aspect of architecture concerns memory or optical data storage. Some of the previous schemes will need to both hold the results of calculations and acquire data from memory in the same parallel format as the logic operations. This requirement is likely to converge upon the already established optical data storage technology; it leads to the requirement for methods of parallel reading and writing without mechanical movement. Theoretically, data retrieval rates can be increased greatly, compatible with the rates implied in the wish list (Table II).

3. Key Devices, Components, and Functional Units

Virtually all architectural schemes, including those in Section 2, require both optical logic planes and spatial light modulators. The wish list gives a useful current measure of performance requirements and, theoretically at least, there seems every reason to believe that this can be achieved. A second class of (passive) devices embraces the optical interconnects needed together with a third class—efficient lasers to provide optical power beams to the required arrays. Progress with computer-generated holographic optical elements, sometimes in the form of binary-phase (Dammann) gratings, has been spectacular. Arrays of beams of up to 200×200 have been achieved. In smaller (16×16) array, ~2% uniformity of intensity [11, 12] has been demonstrated. Fan-out and fan-in can be prescribed and achieved with micro-optical elements [13]. In a more restricted way, focussing to micron dimension is standard commercial practice in optical data storage [14].

3.1. Nonlinear Optical Logic Plane Devices

Although there have been numerous reports of device effects apparently usable for logic and bistable memory, only three have survived the transition to inclusion in optical circuit demonstrations: the *nonlinear Fabry-Perot etalon* (NLFP), the *Self Electro-optic Effect Device* (SEED) and the *Liquid Crystal Light Valve* (LCLV). The latter, when combined with a write mechanism through a CRT, is, of course, an electrically addressed SLM. Here, we shall concentrate on recent circuit achievements on NLFPs in the form of bistable etalons with absorbed transmission (BEATs) and the symmetric SEED (S-SEED). Processing of information, of course, is insensitive to the nonlinear mechanism of the device; it is only the performance that matters. Both SEEDs and BEATs have progressed now to the stage where iterative loop processing with restored logic levels can be achieved and parallel array cascading of up to several hundred beams demonstrated [7, 10, 15].

3.1.1. NLFPs

The original devices in InSb and GaAs date from 1979 [16, 17] and utilised refractive nonlinearity, n_2, caused by optical carrier excitation in these semiconductors with $n_2 \propto \tau/E_G^3$. Neither of these prove convenient for technical reasons: Infrared wavelength and cryogenic temperatures in the case of InSb, despite excellent switching energies and operating powers; too much absorption (and hence, thermal instability) and too short

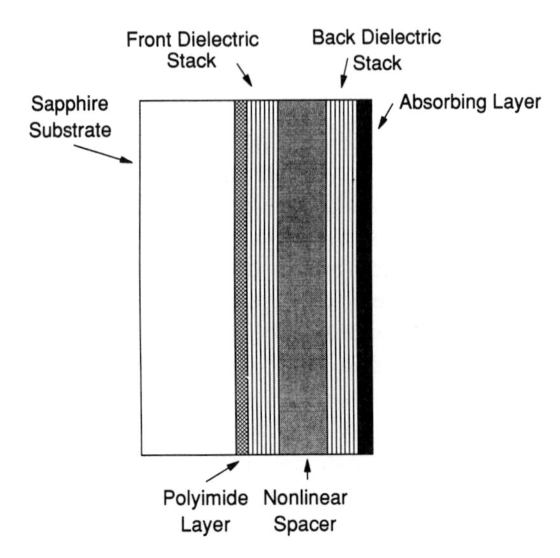

FIG. 2. Schematic of BEAT device.

a carrier lifetime in the case of GaAs. Recent work on GaAs [18] has given a 75% duty cycle with 0.5 s of stability and, with the possibility of longer carrier lifetimes, gives some hope for the future. However, for immediate demonstration, the thermo-optic effects in interference filters, particularly with an absorption layer as developed by Walker [19] (BEAT) (Fig. 2), have yielded more convenient devices with roughly the same operating powers, although rather slower, than InSb. With notable stability, the separation of the exciting power absorption mechanism from resonator structure also allows flexibility for signal beams and quasi-three-port operation.

The ability to operate stably at any part of a nonlinear characteristic, which can be adjusted at will to be single-valued with gain or bistable with memory, allows the use of a biasing beam to hold and a signal beam to switch. This is a threshold device that has proved sufficiently tolerant both to operate circuits and to change controllably the logic function. Separation of signal and hold beams allows restoration of logic levels, and infinite cascadability is obtained when fresh hold beams are applied to successive devices. The operation of a *classical finite state machine*, using these principles, demonstrated this capability [8]. The BEATs use ZnSe as the thermo-optic material and, in convenient cross-section (10 μm dia.), can show milliwatt operating power with μs response time. Use of liquid crystals as the thermo-optic material improves operating power by two orders of magnitude with μW power-switching already demonstrated [20]. Advantages of NLIFs include uniformity, purely optical pixellation (i.e., requiring no transverse fine structure), and the ability to be fabricated for any convenient laser wavelength. Recovery times tend to limit performance.

3.1.2. S-SEEDs

The SEED, by contrast, is a hybrid optoelectronic device in which the nonlinearity is created by an electronic current triggered by the incident light. It uses the electric field modulation of absorption in an exciton characteristic of electrons confined in quantum wells. Within a p-i-n junction, incident light causes a change of voltage that, in turn, operates the electro-optic modulation of absorption occurring in the enclosed multiple quantum well (MQW). To achieve these devices in which light switches or controls light, the p-i-n structure acts as a detector that, working through a bias resistor, changes the voltage across the modulator. It achieves nonlinearity and bistable memory function using a process of increasing absorption, so that its characteristics can be similar to the NLFP. It has no resonator, but the precision of construction of the quantum wells and the sharpness of the exciton line impose tolerance limitations in both fabrication and operating wavelength. If the bias resistor is included in a micro-integration of the detector and modulator, low switching energies are achievable. Contrast levels tend to be rather modest at around 3:1. This device was first reported by Miller in 1983 [21].

An important further development was the symmetric SEED (S-SEED) developed by Lentine *et al* in 1988 [22] (Fig. 3). In this, a second identical SEED is used as the load for the quantum well diode. Thus, there are two diodes in series connected to the voltage supply. It is operated with a beam on each diode. The device proves to be bistable in the ratio of the two beam powers; its importance is that it is insensitive to intensity variation and is able to operate a form of dual rail logic. A further phenomenon is that the device may be operated at various power levels, enabling writing at a low level and reading at a high level. This can be translated into two cycles, in the second of which power gain is achievable; this is known as *time sequential gain* and is useful for circuit cascading purposes. SEED devices have been tested down to switching times ~ 1 ns, but this does not represent the fundamental limit. In circuitry application, devices of 10×10 μm area typically use μW for μs switching times. There is a strict power speed trade-off.

3.2. Operational Demonstration of Vital System Properties

Prior to 1990, there had been no demonstration of moderately—let alone highly—parallel all-optical digital circuits. Since 1982, progressively

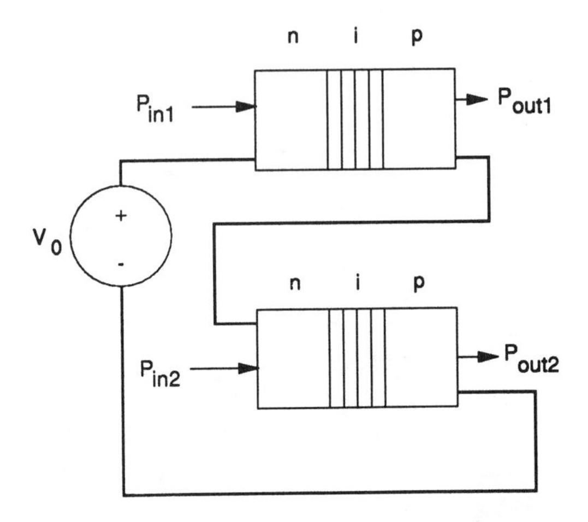

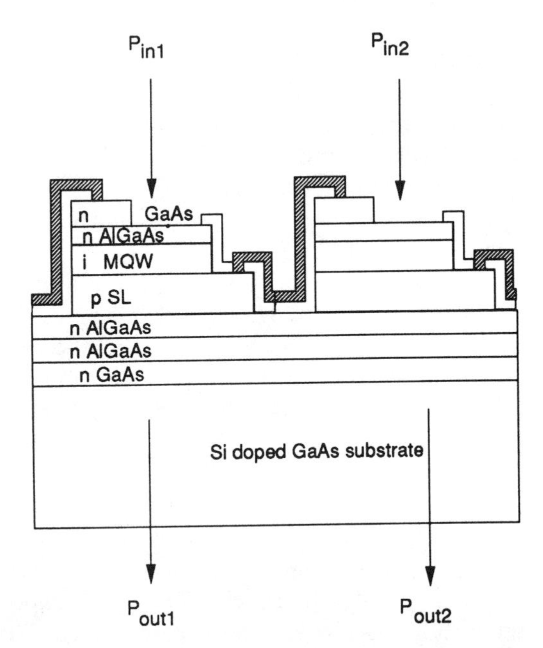

FIG. 3. Symmetric SEED circuit and structure schematic: SL is a short period superlattice used to obtain a high-quality structure, and MQW is the multiple quantum well material.

more complicated single-channel circuits have been used to demonstrate vital functions.

3.2.1. Restoring Logic and Hard Limiting

The logic systems described in this chapter are based upon light intensities. Unlike electronics, where current and voltage are available, we only have one variable in this family of logic devices. Conditioning the input into optical logic gates, therefore, is important. Operationally, the absorbing layer of a BEAT device (Fig. 2) allows the decoupling of the hold or bias power from the switch or signal power—by absorption (or beam angle). The output beam is the reflected hold power beam that produces a hard limiting response (i.e., not dependent on level of signal input when switched). Thus, succeeding stages receive a conditioned input.

3.2.2. Cascadability

This configuration also satisfies the condition for cascadability that the gate must supply more energy than the switching signal and that this be done from a separate power supply.

3.2.3. Prevention of Signal Corruption

An element operated in the hold-and-switch mode described previously delivers sufficient signal to switch a second element, and so on. If included in a loop circuit, measures must be taken to ensure that the optical signal does not circulate more quickly than the element is able to switch. For this, a lock-and-clock three-phase control of holding power can be imposed on a loop containing three elements. The NLFP thus is used in a latching mode to trap the logic signal successively at each stage [8]. The successful operation of a single-channel computational and storage loop represents a primitive proof-of-principle demonstration of an optical finite state machine as it is readily extendible to 2-D arrays. The test results show further that, in practice, indefinitely extendible restoring logic has been achieved and, further, by using one gate in reflection as a NOR gate to achieve inversion, all elements necessary to make a digital optical processor have been demonstrated. These demonstrations were achieved by 1986 using NLFPs , but they are applicable to other forms of logic device.

4. Implementation of Demonstrations

During 1989 and 1990 a number of important circuit demonstrations have been implemented.

4.1. Single Channel Programmable Circuit

We begin by describing the single channel O-CLIP circuit (Fig. 1). This has been implemented at Heriot-Watt and, in part, also at Boeing High Technology Center. The logic gates used have been ZnSe BEATs and the laser power supplies GaAs semiconductor diode lasers operating at 830 nm. The hardware implementation is shown in Fig. 4. In fact, four gates are used to provide both programmable logic (two gates), simulated thresholding (one gate), and lock-and-clock circulation of data (hold-beam phasing

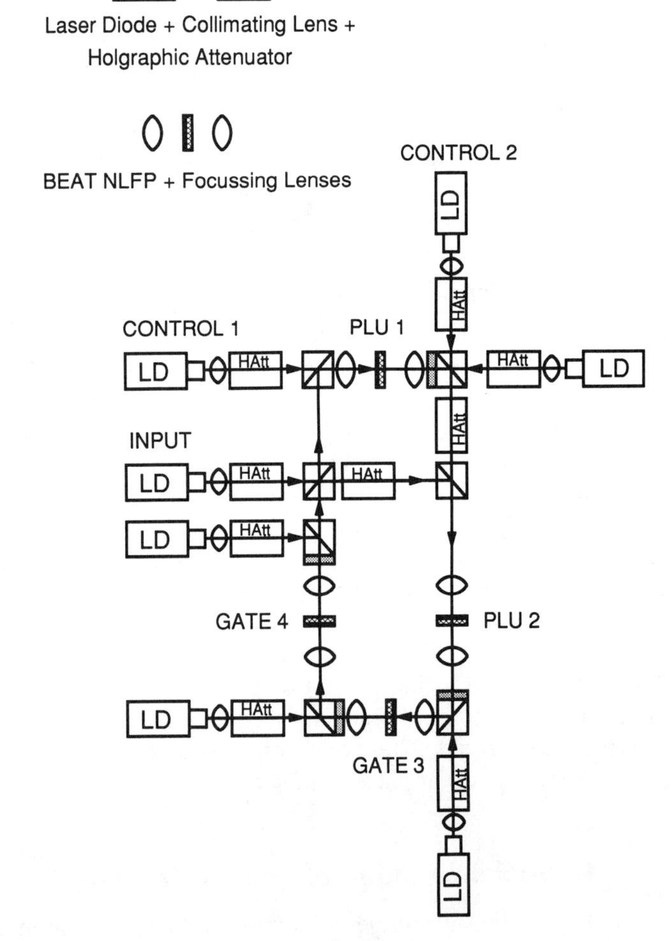

FIG. 4. Details of Optical-CLIP loop circuit.

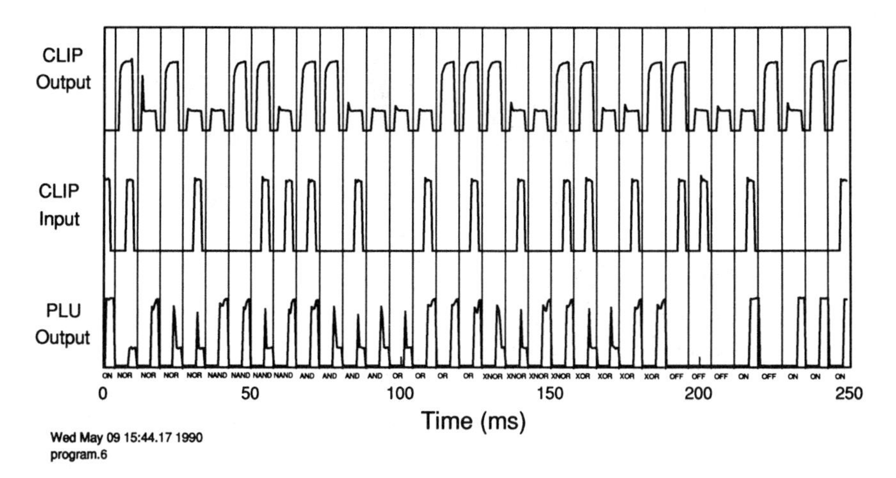

FIG. 5. Dynamic demonstration of all Boolean functions.

in all four gates). Two of the gates receive only holding and information beams; the other two additional control signals to implement the programming. A fifth laser beam supplies the CLIP input to the circuit. This may differ at each cycle. Figure 5 shows a test of this circuit, indicating the satisfactory achievement of all the two-input symmetric Boolean logic functions programmed in sequence through the computer-set control beams from cycle to cycle. In one test, at least 50,000 have been recorded with satisfactory stability.

This demonstration also indicates that the tolerances are adequate for function change. In this case, the hold levels are controlled dynamically and correction procedures could be applied in the event of drift.

4.2. Extension of CLIP Architecture to 2-D Processing

Extension of the circuitry just described to a 2-D array allows us to explore realistic 2-D processing cases. This has been demonstrated by simulation on an electronic DAP. Such an exercise is illustrated in Fig. 6 for the case of noise removal in which each square indicates successive optical logic planes.

A primitive image in the form of a cross, together with some noise, is brought into the optical circuit and subjected to an OR operation. An optical interconnect then imposes the four nearest neighbour pixels of each pixel throughout the image. The resultant image then is thresholded such that the pixel is ON if any of the superimposed pixels are ON. The data then is stored and synchronised through a three-element lock-and-clock

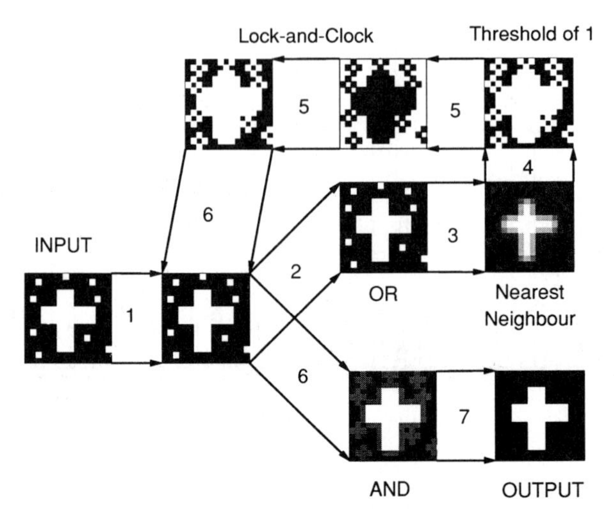

FIG. 6. Successive logic planes in CLIP noise removal.

memory and then brought back into the circuit to be superimposed with the original image. This image is subjected now to the AND function and results in noise removal. It can be seen that noise has been removed in just one complete cycle of the optical CLIP.

The next step in developing a working optical CLIP is to replace the single gates in the iterative processor (Fig. 1) by optical gate arrays. To this end, two 15×15 arrays have demonstrated lock-and-clock transfer of data. The arrays were BEAT devices designed to operate at 1.06 μm, permitting the use of a cw Nd:YAG laser as the optical power source. The uniform multilayer structure was deposited on a 3 μm-thick layer of polyimide spun onto a sapphire heat sink (Fig. 2). The device (with no physical pixellation) was illuminated by focal spot patterns (15 μm dia.) generated from a specially constructed Dammann grating. Figure 7 shows the test transfer of data with greater than 97% accuracy in this first experiment.

4.3. S-SEED Array Cascading

A 4-S-SEED-array loop has been demonstrated using 32-element arrays. Separate interconnect geometries have been tried, including a split/shift pattern and others with devices working just as inverters. Rates of 1.1 MHz with total laser diode power of 9 mW have been reported [10, 23].

4.4. S-SEEDS Switching Networks

A crossover network (Fig. 8), topologically equivalent to a shuffle network, has been implemented optically with S-SEED arrays using 32 inputs.

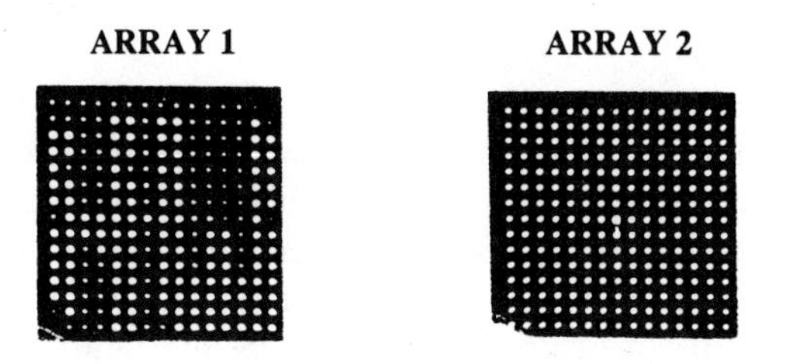

'Image' locked into ARRAY 1. Signals to ARRAY 2 blocked.

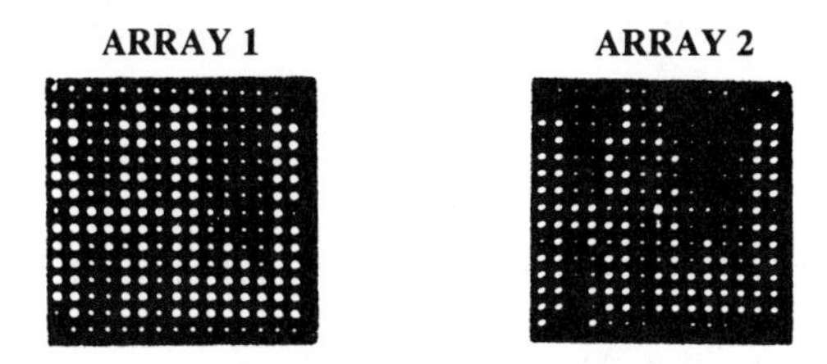

The pattern on ARRAY 1 is transferred to ARRAY 2 with greater than 90% fidelity.

FIG. 7. Digital parallel optical data transfer.

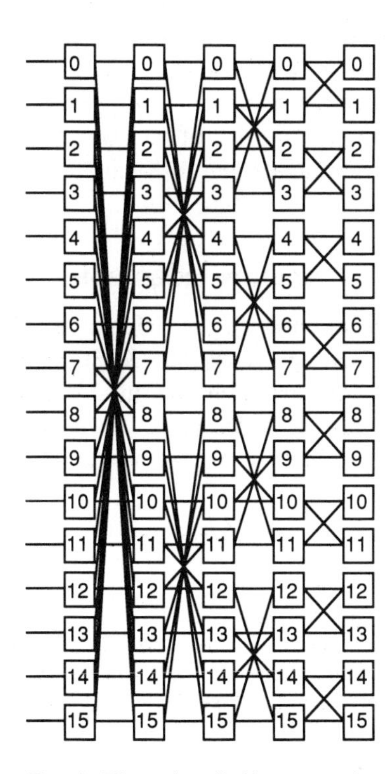

FIG. 8. Photonic switching network.

The S-SEED itself is used as a two-input, two-output node (called a two-module). This module conducts the function OR and passes the results on to two output ports only if the gate is enabled by a clock signal. It can be implemented by using a single S-SEED if an electronically controlled SLM is placed in the path of the clock signals to provide control. A four-stage crossover has been constructed using 8×4 S-SEED arrays. Data from a number of input fibres was switched to any of eight output fibres and showed error-free switching as vertified by oscilloscope traces of signals. A maximum speed of 57.7 kBs^{-1} currently is limited by laser power and optical loss.

5. Conclusion

The progress reviewed here is allowing practical demonstration now of simple realisable computational architectures—in which logic plane speaks to logic plane purely optically—to be achieved. A large number of necessary conditions already have been met in reaching this stage. Both BEATs and SEEDs have proven useful for demonstration and improved devices of various form can be expected in the future.

Increasing the size of arrays is likely to lead to the inevitability of logic error between planes.Methods to correct for this will be important to further progress. A simple exercise problem of noise removal is described. Current work supports the view that architecture and device improvement will go hand in hand now toward proving the technical feasibility of digital optical computing. Demonstrations of increasingly meaningful processing with 16×16 arrays are being prepared for demonstration within the next year.

Acknowledgments

We acknowledge the support of the Science and Engineering Research Council, UK, and of Boeing Aerospace & Electronics Inc. Unpublished results have been made available to us by Robert Craig, Douglas McKnight, and Mohammed Taghizadeh. We acknowledge the advice of colleagues in Edinburgh, Seattle, Naperville, and Holmdel.

References

1. R. W. Keyes, *Optica Acta* **32**, 525, 1985.
2. R. W. Keyes, *Science* **320**, 138, 1985.
3. K. H. Bohle, "Optical Computing," *Proc. 34th Scottish Universities Summer School,* (B. S. Wherrett and F. A. P. Tooley eds.,), 265–280, Edinburg University Press (1988).
4. D. J. Wallace, "Optical Computing," pp. 281–317, Edinburgh University Press, Edinburgh, 1988.

5. M. J. Murdocca, "Digital Design Methodology to Optical Computing," MIT Press, Cambridge, MA 1989.
6. M. J. Duff and J. J. Fountain, "Cellular Logic Image Processing," p. 277, Academic Press, London, 1986.
7. B. S. Wherrett, S. Bowman, G. S. Buller, R. G. A. Craig, J. G. H. Mathew, S. D. Smith, J. F. Snowdon, F. A. P. Tooley, and A. C. Walker, "Design of Optical Computing Systems," *Tech. Digest: Optical Computing 1990*, Kobe, Japan (April, 1990).
8. S. D. Smith, A. C. Walker, F. A. P. Tooley, and B. S. Wherrett "The Demonstration of Restoring Digital Optical Logic," *Nature* **325**, no. 6099, 27–31 (1987).
9. K. H. Brenner, A. Huang, and N. Streibl, "Digital Optical Computing with Symbolic Substitution," *Appl. Optics* **25**, 3054 (1986).
10. T. J. Cloonan, F. B. McCormick, M. J. Herron, F. A. P. Tooley, G. W. Richards, E. Kerbis, J. L. Brubaker, and A. L. Lentine, "A 3D Crossover Switching Network based on S-SEED Arrays," *Proceedings: Photonic Switching Topical Meeting,* Kobe, Japan (April 1990).
11. J. Turunen, A. Vasara, J. Westerholm, G. Jin, and A. Salin, "Optimisation and Fabrication of Grating Beamsplitters," *J. Phys D* **21**, S102–S105 (1988).
12. M. R. Taghizadeh, J. Turunen, A. Vasara, and J. Westerholm, "Binary-Phase Grating Beamsplitters for Very Large Array Generation," Paper TUV17, *Proc. of OSA Annual Meetings*, Orlando, FL (1989).
13. M. R. Taghizadeh, I. R. Redmond, B. Robertson, A. C. Walker, and S. D. Smith, "High Efficiency Holographic Optical Elements for All-Optical Digital Computing," *Proc. Soc. Photo-Opt. Instrum. Eng.* **1136**, 261–268, (1989).
14. G. E. Thomas, "Future Trends in Optical Recording," *Philips Techn. Rev.* **44**, no. 2, 51–57 (1988).
15. B. S. Wherrett, R. G. A. Craig, J. F. Snowdon, G. S. Buller, F. A. P. Tooley, S. Bowman, G. S. Pawley, I. R. Redmond, D. McKnight, M. R. Taghizadeh, A. C. Walker, and S. D. Smith, "Construction and Tolerancing of an Optical-CLIP," *SPIE Proceedings,* OE-LASE, Los Angeles, (January, 1990).
16. D. A. B. Miller, S. D. Smith, and A. Johnston, "Optical Bistability and Signal Amplification in a Semiconductor Crystal: Applications of New Low-Power Nonlinear Effects in InSb," *Appl. Phys. Lett.* **35**, no. 9, 658–660 (1979).
17. H. M. Gibbs, and S. L. McCall, "Optical Bistability in Semiconductors," *Appl. Phys. Lett.* **36**, no. 6 (1979).
18. E. Masseboeuf, O. Sahlén, U. Olin, N. Nordell, M. Rask, and G. Landgren, *Appl. Phys. Lett.* **54**, no. 23, 2290 (1989).
19. A. C. Walker, "Reflection Bistable Etalons with Absorbed Transmission," *Opt. Commun.* **59**, 145–150 (1986).
20. A. D. Lloyd and B. S. Wherrett, "All-Optical Bistability in Nematic Liquid Crystals at 20 μW Power Levers," *Appl. Phys. Lett.* **53**, no. 6, 460–461 (1988).
21. D. A. B. Miller, "Multiple Quantum Well Optical Nonlinearities: Bistability from Increasing Absorption and the Self Electro-Optic Device," *Phil. Trans. R. Soc. Lond. A* **313**, 239–244 (1984).
22. A. L. Lentine, H. S. Hinton, D. A. B. Miller, J. E. Henry, J. E. Cunningham, and L. M. F. Chirovsky, *Appl. Phys. Lett.* **2**, 1419–1421, (1988).
23. M. E. Prise, R. E. LaMarche, N. C. Craft, M. M. Downs, S. J. Walker, L. A. D'Asaro, and L. M. F. Chirovsky, "A Module for Optical Logic Circuits Using Symmetric Self Electrooptic Effect Devices," to be published in *Applied Optics* (May, 1990).

CHAPTER 34

Computing: A Joint Venture for Light and Electricity?

Pierre Chavel

Institut d'Optique (Laboratoire associé au Centre National de la Recherche Scientifique), Centre Universitaire d'Orsay, Orsay, France

1. Introduction

The founders of Fourier optics probably were the first to address clearly the potential relation between optics and information processing: To cite but a few classical works [1], Duffieux expressed imaging as a convolution; Maréchal and Croce and O'Neill suggested use of this relation not only to analyze the physics of imaging, but also to process intentionally the information contained in the object. In the first years of holography, Vander Lugt and Lohmann and his co-workers devised ways to implement almost arbitrary convolutions.

It was soon realized, however, that electric computers had a much larger generality in their way of processing data, and that the absence or, at the least, the scarcity of nonlinear operations in Fourier optics was a handicap. After some preliminary work scattered over 20 years, the idea that, with adequate extensions of existing concepts and components, optics could be used to build a large class of processors—and even to overcome some of the limitations of existing computers—became the subject of active discussion and research around 1980. In part, this resulted from the fact that nonlinear optical elements such as bistable arrays were starting to appear as possible objects for optoelectronic technologies.

Optical computing now is a field of applied science with well-defined objectives. Industrial and public funding contribute to its development, its

ISBN 0-12-289690-4

community has regular meetings whose proceedings present state-of-the-art views of the subject [2], and a few books have been published recently [3, 4, 5].

In his closing remarks at the 1990 International Topical Meeting on Optical Computing in Kobe, Japan, S. Ishihara of Electrotechnical Laboratories, Japan compared optical computing with a treasure on a treasure island, saying that scientists landed on the island but have yet to find the treasure. What exactly is the optical computing community seeking? Is the role of optical computing versus electronic computing clearly defined? Can light supplement electricty in digital computers, or should it actually replace it? To suggest an answer to these questions, we shall examine herein first the processing functions and then the components that are available or being developed, and finally describe and try to assess the architectures that have been proposed so far and, to a still limited extent, demonstrated.

2. The Processing Functions of Light

The purpose of this section is to review the functions that may be performed with the help of light in a computer. Starting from a historical viewpoint, we shall discuss the merits of Fourier optics for computing and its relation to matrix vector multiplication and interconnects; we shall then turn to processing functions involving nonlinear operations, and close the section with a brief mention of optical memory functions.

2.1. Fourier Optics

A short summary of this now classical subject [6] will be useful here to introduce a discussion about the operations that light perform most straightforwardly. Two basic ideas are relevant:

- The first idea is that the physics of the imaging process can be modelled as a convolution operator. More specifically, let us consider any physical system having an input function $f(\mathbf{x})$ and an output function $g(\mathbf{x})$. Here, f and g are physical quantities defined over the space spanned by vector $\mathbf{x}$. A classical mathematical result states that any system that is described by a *linear* and *shift-invariant* relationship between input and output really performs a convolution. Precisely, to a good first approximation, this applies to optics if functions f and g describe the object and the image of some imaging setup, and $\mathbf{x}$ the two-dimensional image space. Fourier optics states that image $g(\mathbf{x})$ is a convolution between the input object $f(\mathbf{x})$ and the system impluse response $R(\mathbf{x})$.
- The second idea is that in the case of coherent illumination by a plane or spherical monochromatic wave, diffraction of the light illuminating ob-

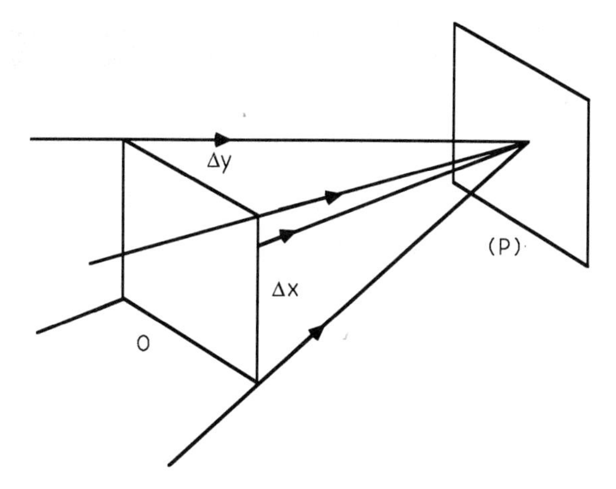

FIG. 1. The optical Fourier transform. O = input object. (P) = Fourier transform plane.

ject $f(\mathbf{x})$ has the effect of creating the Fourier transform of complex amplitude $f(\mathbf{x})$ in any plane conjugate to the source point. Hence, the laws of light propagation allow us to perform, to a good approximation, Fourier transformations of two-dimensional signals.

Let us examine these well-known phenomena from the perspective of computing power, starting with the optical Fourier transform. With reference to Fig. 1, let an input object be illuminated over a finite rectangular area Δx, Δy by a coherent, monochromatic spherical wave of wavelength λ converging on plane (P), a distance d from the object plane. $f(\mathbf{x})$, the diffracting complex amplitude, has a bounded support. The sampling theorem applies to the Fourier plane complex amplitude, so that independent samples can be obtained at regular intervals with periodicity $(\lambda d)/(\Delta x)$, $(\lambda d)/(\Delta y)$. The spatial extent of the Fourier transform of $f(\mathbf{x})$ is, mathematically speaking, infinite. However, the accuracy of diffraction as a means of computing a Fourier transform is limited to small angles. The practical result is that with conventional optics, $1{,}000 \times 1{,}000$ samples typically can be obtained with reasonable accuracy. It is quite important, in practice, to use reasonable order of magnitudes when assessing the computing power of optics; in all further discussions in this chapter, we shall stick to this figure of $1{,}000 \times 1{,}000$; special-purpose optics, such as Fourier lenses in the present case, may be able to increase it by a factor of 10 or so. Since every Fourier sample is a linear weighting of all object points, and since $N = 10^6$ independent samples of the Fourier transform correspond to 10^6 independent samples of the object, this classical operation is worth

$N^2 = 10^{12}$ parallel, weighted interconnects and the same number of additions and multiplications. This in every respect is a very large amount of computation, and its fully parallel implementation by optics deserves mentioning.

Five negative comments, however, are in order here, limiting practical interest of the phenomenon:

- The Fourier transform is a very specialized operation.
- The calculation is performed with analog, not digital, accuracy.
- Light propagation over some distance is required to perform the Fourier transform without requiring very large apertures and, therefore, aberrating beams; so, it does not appear feasible to integrate the optical analog Fourier transform in a very small space, such as that occupied by chips in an electric computer.
- The computing power of a computer should be expressed as throughput, not just number of operations; the problem of parallel input of data and parallel output of results compatible with the speed of calculation is not yet fully solved and extends beyond the domain of Fourier optics.
- The algorithm used by light propagation to perform a Fourier transformation is not optimal; it really uses N^2 additions and multiplications, while it is known that the same calculation can be performed with order $N \log N$ operations or slightly less, making the achievement in terms of number of operations much less impressive than it appears at first glance. Here, however, it can be pointed out that the parallel implementation of fast Fourier transform algorithms requires highly specialized hardware with heavy interconnection loads.

Turning back to the optical implementation of analog convolution for a similar critical assessment, it is clear that if N is the total number of pixels in the input $f(\mathbf{x})$ and n the number of degrees of freedom in the impulse response $R(\mathbf{x})$, optics effectively performs nN operations in parallel. The operation performed is of the form:

$$g(\mathbf{x}) = \int_{-\infty}^{+\infty} f(\mathbf{x}')R(\mathbf{x} - \mathbf{x}')\, d\mathbf{x}'. \tag{1}$$

In the case where sampling is legitimate, which is quite adequate for a comparative evaluation of optical computing and other approaches to computing, the convolution operation of Eq. (1) can be rewritten using integer indices in the form:

$$g_i = \sum f_j R_{i-j}. \tag{2}$$

In Eq. (2), the summation boundaries have not been made explicit for conciseness, but obviously are determined by the bounded supports of the object f and the impulse response R. The minimal number of operations to obtain the result using any processor is the smaller of nN and $N \log N$. In other words, with N again typically of order 10^6, a standard sequential digital computer can perform the convolution faster using two fast Fourier transformations than by direct calculation if n is more than a few tens. In practice, n may range from very small values like $n = 2$ for discrete differentiation to much larger values for applications such as pattern recognition; since parameter n is limited by diffraction and aberrations in the same way as N, 10^6 again is a safe maximum order of magnitude. The discussion about the interest of using optics for convolution, therefore, is quite similar to the previous one about Fourier transformation: An optical convolver can be compared to a massively parallel machine, but is impeded by analog accuracy, input and output limitations, system integration, and lack of generality. The result is that optical signal processing up to now has not matched expectations. Recent evolution in optical computing devices, concepts, and demonstrations address the drawbacks mentioned here and will be considered in the forthcoming sections.

2.2. Matrix-Vector Multiplication

In Eq. (2), the parameters i and j denoting the pixels intentionally were not set in boldface characters that would have indicated a vector notation. It is true that most of the setups using free-space optics work on two-dimensional data. However, we are considering sampled data. Therefore, it is a mere matter of data labelling to use two-dimensional or one-dimensional notation, and while the full use of two-dimensional data *space* is important for the performance of setups, one-dimensional *notation* is more convenient for our present discussion. Parameters i and j being simply integers, it is clear that Eq. (2) is a particular case of the matrix-vector product:

$$g_i = \sum R_{ij} f_j, \tag{3}$$

namely, it corresponds to the product with a vector of a so-called Toeplitz matrix.

A first and important extension of the scope of optical computing, therefore, is the implementation of the general matrix-vector product, or, in other words, the incorporation of full-space variance in the linear relationship between input and output. Numerous experimental configurations, using one-dimensional or sparse two-dimensional data arrangements, have been proposed and this is not a place for a complete review. Therefore, we shall mention only two of them: Figure 2 schematically depicts a cylindrical lens spreading light from pixel i of the one-dimensional

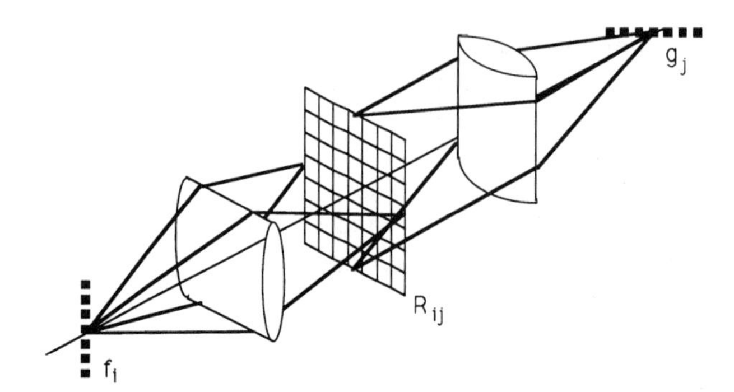

FIG. 2. One optical matrix-vector multiplication setup using a thin mask. Cylindrical optics is used to associate pixels of the one-dimensional input and of the one-dimensional output with rows and columns of the matrix mask.

data array f onto row i of the two-dimensional mask R and a second cylindrical lens collecting light from column j of the mask onto detector j of the one-dimensional result g. This scheme, originally proposed by Goodman [7], shows that:

- The mask R_{ij} may possess the same number N of degrees of freedom of typically 10^6 mentioned earlier for a two dimensional image, but
- Restriction of object format to one dimension will reduce the number of object points to typically 10^3.
  These orders of magnitude apply as well to all other setups using masks or thin holograms for the implementation of matrix R.

The use of thick gratings or holograms, therefore, has been suggested as one way to increase the performances. Psaltis has shown that a setup of the kind sketched in Fig. 3 has the potential for an arbitrary matrix-vector multiplication between $N^{3/4}$ input pixels f_i and $N^{3/4}$ output pixels g_i, the $N^{3/2}$ weights of matrix R being stored in the volume hologram [8]. Using our typical order of magnitude, this should open the way to the storage of 10^9 matrix coefficients, although their individual and accurate control may not be easy.

2.3. Connections

If matrix R is constrained to binary values, zero or one, then it can be understood as an interconnection matrix. Some examples may be useful here: The interconnects networks found commonly in switching applications or in electric parallel processing structures, such as the omega network or the perfect shuffle [9], rely on one-to-one interconnects between an input array and an output array. They are described by permutation

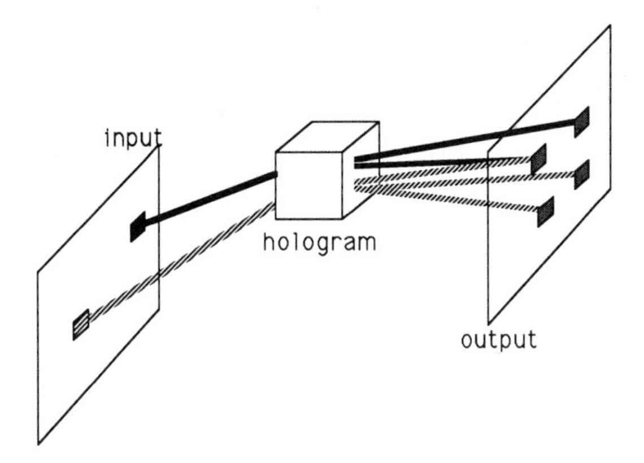

FIG. 3. One optical matrix-vector multiplication setup using a thick hologram. Pixels in the input and output planes are arranged in such a way that the elementary thick gratings encoding each connection never coincide.

matrices. Many switching networks such as the crossbar or the hypercube involve active matrices, including potentially more than one "one" per row and per column, and some control unit for enabling and disabling potential links. In particular, in the case of the crossbar, the most complete network, any element of the matrix may be set to unity at any time by the controller.

It follows from the previous section that optics lends itself nicely to the implementation of interconnect networks. The more elaborate setups are adequate for arbitrary interconnects, and the most classical optical setups are quite efficient for implementing large numbers of shift-invariant interconnects. Comparing light with electricity, this can be seen as a consequence of the capacity of photons to propagate in free space without interaction; it is a clear and intuitive indication, therefore, for the use of light in signal- or data-processing systems in general.

2.4. Binary Computing and Binary Pattern Recognition

The processing operations considered so far all are linear. Some applications, such as broadband spectrum analysis and matched filtering for pattern recognition, have been implemented optically with success for a long time and still are interesting. However, it is important to extend our discussion to nonlinear processing, both because linearity has been recognized as a handicap for the development of optical computing and because no physical law precludes nonlinear devices with attractive performance.

Electric computers rely on binary signal processing. This is dictated mainly by technology because of device reliability and uniformity. Yet, computer history has shown the power of binary signal processing. Some authors have suggested that optics may allow an escape from the rules of binary logic in future computers and the use multivalued logic instead, potential advantages being more compact processors and simpler architectures [10]. Assuming, however, that the same reliability and uniformity issues may prevail for optics as for electronics, we shall concentrate on binary computing.

Let us consider, therefore, one binary computing site labelled i in an array of N sites and let f_i denote its state, zero or one, at some time. Its evolution is governed by some clocking signal and by all the other bits somehow connected to it. We shall call these the neighbors of bit i. The result of the evolution at each clock cycle depends on the state of all binary neighbors, i.e., on the binary pattern present in the neighborhood. Since the only possible values of the result are "zero" and "one," we always may think of the computing operation as the result of a pattern recognition operation on the neighborhood and consider that result "one" denotes the recognition of some match pattern. It may not always be fruitful to consider every binary processing operation as a binary pattern recognition; nevertheless, this always is possible and may be useful in the context of optical computing for two reasons. One the one hand, pattern recognition already is a familiar operation in optics, and on the other hand, recognition of complex binary patterns requires many interconnects and thereby can benefit from one important aspect of optical computing already mentioned.

2.5. Memory

Although computers obviously use displays and optical links, these subjects usually are not considered part of optical computing, both because they pertain to many other fields as well and because they use specific technologies and systems. The same can be said to a large extent of optical memories, and it is a fact that optical disk memories have developed into a domain independent of optical computing for many years. However, drawing sharp distinctions in all these cases would be arbitrary rather than beneficial. Another chapter of this book is devoted to optical memories. It is desirable nevertheless to include in this list of optical computing operations the basic functions of optical memories.

Perhaps the most immediate application of optical bistability could be memory. Functions very similar to dynamic random access memory, static random access memory and permanent memory could be obtained from various classes of bistable devices: If the bistable phenomenon merely consists of an input-output relationship showing hysteresis, but can be observed only for a very short period of time, perhaps nanoseconds, because of thermal effects, the function is analog to dynamic memory; if a holding beam can be used to maintain bistability permanently, the function is that of static memory; finally, if the effect persists even in the absence of a holding beam, the function is that of permanent memory.

Other optical memories are being investigated. Of course, mass storage of information on optical disks has been commercially available now for some time. Recent efforts to make the access of information parallel—for example, by storing information on the disk in the form of holograms [11, 12]—may be relevant to optical computing.

3. Components for Optical Computing

Section 2 listed the functions that can be expected from optics in a processor. Implementation of these functions relies on components. This section, therefore, is devoted to components, both active and passive. Active components are sources, detectors, modulators, and in general, all those whose behavior is modified by light or by electricity; they are comprised, therefore, of nonlinear optical and of optoelectronic components. Conversely, passive components are linear and purely optical.

3.1. Active Components

Figure 4 shows one matrix of optical nonlinear devices [13]. Each device consists of a GaAs/GaAlAs multiple quantum well placed between GaAs/GaAlAs mirrors and has a lateral size of 1.5 μm (right side) or 4 μm (left side).

As mentioned in the introduction, the present research efforts devoted to optical computing are to an extent justified by the realization some years ago of the technological possibility to make such devices. These devices now exist and several other chapters in this book are devoted to them, illustrating the activity of the domain; the reader is referred to them for a physical discussion of their principles and for a recent bibliography. We shall content ourselves here with a few characteristic examples:

- AT & T Bell Laboratories' S-SEED (symmetric self-electro-optic devices) are manufactured in differential switch-pair arrays and their commercial development has been announced recently; pnpn GaAs differential optical switches emitting light in response to incoming light beams have been demonstrated at several places; in general, it is clear that III-V semi-conductor optoelectronic physics and devices have made considerable progress in recent years and that integration in large

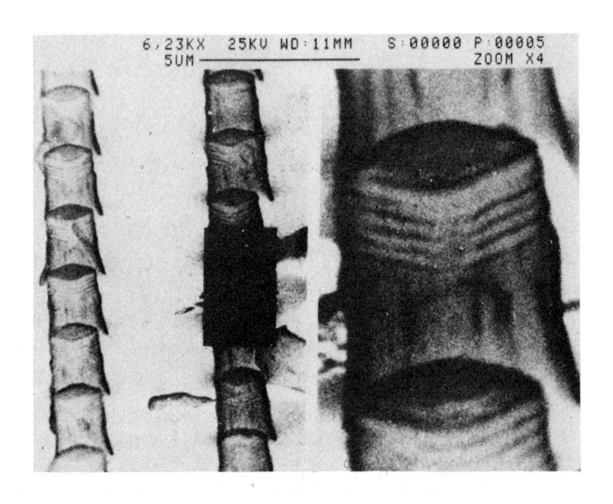

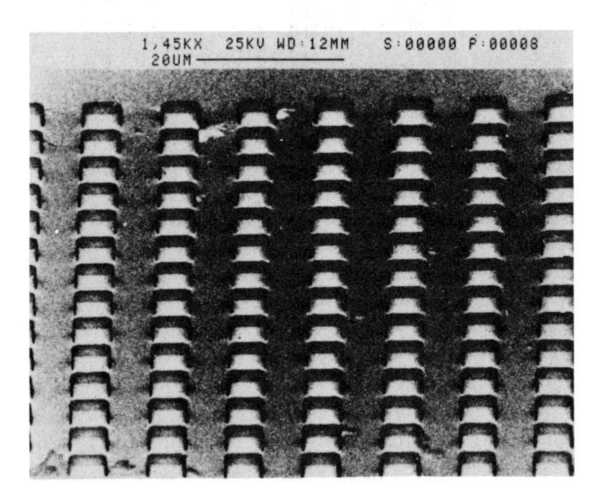

FIG. 4. Portions of two arrays of multiple quantum well nonlinear optical elements (courtesy of J. L. Oudar and R. Kuszelewicz, C.N.E.T., Bagneux, France).

two-dimensional arrays is possible. Present performances already show device areas and response times comparable to those of silicon electronic devices—typically, square micrometers and nanoseconds. The total energy required to observe switching or bistability still is notably larger than the switching energy of a CMOS transistor—typically, picojoules for optoelectronic devices, tens of femtojoules for a commercial

CMOS gate, but these figures are not final. It will be very interesting, therefore, to watch the further progress of the fabrication and integration technologies in the coming years.
In a different category, both optically addressed and electrically addressed liquid-crystal light valves have progressed recently, in particular through the use of smectic ferroelectric media that show a faster response than the nematic liquid crystals used most commonly in display applications. Devices with a space bandwidth product between 10^4 and 10^5 are available commercially. Further progress in the number of pixels can be expected, but it seems unlikely that light valves will attain a degree of compactness, a cycle time, and an energy requirement comparable to those of compound semiconductor devices in the foreseeable future; nevertheless, they may offer attractive characteristics for some applications.

These examples suggest that within a few years, it may no longer be true that optical computing is limited by the lack of active components. Some, if not all, of the functions of optical computing listed in Section 2, therefore, may have a bright future, possibly even in a relatively short time. The issues of system architectures and the integration of the new components in systems becomes relevant. Before we turn to their examination, a word on passive components is appropriate.

3.2. Passive Components

Compared to active components, it may seem that passive components have been in a mature state for quite a long time, and that little progress is required from lenses, mirrors, prisms or gratings. However, system integration already has been mentioned as one important and difficult issue in optical computing. In particular, it will be necessary to provide the new active components both with signal beams carrying two-dimensional information to be processed and with illuminating beams carrying the energy needed for proper operation; also, the data resulting from processing by active components must be carried to the next stage. The most common passive components hardly are suitable for these functions except in bulky demonstration setups. Two important examples will illustrate the issue of integration.

In Section 2.2, we have described setups for matrix-vector multiplication. Scientists at Mitsubishi recently have demonstrated an optical neural chip where the maxtrix-vector product, as suggested on Fig. 5, is performed by putting into close contact a linear array of vertical stripes providing the input signals f_i, a transparent two-dimensional mask, and a linear array of horizontal detector stripes [14]. The performance of this

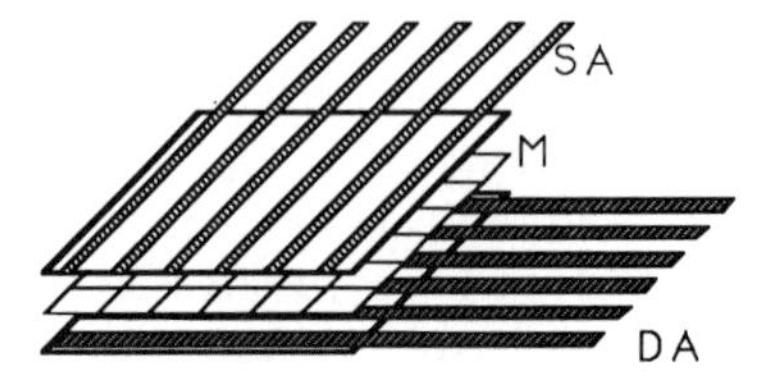

FIG. 5. Schematic principle of an optical neural chip. S.A. = source array; M = matrix of synaptic weights: D.A. = detector array.

chip is limited by diffraction and shadowing effects. While these effects cannot be overcome by standard passive components in a compact configuration, suitable integrated passive components, such as lenslet arrays, can be developed especially for that and similar purposes. These presently are lacking, to a large extent.

Figure 6 illustrates the illumination of an array of optical active devices by a beam carrying energy rather than data. This may be a reading beam testing the state of the devices, or a holding beam setting them to the proper operating point. If the devices are pixellated as is more and more often the case, the configuration of Fig. 6a wastes most of the illuminating beam energy, so that some passive component termed an array illuminator is required, as shown on Fig. 6b. A review of approaches to array illuminators can be found in [15].

Micro-optics, therefore, is a new domain of technical research fostered by the need for using the active elements suitable for optical computing in compact setups and with optimal performances. The components required here are very similar, in fact, to those developed for other applications, such as laser-diode coupling, sensors, and optical switching. Recent progress in hologram arrays and lenslet arrays, as described in other chapters of this book, are among the first results in this domain.

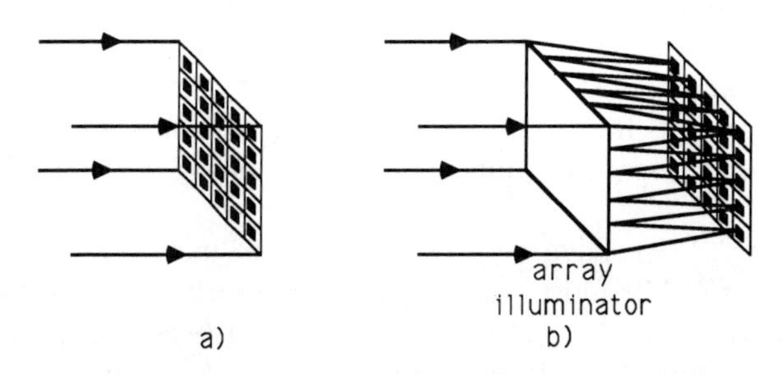

FIG. 6. Illumination of a pixellated array of electro-optic or nonlinear devices: a) by a plane wave; a large fraction of the energy is wasted; and b) by an array illuminator focussing the energy onto the pixels.

4. Computing with Light

In the two previous sections, we have summarized the processing functions that light can be expected to provide, with emphasis on optical interconnections, and we have mentioned the development of components that may allow real achievement of some processing advantage from the use of light beams in processors. In the search for cases where this advantage can be used best, several classes of systems have been put forward. A selection of them is reviewed in this section. The various approaches have little in common except that all emphasize the role of interconnects. We have chosen to examine first some cases where optics is used to boost the performances of otherwise classical electric processors; then, in contrast, propositions for purely or mostly optical processors; and finally, the possibility of intermediate solutions.

4.1. Optical Interconnects

Since it generally is true that optics is attractive for interconnections, the most modest, realistic and, near-term application of optical computing appears to be the use of light beams at the various levels of interconnections inside processors. Before this general idea is put into practice, however, it remains to be seen if limitations in electric computing interconnects are severe enough to justify the use of alternate technologies. This question was asked clearly for the first time in an article by Goodman *et al* in 1984 [16] and since then has been the subject of much work [17].

The answer depends considerably on the interconnection level: on chips, between chips, between boards, and between cabinets. At the latter level, large-bandwidth, fixed links are needed and it is straightforward to look for solutions in the domain of optical fibers, which already are successful—technically speaking, at least—at the levels of local area networks and long distance communications. We shall stick to the practice of including this aspect in optical telecommunications rather than optical computing.

4.1.1. Interconnects Inside Chips

Let us turn now to the innermost interconnection level, that of communications inside a chip, and examine qualitatively the influence of connection length on chip-area budget, time delay, and energy consumption. The area required for an electric interconnection between two transistors obviously increases with distance because of the space devoted to the metal strips, while optical interconnects take up relatively little space, as to a large extent, they can overlap in space without prejudice. Similarly, the time delay of an electrical interconnection increases with distance because of interline capacitances that contribute factors RC in the time delay, while the time of light propagation hardly comes into play at the scale of centimeters for clock frequencies below hundreds of megahertz. Finally, the same capacitances contribute $1/2\ CV^2$ terms in the energy, while the attenuation of a suitably focussed or guided light beam is negligible at the distances considered. Therefore, there must be a distance beyond which an optical interconnect would save area and energy and increase operating frequency on a chip. Some estimations [18] suggest that this critical distance is of the order of centimeters: The issue should be sensible, therefore, in wafer-scale integration, and some cases of the use of optics may even be worth considering already. The most popular example is that of optical clock distribution on the chip [19], which is a case where an external signal, created optically outside the chip and without the need for a feedback signal from the chip to the clock, can be broadcast to many sites on the chip and thereby save space on the chip and avoid clock skew at high cycle frequencies.

Nevertheless, the complexity of combining light detection, focussing or guiding, and, especially, light emission with silicon technology is a considerably limiting factor at this time. Moreover, electric interconnects at the chip level are not considered as severely limited because most of the area of the chip over several layers is available to them. The same is not true at the two intermediate levels of electric interconnections—between chips and between boards—that we shall consider now.

4.1.2. Interconnects between Chips and between Boards

The size of input and output interconnect pins and the distance between pins are limited by several phenomena of practical importance: mechanical resistance of the metallic pins if the chip is to be inserted in a socket, alignment tolerances, and electromagnetic interference between signals traveling in neighboring pins; these set limits to the number of electric interconnects that a chip can support, and impose large interconnect capacitances. With chips containing 10^4 or 10^5 transistors, the restriction of chip-to-chip interconnects to hundreds of pins is an important constraint of system architectures.

The same exact phenomena occur at the level of board-to-board interconnects, where they are even more critical. At frequencies exceeding about 50 MHz, interference phenomena commonly impose the insertion of grounded pins between each pair of active pins, thus decreasing the number of useful channels even further.

The quest for practical solutions involving optics and their fine analysis is progressing at many places [20].

4.2. Optical Computers

As opposed to the term *optical computing*, that may commonly be used in all cases where an optical signal contributes somehow to the operation of a processor, the expression *optical computer* might be used in reference to a processor where all or most of the processing is done by light. As opposed to the previous section, where optics is constrained by the requirements of the prevailing technology and where a demonstrator can be of interest only if its performances are competitive immediately, optical computers are an area where imagination is almost unrestricted and where demonstrators can be interesting, even if very preliminary, because they help develop architectural concepts that still are quite unlikely to be used in commercial processors in the near future. Let us cite a few examples here.

4.2.1. Optical Microprocessors

One approach is to stick to the general operation, if not architecture, of present microprocessors and investigate the possibility of implementing them optically, using existing, commercially available components. P. Guilfoyle is building a 32-bit processor based on multichannel acousto optic modulators [21], where light is used in combination with avalanche photodiodes to make binary decisions of binary pattern recognition; as explained in Section 2.4, this is a general approach to binary computing and can be used to synthesize logical functions of any complexity, such as those of the instruction set of a microprocessor. The processor is planned to operate at a frequency of 100 MHz. Another project of the same category, but relying on fiber optics and lithium niobate-integrated electro-optic switches, is under investigation at the University of Colorado [22].

4.2.2. Neurons

A second approach insist that the large interconnection capability of optics should be used in those processors that require more interconnects, i.e., parallel machines.

In neural processors, emphasis is set on the completeness and generality of the $N \times N$ interconnection set between N processors. The role of optics can be restricted to the implementation of the interconnections—the so-called synaptic matrix-vector product—or it also can include the updating of the synaptic matrix and the neural nonlinearity. Another chapter of this book is devoted entirely to this subject, and the question of rewritable optical interconnects pertains to the subject of photorefractive devices examined in yet another chapter.

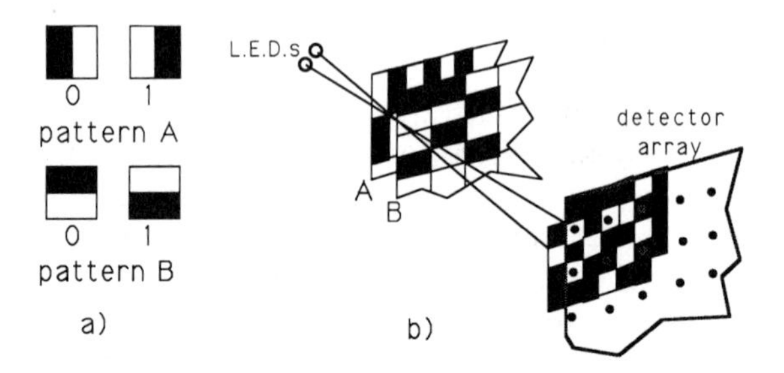

FIG. 7. Optical Parallel Logic Arrays: a) Binary codes for patterns A and B. b) Implementation of the Boolean operation "XOR" using shadow casting.

4.2.3. Cellular Processors and Symbolic Substitution

As mentioned in Section 2, system architecture is simpler for space-invariant interconnection pattterns and larger numbers of cells then can be processed. Several laboratories have investigated setups allowing the achievement of space-invariant, reprogrammable operations on large numbers of pixels and the kinds of processors that can be built with these concepts. Figure 7 shows the principle of the Optical Parallel Array Logic (OPAL), introduced by Tanida and Ichioka [23]. Two binary image planes labelled A and B are encoded as shown in Fig. 7a. A proper positioning of four light sources casts shadows on an array of photodetectors, one photodetector per data pixel shown in Fig. 7b. By switching the sources on or off, any binary Boolean function can be implemented; Fig. 7b, for example, shows the *exclusive or* combination. The system can be made compact and can be extended to perform more complex operations. Therefore, provided some solution for the parallel data input is available, parallel arithmetic is feasible.

Figure 8 shows the principle of *symbolic substitution*, where a binary output array B is formed from a binary input array A by applying a number of substitution rules; for example, in the figure, all occurrences in array A of the match pattern shown on the left side of the inset are searched and replaced in the output by the corresponding left-hand side. The operation is performed for:

- Each match pattern, in parallel over the entire data plane.
- The various match patterns, sequentially in time (or simultaneously on several separated setups).

The results of substitution then are accumulated in the output array. The example of the figure corresponds to one symbolic substitution implementation of binary addition, if it is understood that the process is iterated until convergence. Brenner *et al.* have shown that symbolic substitution

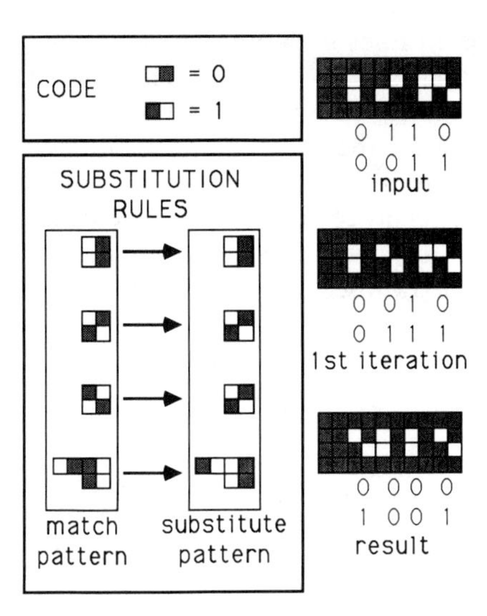

FIG. 8. "Symbolic substitution" of each match pattern shown in the left part of the inset by the corresponding substitute patternshown on the right side results in binary addition.

processors have the power of Turing machines; i.e., they are general binary processors [24]. Symbolic substitution processors are a nice illustration of the idea developed in Section 2.4 that any binary processing can be understood as a binary pattern recognition, and that such a description is well suited to optical computers because optics is a powerful means of performing pattern recognition in parallel.

4.2.4. Discussion and Examples of Cellular Processors

In spite of the fact that cellular processors of that kind have the power of Turing machines, their practical interest often has been questioned. Moreover, even if they can be shown to be attractive, their electric, massively parallel implementations still may outperform the optical ones. This question seems to have remained open. Following are some elements of discussion.

There are at least some well-identified cases where cellular machines can perform useful tasks with full parallelism. We shall cite two examples: *binary image preprocessing* and *fluid dynamic simulations*. The first case is well-known and corresponds to such operations as edge detection, noise cleaning, and skeletonization in binary images. It is described, in general, by theories such as mathematical morphology and binary image algebra [25]. The second case is illustrated by the fact that the statistical properties of a gas, as described by the Navier-Stokes equations, can be simulated completely by collision rules between particles moving on the nodes of a lattice, i.e., by symbolic substitution [26]. This kind of processor is known as a *lattice gas automaton*. These examples obviously correspond to practical problems known for their intensive computation requirements. The demand may justify the construction of such specialized processors.

Whether such parallel processors can be implemented best electrically or optically depends on the state of technology, and the development of the components mentioned in Section 3 may make optics a viable solution. This should be the case, in particular, for processors involving large interconnection patterns because they are difficult to design on all-electronic integrated circuits.

4.3. Computing with Some Optics

Sections 4.1 and 4.2 were devoted to two opposite views of the possible role of optics in computing. In this last section, we describe an intermediate route, where optics is used in combination with electronics, but where the gain brought by optics does not come without the need for revision of the conventional electric architecture. In particular, free space should be left in front of the chips to allow for large amounts of optical inputs and outputs.

One set of possibly interesting architectures derives from the idea of a *smart retina*, where an array of photodiodes is upgraded into an array of processors. The pixels of the incoming image are processed in parallel before being transferred out of the chip. In a smart retina, optics serves only for the parallel input of data to be processed. However, this idea can be combined with other optical functions. Let us mention instruction broadcasting: In many cases of the cellular processors cited earlier—mathematical morphology and lattice gas automata—the interconnections required are restricted to relatively small sizes that electronics can handle straightforwardly. The advantage of using optics at that level, therefore, is questionable; but in such S.I.M.D. machines (single instruction on multiple data), all processors on the chip need to receive the same instruction code at the same time. This is equivalent to a clock signal distribution, except that instruction codes use more bits. Therefore, optical instruction code broadcasting can save space and ease synchronization problems. A holographic array illuminator allowing broadcast of instruction codes using Bragg-matched incoming signals, while at the same time letting Bragg-mismatched data signals go undiffracted through the hologram to the processing chip, is under study. Additionally, compound-semiconductor arrays of the type mentioned in Section 3.1 could be used for parallel output of data, allowing construction of a parallel processor where all the

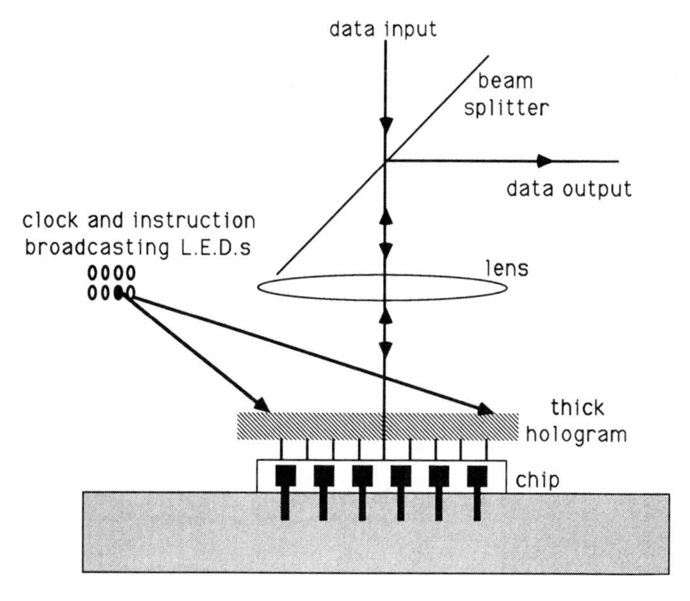

FIG. 9. Project of a hybrid optical-electronic cellular automaton chip. Optics is used for instruction broadcasting through the array illuminator hologram, for parallel data input, and for parallel data output.

local processing would be done electrically and where all parallel signals would be carried optically. (See Fig. 9 and [27]).

Other examples of processors presently being investigated, and combining electronics and optics as intimately as possible, include the parallel distribution of optical random numbers to massively parallel chips for simulated annealing [28], the optical control of synaptic strengths on a neural chip via a photoresistive effect [29], and the U.C.S.D. POEM processors [30].

5. Conclusion

Optical computing is a treasure island, as said earlier. Scientists landed on the island but still have to find out how valuable the treasure really is. It seems safe to predict that many new optoelectronic components will keep appearing and improving because they are interesting aside from their applications in optical computing. It is not really clear, however, whether it is wise to test new architectures on new components or stick to conventional architectures; i.e., whether the new devices will find applications in analog processors, neural processors, cellular automata, microprocessors with little parallelism, or optical interconnects.

These architecture guidelines, summarized in Section 4, probably are fairly well analyzed by now. But are planar waveguides more useful than lenslet arrays for system integration? Is reprogrammability of interconnects using photorefractive crystals a powerful approach? Is it possible to build arrays of components working at megahertz frequency into such massively parallel machines that parallelism will overcome the cycle time handicap, or is there any realism in gigahertz optical processors? Will it prove less difficult to make two-energy optical gate arrays than to integrate the passive elements needed to operate them? Demonstrations of architectures and component integration are needed for the assessment and evolution of these and other concepts. The present time is an exciting period because it is clear that the next few years will bring some answer... and we do not know which.

References

1. For references, see J. W. Goodman, "*Introduction to Fourier Optics and Holography*," McGraw-Hill, San Francisco, 1968.
2. "Optical Computing 87," Lake Tahoe, Nevada *Technical Digest, Optical Society of America*; "Optical Computing 88, Toulon," France, *S.P.I.E Proc.* **963** (1989); "Optical Computing 89," Salt Lake City, Utah *Technical Digest, Optical Society of America*; "Optical Computing 90," Kobe, Japan, *Conference Record, Japanese Society of Applied Physics and S.P.I.E.*, (1990).
3. P. Mandel, S. D. Smith, and B. S. Wherrett, "*From Optical Bistability towards Optical Computing*," Elsevier, Amsterdam, 1987.
4. D. Feitelson, "*Optical Computing: a Survey for Computer Scientists*, M.I.T. Press, Cambridge, MA 1988.
5. R. Arrathoon (ed.), "*Optical Computing, Digital and Symbolic*," Marcel Dekker, New York, 1989.
6. J. W. Goodman, "*Introduction to Fourier Optics*," McGraw-Hill, New York, 1968.
7. J. W. Goodman, A. R. Dias, and L. M. Woody, *Opt. Let.* **2**, 1–3 (1978).
8. D. Psaltis, X. G. Gu, and D. Brady, *S.P.I.E Proc.* **963**, 468–474 (1989).
9. A. A. Sawchuk and B. K. Jenkins, *Proc. S.P.I.E* **143–153** (1986).
10. See, for example, in Ref. 5, chapters by M. Conner and G. Eichman, and chapter 6 by C. Moraga.
11. D. Psaltis, M. A. Neigeld, and A. Yamamura, *Opt. Let.* **14**, 429–431 (1989).
12. A. L. Mikhaelian, *Conference Record, Optical Computing 90*, Kobe, Japan, 193–194 (1990).
13. J. L. Oudar, B. Sfez, R. Kuszelewicz, J. C. Michel, and R. Azoulay, *Phys. Stat. Sol.* (b), **159**, 181–189 (1990).
14. Y. Nitta, J. Hta, K. Mitsunaga, M. Takahashi, S. Tai, and K. Kyuma, *Jap. J. Appl. Phys.* **11**, L2101–L2103 (1989).
15. A. W. Lohmann, W. Lukosz, J. Schwider, N. Streibl, and J. A. Thomas, *OC'88, Proc. S.P.I.E.* **963**, 232–239 (1988).
16. J. W. Goodman, F. I. Leonberger, S.Y. Kung, and R. A. Athale, *Proc. I.E.E.E.* **72**, 850–858 (1984).
17. For a recent collection of papers, see special issue of *Applied Optics,* **29**, no. 8 (March 10, 1990).

18. L. A. Bergman, W. H. Wu, A. R. Johnston, and R. Nixon, *Opt. Engin.* **25**, 1109–1118 (1986).
19. B. D. Clymer and J. W. Goodman, *Opt. Engin.* **25**, 1103–1108 (1986); ibid. **27** (1987).
20. Recent examples include E. E. Frietman, W. van Nifterick, L. Dekker, and T. J. M. Jongeling, *Appl. Opt.* **29**, 1161–1177; J. Jahns, Y. H. Lee, and J. L. Jewell, *Conference Record, Optical Computing 90*, Kobe, Japan, 164–165, (1990).
21. P. S. Guilfoyle and W. J. Wiley, *Appl. Opt.* **27**, 1661–1673 (1988); P. S. Guiloyle, *Proc. S.P.I.E.* **CR35**, 288–309 (1990).
22. H. F. Jordan, *Proc. S.P.I.E.* **CR35**, 266–287 (1990).
23. Y. Ichioka and J. Tanida, *Proc. I.E.E.E* **72**, 787–796 (1984).
24. K. H. Brenner, A. Huang, and N. Streibl, *Appl. Opt.* **25**, 3054–3060 (1986).
25. K. S. Huang, B. K. Jenkins, and A. A. Sawchuk, *Appl. Opt.* **28**, 1263–1278 (1989).
26. J. Taboury, C. Chauve, and P. Chavel, "Optical Computing 88," *S.P.I.E Proc.* **963**, 680–686 (1989).
27. I. Seyd Darwish, J. Taboury, P. Chavel, and F. Devos, *Proc. I.C.O.* **15**, 173–174 Garmisch-Partenkirchen (August, 1990).
28. Ph. Lalanne, H. Richard, J. C. Rodier, P. Chavel, J. Taboury, K. Madani, P. Garda, and F. Devos, *Opt. Commun.*, **76**, 387–394 (1990).
29. E. A. Rietman, R. C. Frye, C. C. Wong, and C. D. Kornfeld, *Appl. Opt.* **28**, 3474–3478 (1989).
30. F. Kiamilev, S. C. Esener, R. Paturi, Y. Fainman , P. Mercier, C. C. Guest, and S. H. Lee, *Opt. Engin.* **28**, 396–409 (1989).

Index